D 28.45
Polly

Engineering Fluid Mechanics

Engineering Fluid Mechanics

THE INTERNATIONAL STUDENT EDITION

by
W. P. Graebel
Professor Emeritus
Dept. of Mechanical Engineering
and Applied Mechanics
The University of Michigan

Taylor & Francis Publishers
New York   London

Published in 2001 by
Taylor & Francis
29 West 35th Street
New York, NY 10001

Published in Great Britain by
Taylor & Francis
11 New Fetter Lane
London EC4P 4EE

Library of Congress Cataloging-in-Publication Data
Graebel, W. P.
    Engineering fluid mechanics / by W.P. Graebel.—International student ed.
        p. cm.
    Includes bibliographical references and index.
    ISBN 1-56032-733-2 (alk. paper)
    1. Fluid mechanics. I. Title.

TA357.G692 2000
620.J'06—dc21   00-049750

Contents

Preface

This text covers the necessary material for an introductory course in fluid mechanics. There is sufficient material presented that it could serve as a text for a second course as well.

The text is designed to emphasize the physical aspects of fluid mechanics and to develop analytical skills and attitudes in the engineering student. Example problems follow most presentations of theory to ensure that the student grasps the implications of the theory and is able to apply it. In topics that involve more than elementary calculations, step-by-step processes outline the procedure used so as to generalize the students' problem-solving skills. To demonstrate the design process beyond the problem-solving techniques, an appendix presents some of the more general considerations involved in the design process.

Elementary fluid mechanics is one of the basic core courses of undergraduate engineering, along with statics, dynamics, mechanics of materials, thermodynamics, and heat transfer. I have endeavored to show linkages to these subjects as well as to elementary physics, to both build on previously learned knowledge and to provide a bridge to courses to be taken in the future.

I have included frequent references to applications throughout the text, as well as an appendix on the history of fluid mechanics. Fluid mechanics is a required subject for many engineering programs, and a student starting such a course is frequently not sure why the subject is of importance. I have found that including such material in my own teaching has enhanced student interest in the subject and resulted in a more appreciative audience.

The subject matter is organized in the following manner:

- Chapters 1 and 2 serve as an introduction to the subject. Terms and concepts are defined and the student is given practice with pressure calculations. A general procedure for attacking engineering problems is suggested.
- Chapter 3 is really the heart of the book. The concept of control volume is presented, along with the fundamental equations of continuity, momentum, and energy. These presentations are for one-dimensional analysis. Chapter 4 extends this theory to three dimensions with the development of the Euler and Navier-Stokes equations. I have included some simple solutions of these equations, which is not traditional for textbooks at this level. I have done so because I feel that without applications, development of the theory leaves the typical undergraduate student with rather

a "so-what" feeling. This chapter can either be presented following Chapter 3 or delayed to a later portion of the course, depending on the instructor's goals. Some instructors may wish to omit it completely.

- Chapter 5 considers the subject of dimensional analysis and provides a road map to the chapters that follow.
- Chapters 6 and 7 develop elementary viscous flow theory and general Reynolds number effects. Chapter 6 deals exclusively with laminar flows, and Chapter 7 with turbulent flows. In pipe flow calculations I have supplemented the traditional Moody diagram with two others, so that the student can solve problems directly and avoid having to deal with trial-and-error solutions of pipe flow problems.
- Chapter 8 deals with open channel flows and Froude number effects. For courses that focus on such flows, it could follow chapter 3 directly.
- Chapter 9 deals with compressible flows and Mach number effects. The material starts with a general discussion of compressibility, then goes to a brief discussion of compressibility effects in liquids before finishing with a discussion of compressibility in gases.
- Chapter 10 gives a summary of measurement techniques suitable for fluid flows, and Chapter 11 discusses aspects of hydraulic machines.
- The concluding chapter points out some of the more advanced topics in fluid mechanics, and indicates to the student the type of courses that might be useful in developing further interest in fluid mechanics.

While at The University of Michigan I have been fortunate to teach a wide variety of engineering subjects to students in engineering mechanics, mechanical engineering, civil engineering, chemical engineering, aerospace engineering, naval architecture and marine engineering, and meteorology and oceanography. This, along with seminars, serving on doctoral committees, and research and consulting activities, has broadened my interests and has given insight into the wide range of applications of fluid mechanics in many areas of engineering. I have tried to include a flavor of many of these applications in my presentation of the material in this book.

No book can suit all students. When I learn a new subject, I find three or four—or more—books dealing with it and study all of them. I find that different authors coming from different points of view help me to find the thread on which to base my own understanding of the subject, to place it in the context of things I am already familiar with. I encourage students to do likewise, and to utilize the library at their institution to the fullest extent.

Many people have influenced my presentation of the material in the book. Certainly the students I have taught have been a great help in teaching me what is effective in teaching. Reviewers of early drafts of the book have also been helpful in their criticisms. I would especially like to thank my wife, June, for her help during the preparation of this book in typing and grammar suggestions, and in her general support and understanding of the effort. I would also like to honor the memory of two cherished people: Chia-Shun Yih, my teacher, colleague, and friend, who first sparked my interest in fluid mechanics, and who taught me much, much more; and Vernon A. Phelps, who broadened my outlook on engineering and suggested new paths to follow. The world is poorer for their absence.

W. P. Graebel

chapter 1

Introduction to Fluid Mechanics

Chapter Overview and Goals

This chapter introduces some of the concepts that we will be using throughout our study of fluid mechanics. We start by defining the term "fluid" and introduce a number of common fluid properties such as mass density, bulk modulus, viscosity, and surface tension. The concepts of stress, absolute and gage pressure, cavitation, Newtonian fluids, and non-Newtonian fluids, together with the no-slip boundary condition, are also introduced and discussed. Several examples of applications are given.

By the end of the chapter, you should be familiar with these concepts, and also with those units of the British gravitational and SI systems of units that are applicable to fluid mechanics. With the help of Appendix A, you should be able to express all quantities in both sets of units.

You should also begin to have a grasp of the magnitudes of the numbers that are reasonable for the various quantities in each set of units, so that you will be able to make judgments as to whether numbers you obtain in calculations are reasonable. The definitions and concepts introduced in this chapter will occur throughout the book; therefore it is important to become accustomed to them.

1. Introduction

Since prehistory, mankind has been interested in being able to predict and/or control how fluids flow. Weather prediction has always been important for agriculture, fishing, and water transportation. Civilizations have started—and ceased—because of the availability of water supplies. The transport of water for agriculture, drinking, and bathing led to such engineering marvels as the aqueducts of the early Romans. Some of these are still in use after more than 2000 years. Control of air flow to decrease erosion of the ground; drag on cars, trucks, and airplanes; and the dispersion of pollutants are all important in our modern lives. Instrumentation for monitoring pressures and flow rates in blood vessels and pressures in the eye has become an important diagnostic tool for medicine.

To resolve the engineering problems that arose in these early attempts to predict and/or control how fluids flow, many people developed individual theories to deal with specific isolated problems even before written history. Starting in the fifteenth century with Newton, and in the sixteenth century with Euler and the Bernoulli family, a general mathematical formulation of fluid mechanics was begun, culminating in the mid-nineteenth century with the work of Navier and Stokes. The latter completed the structure needed for the general mathematical formulation of fluid mechanics. The great scientific advances that were made in that period put the mechanics of fluids on a thorough scientific basis, against which both earlier and later theories and approximations could be tested, and our knowledge and understanding of the flow of fluids increased.

This scientific understanding of how fluids behaved was needed for the technical demands of the industrial revolution and the advanced technology that followed. The development of modern ships and aircraft was possible only because of the general scientific formulations of the nineteenth century, and the application of theory to technology in the twentieth century. Based on these fundamental theories, Orville and Wilbur Wright, Frederick Lanchester, Nicolai Joukowski (also spelled Zhukovskii), and Ludwig Prandtl made modern aviation and the space program possible. The use and behavior of fluid flow in transportation, prediction of circulation in the atmosphere and oceans, power transmission and generation, lubrication, transport of mass and heat, and so many other areas, makes fluid mechanics one of the cornerstones of our modern technological society. It would be difficult to imagine our life today without the myriad ways in which we have applied our knowledge of fluid flow.

As fluid mechanics developed and our knowledge of the behavior of how fluids flow grew, the field became divided into specializations, and various technical areas were given special names. Hydraulics, for example, refers to the flow of liquids in channels, canals, and pipelines. Pneumatics deals with the flow of air, usually in small-diameter tubes. Gas dynamics deals with the high-speed flow of gas when compressibility effects are important. If the fluid density is low enough that means free paths between molecules are large, we speak of rarefied gas dynamics. For ionized gases, we talk of plasma flows, and when in the presence of magnetic fields, magnetohydrodynamics. Meteorologists deal with the flow of air in our atmosphere, while oceanographers are their underwater counterparts. Many other specialities exist, and new ones are still appearing.

2. Definition of a Fluid

All matter exists in one of three phases: liquid, vapor (or gas), and solid. The word "fluid" is used as a general term for the first two of these phases, since the basic mechanical behavior of liquids and gases is very similar. Which phase the matter is in depends on the values of the various thermodynamic variables such as pressure and temperature. Two typical plots showing phase and phase changes when the matter is in static thermodynamic equilibrium are given in Figures 1.1 and 1.2. Figures 1.3 and 1.4 are two-dimensional projections made from Figures 1.1 and 1.2. They show planes of constant mass density drawn through the *critical points* of Figures 1.1 and 1.2. The point labeled "critical point" in these figures corresponds to the point of highest temperature possible for a liquid-vapor mixture to exist in the equilibrium state.

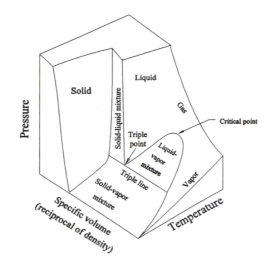

Figure 1.1. Pressure-density-temperature equilibrium surface for a substance that contracts on freezing (e.g., carbon dioxide).

The primary difference between a solid and a fluid is in the strength and type of the molecular bond. A solid is made up of a closely packed molecular structure, where breaking the bonds requires considerable energy. In a fluid the bonds are looser and can be easily broken. A fluid is defined as a substance that will deform and move when a tangential (shear) stress is applied to it, the motion continuing as long as the shear

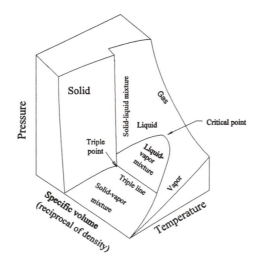

Figure 1.2. Pressure-density-temperature equilibrium surface for a substance that expands on freezing (e.g., water).

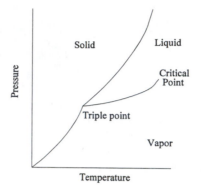

Figure 1.3. A constant density plane passed through the critical point of Figure 1.1.

stress is present. In contrast, when a shear stress is applied to a solid, an initial deformation results, but motion soon stops and does not resume until the shear stress is again changed.

The distinction between fluids and solids is actually not as sharp as is implied in the above paragraph, or perhaps as might be imagined from the equilibrium diagrams of Figures 1.1 and 1.2. Materials such as aluminum can flow like a very viscous liquid when stressed above their elastic limit into the plastic range, and there is a large family of polymer materials that exhibit both solid and fluid behaviors, depending on the level of the applied stress. These polymers are termed "viscoelastic solids" if they are more solid than fluid in their behavior, and "viscoelastic fluids" if they are more fluidlike.

Note that in the definition of a fluid we have spoken only of tangential (shear) stresses, and have not said anything about a fluid's behavior due to applied normal stresses. This is because application of normal stresses to a fluid may or may not cause it to move, depending on the circumstances. For example, in the case of water at rest in a drinking glass, the pressure—a form of normal stress—increases in the water from the minimum pressure (atmospheric pressure) at the top of the water surface to a maximum pressure at the bottom of the glass, yet the water may be completely still.

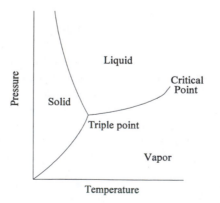

Figure 1.4. A constant density plane passed through the critical point of Figure 1.2.

Conversely, if we have a fluid in a horizontal pipeline and the pressure is different at the two ends of the pipeline, the fluid can be expected to flow toward the end of lower pressure. Attempting to define a fluid by its behavior under normal stresses would therefore lead to a complicated list of qualifications and decisions. On the other hand, under shear stresses, fluids will always flow. Thus defining a fluid by its behavior under shear stresses is unambiguous and avoids difficulties.

From the definition of a fluid it is seen that both the gaseous and liquid phases of matter qualify as fluids. A gas is a molecularly mobile substance, capable of expanding until it completely occupies any container. The molecules of a liquid can also move freely relative to one another, but the intermolecular bonding is stronger, and the molecules are not capable of expanding and completely occupying a container as does a gas. Under the action of gravity a liquid will occupy only the bottom of a container, with a flat horizontal surface on the top separating it from any gases that may be above the liquid. From the point of view of the mechanics of a given flow, it will be seen that even though a gas is in a different phase from a liquid and can be compressed much easier than a liquid, in many instances it is not necessary to distinguish whether we are dealing with a liquid or a gas.

3. The Continuum Hypothesis

To the eye a fluid appears to be a continuous substance. In fact, it is composed of myriads of continuously moving molecules, all interacting and colliding with one another. While to a certain degree it is possible to use the laws of motion to describe what happens to each molecule (this is the subject of kinetic theory, or statistical physics), such an analysis would provide more detail than is normally of engineering interest, and the complexity of the calculation for even simple flows would be much beyond our capabilities.

Instead of using this statistical approach in dealing with the mechanics of fluid flow, it is customary to think of the fluid as if it were made up of a very, very large number of fluid particles. This allows us to concentrate on the macroscopic, or bulk, properties of the flow. As a mathematical idealization we let the size of these fluid particles conceptually shrink to zero, and the number of fluid particles become infinite. In this idealization we are taking mean values over many molecules, and thereby treating the fluid as a ***continuum***; that is, as if it were a homogeneous uniform volume where noncontinuum entities such as molecules, atoms, and the like do not exist. While in fact fluids are aggregations of molecules, the molecules being free to move about and collide with one another, as long as the molecules are separated by distances large compared to their diameters, and the number of molecules per unit volume is sufficiently great that a macroscopic approach is possible, a continuum model of a fluid where we deal with macroscopic quantities rather than microscopic quantities is in fact usually very good.[1]

---

[1] A volume of 10^{-18} cubic meter contains something of the order of 2.7×10^7 molecules at normal temperatures and pressures.

An exception to this continuum model is a rarefied gas, where the mean free path between collisions of molecules may be large. Rarefied gases are encountered in space flight, where the mean free path between molecules can be comparable to the dimensions of a space ship. In such cases, it is necessary to change from the deterministic continuum physics of Newton to statistical or stochastic concepts. While these effects are important in flows in the upper atmosphere and in some of the new experimental areas of power generation, they require a treatment that is so different from other flows that they will not be considered here.

In our continuum hypothesis we will be using mathematical limits where we will let areas and volumes go to zero. Physically, even the smallest volumes we consider must be much larger than the size of the molecules of our fluid for the continuum hypothesis to be reasonable. One way of interpreting the continuum hypothesis is to say that we have replaced the real fluid with a mathematical model that is just as continuous as the fluid appears to our eye. This is a model that is well suited to engineering calculations.

4. Systems of Units

Two major engineering systems of units now coexist in the world. The older "British gravitational system" (sometimes called the U.S. customary system) is still used in much of U.S. industry and commerce, while the newer "International system of units" (colloquially called "metric" units, but more accurately termed "SI" units after its French name "Le Système International d'Unités") is now used in most of the rest of the world. It is important for a student of engineering to become familiar with a particular set of units in order to acquire a physical appreciation of typical orders of magnitudes of quantities. Today and for the foreseeable future it is necessary for students to become familiar with both the British gravitational and SI sets of units, since the U.S. is the lone holdout among industrialized nations in fully adopting the metric system, although the pressures of international trade and the resistance of Europe and Japan to nonmetric usage is putting strong pressure on U.S. industry to convert.

Unfortunately, many units still remain as carryovers from a number of older systems, and will remain long after the entire world has adopted the metric system. Since these units still exist in various books, technical papers, and even in common usage in specialized fields, they must be learned as well. A list of factors useful for converting from one set of units to another can be found in Appendix A.

a. British gravitational system of units

The basic units for the British gravitational system are pounds for force, slugs for mass, feet for length, seconds for time, and degrees Rankine for absolute temperature. The standard gravitational constant at sea level is 32.174 feet per second squared, and by Newton's second law (force = mass times acceleration), the units of mass and force are related. Thus a mass of 1 slug would have a weight (force) of 32.174 pounds at sea level on earth. Also, a force of 1 pound would be needed to accelerate a mass of 1 slug with an acceleration of 1 foot per second squared.

Other British system units that are sometimes used are the pound-mass (instead of slug) for mass and the poundal (instead of pound) for force. A mass of 1 slug is equivalent to a mass of 32.174 pound-mass, and a force of 1 pound is equivalent to a force of 32.174 poundals. These units are summarized in Appendix A. The poundal and the pound-mass will not be used in this text.

Fahrenheit is frequently used as a relative temperature. Absolute temperature (degrees Rankine) is related to relative temperature (degrees Fahrenheit) by

°Rankine = °Fahrenheit + 459.67.

The unit of pressure most often used in the British system is the pound per square inch (psi). In everyday conversation, this is abbreviated by many people to "pounds," a confusing shorthand best avoided.

b. SI system of units

The basic units for the SI system are kilograms for mass, meters for length, seconds for time, and degrees Kelvin for temperature.[2] The (derived) force unit is newtons, where a newton is the force needed to accelerate a mass of 1 kilogram with an acceleration of 1 meter per second squared. The gravitational constant at sea level is 9.807 meters per second squared. A mass of 1 kilogram would consequently weigh 9.807 newtons, and a force of 1 newton would accelerate a mass of 1 kilogram at 1 meter per second squared. Absolute temperature (Kelvin) is related to relative temperature (degrees Celsius) by

Kelvin = °Celsius + 273.15.

Other units (not preferred) that are in use in metricized countries include the kilogram-force, where 1 kilogram-force is equivalent to 9.807 newtons, and the dyne, where 10^5 dynes equals 1 newton. These units also will not be used in this text.

Note that in the SI system of units, even though some of the unit names are proper names of persons, such as newton, celsius, kelvin, stokes, watt, pascal, hertz, and joule, correct practice is to *not* capitalize the unit name, even though capital letters may be used as abbreviations. Also, when giving temperature in degrees Kelvin, the degree sign ° is always omitted, e.g., 10°C = 293.15 K, *not* 293.15°K.

The letter g is customarily used to denote the gravitational acceleration in whatever system of units we are using. Thus, at sea level,

g = 32.174 feet per second squared = 9.807 meters per second squared.

Confusion frequently arises about the term g, for it plays a dual role in mechanics. On the one hand, according to Newton's law of gravitational attraction two masses of mass M and m experience a force of mutual attraction equal to

---

[2]Other fundamental units used in electromagnetics, optics, and geometry include the ampere, mole, and candela, as well as the two supplementary units radian and steradian.

$$F = CmM/R^2,$$

where R is the distance separating their centers and $C = 6.67 \times 10^{-11}$ newton-meters^2 per kilogram^2. The mass of the earth is approximately 5.98×10^{24} kilograms, and its radius is approximately 6.379×10^6 meters. Therefore a mass of 1 kilogram is attracted to the earth at sea level with a force of

$$F = 6.67 \times 10^{-11} \times 5.983 \times 10^{24} \times 1/(6.379 \times 10^6)^2 = 9.807 \text{ newtons.}$$

Our g is therefore equal to CM/R^2, and for this example g represents the gravitational attraction per unit mass. In this use of g the term "gravitational acceleration" really does not apply to g, for if our mass is resting on the surface of the earth, no acceleration actually takes place.

On the other hand, if we hold our 1-kilogram mass above the surface of the earth and release it, the gravitational force $F = 9.807$ newtons still attracts it, and by Newton's law of motion this force must equal the mass times its acceleration. Thus the 1-kilogram mass experiences an acceleration of 9.807 meters per second squared, and the term "gravitational acceleration" for this example is an apt descriptor of g.

A difference in commercial practice between metricized countries and the U.S. is that in metricized countries one typically sells by mass units (kilogram, frequently abbreviated as "kilo"), while in the U.S. one sells by weight units (pound), which implies force. To convert, at sea level we have

1 kilogram = 1/14.593 slug = 0.0685 slug.

A kilogram mass of any substance would thus weigh 0.0685 slug times 32.174 feet per second squared = 2.205 pounds (force), or 9.807 newtons (force).

While the kilogram, meter, second, and degree Celsius are the basic SI units, the SI system goes on to define new names (generally based on the proper names of distinguished scientists of the nineteenth century) for various combinations of the basic units that frequently appear. Consequently we also have the additional units that appear in Table 1.1.

Table 1.1. SI derived units

Quantity	Unit	Relation
force	newton	1 kilogram-mass accelerated at 1 meter per second squared
	kilogram-force	1 kilogram-force = 9.807 newtons
	dyne	10^5 dynes = 1 newton
pressure	pascal	1 pascal = 1 newton per meter squared
	kilopascal	1 kilopascal = 1,000 pascals
work or energy	joule	1 joule = 1 newton-meter
	kilojoule	1 kilojoule = 1,000 joules
power	watt	1 watt = 1 joule per second
frequency	hertz	1 hertz = 1 cycle per second

From an engineering point of view, the size of the base unit for pressure, the pascal, is not very convenient. To appreciate its magnitude, think of an apple that weighs about 1 newton (0.22 pound). If we mash the apple into apple sauce, and spread it evenly over a small table that is 1 meter by 1 meter in size, the small pressure that we have applied to the table has a magnitude of 1 pascal! Thus for pressures whose magnitudes lie in the range of usual engineering interest, the numbers we have to deal with in terms of pascals become quite large. (Atmospheric pressure, for instance, is about 14.7 pounds per square inch in British units, or about 101,325 pascals in the SI system.) For this reason, a unit of pressure close to an atmosphere, the bar (1 bar = 100,000 pascals), is often used. Other attempts to address this issue exist. For instance, it is not unusual to see European tire pressures quoted either in kilogram-force per centimeter squared (1 kilogram-force per centimeter squared = 0.980,66 bar) or in atmospheres (1 atmosphere = 1.0133 bars). We will instead use kilopascals as our basic SI pressure unit for pressure levels in the range of atmospheric and above. Similarly, we will use kilogram and kilojoule in our work rather than gram and joule, since their magnitudes are better suited to engineering calculations.

Even at low pressure levels, again the pascal is not a particularly convenient unit. In many measurements at these levels, such as in biology and medicine (e.g., blood pressure and intraocular pressure), standard practice is to use millimeters of mercury as the pressure unit, where 1 millimeter of mercury (1 mmHg, also denoted as 1 torr, named after Torricelli) equals 0.019337 pound per square inch equals 133.322 pascals.

Up to this point we have been writing out the units in full. Abbreviations as shown in Table 1.2 will be used in the remainder of the text. Also, since one of the principal advantages of the SI system is the use of powers of 10 as multiplication factors, prefix names as given in Appendix A will be used.

Table 1.2. Abbreviations of units

Quantity	SI		British	
	Unit	Unit abbreviation	Unit	Unit abbreviation
length	meter	m	feet	ft
	millimeter	mm	inch	in
time	second	s	second	s
mass	kilogram	kg	slug	slug
			pound mass	lbm
force	newton	N	pound	lb
pressure	pascal	Pa	pound per square inch	psi
	kilopascal	kPa	pound per square foot	psf
work	joule	J	foot pound	ft-lb
	kilojoule	kJ		
power	watt	W	horsepower	hp
	kilowatt	kW		

The sole dimension that has the same units in both systems, time, is the one unit that is not completely decimalized, and that has the most complicated "system." Typically, for time units less than a second, decimal units are used for both British and SI units, and so we will use seconds as the basic time units in most of our calculations. For time intervals greater than a second, the numbers 7 (days in a week), 24 (hours in a day), 28, 29, 30, 31 (days in a month), 52 (weeks in a year), 60 (seconds in a minute, minutes in an hour), and 365 (days in a year) all appear. Our present system is a combination of an early Egyptian system, which divided the day into 24 hours, and a Babylonian system, which introduced the divisions of hours and minutes by 60. Because 24 and 60 had more common factors than 10, and since computers capable of division were not commonly available until the mid-twentieth century, division was simpler in this system.

Over the centuries, a number of attempts have been made to fully decimalize time, but all foundered on the fact that, contrary to the case for length and mass units, a truly decimal time does not fall close to present units. A 20-hour day would come close to our present units, but unfortunately would not be truly decimal. Dividing the day into 100 units gives units that are shorter than is convenient. Metrified time would necessitate changes in time zones, church calendars, political holidays, union contracts, the average work week, in fact a seemingly endless list of our everyday lives. The political and economic problems that would be encountered in such changes are obviously tremendous, and would dwarf any resistance we have already seen in the change to length and mass metrication. This and other reasons made it politically and economically expedient to sacrifice time metrication in order to keep length and mass metrication when metrication took place in the last half of the twentieth century. The net result is that any changes in time units have for a vast multitude of reasons been placed on hold for the foreseeable future.

5. Stress and Pressure

The elementary definition of *stress* is a force divided by the area on which it acts. This is too simple for actual use, for both force and area have direction, and our experience with vectors does not tell us anything about vector division. Stress, strain, rate of deformation, and the moments of inertia all share similar bidirectionalities. (This bidirectional nature of stress is why, when studying elementary mechanics of solids, you used the graphical form of the transformation of axes formula, Mohr's circle, to transfer stress components from one set of axes to another in two dimensions.)

For a more detailed discussion of stress, we consider a two-dimensional surface. We first decompose the force acting on the surface into components normal and tangential to the surface. Therefore we can represent the state of stress on that particular surface as being composed of a ***normal stress component*** (the force component normal to the surface divided by the surface area) together with a ***tangential stress component*** (the force component tangent to the surface divided by the surface area).

More formally, if we consider a small plane area ΔA in a fluid, we can divide the force into a component ΔF_n normal to the area and a component ΔF_t tangent to it. The normal stress is defined in the limit as the area becomes small as

$$\text{Normal stress} = \lim_{\Delta A \to 0} \Delta F_n/\Delta A, \qquad\qquad (1.5.1)$$

and the tangential stress by

$$\text{Tangential stress} = \lim_{\Delta A \to 0} \Delta F_t/\Delta A. \qquad\qquad (1.5.2)$$

Generally, the value we get for the normal stress will depend on the orientation of the area ΔA, that is, on whether the area is horizontal, vertical, or at some angle between these values. Since the decomposition of the force depends on the orientation of the area, we see that if at the same point in space we had chosen an area of the same size but with a different orientation, we would have arrived at quite different values for normal and tangential stress components.

Since stress components have the directions of both the force and the orientation of the area associated with them, they are clearly more complicated than vectors. Quantities that have no directions associated with them (such as density, pressure, and temperature) do not change as we look at different orientations of our coordinate system. Such quantities are called *scalars*, and are said to be *invariant* (unchanged) with respect to the orientation of the coordinate system we use. Other quantities such as force and velocity do have directions associated with them, and we call them *vectors*. Their magnitude and directions are also invariant with respect to the orientation of the coordinate system used, but the components are not. To transform the components of a vector from one coordinate system to another, we use the parallelogram law, or some equivalent representation of it.[3]

When we consider quantities such as stress that have two or more directions associated with them, we stop the process of giving special names such as scalar and vector, and call them nth-order *tensors*, where the order n refers to the number of directions associated with the quantity. Thus a scalar (with zero directionality) is a zero-order tensor, a vector (with one directionality) is a first-order tensor, and stress (with two directions) is a second-order tensor. To decide whether a given quantity is a tensor, we must know how it transforms from one set of coordinates to another. The trivial law of transformation for scalars, the parallelogram law for vectors, and Mohr's circle transformation in two dimensions and its three-dimensional counterpart for stress, strain, rate of deformation, and moments of inertia are all tensor transformation laws appropriate to second-order tensors.

We have already used the term *pressure*, relying on previous understanding of the terms from inflating tires and the like, and even its use in political and social science contexts. A more precise definition is, however, necessary for our study of fluids. In a fluid, pressure is usually the numerically largest portion of the normal stress

---

[3]There are quantities that have directionality, but that are not vectors. An example of this is finite angle rotations. The test of whether a quantity is a vector is—does its components transfer from one coordinate system to another according to the parallelogram law.

components. If the fluid is at rest (***hydrostatic***), or is flowing but has no deformation (***rigid-body motion***), then the value of the normal stress will not depend on the orientation of the area ΔA, and the normal stress is independent of direction. In this case, we call the negative of the normal stress the "pressure p."

Pressure is a scalar, and because of the negative sign involved in the definition, a positive value for the pressure indicates a compressive stress, i.e., a stress pushing on a surface. Normal stresses are by convention considered to be positive when they are associated with tensile forces, i.e., forces pulling on a surface. Fluids, however, can withstand only very small tensile forces, and then only if they are almost absolutely free from impurities such as dirt and dissolved gases.

A more general definition of pressure in a fluid has to distinguish between whether the flow is compressible or incompressible. For compressible flows, we define pressure to be the thermodynamic pressure as given by a state equation such as the perfect gas law. For incompressible flows, there is no equation of state (such as the ideal gas law $p = \rho RT$), and we take the pressure to be the negative of the mean value of the three normal stress components at a point.

The dimensions of pressure are the same as stress, that is, force per unit area. In terms of the British system of units, this becomes pounds per square inch (psi) or pounds per square foot (psf). The SI system, as we have already pointed out, uses either the kilopascal (kPa), which is 1,000 N/m^2, or the bar, which is 100,000 N/m^2.

When pressure is referenced to an absolute vacuum it is referred to as ***absolute pressure***. Most pressure-measuring devices measure pressure relative to some reference pressure, such as atmospheric. Consequently we frequently deal with ***gage pressure***, which is absolute pressure minus reference pressure. Since in most engineering contexts we are interested in pressure differences, or in forces above and beyond those due to atmospheric pressure, working with gage pressure causes no problems. The exception is when we use an equation of state, such as the ideal gas law, where it is necessary to use absolute pressure. In the cases where absolute pressure is used (principally when we study compressible flows and have to use the state equation), we will designate absolute pressure by writing either psia, psfa, or (absolute) after the pressure value. Pressure levels below atmospheric are frequently denoted as suction, or vacuum, pressure. (See Figure 1.5.) Unless otherwise stated, gage pressure is to be assumed as given. Note that

$$p \text{ (absolute)} = p \text{ (gage)} + p \text{ (atmospheric)} \tag{1.5.3}$$

and

$$p \text{ (vacuum)} = p \text{ (atmospheric)} - p \text{ (absolute)}. \tag{1.5.4}$$

The following are examples of these various relative pressures. An atmospheric pressure of 14.7 psi has been used in each example.

$$p \text{ (gage)} = 30 \text{ psi is the same as } p \text{ (absolute)} = 44.7 \text{ psi.} \tag{1.5.5}$$

$$p \text{ (absolute)} = 40 \text{ psi is the same as } p \text{ (gage)} = 25.3 \text{ psi.} \tag{1.5.6}$$

$$p \text{ (vacuum)} = 4.7 \text{ psi is the same as } p \text{ (absolute)} = 10 \text{ psi.} \tag{1.5.7}$$

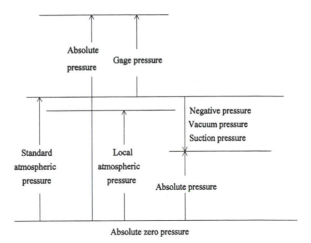

Figure 1.5. Visualization of absolute, gage, and vacuum pressures.

6. Fluid Properties

a. Mass and weight densities

Mass density of a fluid is defined as the mass of the fluid per unit volume. The Greek letter ρ is usually used to denote mass density, more simply called ***density***. The reciprocal of mass density, specific volume v_s, or volume per unit mass, is commonly used in thermodynamics. Mass density and specific volume are state properties of a fluid. That is, they generally depend on the local thermodynamic state, which can be specified by two independent properties such as pressure and temperature. Their values thus can vary with time and place in the fluid.

A formal definition of mass density within the framework of our continuum hypothesis is as follows: Consider a fluid of total mass "m" contained in a volume "V" centered at a point "P" in a fluid. The ratio m/V is the average mass density of the fluid in the volume V. If we consider different volumes centered at the point P, we would possibly find different values for the ratio of mass to volume. If we take the volume so small that there were only a few molecules in it, the ratio would fluctuate rapidly as molecules enter and leave the volume, and we would be beyond our continuum model. However, if we start with a sufficiently large volume and then consider successively smaller volumes surrounding the point P, we would find that the ratio of mass to volume would settle down to a value as long as we did not get down to the molecule size. Thus we define the mass density at the point P as the limit as the volume V shrinks to zero, or formally by

$$\rho = 1/v_s = \underset{V \to 0}{\text{limit}} \, (m/V). \tag{1.6.1}$$

Useful values of mass density to remember are those for freshwater at standard pressure (760 mmHg) and 4°C, which is

$$\rho_{H_2O} = 1.94 \text{ slugs/ft}^3 = 1,000 \text{ kg/m}^3, \tag{1.6.2}$$

and for air at standard pressure and 10°C, which is

$$\rho_{air} = 0.00244 \text{ slug/ft}^3 = 1.26 \text{ kg/m}^3. \tag{1.6.3}$$

It is frequently convenient to deal with the weight density, or *specific weight*, rather than with mass density. Specific weight (denoted by the Greek letter γ) is defined as the mass density times the gravitational constant *g*. Thus

$$\gamma = \rho g \tag{1.6.4}$$

and

$$\gamma_{H_2O} = 62.4 \text{ lb/ft}^3 = 9,807 \text{ N/m}^3 \tag{1.6.5}$$

for freshwater under the standard conditions used for equation (1.6.2), and

$$\gamma_{air} = 0.0765 \text{ lb/ft}^3 = 12.02 \text{ N/m}^3 \tag{1.6.6}$$

for air under standard conditions.

The *specific gravity* of a substance, denoted by SG, is defined as the specific weight of that substance divided by the specific weight of a given fluid under standard conditions. For liquids, water is commonly used as the reference fluid, and consequently, we have the specific gravity of water given as

$$SG_{water} = 1.0 \tag{1.6.7}$$

at standard conditions. For air under standard conditions this would give a value of

$$SG_{air} = 0.00126 \tag{1.6.8a}$$

if water were used as the reference fluid. More commonly for gases, air at standard conditions is used as the reference fluid so that

$$SG_{air} = 1.0. \tag{1.6.8b}$$

(*Note*: Some fields such as chemistry and physics, and some industries such as the petroleum industry, might use other fluids as more convenient reference for their definitions of specific gravity. In this book we will use the term "specific gravity" only when speaking of liquids, and water will be the only reference fluid.)

b. Bulk modulus and coefficient of compressibility

The compressibility of a substance is defined as the amount of small pressure change needed to change a given volume of fluid a given small amount. The property used to describe this is *bulk modulus*. A formal definition of this within the continuum hypothesis is

$$K = -v_s \frac{dp}{dv_s} = \rho \frac{dp}{d\rho},$$ (1.6.9)

where p denotes pressure. The reciprocal of bulk modulus is termed the **coefficient of compressibility**. Bulk modulus has dimensions of force per unit area. Typical values are

$K = 311{,}000$ psi $= 2.14$ GPa (gigapascals) (1.6.10)

for water at 60°F (15.5°C), and

$K = 20.6$ psi $= 0.142$ MPa (megapascals) (1.6.11)

for air.

Example 1.6.1. The compressibility of water
Two cubic feet of water at 60°F is contained within a 10-in-diameter piston. How much force must be applied to change the water volume 1%?

Given: $K = 311{,}000$ psi, $V = 2$ ft^3, $\rho = 1.94$ slugs/ft^3 before application of the force, $\Delta V = -0.02$ ft^3, $p = 0$ gage before the force is applied.
Approximations: Differentials can be approximated by incrementals over a sufficiently small range of the parameters.
Solution: From (1.6.9), $K = -v_s \, dp/dv_s = -V \, dp/dV$ since the total mass of the water does not change. Approximating the differentials by the incrementals gives

$K = -V \, \Delta p/dV = -V \, \Delta p/\Delta V,$

which after solving for Δp becomes

$\Delta p = -K \, \Delta V/V = -311{,}000 \times (-0.02)/2 = 3{,}110$ psi.

The force needed is then

$F = A_{piston} \, \Delta p = \pi \times 5^2 \times 3{,}110 = 244{,}259$ lb.

An important property related to the bulk modulus is the **speed of sound**, or **sonic speed**, defined by

Sonic speed $= \sqrt{K/\rho}.$ (1.6.12)

Using the above values for density and bulk modulus we see that sonic speeds in water and air are

Sonic speed in water $= \sqrt{311{,}000 \times 144/1.94} = 4{,}800$ ft/s $= 1{,}463$ m/s,

Sonic speed in air $= \sqrt{20.6 \times 144/0.00244} = 1{,}102$ ft/s $= 336$ m/s.

For most of the problems we will consider when the fluid is a liquid, the fluid will be treated as if it were incompressible. This is because in most flows the pressure changes involved are small enough, and the bulk modulus big enough, that significant changes in density do not occur.

An important parameter useful for deciding whether fluid compressibility need be considered negligible for a given flow is the **Mach number**, defined as

Mach number = M = local flow velocity/sonic velocity.   (1.6.13)

If the Mach number is less than 0.25 or so, compressibility effects are generally minor, and to a good approximation the flow can be considered to be incompressible, regardless of whether the fluid is a liquid or gas. Note that we are not necessarily assuming that the *fluid* is incompressible, but rather that the *flow* is behaving in an incompressible manner. This is a valid approximation for most physical flow phenomena in liquids, and even for many flows in gases, providing that the flow is such that the Mach number is sufficiently small. For compressibility-associated phenomena such as sound transmission, or flows where the Mach number is greater than 0.25 or so, compressibility effects become important and must be included, even for liquids.

The distinction we have made above between "fluid properties" and "flow properties" is an important one. We will encounter this again and again in our studies, and it is important for our understanding of fluid mechanics that this distinction always is clear.

c. Vapor pressure

A liquid will change to the vapor state at a given temperature when the local pressure equals the saturation pressure, the pressure (a function of the temperature) at which the liquid and vapor states can coexist. The pressure at which this state change takes place is termed *vapor pressure*. It is the pressure at which at a given temperature the liquid "boils." In thermodynamics it is customary to plot the equilibrium state of a substance in a three-dimensional state diagram such as Figures 1.1 and 1.2, where the coordinate axes are pressure, specific volume, and temperature. For any substance there is a distinct portion of the surface above the triple line where the substance consists of a mixture of liquid and vapor. When this portion is projected onto the pressure-temperature plane, giving a plot of saturation pressure as a function of temperature, it is called the *vaporization line*, or alternatively, the *saturation line*. Typical values for water, detailing a portion of the curve in Figure 1.4 between the critical and triple points (the triple point is the end-on view of the triple line as seen in the temperature-pressure plane), are shown in Figure 1.6.

Situations in which fluids can be at vapor pressure include the mercury barometer, where the mercury in the cavity above the liquid mercury column exists as a vapor and is at vapor pressure, and when water is heated to its boiling point. More important, as we shall see in Chapter 3, for liquids it is possible for the local pressure in the fluid to be reduced to vapor pressure by the dynamics of the flow itself, at which point the liquid changes to the vapor phase.

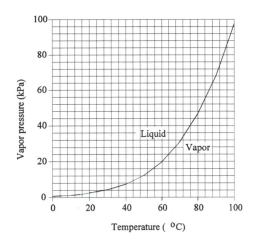

Figure 1.6. Dependence of vapor pressure on temperature for water.

Example 1.6.2. Demonstration of vapor pressure

Take a glass tube of roughly 1/2-in diameter and 8-in length and heat it with a Bunsen burner or propane torch at about midlength. Draw it so that the tube necks down in its middle to about half its original diameter. Attach it to a water faucet with a hose, as shown in Figure 1.7. Open the water faucet only slightly. The flowing water should appear transparent inside the tube. Increase the flow slowly until a cloudy patch just appears at the necked-down region. Explain this cloudy patch.

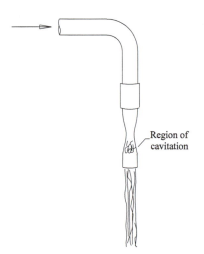

Figure 1.7. Demonstration of cavitation.

As the flow velocity in the necked-down portion of the tube increases, the pressure there decreases due to the faster flow. At the point where the cloudiness appears, the pressure in this region has been reduced to vapor pressure. The cloudiness you see is in fact due to small bubbles of water vapor in the flowing water. If you listen closely, you will hear a hissing noise, which is the noise made by bubbles hitting the tube wall and collapsing. Increasing the flow rate will increase the size of the cloudy region, and also make the noise louder. Downstream of the necked-down region, the cloud of bubbles disappears. This is because the fluid in this region is flowing slower than in the neck, and the pressure is increased over that in the neck, causing the vapor to become liquid again.

The phenomenon described in the above experiment is called *cavitation*. Cavitation can strongly affect the performance of pumps and turbines, to the extent that in extreme cases it can cause mechanical damage to components and even stop the operation of the device. When gas bubbles formed by cavitation strike a solid surface the gas bubble tends to collapse, forming a liquid jet that in turn impacts the solid surface. This can result in erosion of the surface, thereby damaging pumps, turbines, and propellers, and eventually leading to the repair or replacement of parts.

d. Surface tension

At the interface between a liquid and another liquid or a gas, the difference in the molecular structure of the two substances results in an imbalance of the molecular forces. These forces can be quite strong, but they decay very rapidly with distance from the interface. From a continuum mechanical point of view, the interface behaves as if it were a very thin elastic film, or layer, with this film possessing elastic properties. Familiar examples of this interfacial effect are soap bubbles, the rise of oil in wicks and of water in soil, the wetting of paper, the breakup of jets, and the ability of small insects to walk on water.

The amount of force per unit length necessary to deform this interfacial surface is called the *surface tension*, and is denoted by the Greek letter σ. Representative values are given in Appendix B for various fluids.

Example 1.6.3. Measurement of the surface tension of a film
Construct a wire "U" by taking an ordinary wire and bending it 90° twice. Lay a straight wire on top of the U so that it makes a right angle with the two legs of the U. Dip the assembly into a liquid containing a mixture of water and a small amount of detergent, of the type used for washing dishes. (The surface tension of this liquid can be strengthened by adding a little glycerine to the solution.) An approximately rectangular soap film is thus formed. You will have to exert a force to hold the straight wire in place. How is this force related to the surface tension?

In this example, since the liquid film has two surfaces exposed to air, it is helpful to think of the problem as consisting of two separate liquid-gas films that lie next to one another.

If the wire has a length L in contact with the film, the force needed to hold the films in equilibrium will then be

$$F = 2L\sigma,$$

where σ times L is the force on one film, and the 2 takes into account the second film. Generally, this 2 multiplier will occur in a liquid film separating a gas from a gas (e.g., as in a soap bubble blown by a child), where there are actually two films present. A film separating a gas from a liquid (e.g., a bubble in a carbonated beverage) is a single film, and would not have the 2 multiplier.

The surface tension of a liquid is always a function of the solid or fluid with which the liquid is in contact. If a value for surface tension is given in a table for oil, water, mercury, or whatever, and the contacting fluid is not specified, it is safe to assume that the contacting fluid is air. The value of σ can be determined experimentally by stretching a film of fluid and measuring the required force (as in Example 1.6.3), by measuring the shape of a drop suspended under a tube, or by several related methods. However, surface tension can be greatly affected by surface contaminants, and the surface tension of a "clean" fluid (a state that is very difficult to maintain) can have a σ that differs appreciably from that of the same fluid in a "dirty" condition. ***Surfactants*** are chemicals developed to change the surface tension of a liquid for a desired purpose. Detergents and wetting agents are familiar examples of liquids with surfactant properties. When bubbles or films are made for experimental purposes, those made from pure water are very fragile and burst easily. The addition of soap, detergents, and/or glycerine "strengthens" the bubble considerably.

Vapors that may be present in a laboratory (alcohol, for example) can also have an appreciable effect on surface tension. An example of this can be seen in the phenomenon of "wine tears" (Figure 1.8). If a fortified wine (wine to which additional alcohol was added in its manufacture, such as port or madeira) is poured into a wine glass, a thin film of wine can be seen to rise on the side of the glass, forming a thicker bead of wine on the top of the film. Wine droplets form on the bead that then fall back into the glass. The explanation of this phenomenon is that the surface tension of the wine pulls the film up, but the thinness of the film makes it easier for the alcohol in the wine to evaporate, thereby changing the surface tension of the wine in the film. The net effect is that the surface tension of the wine has caused more liquid to be raised than can now be supported, and the wine (now somewhat reduced in alcoholic content) falls back into the glass in the form of "tears."

Another example of surface tension effects can be seen in the behavior of the jet of a fluid with reasonably large surface tension issuing slowly from a horizontal round tube, as in Figure 1.9. The fluid at the outside of the jet is moving slower than the fluid at the center of the jet. Having less momentum, it tends to "fall off" of the main jet. A very thin film of fluid immediately forms and suspends this fallen rope of fluid from the main jet, which it joins further downstream.

When a liquid-gas interface ends at a solid wall α, the ***contact angle*** that the interface makes with the wall (Figure 1.10) is a function of what the two fluids are,

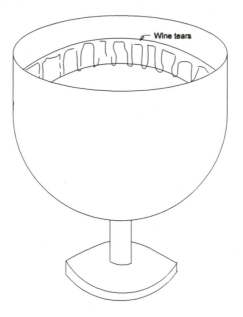

Figure 1.8. "Wine tears" caused by the forces of surface tension, gravity, and evaporation.

and also can depend on the composition of the wall. Contact angle is defined as the angle between the wall and the surface, measured in the liquid from the wall. A contact angle of zero is referred to as **perfect wetting**. A contact angle of 180° is referred to as **complete nonwetting**. If the wall attractive forces (**adhesive forces**) are stronger than the fluid attractive forces (**cohesive forces**), the surface of the fluid is concave upward at the wall, and we say that the fluid "wets" the solid. Water in contact with glass is an example of this. If the fluid attractive forces are stronger than the wall attractive forces, the surface of the fluid is concave downward at the wall, and we say that the fluid does not wet the solid. Mercury in contact with glass is an example of this.

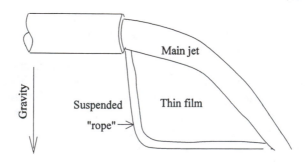

Figure 1.9. Surface tension effects in an exiting laminar jet.

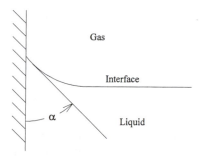

Figure 1.10. Contact angle definition.

For this case of a liquid-gas interface meeting a solid wall, the contact angle is a function of the three materials involved. Contact angles for several different combination of solids and liquids are given in Appendix B.

For a small spherical gas bubble, the relation of surface tension to the various parameters involved can be easily calculated. Consider the forces acting on a hemisphere as in the free-body diagram shown in Figure 1.11. Let p_i and p_o be the pressures inside and outside the bubble and r be the radius of the bubble. If the bubble is immersed in a liquid, then a summation of the pressure and surface tension forces perpendicular to the base of the hemisphere (see the free-body diagram in Figure 1.11) gives

$$(p_i - p_o)\pi r^2 = 2\pi r\sigma,$$

or, upon dividing by the area πr^2,

$$p_i - p_o = 2\sigma/r. \tag{1.6.14}$$

The force on a small area due to the outer pressure p_o acting on the spherical part of the hemisphere is locally perpendicular to the hemisphere. The resultant, or net, force due to p_o is perpendicular to the hemisphere base.

If gas is enclosed in a liquid bubble that in turn is immersed in a gas, the bubble consists of a thin liquid layer surrounded by two gases. Then, since the liquid is

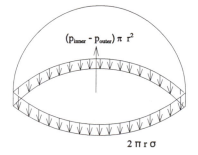

Figure 1.11. Forces acting on a hemisphere-shaped bubble.

exposed to gas on two sides, the bubble has two interfaces (as in Example 1.6.2), and the surface tension force is double that of the previous example. In that case, (1.6.14) becomes

$$p_i - p_o = 4\sigma/r. \tag{1.6.15}$$

Equations (1.6.13) and (1.6.14) are sometimes referred to as the "law of Laplace."

Example 1.6.4. Demonstration of surface tension of a bubble
Take two small tees (the glass tees used to connect tubing that are commonly found in chemistry laboratories work well for this purpose) and connect them with tubing and valves as shown in Figure 1.12. Close the three valves, and then dip the free ends of the tube in a soap solution. (Again, a little glycerine added to the solution will make your bubbles stronger.) By opening the two bottom valves one at a time and blowing into the tees, create two bubbles of different size. Close each of these valves after forming that bubble. If you slowly open the valve connecting the two bubbles, what will happen? Which of the bubbles will grow in size?

The larger bubble will grow, and the smaller one will shrink. According to equation (1.6.15), the larger bubble contains air at a pressure lower than that of the air in the smaller bubble. The greater pressure in the smaller bubble will cause air to flow from the smaller bubble into the larger bubble, increasing its size while at the same time the smaller bubble shrinks. If you continue the experiment long enough the large bubble will burst.

Example 1.6.5. Demonstration of surface tension at a small hole
Take a fine wire mesh and cut and shape it so that it fits snugly over the mouth of a bottle partially filled with water. You should be able to turn the bottle upside down without having any water come out. Inserting a fine wire through the openings in the mesh should

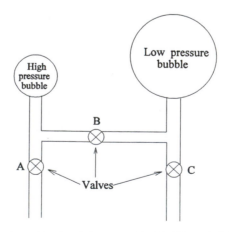

Figure 1.12. Demonstration of Laplace's law, described in Example 1.5.1.

not cause water to flow out. Obviously this effect is due to surface tension. Why is the wire mesh necessary?

The reason for the wire mesh is that each opening of the mesh allows a small confined surface to form, and the surface tension in each of the openings is sufficient to support a small amount of water. The large number of openings in the mesh substantially increases the amount of water the mesh is able to support. (Spreading a little oil on the mesh before performing this experiment keeps the water from wetting the mesh and helps things out.)

CAUTION: Just in case your mesh is not properly fastened to the bottle, it is a good idea to perform this experiment in a location that won't be damaged if the weight of the water is too great for the surface tension forces.

Early Roman literature tells the story of the vestal virgin who was accused of violating her vows. She was condemned to death unless she could transport water in a sieve. She successfully did this, thereby saving her life. Apparently she knew more about fluid mechanics than did her judges.

An important question involving surface tension is whether a drop of liquid placed on the surface of another liquid will remain as a drop because of the surface tension forces or will spread out as a film because of the gravity force. We can find the conditions that divide the two cases by considering the free-body diagram of Figure 1.13. If we assume that the drop stays a drop, then summation of the three surface tension forces gives

$$\sigma_A \sin \alpha - \sigma_{AB} \sin \beta = 0 \qquad (1.6.16)$$

in the vertical direction and

$$\sigma_B - \sigma_A \cos \alpha - \sigma_{AB} \cos \beta = 0 \qquad (1.6.17)$$

in the horizontal direction. With the help of some simple algebra and trigonometry, these can be rewritten as

$$(\sigma_A/\sin \beta) = (\sigma_{AB}/\sin \alpha) = [\sigma_B/\sin(\alpha + \beta)] \qquad (1.6.18)$$

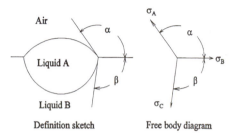

Definition sketch   Free body diagram

Figure 1.13. Angle definitions for Example 1.5.3.

In equations (1.6.18) (there are two equal signs, hence two equations) the surface tensions are known. If we can find values of the angles α and β that satisfy these two equations, then a drop is possible. If we cannot find such values, then spreading will occur.

Example 1.6.6. Surface tension

A drop of water at 25°C (liquid A) is floated on mercury (liquid B). The surface tensions of these liquids measured with respect to air are 0.0756 N/m (σ_A) and 0.513 N/m (σ_B), respectively. What measurements must we make to determine the surface tension σ_{AB} of the interface between fluids A and B?

Given: $\sigma_A = 0.0756$ N/m, $\sigma_B = 0.513$ N/m.

If the diameter of the drop is large compared to the radius of curvature of the drop where the two liquids and the air meet, the problem can be considered to be essentially two-dimensional. From the free-body diagram of Figure 1.13, summation of forces in the vertical (y) and horizontal (x) directions gives

$$\sigma_A \sin \alpha - \sigma_{AB} \sin \beta = 0 \tag{a}$$

and

$$\sigma_B + \sigma_A \cos \alpha + \sigma_{AB} \cos \beta = 0, \tag{b}$$

where the length that has been used is a unit distance into the page. From the first force summation (a), solving for σ_{AB} we have

$$\sigma_{AB} = \sigma_A \sin \alpha / \sin \beta. \tag{c}$$

Using (c) to eliminate σ_{AB} in the horizontal force equation (b), we find after a little algebra and trigonometric substitution that

$$\sin \beta = -(\sigma_A/\sigma_B) \sin(\alpha + \beta) = -0.147 \sin(\alpha + \beta). \tag{d}$$

Thus by measuring the combined angle $\alpha + \beta$, we can calculate ß by equation (d), then subtract β from $\alpha + \beta$ to find α, and use equation (c) to calculate σ_{AB}. Since the number (0.147) multiplying $\sin(\alpha + \beta)$ is less than 1, we are assured that for any combined angle $\alpha + \beta$ that we measure, $\sin \beta$ will lie in the range between -1 and $+1$, and thus no spreading of the water drop upon the mercury will occur. A plot of equation (d) is given in Figure 1.14.

Gas-liquid interfaces appear often in fluid mechanics applications. Rivers, channels, and lakes are common examples, as is water in a drinking glass, washbasin, or bathtub. Such interfaces are called *free surfaces*, "free" in the sense that they are able to easily change in shape. The term "free surface" is used for any gas-liquid interface, regardless of whether surface tension is important to the dynamics of the interface.

The examples we have given for surface tension involve simple algebra for their solution. More complicated geometries, such as the films formed when a wire frame

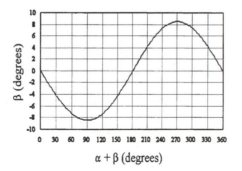

$\alpha + \beta$ (degrees)

Figure 1.14. Results of equation (d), Example 1.6.6.

is dipped into a soap solution, tax mathematical methods to the limit. An approach used is to realize that the surface formed by such a film is a ***minimal surface***. That is, its surface has the smallest area possible subject to the constraint of the wire frame support. Such mathematical problems are referred to as ***Plateau's problem***, and a sizable literature has built up for dealing with them.

e. No-slip condition

Another example of the effect of molecular forces when different molecular structures interact is seen when a fluid is in contact with a solid surface. It is found experimentally in all cases (again with the possible exception of rarefied gases) that the relative velocity between the solid surface and the fluid is zero at the surface of contact. In these regions, locally, the molecular bonding (due to the differing molecules in the solid and the fluid) is extremely strong. While the fluid velocity will vary with distance from the solid, because of the molecular bonding at the surface of the wall, the fluid has zero velocity with respect to the solid. We describe this by saying that the fluid does not slip on a solid boundary.

This ***no-slip condition*** is not an obvious one by any means. If you stand on a steep bank of a rapidly flowing river, it appears as if the water near the bank is flowing as fast as it is in the middle of the river. Similarly, being next to a building wall on a windy day does not provide much shelter from the wind. This is because the flow in such cases is turbulent, and the region in which the flow speed changes to the fast flow to zero velocity is very very thin. The no-slip condition was first proposed by Daniel Bernoulli in 1738 on the basis of several experiments. It was much contested in the nineteenth century, with prominent scientists such as Poisson and Navier contending that slip must take place, and Stokes (after first believing in some form of slip) and Poiseuille contending that experiments showed no slip. Maxwell later carried out calculations based on a molecular model and showed that unless the molecular velocity varies appreciably over the mean free path length, no slip will occur. Today this is the generally accepted state of affairs, with experimental measurements confirming the no-slip condition. While there are still occasional reports by investigators that an exotic fluid has been found that *does* slip, to date careful tests have not verified any of these claims.

Readers interested in a more detailed account of the history of the no-slip condition should read the account in the note at the end of *Modern Developments in Fluid Dynamics*, vol. II, referenced at the end of this chapter.

f. Absolute viscosity

From the definition of a fluid, we can expect that when the fluid is in motion, deformation will occur due to applied stresses. To complete the description of the flow, it is necessary to know the relationship between the stresses and the resulting deformation.

Newton was the first to successfully propose such a relation. He put it in the context of the idealized experiment shown in Figure 1.15. Here the fluid is contained between two large, parallel plates placed a distance b apart. The bottom plate is held stationary, and a force F is applied to the top plate. As a result, the top plate moves with a velocity V. At the lower plate, because of the no-slip condition the fluid velocity will be zero, and at the upper plate it will be V, again consistent with the no-slip condition.

Newton reasoned (we will demonstrate this more carefully in Chapter 6) that between the two plates the velocity will vary linearly with distance from one of the plates, according to

$$v(y) = Vy/b. \tag{1.6.19}$$

If we look at a very small element in the flow that at the first instance of our observation is a rectangle (Figure 1.16), because of the linearly varying velocity this element will start to deform into a parallelogram. By considering the limit as the height of the element becomes very small, it is seen that, with the horizontal velocity component of point A denoted by $v(y)$, then over a time interval Δt point A will move to the right a distance $v \Delta t$. During this same time interval, point B, a distance Δy above A, moves a distance $[v + (dv/dy) \Delta y] \Delta t$ to the right, since it has a slightly different velocity. Thus B has shifted to the right a distance $(dv/dy) \Delta y \Delta t$ relative to A. The shift per unit distance Δy and per unit time Δt is called the ***time rate of deformation***, and in this example is just the velocity gradient dv/dy. Newton hypothesized that this time rate of

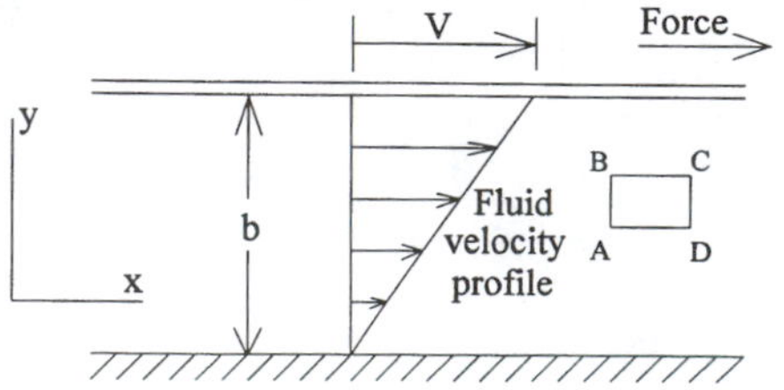

Figure 1.15. Velocity profile for Couette flow between parallel plates.

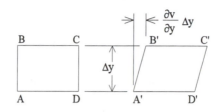

Figure 1.16. Deformation of an infinitesimal element in Couette flow.

deformation is proportional to the shear stress given by the applied force per unit area, or

Shear stress = $\mu \times$ time rate of deformation, $\qquad\qquad$ (1.6.20)

where the time rate of deformation is V/b in Figure 1.15. A fluid obeying a law such as (1.6.20), where all viscous stresses are linearly proportional to the rate of deformation, is called a ***Newtonian fluid***.

The constant of proportionality μ in (1.6.20) is called the ***viscosity***. It is sometimes termed the "dynamic," or "absolute," viscosity. It arises by virtue of the transfer of momentum between molecules, due to both the attractive forces between molecules (the dominant factor affecting the viscosity of liquids) as well as the collisions that occur between the molecules (the dominant factor affecting the viscosity of gases). By this means, adjacent layers of fluid having different velocities (so that there is a velocity gradient) can exchange momentum through the mechanism of viscous shear. This process of transferring momentum is a mechanism that thereby acts to try to smooth out velocity differences. If we were to stop moving the plate in Newton's experiment, viscosity would, over a sufficiently long time, act to bring the fluid to a state of rest.

From (1.6.20) we see that viscosity has the dimensions of force multiplied by time per unit area. In both the British and SI systems of units, there is no special name for the unit of viscosity, and either lb-s/ft^2 or slug/ft-s is used in the British system. In the SI system, either N-s/m^2 or Pa-s is used. Frequently poise, a carryover from the old cgs system, is used, where

1 poise = 0.1 Pa-s = 1/479 lb-s/ft^2 = 0.0020877 lb-s/ft^2. $\qquad\qquad$ (1.6.21)

(The name poise is derived from the proper name Poiseuille, a French physicist-physician who was interested in the flow of blood.)

g. Kinematic viscosity

Very often the viscosity coefficient appears in formulas divided by the mass density. This combination is so useful that it is given its own symbol, ν, and name, ***kinematic viscosity***, where

$$\nu = \mu/\rho. \tag{1.6.22}$$

The name "kinematic" refers to the dimensions of the term, which are length squared per unit time. Thus, only geometric units (kinematic), and not force units (kinetic), appear. Again, no special names have been given to the units, with ft^2/s or m^2/s being preferred, but the older cgs unit stokes is occasionally used (sometimes appearing as "stoke," in an apparent attempt to "singularize" a proper name), where

$$1 \text{ stoke} = 0.0001 \ m^2/s = 1/929 \ ft^2/s. \tag{1.6.23}$$

George Stokes was a famous nineteenth century British fluid dynamicist.

Many other technical units for kinematic viscosity exist. In some cases they are used in conjunction with specific viscometers (devices for measuring viscosity). The Saybolt viscometer, for instance, consists of a container with a hole in the bottom, and the time needed for a given sample size of the fluid to drain out is termed the viscosity in Saybolt universal seconds (SUS, or sometimes SSU). This unit can be converted to other units through a formula of the type

$$\nu = At + B/t, \tag{1.6.24}$$

where A and B are instrument constants (see American National Standard Institute, standard number 211.129-1966). Other units used include the familiar SAE (Society of Automotive Engineers) terminology for motor oils, where the designation indicates not only the viscosity range, but something about the behavior of the oil over a range of temperature. Generally, absolute viscosities do not vary appreciably with pressure unless the pressure becomes very high, such as between the cylinder walls and the pistons of internal combustion engines. Kinematic viscosities, particularly of gases, will vary with pressure because the density varies with pressure. Both viscosities have pronounced variations with temperature. For liquids, increasing temperature reduces the cohesion between molecules, and hence causes the viscosity to decrease. In gases, since viscosity is more due to random motion of the molecules, viscosity increases with temperature.

Typical values for both μ and ν as functions of temperature are found in the plots shown in Appendix B. It is seen that both absolute and kinematic viscosities decrease with temperature for liquids, and increase for gases.

7. Non-Newtonian Fluids

Not all fluids can be described by the simple law for Newtonian fluids (1.6.20). Generally, if the molecular weight of a fluid is of the order of half a million or more, or if the fluid is a nondilute suspension, the fluid behaves as a ***non-Newtonian fluid***, meaning that (1.6.20) may not apply. Suspensions of particles or fibers can also exhibit non-Newtonian effects. Finding an appropriate law to relate applied forces to flow quantities for such fluids is very difficult, since the phenomena encountered in the flows of these fluids can be quite unusual and adequate explanations of the relation of flow to stress state (certainly to the point where it can be codified into a set of equations

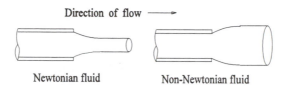

Figure 1.17. Die swell.

that can be used to give a quantitative prediction of such behavior) are not known. The problem is compounded by the fact that many non-Newtonian fluids, when placed in a conventional viscometer, do not exhibit any non-Newtonian effects. This is because in such viscometric flows the non-Newtonian effects appear in the normal stresses, but not the shear stresses.

Several examples of such non-Newtonian flow effects follow.

Die swell. In the formation of plastic rods or filaments, the molten plastic is extruded through a small opening called a "die." When a Newtonian fluid emerges from a die or capillary tube, the diameter of the jet generally can be expected to decrease. When fluids such as polymers are extruded, their diameter may increase as much as 3 or 4 times the original diameter. (See Figure 1.17.) Such fluids exhibit elastic properties as well as viscous properties, and are said to be ***viscoelastic***. A knowledge of the degree of die swell is important in the production of man-made fibers.

Weissenberg effect. When a Newtonian fluid is stirred by a rotating rod, the free surface tends to form a depression near the rod. In the case of a viscoelastic fluid, the liquid at the free surface is seen to climb the rod. (See Figure 1.18.) This climbing phenomenon, called the "Weissenberg effect" after its first describer, K. Weissenberg, can easily be seen when mixing some paints, or in the kitchen when an electric mixer is used to stir dough, egg whites, gelatin, some condensed milks, ketchup, or other "thick" substances with protein structure.

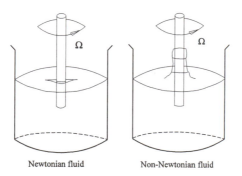

Figure 1.18. The Weissenberg effect.

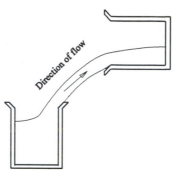

Figure 1.19. Self-siphoning effect.

Self-siphoning effect. When very elastic viscoelastic fluids are poured from one container to another, the fluid can be made to flow uphill by raising the lower container and lowering the upper container. The fluid then can flow up the inside of the (now) upper container, over the edge, and then down into the other container. (See Figure 1.19.) Flows of non-Newtonian fluids are good examples of the dangers inherent in relying too heavily on "physical intuition" in engineering, since effects such as these may lie outside our experience and not within the body of knowledge on which our intuition is based.

Toms effect. When very small amounts (typically 40–50 parts per million, abbreviated ppm) of polymers such as polyethylene oxide (a thickening agent used in water treatment settling tanks) or guar gum (used as a thickening agent in foods) are dissolved in a fluid, the flow rate through a pipeline is increased with no change in pumping pressure, provided that the flow is turbulent. This has been found useful in pumping water on fires in tall buildings, pumping crude oil in long pipelines, and in increasing the speeds of ships, submarines, and torpedoes. The effect is strongly dependent on the concentration of the additive, for when the concentration becomes too large (typically around 1–2%, and sometimes even lower) the liquid becomes a thick gel and its fluidlike properties have all but disappeared. Long-chain molecules can easily be fractured, so if such fluids are subjected to mechanical forces such as in pumps the molecules break down and the effect is lost.

Electrorheologic fluids. A somewhat different class of non-Newtonian fluids was discovered in the late 1940s. These fluids are suspensions of fine particles (typically 0.1–100 μm in diameter) in a nonconducting oil. Under ordinary conditions the suspension behaves as a Newtonian fluid. However, when an electric field is applied, the fluid rapidly turns into a gel-like solid. This process is reversible, the suspension returning to the fluid state when the electric field is removed.

While the exact mechanism of this behavior is still controversial, observation of the fluid under a microscope shows that in the absence of an electric field, the suspended particles are distributed more or less uniformly. When the electric field is applied, the suspended particles become aligned in a chainlike structure. The strength of this chain structure is proportional to the field strength. When the gel is subjected to shear stresses, the particles become separated, but retain their electric charge as long as the electric field is applied. The charge attraction between the particles provides a resistant shear force above that of molecular viscosity.

A solution of corn starch in cooking oil is a simple example of such a fluid. Other oils exhibiting this effect that have been found useful for engineering applications are silicone oil, mineral oil, and chlorinated paraffin. Many of the particles that have been used require the presence of water, thereby limiting the temperature range of their usefulness. However, in 1988 it was found that aluminosilicate ceramic particles operate well without water being present, and thus can be used at higher temperatures than is possible when water is present.

Since the discovery of such fluids, applications that have been suggested include the following: using the electric field as a valve with no moving mechanical parts to control the flow of these fluids; as electrically actuated brakes; for the transmission of mechanical power; as stiffening devices in helicopter rotors or fishing poles. Actual use has awaited the development of fluids capable of functioning at temperatures higher than the boiling point of water. The discovery of anhydrous suspensions with these electrorheologic properties now leaves the door open to a wide variety of engineering applications.

8. Problem-solving Approach

For an engineer to understand fully the very fundamental physical principles of fluid mechanics, it is necessary that he or she be able to apply them. In applying the basic laws that we have just developed, it is good engineering practice to go through a formal procedure and approach for every problem until the procedure becomes second nature. A stepwise procedure useful for both "homework" problems you encounter in school and, later, "real" problems you encounter in engineering practice, is as follows:

What quantity or quantities are being sought?

Make a list of what information is being sought. This list should be short and succinct. This may seem to be an obvious thing to do, but too often it is lost sight of in carrying out the work.

What information is given?

Make a succinct list of which flow quantities are known or given in a given problem. List relevant fluid properties that can be found in tables. Make a rough sketch of the flow geometry, and label points on the sketch that might be useful in your analysis.

List your assumptions

In every problem, you, as the engineer, based on your training and experience, must make certain assumptions about the parameters in a given problem. Writing them down in a list, with a few words of explanation as to the range of validity of that particular assumption, can help you avoid later difficulties.

Draw a free-body diagram or control volume

The free-body diagram or control volume is crucial to your analysis and is the first real step in the analysis phase of the problem. You should be familiar with the free-body-diagram concept from earlier courses in mechanics. It isolates the body from the surroundings and defines external and internal forces. Control volumes, introduced in Chapter 3, are a variation of the free-body concept, where fluid is allowed to enter and leave a fixed volume. In this chapter and in Chapter 2, free-body diagrams are the appropriate tool, since the fluid is not in motion. Starting with Chapter 3, the control volume is the appropriate analytical tool since the fluid is in motion. Free-body diagrams and control volumes are analytical tools that have been developed over the years especially for the analysis of engineering problems in mechanics and thermo-dynamics.

It is often tempting to omit the step of drawing a control volume or a free-body diagram as perhaps being "too routine" or "too obvious." Recognize that this step represents **your** engineering analysis of the problem. It is **your** chief contribution to the solution of the problem, and sets forth **your** thoughts in a clear and definite form. It should be the starting point of your analysis, from which you establish the equations that govern your flow. Use the control volume to test and retest your thinking.

The coordinate axes that you have chosen to use should also be shown on the control volume or free-body diagram. It is up to **you** to define the coordinate system that **you** are going to use in a given problem. Some choices for a coordinate system might be more helpful than others, and might provide simplifications in later steps. Be flexible enough to change your coordinate system—and perhaps the control volume—and start over if further study indicates that the change would be to your advantage. Indicate vector quantities (velocities and forces) with arrows, showing how they act on the surfaces of your control volume/free-body diagram. The work you do in this step is fundamental to your analysis. If you don't get this part correct, it is unlikely that further effort will be successful. This cannot be emphasized too strongly.

Apply the conservation laws

With the control volume/free-body diagram in front of you, use the principles of conservation of mass, momentum, and energy to find the unknowns. In Chapter 2 we will consider problems with only very simple flows, so the only one of these three laws we will need is the reduced form of the momentum conservation law (Newton's third law) that says that the sum of forces is either zero or known. In problems in later chapters, the mass continuity principle will almost always have to be used in a problem involving flow, but sometimes only one of the energy and momentum principles need be used, depending on what information is given and what is being sought. While acquiring proficiency in applying these principles, it is a good idea to look at all three laws, to see which laws provide, or require, certain information.

Be sure that when you write your equations, you use the coordinate axes you previously showed, plus the directions associated with the various vector quantities as shown by your arrows. All known quantities must be shown by the arrows in the proper direction. If you assume the wrong direction for an unknown quantity, a minus sign will tell you the correct direction. If your choice of vector direction differs between control volume/free-body diagram and your equations, errors are almost sure to result.

If you find that you do not have enough information to use these conservation laws, go back to "What quantities are being sought" to see if you have missed something. For instance, if the flow is horizontal in a given region, then the vertical acceleration is zero and the pressure gradient in the vertical direction will be hydrostatic. Therefore the pressure becomes a known quantity. If the flow is basically a converging flow, then perhaps energy losses can be neglected in that region. Wall friction is sometimes negligible with respect to pressure forces if mixing and/or flow changes take place away from the wall. Perhaps use of a different control volume would eliminate some of the unknowns and make your problem simpler. Experience will develop these needed skills.

Reexamine the problem and your solution

Having obtained an answer, check it against the assumptions you have made. Do your assumptions still seem reasonable? How reasonable is the magnitude of your answer? Does it check with your physical intuition? (*Note:* Physical intuition is a combination of experience and knowledge. Sometimes, lacking one or the other, physical intuition is not a good guide. We have to work at developing it, and we gain experience through doing problems. Many times an engineer is asked for a quick answer based on experience and judgment, without the luxury of a complete analysis. Your response in such a situation can affect your career, in either a positive or a negative way.)

In the problems that appear in the rest of this book, analyze each of the results using these six steps. Try to understand why the particular control volume was used and why certain assumptions were made. You will find such an organized analytical procedure useful in studying other engineering subjects as well.

Suggestions for Further Reading

ASME Orientation and Guide for Use of SI Units, Guide no. SI-1, American Society of Mechanical Engineers, New York, 1982.

Astarita, G., and G. Marrucci, *Principles of Non-Newtonian Fluid Mechanics*, McGraw-Hill, New York, 1974.

Bird, R. B., R. C. Armstrong, and O. Hassager, *Dynamics of Polymeric Liquids*, vol. **1**: *Fluid Mechanics* and vol. **2**: *Kinetic Theory* (with C. F. Curtiss), Wiley, New York, 1987.

Boys, C. V., *Soap Bubbles and the Forces Which Mould Them*, Doubleday Anchor, Garden City, NY, 1959.

Coleman, B. D., H. Markovitz, and W. Noll, *Viscometric Flows of Non-Newtonian Fluids*, Springer-Verlag, New York, 1966.

Courant, R., and H. Robbin, *What is Mathematics?* Oxford University Press, Oxford, 1996.

Fredrickson, Arnold G., *Principles and Applicatons of Rheology*, Prentice-Hall, Englewood Cliffs, NJ, 1964.

Goldstein, S., *Modern Developments in Fluid Dynamics*, vols. **1** and **2**, Dover, New York, 1965 (originally published in 1938).

Isenberg, C. *The Science of Soap Films and Soap Bubbles*, Dover, New York, 1992.

Vincenti, Walter G., *What Engineers Know and How They Know It*, The Johns Hopkins University Press, Baltimore, 1990.

Problems for Chapter 1

Definition

1.1.   (Multiple choice.) A fluid is a substance that

(a) always expands until it fills any container.

(b) is practically incompressible.

(c) cannot be subjected to shear forces.

(d) will not remain at rest under the action of shear forces.

(e) has the same shear stress at every point regardless of its motion.

Systems of units

1.2.   (Multiple choice.) Dynamic viscosity has the dimensions of

(a) $FL^{-2}T$.   (b) $FL^{-1}T^{-1}$.   (c) FLT^{-2}.   (d) FL^2T.   (e) FLT^2.

1.3.   Convert the following properties to appropriate SI units.

(a) Specific weight = 62.29 lb/ft^3.

(b) Mass density = 1.936 slugs/ft^3.

(c) Viscosity = 2.05 lb-s/ft^2.

(d) Kinematic viscosity = 1.059 ft^2/s.

(e) Surface tension = 0.005 lb/ft.

(f) Bulk modulus = 320,000 psi.

1.4.   Convert the following properties to British units.

(a) Specific weight = 2,100 N/m^3.

(b) Mass density = 997.8 kg/m^3.

(c) Viscosity = 98.154 Pa-s.

(d) Kinematic viscosity = 0.09838 m^2/s.

(e) Surface tension = 0.073 N/m.

(f) Bulk modulus = 2.206×10^6 kPa.

Pressure, stress

1.5.   (Multiple choice.) Select the correct statement.

(a) Local atmospheric pressure is always below standard atmospheric pressure.

(b) Local atmospheric pressure depends only upon the elevation above sea level of a locality.

(c) Standard atmospheric pressure is the mean local atmospheric pressure at sea level.

(d) A barometer reads the difference between local and standard atmospheric pressure.

(e) Standard atmospheric pressure is 34 in of mercury (absolute).

1.6. (Multiple choice.) Select the three pressure intensities that are equivalent.

(a) 10.0 psi, 23.1 ft of water, 4.91 in of mercury.

(b) 10.0 psi, 4.33 ft of water, 20.3 in of mercury.

(c) 10.0 psi, 20.3 ft of water, 23.1 in of mercury.

(d) 4.33 psi, 10.0 ft of water, 20.3 in of mercury.

(e) 4.33 psi, 10.0 ft of water, 8.83 in of mercury.

1.7. Convert the following gage pressures to absolute pressures. Your answers should be in psia if the gage pressure is in British units, and in kilopascals if the gage pressure is in SI units.

(a) 25 bars (gage).

(b) 123 kPa (gage).

(c) 100 psig.

(d) 20 in of mercury (gage).

1.8. Convert the following absolute pressures to gage pressures. Your answers should be in psia if the absolute pressure is in British units, and in kilopascals if the absolute pressure is in SI units.

(a) 25 bars (absolute).

(b) 123 kPa (absolute).

(c) 100 psia.

(d) 20 in of mercury (absolute).

1.9. A European tire manufacturer suggests that its tires should be inflated to a pressure of 1.9 kilogram-force/cm^2. What is this pressure in kPa and in psi?

1.10. Large hovercrafts have been used to transport passengers and vehicles across the English Channel since the 1960s. These SRN-4 hovercrafts are the largest civilian hovercrafts ever used. They weigh about 300 tons unloaded, and can carry 384 passengers and 50 vehicles. They are 91 ft wide and 185 ft long, and can travel at speeds up to 60 knots. They are supported by a cushion of air provided by large fans, and held in place by 20 tons of flexible skirts 15 ft deep.

(a)   Estimate the pressure needed to support an unloaded hovercraft.

(b)   Using average weights of 175 lb for passengers and 4,000 lb for vehicles, find much additional pressure is needed for the fully loaded condition?

1.11. A U.S. tire manufacturer recommends a tire inflation pressure of 35 psi for a 4,500-lb vehicle. Assuming the weight is carried equally by all four tires, what is the area of the tire "footprint" and the average pressure exerted on the ground?

1.12. Ice skates function by creating pressure on the ice and locally melting it. If a skate blade is 11 in long and 1/8 in wide, what is the average pressure for a skater weighing 140 lb and standing on one skate?

1.13. Passenger elevators for two- or three-story buildings often use a hydraulic cylinder as the lifting device. If the elevator cage weighs 1,650 lb, the maximum passenger load is to be 1,200 lb, and the cylinder has a diameter of 10 in, what pressure is necessary to operate the elevator?

1.14. A hydraulic press uses a hydraulic cylinder and piston to provide the force. The piston has a diameter of 0.9 in, and the press has a capacity of 4,000 lb. What is the pressure needed in the cylinder?

Densities

1.15. A liquid of volume 3.5 m^3 has a mass of 5,500 kg. What is its specific gravity? Its specific weight?

1.16. A barrel contains 55 gallons (gal) of oil at a specific gravity of 0.82. What is the weight of oil in the barrel?

1.17. What is the specific gravity (referenced to air) of air at a pressure of 86% atmospheric? Assume that density varies linearly with pressure.

1.18. What is the density of air at 20°C and a pressure of 1 MPa? (*Hint*: Air can be considered to be an ideal gas, obeying the law $p = \rho RT$. Here p and T are absolute pressure and temperature.)

1.19. The gas constant R for an ideal gas is equal to a universal gas constant divided by the molecular weight of the gas. Find this universal gas constant by averaging the products of R and molecular weight in the Appendix B table entitled Approximate physical properties of some common gases, on page 592.

1.20. A gas has a specific volume of 450 ft^3/slug. Find the mass density and specific gravity, the latter referenced to air.

1.21. If at a particular location an object is found to have a mass of 2.5 kg and a weight of 23.5 N, what is the local acceleration of gravity?

1.22. An oil has a specific gravity of 0.815. What is the mass density of the oil? What would be the weight of 55 gal of this oil?

1.23. An open barrel with a diameter of 70 mm and a height of 1.7 m contains water. At 10°C water completely fills the barrel to the lip. If the water temperature is increased 40°C, how much water spills from the barrel?

1.24. What is the weight of a cubic meter of 0.85 specific gravity oil on a planet with a gravitational acceleration of 6.3 m/s^2?

Bulk modulus

1.25. Estimate the bulk modulus of a substance from the following data:
@ p = 200,000 psia, ρ = 1.86 slugs/ft^3;   @ p = 225,000 psia, ρ = 1.90 slugs/ft^3.

1.26. A closed cylinder-piston combination with cross-sectional area 1.4 in^2 contains air at 150,000 psig. How much additional force must be provided to the piston to decrease the volume another 1%?

1.27. A closed cylinder-piston combination with cross-sectional area 1.4 in^2 contains water at 150,000 psig. How much additional force must be provided to the piston to decrease the volume another 1%?

1.28. A storage vessel with a diameter of 400 mm contains 1,000 mm of water and 300 mm of kerosene at 20°C when the vessel is at atmospheric pressure. The pressure is increased to 1 MPa. What are the new heights of the two liquids? K (kerosene) = 1,430 MPa; K (water) = 2,240 MPa.

1.29. Water at atmospheric pressure completely fills a spherical tank with a diameter of 5 ft. The tank is open to the atmosphere. The water is heated from 50°F to 150°F. Neglecting any changes in the tank measurements, how much water will spill from the tank?

1.30. Using the values for sonic speed and density in the U.S. Standard Atmosphere table in Appendix B, compute the bulk modulus of air at 20°C and 5,000 m.

1.31. Compute the speed of sound for oxygen at 100°C. Assume oxygen behaves as an ideal gas.

Mach number

1.32. A plane is flying at an elevation of 10 km and a speed of Mach 1.35. What is the plane's speed in terms of nautical miles?

1.33. A plane is flying at an elevation of 26 km and a speed of 460 m/s. What is the corresponding Mach number?

1.34. Use the ideal gas law to find the speed of sound in oxygen at 110°C.

Vapor pressure

1.35. At standard atmospheric pressure water boils at 100°C. What pressure does water boil at when the atmospheric pressure is 10% below standard atmospheric pressure? (*Hint:* Use Appendix B to find the temperature corresponding to that vapor pressure.)

1.36. What is the boiling temperature of water in cities at mile-high elevation (Denver, for instance)? (*Hint:* Use Appendix B to find atmospheric pressure at the proper elevation. Then, using the table for vapor pressure of water, find the temperature corresponding to that pressure.)

1.37. Water at 20°C flows through a pipeline containing a converging-diverging nozzle such as in Example 1.6.2. The pressure and velocity in the pipe upstream from the nozzle are $p_0 = 30$ psig and $V_0 = 2.5$ ft/sec. The pressure in the nozzle is given by $p = p_0 + 0.5\rho(V_0^2 - V^2)$, where V is the local fluid velocity. At what value of V will the water vaporize?

1.38. Water is placed in a container kept at 20°C. Air is pumped out of the container until the air starts to vaporize. What is the air pressure in the vessel at that point?

1.39. A mercury barometer is used to measure atmospheric pressure. The value of this pressure is given by $p_{atmospheric} = \gamma_{Hg}h + p_{vapor}$, where h is the height of the mercury column, γ_{Hg} is the specific weight of mercury, and p_{vapor} is the vapor pressure of mercury. On a day where the temperature is 25°C the height h is 760 mm. What percent error would be made by neglecting p_{vapor} in the above equation?

Surface tension

1.40. A small bubble is trapped in a liquid-filled tube that is 2 mm in diameter. If the diameter of the bubble is the same as the diameter of the tube and the pressure change across the bubble is 0.9 lb/ft^2, what is the surface tension of the fluid?

1.41. Two very long parallel clean glass plates are spaced 3 mm apart. If the space between them is open to the atmosphere, how high will water rise between the two plates? Neglect any effects near the edges of the plates.

1.42. An air bubble of 3-mm diameter floats in a liquid with a surface tension of 0.08 N/m. What is the pressure inside the bubble?

1.43. An air-filled bubble of 3-mm diameter falls in air. The surface tension of the bubble liquid is 0.08 N/m. What is the pressure inside the bubble?

1.44. A liquid-filled bubble of 3-mm diameter falls in air. The surface tension of the bubble liquid is 0.08 N/m. What is the pressure inside the bubble?

1.45. An air-filled bubble of 1 mm in diameter rises in a drinking glass containing water. The surface tension of the bubble liquid is 0.0728 N/m. What is the pressure inside the bubble?

1.46. A soap bubble of 3-in diameter is found to have an internal pressure of 0.001 psig. What is the surface tension of the soap?

1.47. A clean glass tube open at both ends has an ID of 6 mm and an OD of 12 mm. It is inserted vertically into a container of mercury at 20°C.

(a) How high does the mercury rise in the tube?

(b) What is the magnitude and direction of the force that the mercury exerts on the inside of the tube?

(c) What is the magnitude and direction of the force that the mercury exerts on the outside of the tube?

1.48. An ordinary glass tube open at both ends has an ID of 6 mm and an OD of 12 mm. It is inserted vertically into a container of impure water at 20°C.

(a) How high does the water rise in the tube?

(b) What is the magnitude and direction of the force that the water exerts on the inside of the tube?

(c) What is the magnitude and direction of the force that the water exerts on the outside of the tube?

1.49. A clean wire is bent into a U-shaped frame. The legs of the U are separated by 20 mm. A straight wire is held in place perpendicular to the legs of the U, and the assembly is dipped into a container of glycerine at 20°C. After the assembly is removed from the glycerine, a film is seen to be attached to the interior of the assembly. What force is required to hold the straight wire in place?

Viscosity

1.50. (Multiple choice.) Newton's law of viscosity relates

(a) pressure, velocity, and viscosity.

(b) shear stress and rate of angular deformation in a fluid.

(c) shear stress, temperature, viscosity, and velocity.

(d) pressure, viscosity, and rate of angular deformation.

(e) yield shear stress, rate of angular deformation, and viscosity.

1.51. Two parallel plates 30 cm long and 20 cm wide are separated by 20 mm. Between them is water at 30°C. The bottom plate is held stationary and the top plate is moved at a velocity of 40 cm/s in the direction of the 30-cm length. What force is necessary to move the plate at this speed?

1.52. A metal disk of diameter 100 mm is encased in a housing such that the clearance between the disk and housing is 3 mm on each side of the disk. The housing-disk combination is filled with water at 20°C. What torque is required to turn the disk at 100 rpm? [*Hint:* The fluid velocity can locally be considered to vary linearly between the housing velocity and the disk velocity. The local velocity on the disk will vary linearly with the distance from the axis of rotation; thus $v = (r\Omega)(y/h)$. Integration is required to find the moment. Remember to include both sides of the disk.]

1.53. In Chapter 6 we will see that the velocity distribution within a circular pipe of radius a is given by $v = V_{max}\,(1 - r^2/a^2)$, where r is measured from the pipe centerline. For $V_{max} = 3$ in/s, $a = 0.4$ in, and water at 70°F the fluid, what is the volumetric discharge Q? What is the shear stress on the wall? What force does the flow exert on a 12-in length of pipe?

1.54. A plate weighing 80 lb and having a length of 40 in and a width of 8 in slides down a 10° incline on a 0.001-in-thick film of 20°C water. What is the velocity of the plate once it reaches steady state?

1.55. A fluid with a viscosity of 0.001 Pa-s is subjected to a shear stress of 5 Pa. What is the rate of deformation in the fluid at this point?

1.56. A cubical block weighing 10 lb and measuring 8 in on a side sits on a table top with a coefficient of friction of 0.3. It is desired to slide the block in a direction parallel to one of its edges. Compute the force needed to move the block at a constant speed for each of these 3 cases:

(a) Dry friction

(b) Sliding on a 0.001-in-thick layer of air at 20°C

(c) Sliding on a 0.001-in-thick layer of water at 20°C

1.57. A very large plate is placed equidistant between two vertical walls. The 10-mm spacing between the plate and each wall is filled with kerosene at 30°C. What force per unit plate area is required to move the plate upward at a speed of 35 mm/s?

1.58. A 25-mm-diameter shaft is rotated in a 300-mm-long sleeve containing SAE 50 oil. The clearance between the shaft and the sleeve is 0.6 mm all the way around the shaft. What torque is required to rotate the shaft at a speed of 1,800 rpm?

1.59. A cone-and-plate viscometer is used to measure the viscosity of "thick" fluids. This viscometer consists of a stationary flat plate with a rotating cone (angular velocity Ω). The angle θ (in radians) between the plate and cone is usually very small, so that the velocity in the fluid is given by $v \approx (r\Omega)(z/r\theta) = \Omega z/\theta$ Find the expression for the torque exerted on the plate of radius R.

Non-Newtonian fluids

1.60. A Bingham fluid is one for which a yield stress τ_{yield} must be exceeded before flow occurs. Paints behave as Bingham fluids (otherwise they would flow off of vertical walls), as do certain polymers (e.g., Carbopol, a carboxy vinyl polymer). For such fluids in simple shearing flows, Newton's viscosity law is replaced by

$\tau = \tau, v = $ constant $\quad$ for $|\tau| \leq \tau_{yield}$,

$\tau = \pm \tau_{yield} + \mu \, dv/dy \quad$ for $|\tau| \geq \tau_{yield}$.

The plus sign in the above expression is used for flows where the velocity gradient is positive, the minus sign being used otherwise.

A Bingham fluid with $\tau_{yield} = 0.025$ psf and $\mu = 0.003$ lb-s/ft^2 is flowing between two parallel plates spaced a distance of 0.01 ft. The top plate moves with a velocity of 1.5 ft/s, while the bottom plate remains stationary. The shear stress everywhere in the fluid is measured to be 0.04 lb/ft^2. What is the discharge per unit width?

1.61. The Bingham fluid of the previous problem now flows in a circular pipe of radius 0.01 ft. For this flow the shear stress now varies from zero on the pipe centerline to –0.04 psf at the pipe wall. The velocity is in the form of "plug flow." That is, the profile is given by

$v = V \quad\quad\quad\quad\quad\quad$ for $0 \leq y \leq a$,

$v = V[1 - (y - a)/(0.01 - a)] \quad$ for $a \leq y \leq 0.01$,

where a is the radius at which the yield stress is reached. Find a, V, and the discharge.

chapter 2

Hydrostatics and Rigid-Body Motions

Chapter Overview and Goals

The subjects of hydrostatics and rigid-body motions are special cases of fluid dynamics where the fluid is at rest, is undergoing a constant linear acceleration, or is being rotated at a constant angular velocity. Our study of these topics will introduce the concepts of surface and body forces, pressure prism, pressure centroid, and pressure center. The theory we will introduce is the law of hydrostatics, which is a special case of the momentum equation (Newton's second law).

As applications of our theory, we consider manometers (used for measuring pressures and pressure differences), forces on plane and curved surfaces, linear uniform acceleration, and constant angular rotation. At the end of your study you should understand these subjects sufficiently to work problems covering these topics.

While this chapter is a special case of the theory presented in Chapter 3, it is singled out for special treatment so that you can concentrate on the physical principles and methods of analysis, which are clearer in these simpler cases.

1. The Hydrostatic Equation

Hydrostatics is the study of the pressure field in a fluid either at rest or moving with a constant and uniform velocity so that the fluid is in rigid-body motion. There is no relative deformation of the fluid, so that all stresses due to viscosity vanish. On any surface that we examine in a static fluid, all stresses locally act normal to that surface. Furthermore, the stresses are all compressive. That is, they act to increase the density of the fluid. The standard convention for stress is that normal stresses are positive if they are tensile, that is, if they are acting to stretch a fluid element. In hydrostatics, the stresses are almost always compressive.[1] The negative of each of the normal stresses

---

[1] In most situations, liquids cannot withstand tensile stresses, and in fact will change to the vapor phase as the vapor pressure is reached. However, if extreme care is taken to remove all dissolved gases and contaminants, the stress can be adjusted so that regions of the liquid are in tension.

is equal to the same quantity, called the "pressure." Within the fluid, the pressure will vary in the direction of gravity if the fluid is at rest, or in the direction of the apparent gravity (the vector resultant of gravity and acceleration) if the fluid is in rigid-body motion.

To investigate the nature of pressure variations in a static fluid, we isolate an arbitrarily shaped volume of fluid and sum the forces acting on that volume of fluid. We have a choice of selecting either a finite (Figure 2.1) or an infinitesimal (Figure 2.2) size volume. Both free-body diagrams have their uses and advantages. We will start with the finite volume free body, and then consider the infinitesimal free-body diagram.

The force system acting on this free body is made up of *surface forces*, or stresses, and *body forces*. The surface forces are due to contact of the fluid contained within our free body with the surrounding fluid. In the case of hydrostatics, these forces are transmitted only through normal forces. The shear forces that would arise from viscosity and relative motion in the fluid are all zero. Body forces (sometimes called volume forces) are due most commonly to the gravitational attraction between the fluid and the earth, but they can be caused by other effects as well.[2] In hydrostatic situations, it is the body forces that cause pressure variations in the fluid.

The free-body diagram shown in Figure 2.1 is drawn with gravity acting in a general direction. The gravity vector $\mathbf{g}$, with a magnitude on earth of 32.174 ft/s^2 or 9.807 m/s^2, depending on the units used, has components (g_x, g_y, g_z) along the xyz coordinate system. The body force (gravity) exerts a force in the direction of gravity and is equal to the mass density times the magnitude of $\mathbf{g}$ times the volume, i.e., $\rho\mathbf{g}$ times volume). The net force acting on our free body due to the body force terms is then, with $\mathbf{g}$ directed toward the gravitational force,

$$\mathbf{F}_B = \iiint_V \rho\mathbf{g}\, dV.$$

The net surface force is given by

$$\mathbf{F}_S = \iint_S -p\mathbf{n}\, dS.$$

Here $\mathbf{n}$ is the local unit vector acting normal the surface dS and pointing outward from the volume. The force $p\, dS$ is pushing on the surface in the opposite direction of $\mathbf{n}$, hence the minus sign. Since the free body must be in static equilibrium, the sum of the forces must be zero, or

$$0 = \mathbf{F}_B + \mathbf{F}_S = \iiint_V \rho\mathbf{g}\, dV - \iint_S p\mathbf{n}\, dS. \tag{2.1.1}$$

---

[2] Some such effects are the rotation of a fluid, giving centripetal acceleration that acts like a body force, and magnetic or electric fields, which in liquid metals can exert body forces of attraction or repulsion.

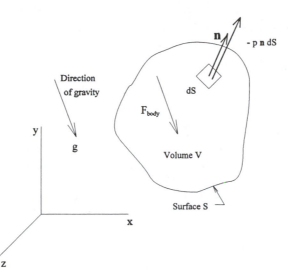

Figure 2.1. Free-body diagram of a finite size arbitrary shape.

Equation (2.1.1) tells us then that any variation of pressure in our fluid is caused by gravity, but because one of the terms is a volume integral and the other is a surface integral, it is difficult to see how the two integrals can be brought together. Some of you may be familiar with the divergence theorem, which states that for any scalar Q the following relation holds:

$$\iint_S Q\mathbf{n}\,dS = \iiint_V \nabla Q\,dV,$$

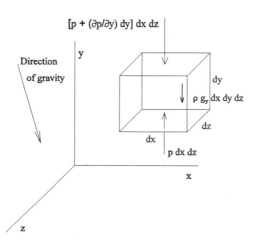

Figure 2.2. Free-body diagram of infinitesimal size volume showing forces acting in the y direction.

where the del operator ∇ is defined by

$$\nabla Q \equiv \mathbf{i}\,\frac{\partial Q}{\partial x} + \mathbf{j}\,\frac{\partial Q}{\partial y} + \mathbf{k}\,\frac{\partial Q}{\partial z}.$$

Then we can rewrite (2.1.1) as

$$0 = \mathbf{F}_B + \mathbf{F}_S = \iiint\limits_V (\rho\mathbf{g} - \nabla p)\,dV. \qquad (2.1.1a)$$

Since the choice of the volume included in the free-body diagram is entirely arbitrary, the integral can vanish for all volumes only if the integrand itself is zero. Thus we have

$$0 = \rho\mathbf{g} - \nabla p. \qquad (2.1.1b)$$

For students not familiar with the divergence theorem, the same results can be obtained with somewhat simpler mathematics if we choose an infinitesimal rectangular volume dx by dy by dz. A free-body diagram of such a volume is shown in Figure 2.2. To determine the forces acting on the free body, we consider an infinitesimal rectangle dx by dy by dz as shown in Figure 2.2. The body force, that is, the weight of the element, has components given by

$$F_{bx} = g_x\rho\,dx\,dy\,dz, \qquad (2.1.2a)$$

$$F_{by} = g_y\rho\,dx\,dy\,dz, \qquad (2.1.2b)$$

$$F_{bz} = g_z\rho\,dx\,dy\,dz. \qquad (2.1.2c)$$

On each of the six surfaces of our rectangle, the surface force is equal to the pressure on that surface times the area. This force acts in a direction normal to the area and, for positive pressures, is directed toward the interior of our rectangle. We denote by p the pressure acting on the three surfaces that face in the negative x, y, and z directions. Since there can be pressure variation in the fluid, the pressure on the other three surfaces is as follows:

Pressure on surface facing in plus x direction is $p + (\partial p/\partial x)\,dx$,

Pressure on surface facing in plus y direction is $p + (\partial p/\partial y)\,dy$,

Pressure on surface facing in plus z direction is $p + (\partial p/\partial z)\,dz$.

The argument for writing these pressures in this manner is as follows. Consider first the variation of pressure in the x direction. We have let the pressure on the surface facing in the negative x direction be denoted by p. On the opposing face the average pressure will be different, but only by a small amount since the two surfaces are separated by an infinitesimal distance. Expanding the pressure in a Taylor series in x about the rearward-facing surface and keeping only the first two terms, we find the pressure acting on the face whose normal points in the positive x direction to be $p + (\partial p/\partial x)\ dx$. The same argument holds in the other two directions.

The net force on the left-facing surface is then $p\ dy\ dz$, and on the right-facing surface it is $[p + (\partial p/\partial x)\ dx]\ dy\ dz$. Summing the forces acting on the rectangular element the net force acting in the x direction on the body due to surface forces is found to be

$$dF_{x\ \text{surfaces}} = p\ dy\ dz - \left(p + \frac{\partial p}{\partial x}\ dx\right) dy\ dz = -\frac{\partial p}{\partial x}\ dx\ dy\ dz. \tag{2.1.2d}$$

Similar summations in the y and z directions give

$$dF_{y\ \text{surfaces}} = p\ dx\ dz - \left(p + \frac{\partial p}{\partial y}\ dy\right) dx\ dz = -\frac{\partial p}{\partial y}\ dx\ dy\ dz, \tag{2.1.2e}$$

$$dF_{z\ \text{surfaces}} = p\ dx\ dy - \left(p + \frac{\partial p}{\partial z}\ dz\right) dx\ dy = -\frac{\partial p}{\partial z}\ dx\ dy\ dz. \tag{2.1.2f}$$

To have static equilibrium, the net sum of body and surface forces acting on our free body must be zero. The result is

$$dF_{x\ \text{surfaces}} + dF_{x\ \text{body}} = -\frac{\partial p}{\partial x}\ dx\ dy\ dz + \rho g_x\ dx\ dy\ dz,$$

$$dF_{y\ \text{surfaces}} + dF_{y\ \text{body}} = -\frac{\partial p}{\partial y}\ dx\ dy\ dz + \rho g_y\ dx\ dy\ dz,$$

$$dF_{z\ \text{surfaces}} + dF_{z\ \text{body}} = -\frac{\partial p}{\partial y}\ dx\ dy\ dz + \rho g_z\ dx\ dy\ dz.$$

Upon dividing these three equations by the volume $dx\ dy\ dz$ of the rectangle, we are left with

$$-\frac{\partial p}{\partial x} + \rho g_x = 0, \tag{2.1.3a}$$

$$-\frac{\partial p}{\partial y} + \rho g_y = 0, \tag{2.1.3b}$$

$$-\frac{\partial p}{\partial z} + \rho g_z = 0. \tag{2.1.3c}$$

Students familiar with the del operator will recognize that these three vector-component equations can be written in the vector form

$$-\boldsymbol{\nabla}p + \rho\mathbf{g} = 0. \tag{2.1.3}$$

Equation (2.1.3) is the same result as equation (2.1.1b), as of course it should be. It tells us that the pressure gradient is in the direction that $\mathbf{g}$ acts, and that the pressure gradient has a magnitude of ρg, which is the specific weight. The *isobars*, that is, the surfaces along which the pressure is constant, according to (2.1.3) are surfaces that locally are perpendicular to the direction of gravity.

In our analysis using the infinitesimal free-body diagram, you may have wondered why we could let the pressure on the three faces facing in the negative coordinate direction all have the same value. Note that in our force summations the values of the pressure itself dropped out and we are left with only the changes in pressure. Therefore, no matter what we call the pressure on the three surfaces, they would drop out in our final result.

You may also have wondered why we didn't take account of the variation of pressure over the surfaces of the rectangle, or why we didn't include higher order terms in the Taylor series expansions. Again, if we did this we would find that the additional terms would all be of higher order in dx, dy, and dz, and would vanish in the limit as the volume got smaller and smaller. Therefore our results would be unchanged, and we would not have gained any additional information.

To find the pressure as a function of position in the fluid, it is necessary to integrate equation (2.1.3). For our convenience and easier physical understanding we will orient our axes so that gravity acts in the negative y direction. With this choice, $\mathbf{g}$ has the components $(0, -g, 0)$. Integration of (2.1.3) with $\rho g = \gamma$, the specific weight, results in

$$p = -\int \gamma \, dy + C, \tag{2.1.4}$$

where C is a constant of integration and γ is the specific weight. To complete the determination of pressure, it is necessary to know how the specific weight varies with height and how to determine the constant of integration.

An important special case of hydrostatics is when the specific weight is constant, which is generally valid for liquids unless under extreme pressures, and which may be used for gases when the specific weight times the total height variation is negligible compared to the nominal pressure. In this case

$$p = -\gamma y + C,$$

where C is a constant of integration. By setting y to zero in this equation it is seen that C is the value of the pressure at $y = 0$. Denoting this value by p_0, we have

$$p = p_0 - \gamma y. \tag{2.1.5}$$

Consequently, as we move down (in the direction of gravity) in the fluid, the pressure increases from the value p_0 at the origin of our coordinate system at a rate of γ per unit length.

Example 2.1.1. Pressure variation in a constant density perfect gas

What is the atmospheric pressure at an elevation of 4,000 m, assuming that the density of the air is constant and that at the earth's surface the temperature is 20°C and the pressure is 760 mmHg?

Sought: Atmospheric pressure at 4,000 m.

Given: Atmospheric pressure measured at sea level is 760 mmHg. This pressure is equivalent to

$$p_0 = 760 \text{ mmHg} = 760 \times 133.322 = 101.32 \text{ kPa}.$$

The temperature 20°C is equal to $273.15 + 20 = 293.15$ K. The gas constant for air is, from Appendix B, $R = 0.2869$ kJ/kg-K.

Assumptions: Assume constant mass density, ideal gas behavior, static air.

Solution: Atmospheric pressure p_0 at the earth's surface is 101.32 kPa and the temperature is 293.15 K. Since air behaves as an ideal gas, the air density at the earth's surface can be found from the ideal gas equation,

$$\rho_0 = p_0/RT = 101{,}320 \text{ (N/m}^2)/[286.9 \times (\text{N} - \text{m/kg} - \text{K}) \times 293.15 \text{ K}] = 1.204 \text{ kg/m}^3.$$

The specific gravity γ of the air is then given by

$$\gamma = 9.8 \times 1.204 = 11.8 \text{ N/m}^3.$$

The air pressure at 4,000 m is given by equation (2.1.5) as

$$p = p_1 - \gamma y = 101{,}320 - 11.8 \times 4{,}000 = 101{,}320 - 47{,}200 = 54{,}120 \text{ Pa} = 54.1 \text{ kPa}$$

$$= 54{,}120/133.322 = 405.9 \text{ mmHg}.$$

Other important special cases of hydrostatic variation are observed in the atmosphere, where the specific weight of air varies with elevation. Depending on the elevation we are at in the atmosphere, two idealized cases are of interest:

Case 1. Isothermal perfect gas. The state equation for a perfect gas, with R as the gas constant, T the absolute temperature, and p the absolute pressure, is

$$p = \rho RT. \tag{2.1.6}$$

In the isothermal case, solving for the density gives

$$\rho = p/RT,$$

where RT = constant. Then, with the y coordinate taken to be positive upward from the surface of the earth (away from the direction of gravity), we have

$$dp/dy = -gp/RT.$$

We have replaced the partial derivative by a total derivative here since p varies only with the vertical coordinate.

Separating variables gives

$$dp/p = -(g/RT)\,dy,$$

which, upon integration, gives

$$\ln p = -gy/RT + D,$$

where D is a constant of integration. If we let $p = p_1$ and $\rho = \rho_1$ at $y = y_1$, then we obtain

$$C = \rho_1/p_1 \quad \text{and} \quad D = \ln p_1 - gy_1 C,$$

or finally,

$$p = p_1 \exp [-\gamma_1(y-y_1)/p_1]. \tag{2.1.7}$$

Example 2.1.2. Pressure variation in an isothermal perfect gas

What is the atmospheric pressure at an elevation of 4,000 m, assuming that the air is isothermal and that at the earth's surface the temperature is 20°C and 760 mmHg?

Sought: Atmospheric pressure at 4,000 m.

Given: Atmospheric pressure measured at sea level is 760 mmHg. This pressure is equivalent to

$$p_1 = 760 \text{ mmHg} = 760 \times 133.322 = 101.32 \text{ kPa}.$$

The temperature 20°C is equal to $273.15 + 20 = 293.15$ K. The gas constant for air is, from Appendix B, $R = 0.2869$ kJ/kg-K.

Assumptions: Assume isothermal conditions, ideal gas behavior, static air.

Solution: Atmospheric pressure p_1 at the earth's surface is given as 101.32 kPa and the temperature is 293.15 K. Since air behaves as an ideal gas, the air density at the earth's surface can be found from the ideal gas equation,

$$\rho_1 = p_1/RT_1 = 101.32/0.2869 \times 293.15 = 1.205 \text{ kg/m}^3.$$

The specific gravity γ_1 at the earth's surface is then given by

$$\gamma_1 = 9.8 \times 1.205 = 11.81 \text{ N/m}^3,$$

and the air pressure at 4,000 m is given by equation (2.1.7) as

$$p = 101.32 \exp (-11.81 \times 4,000/101,320) = 63.56 \text{ kPa}$$

$$= 63,560/133.322 = 476.7 \text{ mmHg}.$$

Case 2. Temperature varies linearly with elevation. In this case, we have

$$T = T_0 + \beta y, \tag{2.1.8}$$

where $\beta = dT/dy$ is called the *lapse rate*. Inserting (2.1.8) into (2.1.6) we obtain

$$\rho = p/R(T_0 + \beta y).$$

From (2.1.3) this leads to

$$dp/dy = -\rho g = gp/R(T_0 + \beta y).$$

Separating variables yields

$$dp/p = -g \, dy/R(T_0 + \beta y),$$

which, upon integration, gives

$\ln p = -(g/R\beta) \ln (T_0 + \beta y) + D,$

where again D is a constant of integration. Letting $T = T_0$ and $p = p_0$ at $y = 0$, we get for our constant of integration

$D = \ln p_0 + (g/R\beta) \ln T_0.$

Putting the expression for D into the expression for p, we finally have

$$p = p_1 \left(\frac{T_0 + \beta y}{T_0} \right)^{-g/R\beta}. \tag{2.1.9}$$

Example 2.1.3. Pressure variation in an ideal gas whose temperature varies linearly with elevation

What is the atmospheric pressure at an elevation of 4,000 m, assuming that the pressure varies linearly with temperature, that at the earth's surface the temperature and pressure are 20°C and 760 mmHg, and that at 4,000 m the temperature is 40°C colder than the surface of the earth?

Sought: Atmospheric pressure.
Given: Atmospheric pressure measured at sea level is 760 mmHg. This is equivalent to
$p_0 = 760$ mmHg $= 760 \times 133.322 = 101.32$ kPa.
The temperature of 20°C is equal to $273.15 + 20 = 293.15$ K.
The gas constant for air is, from Appendix B, $R = 0.2869$ kJ/kg-K.
Assumptions: Assume linear dependence of temperature on height and ideal gas behavior for air.
Solution: From (2.1.8) we have

$\beta = (T_1 - T_0)/y_1 = -40/4{,}000 = -0.001$ K/m.

Given that the earth's surface pressure and temperature are given as 101.32 kPa and 293.15 K, and that air behaves as an ideal gas, the air density at the earth's surface is found from the ideal gas equation to be

$\rho_0 = p_0/RT_0 = 101.2/0.2869 \times 286.55 = 1.23$ kg/m^3.

From (2.1.9),

$$p = 101.2 \left(\frac{293.15 - 0.001 \times 4{,}000}{293.15} \right)^{9.8/293.15 \times 0.001}$$

$$= 101.32 \times 0.986^{33.43} = 63.24 \text{ kPa} = 63{,}240/133.322 = 474 \text{ mmHg}$$

In actuality, the temperature variation in the atmosphere varies in a much more complex manner than is given by either of these models. To provide a standard for comparing the performance of airplanes or aerospace vehicles, a standard atmosphere has been agreed upon. Its temperature decreases linearly with elevation for the first 11 km, becomes isothermal for the next 9.1 km, and then increases or decreases linearly in various successive higher layers. Values for the U.S. Standard Atmosphere are given in Appendix B.

2. Manometers

An important device for measuring pressure differences is the ***manometer***, an instrument that consists primarily of a transparent column containing one or more liquids that do not mix. This is a device that is simple both in concept and in use, is inexpensive, and requires no calibration. As a result, it can be used as a standard for calibrating other devices. Its main disadvantages are:

1. It is extremely important to remove all air or gas bubbles in the liquid-filled columns of the manometer. (Other pressure-measuring devices may also have this requirement.)
2. Reading of a manometer from a remote location is difficult.
3. The length of the manometer becomes awkward when large pressures are to be measured.
4. Because of the inertia (mass) of the fluid in the tubing connecting the manometer to the regions where pressure is to be measured, the manometer can respond only to slowly varying pressure changes of the order of 100 Hz or less.

Several manometer arrangements for measuring pressure in vessels or pipelines are shown in Figure 2.3. (The circles in the figure represent either vessels or cross sections of pipes.) Configuration a can be used when dealing with a liquid that is at a pressure above atmospheric pressure, configuration b when the liquid is at a pressure less than atmospheric pressure, and configuration c when we are measuring the pressure in a

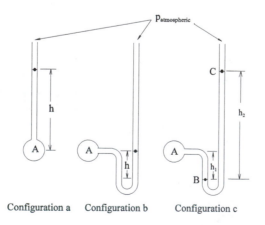

Configuration a   Configuration b   Configuration c

Figure 2.3. Various configurations of manometers connected to pipelines at *A*.

gas. The procedure in analyzing all these cases is the same, and cases a and b are seen to be special cases of c. We will examine the more complicated case c: you can then verify yourself that for case a, the gage pressure is

$$p_A = \gamma h,$$

and for case b it is

$$p_A = -\gamma h.$$

For case c, between points A and B there is a fluid with specific weight γ_1. Between B and C there is a different fluid with specific weight γ_2. For static stability so that there is no overturning of the fluids, γ_2 must be greater than γ_1. Since B is lower than A, according to our hydrostatic law its pressure will be greater than A by an amount equal to the vertical height h_1 times the specific weight of the fluid. The fact that A and B differ in their horizontal positions does not affect this, since there is no horizontal body force and hence no horizontal pressure change. Therefore the pressure at B in terms of the pressure at A is

$$p_B = p_A + \gamma_1 h_1. \tag{2.2.1}$$

At the interface B, the pressure will be the same in the two fluids, providing surface tension effects are negligible. (This can usually be accomplished by using tubes having a diameter of at least 0.5 in.) Then the pressure at C will be less than the pressure at B by an amount equal to the height h_2 times the specific weight of this fluid. Since as shown in the figure C is at atmospheric pressure, we have

$$p_C = p_B - \gamma_2 h_2 = 0 \text{ gage pressure.} \tag{2.2.2}$$

Putting (2.2.1) and (2.2.2) together yields

$$p_A = \gamma_2 h_2 - \gamma_1 h_1. \tag{2.2.3}$$

As it stands, equation (2.2.3) is awkward to use, since it requires reading two heights to get one pressure reading. This can be avoided in practice by setting the scale on which h_2 is read so that it reads zero when p_A is at some reference level, say, p_{A0}. Then

$$p_{A0} = \gamma_2 h_{20} - \gamma_1 h_{10}. \tag{2.2.4}$$

If the pressure is increased to a value p_A, then the interface B will move down an amount Δh, and C must move up an amount Δh, if the two legs of the manometer are of the same cross-sectional area. (If the legs are of different cross-sectional areas, then the product $A \, \Delta h$ must be the same in each leg.) Then the relationships become

$$h_1 = h_{10} + \Delta h \tag{2.2.5a}$$

and

$$h_2 = h_{20} + 2 \, \Delta h \tag{2.2.5b}$$

so that

$$p_A = \gamma_2 (h_{20} + \Delta h) - \gamma_1 (h_{10} + 2 \, \Delta h) = p_{A0} + \Delta h (\gamma_2 - 2 \gamma_1). \tag{2.2.6}$$

Consequently, we have to read only the single change in height Δh and compute the pressure by equation (2.2.6).

Example 2.2.1. Manometers

The fluid in the pipeline of Figure 2.3a has specific gravity 0.8, and the height h is 18 in. What is the pressure in pipeline A?

Sought: Pipeline pressure at point A.

Given: The specific weight of the fluid is $0.8 \times 62.4 = 49.92$ lb/ft^3. The pressure at the open end of the manometer tube is atmospheric.

Assumptions: The manometer fluid is incompressible and of constant density. Pressure variation due to the column of air above the manometer liquid can be neglected since air density is small and the quantity $\gamma_{air}\ \Delta y_{air}$ therefore is also small compared to other pressure differences in the manometer.

Figures: See Figure 2.3a.

Solution: Since the top of the manometer is open to the atmosphere, the pressure there is zero (gage). The pressure in the manometer tube increases in the downward direction, according to specific weight times distance down from the free surface. Thus

$$p_A = \rho g h = 49.93 \times (18/12)$$

$$= 74.88 \text{ psf gage} = 0.52 \text{ psi gage}$$

$$= 74.88/62.4 = 1.2 \text{ in of water}$$

Many times, especially for low pressure values, it is convenient to express the pressure in terms of how high the fluid were to rise in a manometer of the type of Figure 2.3a, with water as the manometer fluid. The dimensions of this pressure is then inches of water. All pressures shown are gage pressures, measured with respect to the atmosphere.

Example 2.2.2. Manometers

The fluid in the pipeline of Figure 2.3b is water, and the height is 3 in. What is the pressure p_A?

Sought: Pipeline pressure at point A.

Given: The specific weight of water is 62.4 lb/ft^3. The pressure at the open end of the manometer tube is atmospheric.

Assumptions: The manometer fluid is incompressible. Pressure variation in the air above the manometer liquid can be neglected since air density is small and the quantity $\gamma_{air}\ \Delta y_{air}$ therefore is also small compared to other pressure differences in the manometer.

Figures: See Figure 2.3b.

Solution: The pressure at the air-water interface is zero gage. Since this interface is below point A, p_A must be less than atmospheric. Thus

$$p_A = -62.4 \times (3/12) = -15.6 \text{ psf gage} = -0.1083 \text{ psi gage}$$

$$= -3 \text{ in of water.}$$

Again, all pressures are gage pressures since they are measured relative to the atmosphere.

Example 2.2.3. Manometers

The pipeline of Figure 2.3b contains water. The manometer level is 30 cm above the centerline of the pipe. What is the pressure in the pipe?

Sought: Pipe pressure on centerline.
Given: The specific weight of the fluid is 62.4 lb/ft^3.
The pressure at the open end of the manometer tube is atmospheric.
Assumptions: The manometer fluid is incompressible. The pressure variation in the air above the manometer liquid can be neglected since air density is small and the quantity $\gamma_{air}\,\Delta y_{air}$ therefore is also small compared to other pressure differences in the manometer.
Solution: Since the free surface of the manometer fluid is above the centerline of the pipe, the pipe pressure is greater than atmospheric pressure by 30 cm of water. Since the mass density of water is 1,000 kg/m^3, the pipe pressure must be greater than atmospheric pressure by

$$\rho gh = 1{,}000 \times 9.807 \times 0.30 = 2{,}942.1 \text{ Pa.}$$

Therefore the pressure in the pipeline is 2.942 kPa gage pressure, which is equivalent to 0.2942 bar gage pressure.

Example 2.2.4. Manometers

The fluid between A and B in the manometer of Figure 2.3c has a specific gravity of 0.9, and the fluid between B and C has a specific gravity of 1.2. Height h_1 is 7 in, and height h_2 is 15 in. What is the pressure at A?

Sought: Pressure at point A.
Given: The specific weight of the fluid is $0.9 \times 62.4 = 56.16$ lb/ft^3. The pressure at the open end of the manometer tube is atmospheric.
Assumptions: The manometer fluid is incompressible. The pressure variation in the air above the manometer liquid can be neglected.
Figures: See Figure 2.3c.
Solution: The pressure at C is zero gage since the manometer is open to the atmosphere. The pressure at B is

$$p_B = (1.2 \times 62.4) \times (15/12) = 93.6 \text{ psf gage.}$$

The pressure at A is

$$p_A = p_B - (0.9 \times 62.4) \times (7/12) = 93.6 - 32.76 = 60.84 \text{ psf gage}$$

$$= 0.4225 \text{ psi gage} = 0.975 \text{ in of water.}$$

Again all pressures are gage pressures since they are read relative to the atmosphere.

The manometers shown in Figure 2.3 are the simplest manometer configurations. Manometers can also be connected so that the open end of the manometers in Figure 2.3 are attached to a reference pressure that is different from atmospheric, so as to obtain pressure differences with respect to some value other than atmospheric pressure. Also, to simplify the reading of the heights in a manometer, some manometers are made with one leg of considerably greater cross-sectional area than the other, and one leg may even consist of an opaque reservoir. Usually when such manometers are constructed, the scale by which the manometer is read is "stretched" to take into account this area ratio. Therefore the level in the section with larger cross-sectional area need not be read, making reading of the manometer easier.

When the pressures to be read by a manometer are small, the manometer is often designed to increase the accuracy to which the height may be measured. The *inclined manometer* (Figure 2.4) does this by putting one leg at an angle θ with the horizontal, and reading the interface with a scale also mounted at this angle. The desired vertical height is this reading times the sine of the angle θ. When θ is made small, the precision of reading is thereby increased.

Example 2.2.5. Inclined manometer

The tube in the manometer of Figure 2.4 is inclined 30° with the horizontal. The vertical leg of the manometer is connected to a pressure p_a in a gas, and the manometer fluid is water. The scale in the inclined leg is set so that when the water level in the vertical leg is at the same level as the level in the inclined leg, the scale reading is zero. If the reading along the scale to the free surface of the manometer liquid is 30 in, find the pressure in the gas.

Sought: Pressure in gas.

Given: The specific weight of the manometer fluid is 62.4 lb/ft^3. The pressure at the open end of the manometer tube is atmospheric.

Assumptions: The manometer fluid is incompressible. The pressure variation in the air above the manometer liquid can be neglected since air density is small and the quantity $\gamma_{air} \Delta y_{air}$ therefore is also small compared to other pressure differences in the manometer.

Figures: See Figure 2.4.

Solution: Since the open end of the manometer is above the level of the gas that is in contact with the manometer liquid, the pressure in the gas is greater than atmospheric.

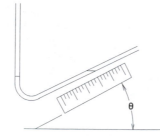

Figure 2.4. Inclined manometer.

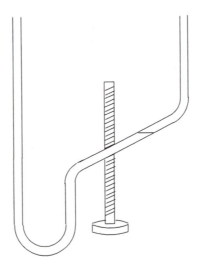

Figure 2.5. Micromanometer.

This pressure is therefore simply the vertical height difference of the manometer liquid, given by

$\rho g h \sin \theta = 62.4 \times (30/12) \times \sin 30° = 78$ psf gage $= 0.542$ psi gage.

Since the inclined tube of the manometer is open at the upper end, all pressures are relative to atmospheric pressure.

Micromanometers (Figure 2.5) mount a short piece of the manometer leg so that it is horizontal. This short piece can be moved vertically by means of a fine screw. The screw is adjusted so that the interface lines up with scribe marks on this short leg. This height is measured accurately by a dial gage, a scale with a vernier, or some similar arrangement. This will yield very precise results, provided the pressures do not fluctuate appreciably. Optical devices (e.g., magnifiers, telescopes) to improve the precision of measurement are also used.

All the above techniques, however, require an operator at the manometer to take readings. To allow remote readings and/or a continuous record of the pressure, transducers that develop an electrical signal proportional to the pressure are used. Resistance- or capacitance-measuring devices can be used to measure the deformation of a diaphragm exposed to the pressure. Also, semiconductor circuits exist whose outputs are particularly sensitive to pressure levels imposed on the semiconductor.

3. Rise of Liquids Due to Surface Tension

In tubes or slots of very small cross-sectional area, surface tension at liquid-gas or liquid-liquid interfaces can cause liquids to rise to considerable heights. Since the fluid

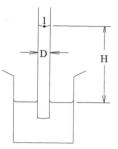

Figure 2.6. Capillary tube inserted into a liquid.

is at rest, our hydrostatic equations together with the surface tension laws are well suited to determining how high the fluid can rise.

Consider first one of the simplest cases, a round tube of diameter D. If we insert this tube vertically into a liquid, the fluid will rise to a height H, as seen in Figure 2.6. If the free surface of the liquid is at atmospheric pressure, then the pressure at point 1 must also be atmospheric. Since the fluid is not in motion, there can be no shear stress at the walls acting to assist in supporting the liquid. The vertical forces acting on the column of the liquid held up by surface tension must then be as in the free-body diagram shown in Figure 2.7, where α is the contact angle and σ is the surface tension. We have taken H as the average height and used gage pressure only, since the atmospheric pressure acts at the top and the bottom of our free-body diagram, and hence cancels out. The force balance then involves only the weight of the column and the surface tension forces. Summing forces, we have

$$-\rho g(0.25\pi D^2 H) + \sigma \cos \alpha \, (\pi D) = 0, \tag{2.3.1}$$

giving as the height to which the fluid will rise

$$H = 4\sigma \cos \alpha/(\rho g D). \tag{2.3.2}$$

We see that if the fluid wets the walls of the tube ($\alpha < 90°$, e.g., water in glass tubes), then the fluid rises in the tube, and the height of the rise is inversely proportional to

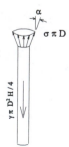

Figure 2.7. Free-body diagram of a capillary tube.

the tube diameter. Similarly, if the fluid does not wet the walls of the tube ($\alpha > 90°$, such as mercury in glass tubes, and water in tubes with walls coated in paraffin wax), the liquid is depressed in the tube.

Note that equation (2.3.2) tells us that, since σ is numerically small, D must also be small for H to be an appreciable value. Tubes of small diameter are called *capillary tubes*, or literally, tubes of hairlike, or thin, diameter. Very small blood vessels are also called capillaries.

The case of two vertical flat walls placed at a small angle with one another gives a result that also can easily be demonstrated in a simple experiment. We can use the same free bodies as in the previous example, but now we consider the figure to represent a two-dimensional case rather than a circular one. Summing forces per unit width into the page, we have

$$-\rho g H D + 2\sigma \cos \alpha = 0, \tag{2.3.3}$$

which upon solving for H gives

$$H = 2\sigma \cos \alpha/(\rho g D). \tag{2.3.4}$$

Equation (2.3.4) is seen to differ from equation (2.3.2) by only a factor of 2.

To perform the above experiment, take two flat glass sheets and a thin spacer, perhaps a rod or wire about a millimeter or two in diameter. The actual value is not important. Clamp the plates together with a rubber band so that two edges of the glass sheets are in tight contact, with the spacer separating the other edges. The spacer should be parallel to the two edges in contact with one another. The two plates thus form a wedge, with the spacing D varying linearly with the distance from the contacting edges. According to (2.3.3) the product HD will be constant. Thus when the plates are inserted into the fluid, the free surface of the liquid will rise (or be depressed, depending on α) and form a hyperbola as seen in Figure 2.8a. It is important for a successful experiment that the glass plates be clean.

For the case where the fluid wets the surface and therefore rises, $\alpha > 90°$, and the pressure immediately under the free surface is atmospheric pressure minus $\rho g H$. If H is too great, this pressure will become equal to the vapor pressure of the fluid, and so we have a practical limit as to how high a liquid can rise under surface tension (approximately 33 ft, or 10 m). Surface tension is occasionally invoked as the reason that sap rises in trees. If that were strictly true, the above suggests that we would have no trees greater than this height. The reasons why trees can be greater than this height are that in portions of the tree the sap is actually in tension, and that the sap-containing passages in the tree are capillary tubes that at their upper ends are closed by leaves. The leaf acts as a thin membrane that has effects similar to surface tension at a free surface, in that it allows for the existence of a pressure difference across it. This pressure difference is, however, related to parameters other than surface tension. Similar membranes exist throughout the passageways of the plant, so in fact the situation is much more complicated than encountered in our simpler fluid examples.

Example 2.3.1. Surface tension and hydrostatics

Two rectangular glass plates are taped together so that they touch on two of their vertical edges, as described in the text. A 3-mm-diameter rod is placed vertically between the two plates, 9 cm from the taped edges. The space between the plates thereby forms a wedge. If the y coordinate is vertical, and $x = 0$ at the taped edge, the spacing varies with x according to $D(x) = 0.003x/0.09 = x/30$.

If the plates are now placed in a shallow dish containing a liquid of specific gravity 0.8, contact angle 25°, and surface tension 0.07 N/m, the liquid is observed to rise in the wedge. What is the shape of the liquid-air interface, if $y = 0$ at the liquid level in the dish?

Sought: Equation of liquid-air interface.

Given: The liquid specific weight is $0.8 \times 9{,}807 = 7{,}846$ kg/m^3. The surface tension and contact angle are 0.07 N/m and 25°, respectively.

Assumptions: The liquid is incompressible and of constant density. The fluid is not in motion.

Figures: See Figure 2.8a.

Solution: The pressure at the liquid-air interface in Figure 2.8a is atmospheric. We can draw the free-body diagram of unit thickness as shown in Figure 2.8b. We sum forces in the vertical direction, and note that, since atmospheric pressure acts on all sides of our free body, it has no net effect. Thus it is appropriate to work with gage pressure. The result is, per unit length into the paper,

$$2\sigma \cos \alpha \quad - \quad \gamma DH = 0.$$
(surface tension force)   (weight)

Solving for H gives

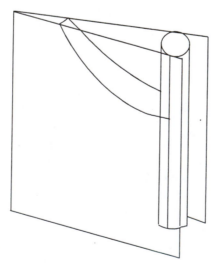

Figure 2.8a. Liquid between two nearly parallel plates.

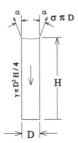

Figure 2.8b. Free-body diagram for Figure 2.8a.

$H = 2\sigma \cos \gamma/(\gamma D)$.

Inserting our previous expression for D along with the numerical values of the various quantities,

$H = 2 \times 0.07 \cos 25°/7847(x/30) = 4.85 \times 10^{-4}/x$.

Since H is proportional to $1/x$, the free surface is a hyperbola.

Example 2.3.2. Surface tension and hydrostatics

A small capillary tube of 2-mm ID is inserted into a fluid with a surface tension of 0.07 N/m and a contact angle of 25°. How high does the fluid rise in the capillary tube?

Sought: Height to which fluid rises in a capillary tube.
Given: The liquid specific weight is $0.8 \times 9{,}807 = 7{,}846$ kg/m³ The surface tension and contact angle are 0.07 N/m and 25°, respectively.
Assumptions: The liquid is incompressible and of constant density. The fluid is not in motion.
Figures: See Figure 2.7.
Solution: From the free-body diagram of Figure 2.7,

$\sigma\pi D \cos \alpha - \gamma\pi H D^2/4 = 0$.

(This is different from the previous example, because there we neglected the curvature in the plane parallel to the paper. Here it is important to include it.) Thus

$H = 4\sigma \cos \alpha/\gamma D = 4 \times 0.07 \cos 25°/(0.8 \times 9{,}807) \times 0.002 = 0.016$ m.

The shape of a free surface acted upon by surface tension is given by the general equation

$$p_1 - p_2 = \sigma (1/R_1 + 1/R_2),\tag{2.3.5}$$

where p_1 and p_2 are the pressures on the two sides of the surface and R_1 and R_2 are what are called the "principal radii of curvature" of the surface. Three special cases of this provide the solutions we have discussed so far: $R_1 = R_2 = \infty$, a flat plane, with surface tension having no effect on the surface; $R_1 = R_2 = r$, a sphere of radius r, and

(2.3.5) reduces to (1.6.14); $R_1 = \infty$, $R_2 = r$, a circular cylinder of radius r, where (2.3.5) differs from (1.6.14) by a factor of 2.

The calculation of other shapes of a free surface is generally a complicated job mathematically, since the equations for principal radii are highly nonlinear. One of the easier cases involves the solution of a two-dimensional surface, say, a liquid near a wall. In that case one of the radii of curvature is infinite (parallel to the wall). For the other, with x being the distance from the wall and y the elevation measured from the level far from the wall, the principal radius of curvature is given by a standard calculus formula as

$$\frac{1}{R} = \frac{\dfrac{d^2y}{dx^2}}{|1 + (dy/dx)^2|^{3/2}} = \frac{p_1 - p_2}{\sigma} = \frac{\gamma y}{\sigma} \tag{2.3.6}$$

since the pressure under the interface varies hydrostatically. Equation (2.3.6) can be integrated by first multiplying both sides by $(dy/dx)\, dx$. Rearrangement then gives

$$\frac{dy}{dx}\, d\, \frac{\left(\dfrac{dy}{dx}\right)}{\left|1 + \left(\dfrac{dy}{dx}\right)^2\right|^{3/2}} = \frac{\gamma y}{\sigma}\, dy,$$

or after integration,

$$\frac{-1}{\left|1 + \left(\dfrac{dy}{dx}\right)^2\right|^{1/2}} = \frac{\gamma y^2}{2\sigma} + A,$$

where A is a constant of integration. Since y and dy/dx both go to zero far from the wall, the constant $A = -1$.

Solving the above equation for dy/dx gives

$$\left(\frac{dy}{dx}\right)^2 = \frac{1}{(1 - \gamma y^2/2\sigma)^2} - 1, \tag{2.3.7}$$

which again can be integrated with the help of tables, giving

$$x = -\frac{\sqrt{\sigma}}{\sqrt{\gamma}}\, \cosh^{-1}\frac{2\sqrt{\sigma}}{y\sqrt{\gamma}} + \sqrt{\frac{4\sigma}{\gamma} - y^2} + B,$$

where B is a second constant of integration. We can choose B by saying that at the wall the elevation is h ($y = h$ at $x = 0$), giving

$$x = -\frac{\sqrt{\sigma}}{\sqrt{\gamma}} \cosh^{-1} \frac{2\sqrt{\sigma}}{y\sqrt{\gamma}} + \sqrt{\frac{4\sigma}{\gamma} - y^2} + \frac{\sqrt{\sigma}}{\sqrt{\gamma}} \cosh^{-1} \frac{2\sqrt{\sigma}}{h\sqrt{\gamma}} - \sqrt{\frac{4\sigma}{\gamma} - h^2}. \tag{2.3.8}$$

The elevation h at the wall is commonly given from knowing the contact angle at the wall. In that case, from (2.3.7),

$$\pm \cot \theta = \frac{1}{1 - \gamma h^2/2\sigma}, \tag{2.3.9a}$$

or after solving for h,

$$h^2 = 2\sigma \, (1 \mp \tan \theta)/\gamma. \tag{2.3.9b}$$

The upper sign is chosen if $\gamma h^2/2\sigma > 1$, the lower sign being used otherwise .

The surface tension of fluids is often used in industrial processes. For example, ink jet printers rely on the surface tension of the ink to form spherical drops. While drop-on-demand is the technique most used in home printers, heavy-usage printers may use a continuous jet of ink that breaks up into droplets under the action of surface tension. Wavy disturbances form on the ink jet, with disturbances of wavelength 4.508 times the diameter of the jet being the ones that grow fastest, pinching the continuous jet into separated droplets. The volume of the cylinder that forms a droplet is $\pi d_{\text{jet}}/4$ times the wavelength. Equating this volume to the volume of a spherical drop gives the drop diameter as $d_{\text{drop}} = 1.891 d_{\text{cylinder}}$.

4. Forces on Surfaces

In many applications the principal result desired from hydrostatics is the resultant force due to a fluid pressure acting on a surface. The line of action of this resultant may also be desired. This would be true for instance if we were interested in finding the force on a dam, bridge pier, or other such structure. So as to make force and moment calculations simpler, we would like to replace the distributed pressure force by a single concentrated force. This replacement is to be done so that the concentrated force is *statically equivalent* to the distributed force. That is, the sum of forces due to the distributed pressure force in any direction is the same as that due to the concentrated force, and the sum of moments due to the distributed pressure force about any point is the same as that due to the concentrated force about the same point.

To accomplish this, we start off by considering an infinitesimal area of size dA. (See Figure 2.9.) This area has a directionality depending on its orientation. To denote this, we let **n** be a unit vector (a vector with length 1, and no dimensions) that is perpendicular to and pointing toward the area. Then the magnitude of the force on the area is the local value of the pressure times the area, and the direction of the force on the area is the same as the direction of **n**. From this we have

$$d\mathbf{F} = p\mathbf{n} \, dA. \tag{2.4.1}$$

Summing forces over many of these infinitesimal areas, we have

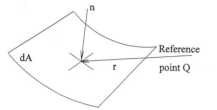

Figure 2.9. Infinitesimal surface used to find force on an area.

$$\mathbf{F} = \iint p\mathbf{n}\, dA. \tag{2.4.2}$$

In general, p will vary with the vertical depth according to equation (2.1.1), and if the surface is nonplanar, $\mathbf{n}$ will also vary in a fashion known from the geometry of the area.

The moment due to the pressure can be found in a similar fashion. If a fixed reference point Q is chosen and $\mathbf{r}$ is the distance measured from this reference point to the area dA, then the moment due to the force $d\mathbf{F}$ acting on dA is

$$d\mathbf{M} = \mathbf{r} \times d\mathbf{F} = \mathbf{r} \times \mathbf{n}\, p\, dA. \tag{2.4.3}$$

Again, summing over the total area we have

$$\mathbf{M} = \iint \mathbf{r} \times \mathbf{n} p\, dA. \tag{2.4.4}$$

When this integration is carried out, $\mathbf{r}$ as well as $\mathbf{n}$ and p will vary with position on the surface.

To find the line of action of the resultant force, remember that its moment must be equal to that given by equation (2.4.4), the moment being about the same reference point Q as used in the computation of $\mathbf{r}$. Letting $\mathbf{r}_{cp}$ be the distance from that reference point to the line of action of the force, from the usual definition of a moment this means that

$$\mathbf{M} = \mathbf{r}_{cp} \times \mathbf{F}. \tag{2.4.5}$$

If $\mathbf{M}$ and $\mathbf{F}$ are both found from computation of the integrals, then $\mathbf{r}_{cp}$ can be found from the solution of equation (2.4.5). Note that this solution is not unique. As can be seen from equation (2.4.5), adding a vector to $\mathbf{r}_{cp}$ that is parallel to the resultant force $\mathbf{F}$ will not change the moment. From the point of view of statics, this nonuniqueness is of no importance. If, however, we wish to make this determination unique, we do so by using the special point where this line of action of the force intersects the surface. This point is called the ***center of pressure***.

Carrying out the integrals given by equations (2.4.2) and (2.4.4) can be tedious and messy problems in geometry and calculus for nonplanar and nonrectangular shapes. We will consider some special cases in which the integration simplifies. In some cases rather than attacking the integration directly, it is easier to reconsider the problem from a different point of view. For complicated surfaces, integration may be simplest if it is carried out numerically.

a. Plane surfaces

When the surface is a plane, the normal vector **n** is a constant and can be brought out of the integral sign. In this case the magnitude of the force is

$$F = \iint p \, dA \qquad (2.4.6)$$

and its direction is perpendicular to the area. The pressure will vary with the coordinate perpendicular to the surface (unless the surface is horizontal) and, if the density is constant, will be of the form

$$p = p_0 - \gamma y \qquad (2.4.7)$$

The force then is

$$F = \iint (p_0 - \gamma y) \, dA = (p_0 - \gamma y_c)A = p_c A, \qquad (2.4.8)$$

where, since the centroid of the area is defined by

$$y_c = \iint y \, dA / \iint dA, \qquad (2.4.9)$$

it is seen that y_c is the y position of the centroid of the area A and p_c is the value of the pressure at that centroid. Thus if the area is simple enough in shape that we know the location of its centroid, equation (2.4.8) provides us with a convenient and quick way of determining the resultant force.

Example 2.4.1. Force on a circular area
Find the force that water exerts on one side of a circle of diameter $D = 18$ in, inclined at an angle of 30° with the horizontal. The center of the circle is submerged a distance $h = 3$ ft into water.

Sought: Force exerted by water on a submerged circular area.
Given: The specific weight is 62.4 lb/ft^3.
Assumptions: The liquid is of constant specific weight and the fluid density is constant. The fluid is at rest.
Solution: Since the area is a flat surface, (2.4.8) holds. Then

$$p_c = \rho g h = 1.935 \times 32.17 \times 3 = 186.7 \text{ psf gage}$$

and

$$F = (\rho g h)\pi D^2/4 = 186.4 \times 3.14159 \times (1.5)2/4 = 329.4 \text{ lb}.$$

Note that atmospheric pressure has not been included in the calculation, only gage pressure. Atmospheric pressure exists on all sides of the surface, and therefore has a zero net effect.

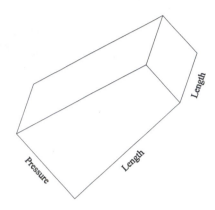

Figure 2.10. Pressure prism.

Another concept that is useful for determining resultant force is that of the **_pressure prism_**. Imagine a geometrical shape in the form of a prism with a base area equal to the given cross-sectional area and a local height proportional to the local value of the pressure. (See Figure 2.10.) This is the pressure prism, and its "volume" is the magnitude of the force. When we take a moment of this volume using equation (2.4.4), we are really finding the centroid of the volume, that is, the centroid of the pressure prism. Therefore the pressure center is a point on the intersection of the area with a line drawn perpendicular to the area and passing through the centroid of the pressure prism. (Note that, unless the surface is horizontal, this centroid is different from the centroid of the area used in computing the force.) Unless an extensive table of geometrical shapes and their centroids is available, the pressure prism concept is useful principally for rectangular areas with one edge parallel to a free surface. It is not particularly useful for curved surfaces, where the pressure prism "folds in" upon itself.

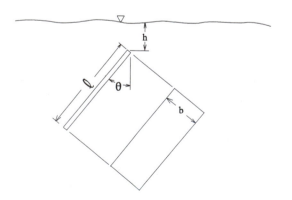

Figure 2.11a. Sketch for a rectangular surface under a free surface.

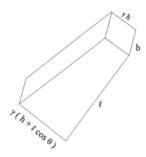

Figure 2.11b. Pressure prism for surface of Figure 2.11a. The height is pressure, the base is area.

Example 2.4.2. Force on a flat plate

Find the force that water exerts on one side of a flat plate b by w submerged so that one edge (that of length b) is parallel to the free surface and a distance h below it. The plate makes an angle θ with the vertical. Also find the location of the center of pressure.

Sought: Force exerted by water on one side of a flat plate and the center of pressure.

Given: The specific weight is 62.4 lb/ft^3.

Assumptions: The liquid is of constant specific weight. The fluid is at rest.

Figures: See Figure 2.11a.

Solution: A side view of the pressure prism is shown in Figure 2.11b. The force can be found either by finding the volume of the pressure prism (average value of the pressure times the base area) or by the pressure at the centroid of the rectangle, to be

$$F = \gamma(h + 0.5w \cos \theta) \, wb.$$

To find the center of pressure, it is necessary to find the centroid of a trapezoid. Using the formula

$$s_c = \Sigma s_{ci} A_i / A_{\text{total}},$$

we can conveniently divide the trapezoid into a rectangle (w by γh) and a right triangle (the two perpendicular sides being w by $\gamma w \cos \theta$). The centroid of a rectangle is at the midpoint, and for the triangle it is 2/3 of the distance from the top end. Measuring s_c from the top end, we find

$$s_c = [w\gamma h w/2 + (2/3) \times (1/2)w\gamma w^2 \cos \theta] / [\gamma h w + (1/2)\gamma w^2 \cos \theta]$$

$$= w(h/2 + w \cos \theta/2)$$

The pressure distribution is therefore statically equivalent to a single concentrated force of magnitude F acting a distance s_c from the top edge of the plate.

Example 2.4.3. Force on a flat plate

Repeat Example 2.4.2 using the integration formulas.

Sought: Force exerted by water on one side of a flat plate and the center of pressure.

Given: The specific weight is 62.4 lb/ft^3.

Assumptions: The liquid is of constant specific weight. The fluid is at rest.

Solution: Using s as the distance measured along the plate from the top edge, we have as the pressure distribution

$$p = \gamma(h + s \cos \theta)$$

and

$$dA = b \, ds.$$

Putting these together, we have

$$F = \int_0^w (h + s \cos \theta)b \, ds = \gamma(h + w \cos \theta/2)wb,$$

$$M = \int_0^w s\gamma(h + s \cos \theta)b \, ds = \gamma w^2(h/2 + w \cos \theta/3)b,$$

$$s_c = M/F = w(h/2 + w \cos \theta/3)/(h + w \cos \theta/2).$$

This result coincides with that of the previous example, as of course it should.

Example 2.4.4. Hydraulic brakes

An important application of hydrostatics is the transmission of power in hydraulic devices. Figure 2.12 shows a typical automobile braking system. The driver applies a force F_D to the brake pedal. This force is transferred to the piston in the master brake cylinder. What force is transmitted to the brakes?

Sought: Force transmitted to brakes.

Given: The areas of the various cylinders and the lever arms.

Assumptions: The brake fluid is of constant specific weight. The fluid is at rest.

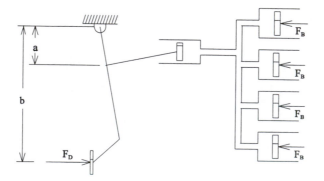

Figure 2.12. Vehicle hydraulic brake system.

Figures: See Figure 2.12.

Solution: By elementary statics, the force applied to the master cylinder is $(b/a)F_D$. If A_m is the area of the master cylinder piston, the resulting pressure in the hydraulic fluid is

$$p = F_D b / a A_m.$$

This pressure is transmitted equally to all wheel cylinders; hence

$$F_B = p A_w = F_D b A_w / a A_m,$$

where A_w is the area of the wheel piston. (Often this area is nearly the same as the master cylinder area.) The brake shoe is then pressed against the brake disk or drum with a force F_B.

Typically in both disk and drum brakes the wheel cylinders are double-acting, so that each end produces a force F_B. Note that this does not affect p or F_B.

In an automobile with "power brakes," there is an intermediate element in the hydraulic line that acts as a pressure multiplier, so that when the engine is running the pressure in the wheel cylinders is greater than in the master cylinder.

b. Forces on circular cylindrical surfaces

Another case that can be treated in a simple manner is a surface made up of a portion of a circular cylinder as shown in Figure 2.13. In this case, rather than finding the forces on the surface by integration of the pressure, it is useful to use a different attack. This will give us still another approach to these problems, and also will aid us in the physical interpretation of our results.

Consider for example the quarter circle between A and B in Figure 2.14. At A the pressure is γh, and at B it is $\gamma(h + a)$. To analyze the problem, consider the sum of forces acting on the volume of fluid in contact with the surface AB and bounded by planes AC and BC. The fluid above the quarter circle exerts a uniform pressure γh on the surface AC, which results in a downward force

$$F_{AC} = \gamma hab. \tag{2.4.10}$$

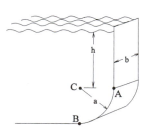

Figure 2.13. Liquid above a curved surface.

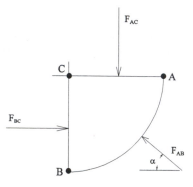

Figure 2.14. Free-body diagram for Figure 2.13.

The fluid to the right of plane BC exerts a pressure that varies from γh to $\gamma(h + a)$. The resultant horizontal force (average pressure times area) is

$$F_{BC} = \gamma(h + a/2)ab. \tag{2.4.11}$$

The weight of the fluid in our free-body diagram is

$$W = \gamma \pi a^2 b/4,$$

and F_{AB} is the force of the surface on the fluid, which by Newton's third law is equal in magnitude and opposite in direction to the force of the fluid on the surface. The angle α is the inclination of this force with the horizontal. Summing forces in the horizontal and vertical direction gives

$$F_{BC} - F_{AB} \cos \alpha = 0,$$

$$-F_{AC} - W + F_{AB} \sin \alpha = 0.$$

The result is

$$\tan \alpha = (h + \pi/4)/(h + a/2),$$

$$F_{AB} = \gamma ab \sqrt{(h + \pi a/4)^2 + (h + a/2)^2}.$$

The vertical component of F_{AB} in this case is seen to be the weight of the fluid above the surface AB.

To find the line of action of the resultant force, we take moments about point C. The moment arm for the force F_{AC} is $a/2$, with a clockwise moment. The force F_{BC} is conveniently broken up into the force due to the uniform pressure γh (moment arm $a/2$) and that due to the linearly varying pressure with moment arm $2a/3$. (Recall that the centroid of a triangle is located 2/3 of the distance from a corner.) These forces give counterclockwise moments. The centroid of a quarter circle is at $4a/3\pi$, which is the lever arm of the weight A. Summation of moments then gives

$$0 = (\gamma hab)a/2 - [(\gamma hab)a/2 + (1/2)\gamma a^2 b](2a/3) - (1/4)(\gamma \pi a^2 b)(4a/3\pi) + F_{AB}d,$$

or

$$0 = F_{AB}d.$$

Thus we conclude that $d = 0$, and the resultant force acts through C. We could have anticipated this result by noting that on every infinitesimal section of the quarter circle the force due to the pressure passes through C.

Example 2.4.5. Force on a cylindrical surface
Repeat the problem described above by the integration method.

Sought: Force on a submerged cylindrical surface by integration.
Assumptions: The liquid has constant specific weight. The fluid is at rest.
Figures: See Figure 2.15.
Solution: In terms of the variable angle θ (see Figure 2.15), we have

$$p = \gamma(h + a \sin \theta),$$

$$dA = ba\, d\theta,$$

$$\mathbf{n} = \mathbf{i} \cos \theta - \mathbf{j} \sin \theta;$$

therefore

$$\mathbf{F} = \int_0^{\pi/2} \gamma(h + a \sin \theta)(\mathbf{i} \cos \theta - \mathbf{j} \sin \theta)ba\, d\theta$$

$$= \gamma ab[\mathbf{i}(h + a/2) - \mathbf{j}(h + \pi a/4)].$$

Taking moments about C with $\mathbf{r} = a(\mathbf{i} \cos \theta - \mathbf{j} \sin \theta)$ yields

$$\mathbf{M} = \int_0^{\pi/2} p\, \mathbf{r} \times \mathbf{n}\, a\, d\theta.$$

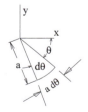

Figure 2.15. Infinitesimal area for integration procedure.

Therefore the line of action of the resultant force passes through the point C, and has the same inclination as the force **F**.

When $h = 2$ ft, $a = 3$ ft, $b = 18$ in, and the fluid is water,

$$\mathbf{F} = 62.4 \times 3 \times 1.5[\mathbf{i}(2 + 0.5 \times 3) - \mathbf{j}(2 + p \times 3/4)] = (982.8\mathbf{i} - 1{,}223.2\mathbf{j}) \text{ lb.}$$

c. Buoyancy forces

The resultant pressure force on a floating or submerged body is called the ***buoyancy force***, and historically it is associated with the Greek philosopher Archimedes. To see the magnitude of this force, we consider a body of arbitrary weight distribution submerged in a fluid and draw a rectangular box with dimensions a, b, c around it to include the body. For a free-body diagram we will take just the fluid inside the box (Figure 2.16). The forces acting on this fluid are the pressure forces on the six sides of the box, the weight of the fluid, and the force that the body exerts on the fluid. To satisfy equilibrium requirements in the horizontal direction, the net horizontal pressure forces must be equal and opposite in direction and will therefore cancel. Therefore the net force of the fluid on the body must be vertical. Letting V be the volume of the body and summing forces in the vertical direction, we have

$$0 = -\gamma h ab + \gamma(h + c)ab - \gamma(abc - V) - F_{\text{body}}.$$

After canceling common terms, we are left with

$$F_{\text{body}} = \gamma V. \tag{2.4.12}$$

We see that the force of the body on the fluid is downward and is equal to the weight of fluid displaced by the body. The force of the fluid on the body is equal and opposite to this. Thus ***the body is "buoyed up" by a force equal to the weight of the fluid it displaces.*** While we proved this for a fully submerged body, the result is true for a floating body as well. Taking moments would show that this buoyancy force acts through the centroid of the displaced fluid. This is called the ***center of buoyancy***. (The

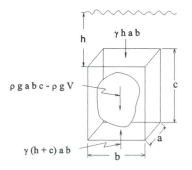

Figure 2.16. Free-body diagram for a fully submerged floating body.

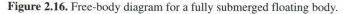

centroid of the displaced fluid is not necessarily the same as either the center of mass or the centroid of the body, since the body may have a nonuniform mass density.)

Example 2.4.6. Buoyant body

A stick of wood with specific gravity 0.4 is anchored to the bottom of a lake (Figure 2.17). Its dimensions are 2 in by 4 in by 9 ft, and the lake is 2 ft deep at this point. Find the angle θ at which the stick floats, assuming that the 4-in dimension is horizontal.

Sought: Angle at which a partially submerged tethered stick floats.

Given: The specific weight of the water is 62.4 lb/ft^3. The specific weight of the wood is $0.4 \times 62.4 = 24.96$ lb/ft^3

Assumptions: The specific weight of the water and the wood are each constant. The fluid is at rest. The length of the submerged portion of the stick is l.

Figures: See Figure 2.17.

Solution: The weight and buoyancy forces are

$$W = 0.4\gamma \times (2/12) \times (4/12) \times 9 = 0.2\gamma$$

and

$$B = \gamma \times (2/12) \times (4/12) \times l = \gamma\, l/18.$$

Taking moments about the anchor point to eliminate the unknown anchor force,

$$0 = B \times [(\,l/2)\cos q] - W \times 4.5 \cos \theta$$

$$= (\gamma l/18) \times [(l/2)\cos\theta] - 0.2\gamma \times 4.5 \cos = (l^2/36 - 0.9)\,\gamma \cos \theta.$$

From this we conclude that there are two possible mathematical solutions

$$l = \sqrt{0.9 \times 36} = 5.692 \text{ ft} \qquad \text{and} \qquad \theta = 90°.$$

The first of these gives

$$\theta = \sin^{-1}(\text{depth}/l) = \sin^{-1}(2/5.692) = 20.6°,$$

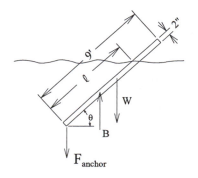

Figure 2.17. Force on a floating stick.

which is a physically possible result. Since the depth is given as 2 ft, $\theta = 90°$ is not physically possible, because this position would be an unstable one for the stick. (If the lake depth were greater than 5.692 ft, the correct choice would be $\theta = 90°$ and the stick would be vertical.)

Exercise 2.4.7. Hot-air balloon

Estimate the temperature of the gas in the envelope needed for the hot-air balloon shown in Figure 2.18 to be able to carry a passenger load of 600 lb. The balloon envelope volume is 77,500 ft^3; the envelope weighs 200 lb; and the passenger basket, burner, and fuel weigh 175 lb.

Given: The fluid is air, volume displaced is 77,500 ft^3, unloaded weight is 200 + 175 = 375 lb, loaded weight is 975 lb, ambient temperature is 70°F.

Assumptions: Air behaves as an ideal gas; the pressure inside the balloon is ambient pressure.

Figures: Figure 2.18.

Solution: The buoyancy forced must equal the weight of the balloon. Therefore

$$W_B = gV\rho_0.$$

From the ideal gas law $p = \rho RT$,

$$\rho_0 T_0 = \rho_1 T_1 = p/R = 14.7 \times 144/53.3 \times 32.2 = 1.2334.$$

Figure 2.18. Single-passenger hot-air balloon. (Photo courtesy Mr. Bob Bowers, *The Ultimate Balloon Adventure*, Las Vegas.)

Thus for the ambient air,

$\rho_0 = 1.2334/(460 + 70) = 0.002327$ slug/ft^3,

and the weight of the displaced air (buoyancy force) is

$W_B = Vg\rho_0 = 77,500 \times 32.2 \times 0.002327 = 5,807$ lb.

The total weight of the balloon is

$W_{total} = W_{envelope} + W_{basket\ etc.} + W_{heated\ air} + W_{passengers} = W_B$ by statics.

When no passengers are present,

$W_{heated\ air} = W_B - W_{envelope} - W_{basket\ etc.} = 5,807 - 200 - 175 = 5,432$ lb

$$= Vg\,\rho_{heated\ air}.$$

Then $\rho_{heated\ air} = 5,432/Vg = 5,432/(77,500 \times 32.2) = 0.002177$ slug/ft^3 and $T = 1.2334/0.002177 = 567°R = 107°F$.

For the balloon with passengers,

$W_{heated\ air} = W_B - W_{envelope} - W_{basket\ etc.} - W_{passengers} = 5,807 - 200 - 175 - 600$

$$= 4,832 = Vg\rho_{heated\ air}.$$

Then $\rho_{heated\ air} = 4,832/Vg = 4,832/(77,500 \times 32.2) = 0.001936$ slug/ft^3 and $T = 1.2334/0.001936 = 637°R = 177°F$.

The temperatures so calculated are average gas temperatures in the balloon envelope. Near the burner, gas temperatures are typically in the range 150–250°F.

The dimensions and weights used here are for a small "sport" balloon. Larger sizes with greater capacities are available. The numbers used here are courtesy of Mr. Bob Bowers, *The Ultimate Balloon Adventure*, Las Vegas, NV.

d. Stability of submerged and floating bodies

The stability of floating and submerged bodies has to do with whether the submerged or floating body returns to its initial position after it has been disturbed. The stability of these bodies is determined by the relative positions of the center of buoyancy and the center of gravity of the body. Consider the body shown in Figure 2.19. When the center of gravity and center of buoyancy lie along the same vertical line as in Figure 2.19a, summation of forces states that the buoyancy force equals the body weight, and the moment about any point is zero. In your imagination, draw a line connecting these two centers on the body. Suppose now that the body is turned through a small angle θ. The center of gravity of the body remains fixed, but the center of buoyancy will shift to a new position. Draw a vertical line from the new center of buoyancy so that it intersects the original line you drew. The point of intersection is called the ***metacenter***, and the distance from the metacenter to the center of gravity is called the

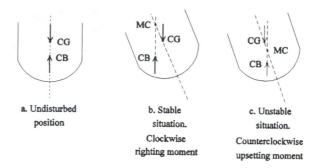

Figure 2.19. Stability considerations for a floating body.

metacentric height. If the metacenter is above the centroid of the body, the buoyancy force and body weight combine to give a clockwise couple, as shown in Figure 2.19b. This couple acts in a direction such as to restore the body to its original position, and the body is considered to be stable. If the metacenter is below the centroid, the couple is in the counterclockwise direction and acts to increase the disturbance, as in Figure 2.19c. The body in this case is said to be *unstable*.

It is possible for both the centroid of the body and the center of buoyancy to change positions for a given body. Suppose for example that you are sitting in a canoe floating on a lake on a calm summer day. If you start rocking the canoe while remaining seated, the centroid of the canoe remains fixed with respect to the canoe. But if you rock the canoe through ever-increasing angles, the center of buoyancy moves farther from its original position. While the canoe may be stable for small angles, there may well be a sufficiently large angle at which the canoe tips.

You can repeat the above experiment, this time standing up in the canoe. By so doing you have raised the centroid of the canoe/passenger combination. You should also find that the canoe will tip at a much smaller angle of rocking. (Remember, I said to do this on a calm summer day, and be sure to wear a life preserver.)

Some U.S. Coast Guard rescue vessels have the capability of being rolled over completely by rough seas, and then automatically righting themselves in several seconds. This is achieved by having air tanks placed high within the ship hull.

As you might imagine, computation of the metacentric height for complicated shapes like ships can be quite tedious, and must be done for a wide variety of angles. The problem is one for which procedures have been established in the literature of naval architecture. The example that follows deals with the simplified case where the angle of tip is very small.

Example 2.4.8. Stability of a floating body

A barge of uniform rectangular cross section has length L, width W, and height H. The centroid of the barge and its cargo is a fixed distance a from the bottom of the barge. The position of the centroid always remains fixed with respect to the barge. The waterline is a distance b from the bottom of the barge. (See Figure 2.20.) The barge floats in seawater

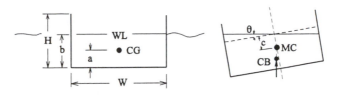

a. Undisturbed position   b. Tipped position

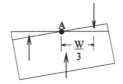

c. Change in buoyancy force.

Figure 2.20. Buoyancy of a barge.

(γ = 64 lb/ft^3). For small angles of tilt, find the metacentric height. Under what conditions is the barge stable?

Sought: Location of metacenter, stability criterion for the barge.
Given: Dimensions of barge L, W, H. Location of centroid and water line a, b. Density of water.
Assumptions: Small angle of tilt, uniform water density, centroid position fixed with respect to the barge.
Figures: Figure 2.20a, b, and c.
Solution: The weight of the barge B must be the weight of the displaced water; thus

$$B = \gamma \times L \times W \times b.$$

When the barge is flat in the water the center of buoyancy is located at a distance $b/2$ from the bottom of the barge. When the barge is tipped through a small angle θ as shown in Figure 2.20b, the new position of the center of buoyancy can be found by summing the moments due to the original rectangular block of water plus the two triangular areas of newly displaced and newly released water. Each triangular area has a volume equal to

$$\text{Volume} = (1/2) \times (W/2) \times (W \tan \theta/2) \times L \approx W^2 L\, \theta/8.$$

Their moment arm about the centerline is, for small angles θ, approximately $W/3$. To replace this by an equivalent force system as in Figure 2.20c, the force must be γbWL and located a distance c above the point A.

To determine c, we sum moments in Figure 2.20b and c about point A due to the resultant force and due to the new buoyant forces and equate them. Thus

$$\Sigma M_A = B \times c \sin \theta \approx \underbrace{B \times (b/2) \sin \theta}_{\text{rectangular block}} - \underbrace{2 \times \gamma \times (W^2 \times \theta/8) \times (W/3)}_{\text{triangular blocks}},$$

the 2 appearing in the last term because of the two triangular volumes. Dividing by the buoyant force and making the small angle approximation for the sine, we are left with

$c = b/2 - W^2/12b.$

Therefore the metacentric height is

$b - c - a = W^2/12b + b/2 - a.$

To be stable, this height must be positive, or

$W > \sqrt{12b(a - b/2)}.$

For floating objects whose cross sections change along their length, stability of the object can be determined by a simple formula. Figure 2.21 shows a cross section of a more generally shaped body displaced through a small angle θ. Compared to the undisplaced body, the center of buoyancy B has moved a distance r to the left in this figure, and is now designated as B'. The magnitude of the buoyancy force is still equal to its original value as the weight of the displaced water. We see that the triangular wedge of water on the left side gives an upward buoyancy force F in addition to the undisplaced buoyancy, and the triangular wedge of water on the right side gives a corresponding downward buoyancy force F. They are separated by a distance s and constitute a clockwise couple of magnitude sF.

We can determine the magnitude of this couple by integration of the pressure, resulting in sF over the cross-sectional area of the body at the water line. From Figure 2.21 we have

$$sF = \int x(x \tan \theta \, \gamma) \, dx \, dy = \gamma \tan \theta \int x^2 \, dx \, dy = \gamma \tan \theta \int x^2 \, dA = \gamma \tan \theta \, I_{yy},$$

where y is perpendicular to the plane of the paper, A is the water line cross-sectional area, and I_{yy} is the second moment of this area about the y axis. (If you have taken a course in the mechanics of materials, you encountered similar I_{yy} in the bending of beams.)

From trigonometry, h, the height of the metacenter above the undisplaced center of buoyancy, is seen to be related to r by

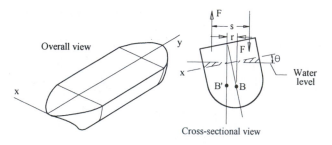

Figure 2.21. Analysis of a floating object of nonuniform cross section.

$r = h \sin \theta$.

To have the summation of moments about B' equal to zero means that

$rW = sF$,

or

$hW \sin \theta = \gamma \tan \theta \, I_{yy}$.

Solving for h gives

$h = \gamma \tan \theta \, I_{yy} / W \sin \theta = \gamma I_{yy} / W \cos \theta$.

Letting $V = W/\gamma$, denote the volume of liquid displaced by the body, and remembering that θ is small so that its cosine is approximately 1, the above simplifies to

$h \approx I_{yy} / V$.   (2.4.13)

Example 2.4.9. Stability for small angular disturbances
Repeat Example 2.4.8 using equation (2.4.13).

Solution: The displaced volume is bWL. The second moment of area of a rectangle is $I_{yy} = (1/12) \, W^3 L$. Therefore, for small angles, the distance between the center of buoyancy and the metacenter is a height

$h = (1/12) \, W^3 L / bWL = W^2/12b$

above the undisturbed center of buoyancy. The metacentric height is obtained by adding the distance between the undisturbed center of buoyancy and the center of gravity, obtaining

$h + b/2 - a = W^2/12b + b/2 - a$,

the result obtained in the previous analysis.

In practice, submerged floating bodies are seldom truly stable due to temperature variations, which in turn lead to density variations. You may have seen in the stores what is termed "Galileo's Liquid Thermometer." It consists of five or more glass balls floating in a liquid-filled transparent tube. The balls are all weighted so that they have slightly different specific gravities, and are marked with a number denoting temperature. The temperature range covered is usually around 16°F (10°C). As the room temperature rises slightly, say, from 74 to 76°F, the water density decreases, and a ball with the number 76 will fall to the bottom while the ball with the number 74 will move from the top of the tube to the center. This thermometer illustrates the sensitivity of floating bodies to slight temperature changes, since the balls generally are seen to be more or less constantly changing their positions.

Other examples of the sensitivity of the stability of submerged bodies are hot air balloons, airships, and submarines. They all can change their depth or height by changing the amount of ballast they have taken on, but to maintain constant attitude some forward motion along with continuous adjustment of fins is also necessary.

5. Rigid-Body Acceleration

In fluid statics we considered the case where the fluid was completely at rest. Because there is no relative motion between fluid particles, all viscous effects vanish, simplifying the calculations considerably. This simplification also holds when the fluid is being accelerated uniformly at a constant rate, so that at a given instant of time, the velocities of all fluid particles are the same. In this case, considering the same free body as in Figure 2.2, the sum of forces will be exactly the same as we found in deriving equations (2.1.4), (2.1.8), and (2.1.9), but instead of setting the sum of forces in any one direction to zero, they will be equal to the mass of the element ($\rho\,dx\,dy\,dz$) times the appropriate component of acceleration. From this we obtain equations (2.1.4), (2.1.5), and (2.1.6), but modified by the addition of an extra term in each. For instance, if **g** acts in downward in the y direction,

$$\frac{\partial p}{\partial y} = -\gamma - \rho a_y = -\rho(g + a_y), \tag{2.5.1}$$

$$\frac{\partial p}{\partial x} = -\rho a_x, \tag{2.5.2}$$

$$\frac{\partial p}{\partial z} = -\rho a_z \tag{2.5.3}$$

The isobars (surfaces of constant pressure) can be found by integrating,

$$dp = \frac{\partial p}{\partial x}\,dx + \frac{\partial p}{\partial y}\,dy + \frac{\partial p}{\partial z}\,dz = -\rho a_x\,dx - (\gamma + \rho a_y)\,dy - \rho a_z\,dz. \tag{2.5.4}$$

When the mass density and accelerations are constant, (2.5.4) becomes

$$dp = -\rho[a_x\,dx + (g + a_y)\,dy + a_z\,dz] \tag{2.5.5}$$

and the integrands are all constants. Then the equations can be integrated to give

$$p = p_0 - \rho[xa_x + y(g + a_y) + za_z], \tag{2.5.6}$$

where p_0 is the pressure at the origin of the coordinate system. Therefore the isobaric surfaces satisfy the equation

$$xa_x + y(a_y + g) + za_z = \text{constant}, \tag{2.5.7}$$

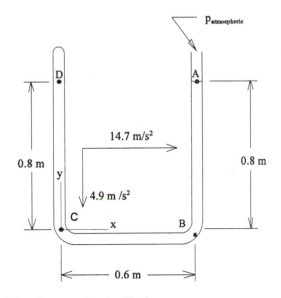

Figure 2.22. Conditions in an accelerating U-tube.

and the isobaric surfaces are planes tilted with respect to the horizontal.

Example 2.5.1. Accelerated fluid
The water-containing U-tube shown in Figure 2.22 is being accelerated to the right with an acceleration of 14.7 m/s^2, and downward with an acceleration of 4.9 m/s^2. Find the pressures at B, C, and D, given that A is open to the atmosphere. Also find the slope of the isobars.

Sought: Pressures at points B, C, D.
Given: The specific weight of the water is 62.4 lb/ft^3. Its mass density is 62.4/32.17 = 1.94 slugs/ft^3.
Assumptions: The specific weight and density of the fluid are constants. The fluid is being accelerated to the right with a constant acceleration.
Figures: See Figure 2.22.
Solution: We take the origin at C to avoid having negative values for x and y. Since we have

$$a_x = 1.5\,g, \qquad a_y = -0.5\,g, \qquad a_z = 0, \qquad \text{and} \qquad p_0 = p_C,$$

we then find from (2.5.6) that

$$p = p_C - \gamma(1.5x + 0.5y).$$

We know that $p_A = 0$, so inserting $x = 0.6$, $y = 0.8$ into the above equation for the pressure gives

$p_A = p_C - \gamma(1.5 \times 0.6 + 0.5 \times 0.8) = 0.$

Solving this for p_C, we have

$p_C = 1.3\gamma = 1.3 \times 62.4 = 81.1 \text{ psf} = 0.56 \text{ psi}.$

Then we obtain

$p_B = p_C - \gamma(1.5 \times 0.6) = 0.4\gamma = 24.96 \text{ psf} = 0.173 \text{ psi}$

and

$p_D = p_C - \gamma(0.5 \times 0.8) = 0.9\gamma = 56.2 \text{ psf} = 0.39 \text{ psi}$

by inserting, respectively, $x = 0.6$, $y = 0$ and $x = 0$, $y = 0.8$ into the above expression for the pressure.

On the isobars, since the pressure is constant, $1.5x + 0.5y$ must also be constant. Thus the isobars have a slope of $\tan^{-1}(-1.5/0.5)$, and so they slope downward at an angle of $71.57°$. Notice that it is very important that the pressure be known at one point in this problem. If the tube had been sealed, during the startup of the motion many things could have happened to the tube (temperature changes, dents in the container, etc.) that would have changed the conditions from what they were at the time the tube was filled with liquid. Knowledge of the complete history of handling the tube would be necessary if the pressure at A had not been established by leaving an opening here.

---

6. Rigid-Body Rotation

A further case of rigid-body motion that is of interest is the case of rigid-body rotation of a fluid. If a fluid is rotated at a constant angular speed, after the startup transients have died out there will be no deformation of a fluid element, and therefore, viscous forces again will not come directly into play. (They do, however, come in indirectly, for without viscosity the fluid would slip at the boundaries and would remain at rest.) Each point in the fluid will travel on a circular path centered on the axis of rotation. The velocity of a fluid particle is equal to the radius of the circle times the angular velocity. The acceleration the fluid particle experiences is of the centripetal type, with magnitude equal to (velocity)^2/radius and directed toward the axis of rotation.

Thus, in addition to the hydrostatic pressure gradients given by equations (2.1.1) and (2.1.2), there are centripetal accelerations. To help our physical intuition it is convenient to think of this acceleration as an additional "variable gravity." Since these accelerations are directed radially and vary with the radius, they must be balanced by a nonconstant pressure gradient in the radial direction. If Ω is the angular velocity of the fluid, the radial pressure gradient must be of the form

$\partial p/\partial r = \rho r \Omega^2,$ $\hspace{3cm}$ (2.6.1)

while the vertical pressure gradient remains as the hydrostatic form

$\partial p/\partial y = -\gamma$ $\hspace{3cm}$ (2.6.2)

provided that the axis of rotation is vertical.

We write

$$dp = (\partial p / \partial r)\, dr + (\partial p / \partial y)\, dy = \rho r \Omega^2\, dr - \gamma\, dy,$$

and integration of the above for constant mass density and angular velocity gives

$$p = p_0 - \gamma y + 0.5 \sigma r^2 \Omega^2. \tag{2.6.3}$$

The isobaric surfaces are found by letting p in (2.6.3) be constant. These surfaces are seen to be parabolas of revolution, of the form

$$y = r^2 \Omega^2 / 2g + \text{constant.} \tag{2.6.4}$$

Note that in using equations (2.6.3) and (2.6.4), r **must** be measured from the axis of rotation.

The above results are used in the casting of large mirrors for telescopes used in astronomical observatories. Molten glass disks are rotated at such a speed that when they solidify they have the desired parabolic shape.

Example 2.6.1. Rigid-body rotation—circular cylinder rotated about its axis
A circular cylinder of radius a and initially filled to a depth h is rotated about its vertical axis. (See Figure 2.23.) Find the shape of the free surface so that the volume of liquid under the free surface is equal to the initial volume of fluid. How fast must the cylinder be rotated for a dry spot to appear at the bottom of the container?

Sought: Rotational speed at which a dry spot first appears at the bottom of a rotating tank.
Given: Initial fluid volume is $\pi a^2 h$.

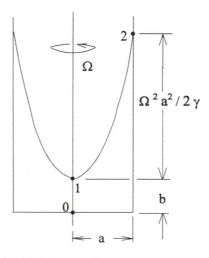

Figure 2.23. Rotating tank with fully wetted bottom.

Assumptions: The fluid is of constant density. The angular velocity is constant.

Figures: See Figure 2.23.

Solution: Using a coordinate system that has the y axis coinciding with the axis rotation and taking the origin of the coordinate system at the bottom of the cylinder, we find the pressure before rotation starts to be given by

$$p = \gamma(h - y) \text{ (gage)}.$$

When the cylinder is rotated, the pressure is given by

$$p = p_0 - \gamma y + 0.5\, \rho r^2 \Omega^2.$$

If the depth at the center of the container is b, then since $p(0, b) = 0$, we get

$$p_0 = \gamma b.$$

We can find b in terms of the rotational speed by making use of the fact that the mass of the fluid in the container remains constant. Referring to Figures 2.23 and 2.24, using the fact that the volume of a paraboloid of revolution is one-half the volume of the circumscribing cylinder, and that on the free surface

$$p = 0 = \gamma(b - y + r^2\Omega^2/2g),$$

then

$$\rho\pi a^2 h = \rho\pi a^2(b + \Omega^2 a^2/4g),$$

and upon solving for b we find that

$$b = h - \Omega^2 a^2/4g.$$

When

$$\Omega = \Omega_0 \equiv \sqrt{4gh/a^2},$$

a dry spot appears on the bottom of the cylinder that grows as the angular speed is increased further.

At speeds greater than Ω_0, the computation of p_0 and b proceeds in a similar fashion, providing we let a portion of the free surface extend below the bottom of the container as in Figure 2.24. It is no longer convenient to use the origin of the coordinate system to find p_0. Instead, we take the origin at the bottom of the container on the axis of rotation and, since the pressure at $r = a_0$ and $y = 0$ is zero, we have

$$0 = \gamma(b + a_0 2\Omega^2/2g).$$

Doing the mass calculations, we have

$$\rho\pi a^2 h = \rho\pi[(\Omega^2 a^2/2g + b)a^2/2 - ba_0 2/2 - (a^2 - a_0 2)b].$$

Eliminating b between these two equations, we have

$$(a_0/a)^4 - (a_0/a)^2 + 4gh/a^2\Omega^2 - 1 = 0,$$

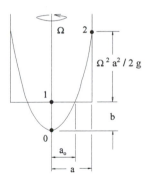

Figure 2.24. Rotating tank with partially wetted bottom.

or

$$a_0^2 = a_2(1 + \sqrt{5 - 16gh/a^2\Omega^2})/2 = a^2(1 + \sqrt{5 - 4\Omega_0^2/\Omega^2})/2$$

and then

$$b = -a_0^2\Omega^2/2g.$$

The depth of fluid on the cylinder wall is $\Omega^2 a^2/2g$. The cylinder must be at least this tall for these results to be valid.

Notice that in solving this problem we used the following facts:

1. The mass of the fluid remains constant throughout the process.
2. We know the pressure at one point ($r = 0$, $z = b$ or $r = a_0$, $z = 0$).
3. We know the equation of the free surface.
4. We know the general equation for pressure distribution.

While all problems involving either rectilinear acceleration or uniform rotation may not require all this input, it should be expected in doing problems that all available information has somehow to be used and incorporated into that process.

Suggestions for Further Reading

Bertholet, M. M., "Sur quelques phenomenes de dilatation forcee des liquides," *Ann. Chim. Phys.*, vol. **3**, pages 30 and 232, 1850.

Dixon, H. H., and J. Joly, "On the ascent of sap," *Ann. Bot.*, vol. **8**, pages 468–470, 1894.

Hayward, A. T. J., "Negative pressure in liquids: Can it be harnessed to serve man?" *Am. Sci.*, vol. **59**, pages 434–443, 1971.

Hayward, A. T. J., "Mechanical pump with a suction lift of 17 m[eters]," *Nature*, vol. **225 (5230)**, pages 376–377, 1970.

Scholander, P. F., "Tensile water," *Am. Sci.*, vol. **60**, pages 584–590, 1972.

Zimmerman, M. H., "Sap movement in trees," *Biorheol.*, vol. **2**, pages 15–27, 1964.

Problems for Chapter 2

Manometers

2.1. A mercury column barometer reads 775 mmHg at 20°C. What is the atmospheric pressure? How significant to the accuracy of your calculation is the correction for mercury vapor pressure at the top of the column?

2.2. For the manometer shown in the figure, what are the gage pressures at points A, B, C, and D? Give your answers in psfg.

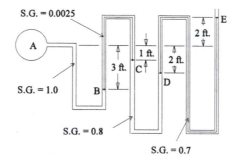

Problem 2.2

2.3. A manometer is connected between two pipelines, shown as A and B in the figure. What is $p_A - p_B$, expressed as feet of water?

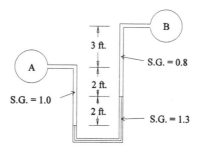

Problem 2.3

2.4. For the manometer reading shown in the figure, what is the pressure difference $p_A - p_B$?

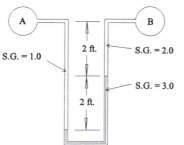

Problem 2.4

2.5. The manometer shown in the figure connects two pipelines A and B. What is the pressure difference $p_A - p_B$?

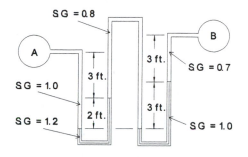

Problem 2.5

2.6. Two manometers as shown in the figure connect 2 pipelines A and B. Calculate the pressure difference $p_A - p_B$.

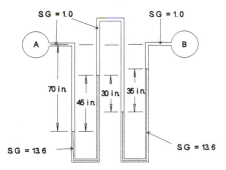

Figure 2.6

2.7. For the manometer shown in the figure, what is the pressure difference between points A and B? State your answer both in psf and ft of water. Which point is at the higher pressure?

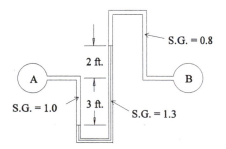

Problem 2.7

2.8. Find the pressure difference $p_1 - p_4$ for the manometer shown in the figure. As intermediate results, give the pressures at p_2 and p_3 in terms of p_4 also.

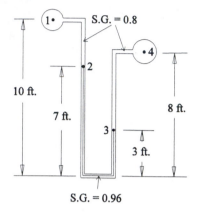

Problem 2.8

2.9. The manometer shown in the figure is used to measure the pressure difference between two pipelines A and B, each containing water. The manometer fluid has a specific gravity of 1.75, and the elevation difference between the two pipes is 0.5 m. The difference between the two meniscuses is 0.3 m. What is the pressure difference $p_A - p_B$?

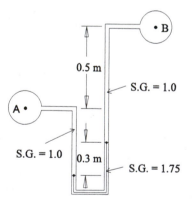

Problem 2.9

2.10. Find the pressure difference $p_A - p_B$ in psi for the situation shown in the figure. $SG_A = 1.5$, $SG_B = 2.2$, and $SG_C = 3$.

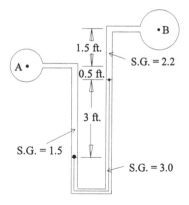

Problem 2.10

2.11. A glass capillary tube with an inner diameter of 2 mm is inserted into a beaker of impure water at 20°C. How high will the water rise in the tube? What is the pressure in the water immediately under the curved free surface?

2.12. Water rises 3 in in a clean glass capillary tube dipped into a beaker of water. What is the inner diameter of the capillary tube?

2.13. A manometer is constructed by inserting a clean glass tube with an inner diameter of 2 mm into a pipeline containing 20°C water. If the actual pressure in the water line is 3 in of water, what pressure will the manometer column show?

2.14. Two parallel clean glass plates separated by 1 mm are placed vertically into a container of water. How high will the water rise between the plates?

2.15. The plates in the previous problem are arranged so that they touch at one end, and 50 mm away are separated by 1 mm. With x measured from the line where the plates touch, what is the equation of the air-water interface?

2.16. A closed water faucet has a small drop of water suspended from its exit. A pressure gage just above the drop registers a pressure of 1.2 Pa. Determine the radius of curvature of the water at the center of the drop. The water is at 25°C and the inner diameter of the faucet is 15 mm.

2.17. A U-shaped wire is formed with arms pointing down of length 2 cm, separated by 10 cm. The wire is attached to a spring that in turn is fastened to a scale. The spring constant is determined by adding a 3-g mass to the wire, which causes the spring to extend an additional 4.2 cm. When the U-shaped wire is dipped into a beaker of water and the beaker is slowly lowered, a 2 by 10 cm water film forms on the wire. The film breaks when the spring is stretched 2 mm from the start position. What is the surface tension of the water?

2.18. A fine capillary tube immersed in a liquid-containing beaker draws the liquid to a height of 50 mm. A student suggests that if the tube was immersed deeper into the liquid, so that less than 50 mm of the tube was above the free surface, a fountain would be formed at the top of the capillary, with fluid being pumped continuously. Explain what actually would happen.

2.19. A wire-frame circular cylinder is formed by joining 2 wire rings, each of 20-mm diameter. When dipped into a soap solution a bubble is formed consisting of a cylinder that joins two spherical ends on the rings. What is the radii of curvature of the spherical ends? Consider the diameter of the circular cylinder to be constant.

2.20. A large-diameter clean glass beaker is half-filled with turpentine at 20°C. What is the elevation rise due to the surface tension of the liquid at the wall?

Forces on plane surfaces

2.21. For the surface *ABCD* as shown in the figure, with water being the fluid and using the given coordinate system, find the pressures at *B*, *C*, *D* in terms of gage pressure. Give the three scalar equations that describe the pressure variations on surfaces *AB*, *BC*, and *CD* in terms of distance measured from *D*. Compute the force per unit width on surfaces *AB*, *BC*, and *CD*. Determine the center of pressure for each of the forces on surfaces *AB*, *BC*, and *CD*. What is the net force per unit width (excluding atmospheric pressure) on the surface *ABCD*? What is the resultant moment per unit width on *ABCD* taken about point *D*?

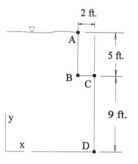

Problem 2.21

2.22. For the surface *ABCD* (shown in the figure) submerged in seawater, find the *x* and *y* components of the resultant force per unit width. Give the location of the point where the line of action of the resultant intercepts the *x* axis to the right of *A*.

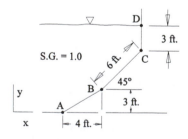

Problem 2.22

2.23. Determine the horizontal and vertical forces on the gate *A* in the figure due to the fluid. Include the effect of the pressure at the top of the gate, but not atmospheric pressure. What couple at *A* is required to hold the gate closed?

2.24. Find the force on the surface *AB* in the figure. Fluid *A* is water, fluid *B* has a specific gravity of 0.8, and fluid *C* a specific gravity of 1.7. The manometer is open to the atmosphere on the right upper end. The dimension of the container into the paper is 0.7 m. What is the pressure at points *A* and *B*? What is the force of the fluid on surface *A*?

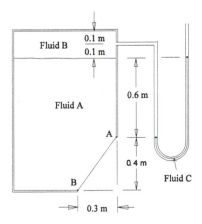

Problem 2.24

2.25. Find the resultant force on the area *A* shown in the figure. Where does the force act? The width of the surface is 7 ft and the fluid is water.

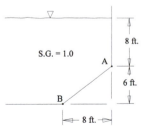

Problem 2.25

2.26. What is the net force on the gate shown in the figure due to the water? The gate is 7 ft wide. How large a vertical force *F* is needed to raise the gate?

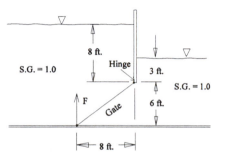

Problem 2.26

2.27. Give the magnitude and direction of the force per unit width acting on the surface A in the figure. How far from A does the line of action of the force intersect the surface?

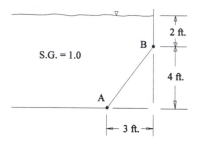

Problem 2.27

2.28. The box shown in the figure has a dimension of 6 ft into the paper. What is the force exerted on side AB, which is 5 ft by 6 ft (into the paper).

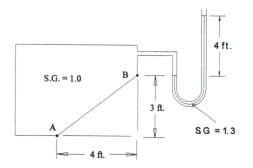

Problem 2.28

2.29. Find the resultant force per unit width on one side of the plane surface AB in the figure. What is the slope of the line of action of this force? How far from B does the line of action intersect the surface AB? SG = 0.85.

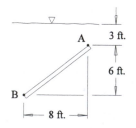

Problem 2.29

2.30. A vertical wall 4 ft high and 4 ft wide stands in saltwater to a depth of 12 ft. ($\gamma = 64$ lb/ft^3.) What is the total force on the wall due to the water acting on one side of the wall? What is the moment about the bottom of the wall?

2.31. For the system shown in the figure, what is the gage pressure at point *B*? At point *A*? If the gate is 5 ft wide, what is the horizontal force component acting on the gate? What is the vertical force component? What couple at *A* would be required to hold the gate closed?

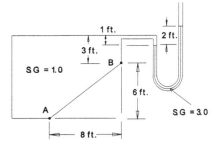

Problem 2.31

2.32. If the flat gate in the previous problem were replaced by a curved gate passing through the same points *A* and *B*, would the horizontal force be different? Would the vertical force be different? State how much they would be different depending on the geometrical shape of the curved gate in terms of the amount of fluid contained.

2.33. Find the force per unit width (into the paper) due to the water acting on the surface *ABC* in the figure. Give the point of intersection of the line of action of this force with a vertical line through *C*.

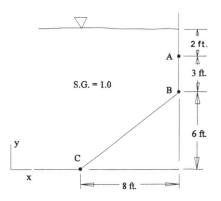

Problem 2.33

2.34. What is the force per unit width (give horizontal and vertical components) of the fluid on the gate *BC* in the figure, given p_A = 20 ft of water. State whether the horizontal components act to the right or to the left, and whether the vertical components are up or down. If the gate is hinged at *B*, what vertical force *F* at *C* is needed to keep the gate closed?

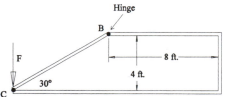

Problem 2.34

2.35. A rectangular surface 6 ft wide and 10 ft long is submerged in water as shown in the figure. What is the force per unit width exerted by the water on one side of the surface? What is the reading h on the manometer?

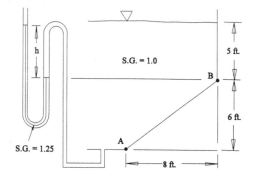

Problem 2.35

2.36. Find the total force per unit width due to the water on one side of the surface ABC in the figure. Give your answer in terms of x and z components. What is the moment of this force about A?

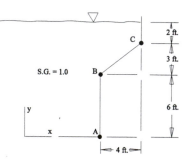

Problem 2.36

2.37. For the rectangular surface AB shown in the figure, using a specific weight of 62.4 lb/ft^3, what is the gage pressure as a function of x and y? Find the force per unit width (into the paper) acting on one side of the surface. Find the center of pressure.

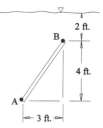

Problem 2.37

2.38. Find the force per unit width due to the water acting on the surface *AB* in the figure. How far from the bottom of the surface measured along *AB* does the resultant force act?

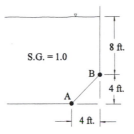

Problem 2.38

2.39. If the flat surface in the previous problem were replaced by a circular surface of radius 4 ft, what would be the change in your answers?

2.40. Find the center of pressure for the rectangular gate *AB* in the figure.

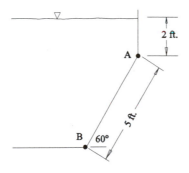

Problem 2.40

2.41. Find the force per unit width exerted by the fluid on the wall *AB* of the figure if the density of the fluid is given by $\rho = 2(1 - 0.1z)$, where z is measured from the bottom. What moment per unit width does the fluid exert on the wall about *A*?

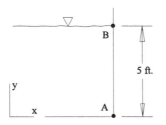

Problem 2.41

2.42. For water as the fluid in the figure find the following: the horizontal component of force on one side of *AB*; the vertical component of force on one side of *AB*; the point of application (center of pressure) of the force, given as a distance along *A* as measured from *A*.

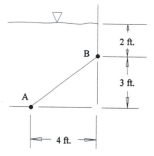

Problem 2.42

2.43. For the configuration of the figure, calculate what the weight of the 3-ft-long gate should be to keep it closed. The density of the oil is 0.8 that of water. The width of the gate is 5 ft.

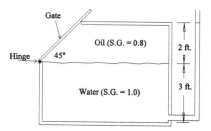

Problem 2.43

2.44. What is the total force on wall *ABC* in the figure due to the water? What is the moment of this face about *A*?

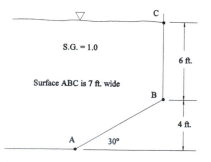

Problem 2.44

2.45. Oil (SG = 0.8) and water are standing in a tank as shown in the figure. What horizontal force *F* per foot of width is required to hold the gate in equilibrium? Neglect the weight of the gate.

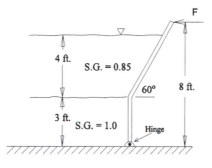

Problem 2.45

2.46. Find the horizontal and vertical components of the force exerted on gate *AB* in the figure per unit width of gate. How far from point *B* does the resultant force act?

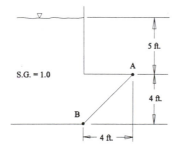

Problem 2.46

2.47. What force is required to start to lift the 100-lb cube shown in the figure. The cube initially makes a tight seal between the oil and water.

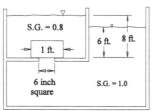

Problem 2.47

2.48. The slanted gate *AB* in the figure is a uniform plate weighing 20,000 lb that separates two bodies of freshwater in a channel 8 ft wide. As the water is drained from the left side of the gate, the reaction at *B* decreases. Determine the depth *h* at which this reaction is zero.

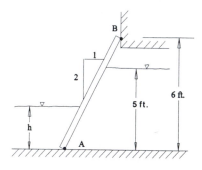

Problem 2.48

2.49. Find the force exerted on the plane rectangular surface *AB* in the figure (10 ft wide into the paper).

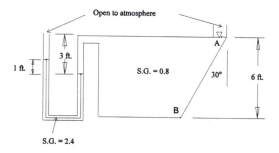

Problem 2.49

2.50. Find the force per unit width on the floor of the oil container in the figure due to the 3 fluids (water, air, and oil). The floor area is 5 ft^2.

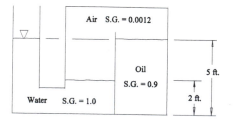

Problem 2.50

2.51. The fluid shown in the figure is saltwater, whose specific weight varies linearly with the depth. Find the force of the fluid on the gate per unit width, and its line of action. $SG_A = 1.0$, $SG_B = 1.5$.

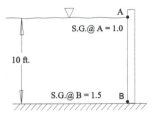

Problem 2.51

2.52. What vertical force P is needed to hold the hinged (at B) gate BC in the figure closed? The fluid is water, and the gate is 8 ft wide into the paper.

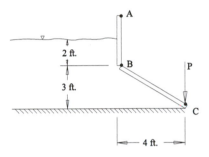

Problem 2.52

2.53. The hinged gate (hinged at its top) in the large box in the figure can only rotate clockwise. At what manometer reading h will the gate first open?

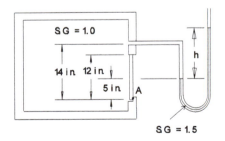

Problem 2.53

Forces on circular cylindrical surfaces

2.54. What is the force per unit width on the surface AB in the figure? Give the coordinates on the point on the line of action where the resultant force intersects the surface AB. $SG = 1$.

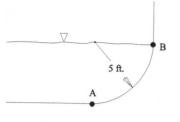

Problem 2.54

2.55. The gate shown in the figure has the shape of a segment of a circle. It is 30 ft wide and pivoted about the circle center. Calculate the magnitude of the horizontal and vertical components of the force on the gate. The pivot is at the same elevation as the water surface. The gate has a radius of 10 ft.

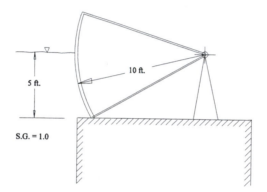

Problem 2.55

2.56. What is the moment M needed to keep the gate A in the figure closed? The gate width is 3 ft.

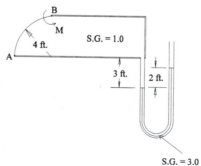

Problem 2.56

2.57. Find the resultant force on one side of the surface *ABC* in the figure. Give your answer in terms of horizontal and vertical components. What is the slope of the line of action of this force? How far above *O* on the line through *OC* does the line of action pass?

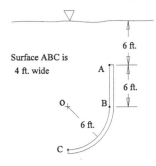

Surface ABC is
4 ft. wide

6 ft.

6 ft.

6 ft.

A

B

O

C

Problem 2.57

2.58. A hemispherical shell 4 ft in diameter is attached with the flat circular edge flush to the vertical inside wall of a tank containing water. If the center of the shell is 6 ft below the water surface, what are the magnitudes of the vertical and horizontal hydrostatic force components on the shell? (Use $\gamma = 62.4$ lb/ft^3 for water.)

2.59. If the gate shown in the figure is 10 ft wide and weighs 15,000 lb with its center of gravity at *A*, find the force at stop *B*.

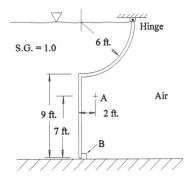

Hinge

S.G. = 1.0

6 ft.

9 ft.

7 ft.

A

2 ft.

Air

B

Problem 2.59

2.60. Determine the force on the surface *ABC* in the figure. How far above *A* is the line of action of the force?

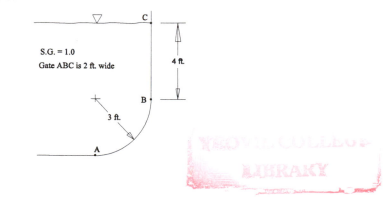

S.G. = 1.0

Gate ABC is 2 ft. wide

C

4 ft.

B

3 ft.

A

Problem 2.60

2.61. A 4-in-diameter circular opening in the side of a tank is closed by a conical stopper as in the figure. The centerline of the stopper is 16 in below the air-water interface. What is the total force exerted on the stopper by the fluid?

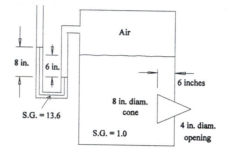

Problem 2.61

2.62. Find the horizontal and vertical components of the force per unit width exerted on gate *A* in the figure.

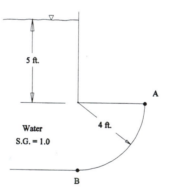

Problem 2.62

2.63. A metal plug of diameter 2 in and weighing 2 lb is fitted into an opening dividing two fluids as shown in the figure. The plug has a slight taper so that it cannot be forced

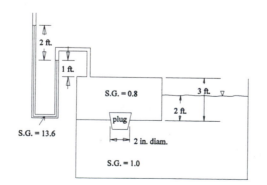

Problem 2.63

downward through the hole. Under the conditions shown, what force would be required to raise the plug? Neglect the height of the plug.

2.64. The corner of a tank consists of a surface of an octant of a sphere with radius of 5 ft. Calculate the total force (and direction) on this spherical surface when the water depth is 10 ft above the center of the sphere.

2.65. Calculate the magnitude of the x and y force components and the center of pressure for the quarter circle gate shown in the figure. The fluid is water and the gate width is 10 ft.

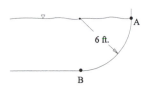

Problem 2.65

Buoyancy forces

2.66. The tank containing water shown in the figure is held closed by a 250-lb lid on top of a hole 3 ft in diameter. The open manometer tube in the top reads a pressure of 20 ft of water. If a large block of concrete ($\gamma = 150$ lb/ft^3) is hung from the lid, what is the smallest volume of concrete needed to keep the lid closed?

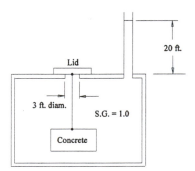

Problem 2.66

2.67. One end of a 25-lb pole 1/4 ft in diameter and 18 ft long is anchored with 10 ft of light cable to the bottom of a freshwater lake where the depth is 15 ft. Find the angle that the cable makes with the vertical and the length d of the pole protruding above the water.

2.68. A cylindrical wooden rod 12 ft long and 1/2 in in diameter is supported at its upper end by a string A as in the figure. The lower part of the rod is submerged in water. Calculate the angle that the string makes with the vertical; and the angle the rod makes with the horizontal. Use SG$_{\text{wood}} = 0.9$.

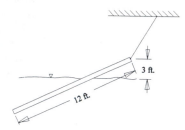

Problem 2.68

2.69. The cylindrical tank shown in the problem figure is 4 ft in diameter and 6 ft high. The tank is initially filled with air at 15 psi and 60°F. Seawater ($\gamma = 64$ lb/ft^3) enters the tank through an opening at the bottom as the tank is slowly lowered. Assume that no air leaks out of the tank and that it is compressed isothermally. What size of solid-steel cube is required to hold the tank 100 ft under the surface, if the tank weighs 50 lb and steel weighs 500 lb/ft^3?

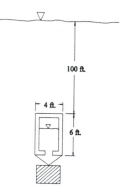

Problem 2.69

2.70. A hot-air balloon envelope has a volume of 200,000 ft^3 and weighs 350 lb. The passenger basket, burner, and fuel weigh 325 lb. What must be the average temperature of the air in the envelope on a 10°C day to carry a passenger load of 1,300 lb?

2.71. Concrete canoe races have become popular on college campuses. A typical canoe can be approximated as a rectangle 12 ft long, an average 18 in wide, and with a side height of 2 ft. If the canoe is to carry a load of 400 lb and 15 in of the canoe is to be submerged, what is the thickness of the concrete shell? (Concrete weighs 150 lb/ft^3.)

2.72. Archimedes is said to have used his buoyancy principle to determine the purity of gold in a crown. (Pure gold has a density of 19,300 kg/m^3.) A certain crown has a mass of 36 kg in air, and 34 kg when fully submerged in water. What is the volume and density of the crown?

2.73. Hydrometers are used for determining the specific gravity of liquids. They consist of an approximately spherical glass bulb with a cylindrical stem. The bulb contains small lead shot for stability and the stem contains a paper scale. The specific gravity is found by reading the scale at the free surface level. A particular hydrometer is constructed with a stem diameter of 6 mm. When immersed in a fluid of specific gravity of 1.6, only the bulb is submerged. When immersed in a liquid of specific gravity 1.2, 75 mm of the stem is also submerged. What is the volume of the bulb?

2.74. A container of water with an internal diameter of 125 mm rests on a scale. A steel cylinder ($\rho_{steel} = 7,860$ kg/m^2) 50 mm in diameter is inserted vertically into the water so that the bottom of the cylinder is immersed 70 mm into the water. The cylinder does not touch the container. How much does the reading of the scale change due to this immersion? Assume that no water spills out of the container.

Stability of submerged and floating bodies

2.75. A rectangular barge 100 ft long, 30 ft wide, and with a draft (the depth of the submerged hull) of 6 ft floats in freshwater. What is the weight of the barge? If the barge tilts through 20° about its longitudinal axis, what is the greatest height above the hull bottom the center of gravity can be for the barge to be stable?

2.76. A wooden cylinder (SG = 0.6) with a diameter of 0.2 m and a length of 0.45 m floats in a fluid having a SG of 1.35. Will the cylinder float upright? Suppose a lead weight with a diameter equal to that of the cylinder is added to the bottom of the cylinder. What would be the minimum thickness of the lead to ensure stability to small disturbances?

2.77. A hollow cubical box is made from 2-ft by 2-ft squares of steel plate. The box weighs 400 lb. It is placed with sides vertical into a container of oil (SG = 0.85). Where is the metacentric height of the box? Is this position stable?

2.78. An inflated tire inner tube is in the shape of a torus with an outer diameter of 1.5 m and an inner diameter of 0.9 m. It is weighted so that its center plane is at the water level. Find the location of the metacentric height. (*Hint:* For a solid circle, $I_{circle} = \pi d^4/64$. For a ring, I will be the difference between the I's for the outer and inner circles.)

2.79. The waterline of a 12-ft-long canoe can be approximated by a rectangle 2 ft by 6 ft, with two triangles (2-ft base by 3-ft height) at each end. The canoe and occupant weighs 250 lb. What is the metacentric height? Comment on the stability of the canoe.

Rigid-body acceleration

2.80. A cubical box 1 ft on an edge is open at the top and two-thirds filled with water. If the fluid were accelerated to the right, how great an acceleration is possible before fluid spills from the box?

2.81. If the box in the previous problem is accelerated vertically downward at the rate 16.1 ft/s, what is the force exerted on the bottom of the box?

2.82. The container shown in the figure is accelerated to the left an amount 8.05 ft/s and upward an amount 64.4 ft/s. Water is in the box. Find the pressures at points A, B, C, and D.

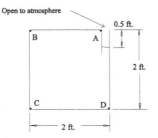

Problem 2.82

2.83. The 2-ft by 2-ft by 2-ft box shown in the figure is accelerated to the right at a rate of 16.1 ft/s. What are the forces on the bottom, sides, and top?

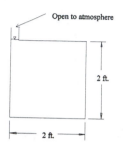

Problem 2.83

2.84. A container of water is being accelerated downward at a rate of 16.1 ft/s and to the right at a rate of 64.4 ft/s. If the box is a 4-ft cube, and was initially one-eighth filled, what is the slope of the free surface? Sketch the pressure distribution along the bottom of the box, giving values of the pressure at the end points of the box.

2.85. The box shown in the figure is accelerated upward at a rate of 16.1 ft/s and to the right at an unknown amount. Find the horizontal acceleration. What are the pressures at points A and B?

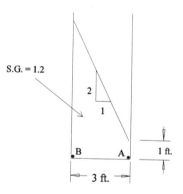

Problem 2.85

2.86. An initially full circular container 2 ft in diameter and 5 ft high is accelerated horizontally so that half of the fluid spills out. What is the magnitude of the acceleration?

2.87. If the cylinder in the previous problem was instead rotated about its axis of symmetry at a constant angular rate, the fluid just reached the top of the cylinder, and one-third of the area of the bottom was dry, what would be the angular rate of rotation?

2.88. The box shown in the figure is accelerated an amount 8.05 ft/s to the left and downward an amount 16.1 ft/s. The box is filled with water ($\gamma = 62.4$ lb/ft^3) and the fluid in the manometer is oil (SG = 0.8). The pressure gage in the upper-left corner reads 0.5 psig. What is the height h?

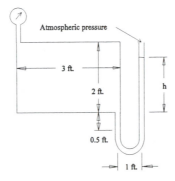

Problem 2.88

2.89. A closed cubical box 2 ft on each edge is half-filled with water and the other half is filled with oil having a specific gravity of 0.75. When it is accelerated upward at 16.1 ft/s^2, what is the pressure difference between the top and the bottom in lb/ft^2?

2.90. The box shown in the figure is half-filled with water. If it is accelerated to the right with an acceleration of 16.1 ft/s and accelerated downward at 24.15 ft/s, what are the pressures at points A and B?

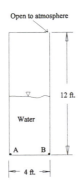

Problem 2.90

2.91. If the container of fluid shown in the figure is being accelerated to the right with an acceleration of $g/2$, and accelerated downward with an acceleration of $g/4$, determine an equation giving pressure at any point in the box in terms of x and y.

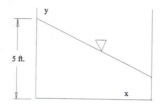

Problem 2.91

2.92. A rectangular tank 0.3 m on a side and 0.5 m deep containing water is accelerated to the left with an acceleration of 14.7 m/s^2 and upward an amount 4.9 m/s^2. Initially it is filled to a depth of 0.2 m. What is the slope of the free surface? Is the slope up to the left or down to the left? What is the pressure at the bottom-right corner? What are the depths at the left and right sides of the tank?

2.93. The box of the figure in Problem 2.82 is accelerated to the right at a rate of 16.1 ft/s^2 and upward at a rate of 40.25 ft/s^2. Give the gage pressures at points A, B, C, D.

2.94. A cubical box 1 ft on an edge is open at the top and two-thirds filled with water. If the fluid were accelerated to the right, how great an acceleration is possible before fluid spills from the box?

2.95. If the box in the previous problem were instead accelerated vertically downward at the rate of 16.1 ft/s, what would be the force on the bottom of the box?

2.96. An open cubical tank of seawater ($\gamma = 64$ lb/ft^3) is accelerated to the right at an acceleration of 16.1 ft/s. The tank is 2 ft on a side and initially was filled with water. What is the net force exerted by the water on the right wall of the tank? What is the force exerted by the water on the bottom of the tank?

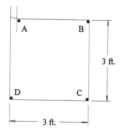

Problem 2.96

Rigid-body rotation

2.97. The U-tube shown in the figure is rotated about the vertical axis z-z. If the fluid is water, how fast must it be rotated until a vacuum is reached at some point in the fluid? Assume that the fluid does not vaporize first. Take $\rho_{atmos} = 15$ psia.

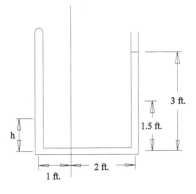

Problem 2.97

2.98. The box shown in the figure is rotated about the side *ABC*. If the distance *AB* = 0.8 ft and *DE* = 3 ft, what is the angular rate of rotation? What is the pressure at *E*?

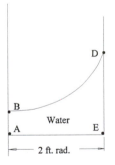

Problem 2.98

2.99. A parabolic mirror having the shape $z = ar^2$ has a focal point at $(1/4a, 0)$. An astronomical telescope is to be constructed having a mirror 30 ft in diameter with a focal length of 40 ft. At what speed should the molten glass be rotated?

2.100. Water contained in a small-diameter tube is rotated about a vertical line at 240 rpm. The closed bottom of the tube is on the axis of rotation, and the tube is inclined 30° with the axis of rotation. If the length of the water column is 0.35 m, what is the pressure at the tube bottom?

2.101. A container of water spinning at 100 rpm is accelerated downward at an acceleration of 0.3 g. The depth is 3 in on the axis of rotation. Find the pressure on the container bottom both on the axis of rotation and 3 in radially outward from it.

2.102. Skim milk can be produced by rotating whole milk in a centrifuge. The fat separates from the skim milk and, being of lighter density, floats on top. If a container of milk 6 in in radius is rotated at 50 rpm about its vertical centerline, and the layer of cream is observed to be 0.5 in thick on the axis of rotation, what is the equation of the interface between the cream and the skim milk? Choose your origin on the axis of rotation at the free surface at the top of the cream. (*Hint:* The interface will be a surface of constant pressure.)

2.103. Casting of thin cylindrical shells can be accomplished by rotating a cylinder sufficiently fast that the liquid clings to the cylinder side. If it is desired to cast such a cylinder with a 3-in outer diameter, a length of 1 ft, a thickness of 0.25 in, and a specific gravity of 0.9, what must be the rotational speed so that the thickness variation over the length is less than 20%?

2.104. The U-tube shown in the figure for Problem 2.97 contains oil having a SG of 0.85. It is rotated about the vertical axis z-z at an angular velocity of 25 rpm. If point A at $r = 1$, $z = h$ has the same pressure as point B at $r = 2$, $z = 1.5$, what is the height h?

2.105. A rectangular box of length 0.4 ft, width 0.2 ft, and height 0.6 ft is half-filled with water. If the box is rotated about the vertical centerline, at what speed will the water spill out of the box?

2.106. A vertical cylindrical tank of diameter 5 in is filled with an oil (SG = 0.9) to a depth of 10 in. If it is rotated about its centerline, at what speed will a dry circle of diameter 4 in appear on the bottom of the tank? What is the maximum pressure in the oil at this speed?

chapter **3**

Fluid Dynamics

Chapter Overview and Goals

In this chapter we introduce a number of flow properties as well as the concepts of discharge and incompressibility. The ideas of control surfaces and control volumes suited to the analysis of fluid problems are introduced along with that of the material time derivative. The latter is used in computing acceleration and in finding the rate of change of flow and fluid quantities such as mass density.

The fundamental equations of fluid mechanics are developed in one-dimensional form by use of the concepts of conservation of mass, linear momentum, angular momentum, and energy. These four laws are expressed in forms suitable for control volume analysis. The use of translating and rotating coordinate systems is demonstrated and applied in several problems.

The basic equations developed in this chapter are fundamental to all fluid flow problems. To understand how they are applied in problems and also to gain an understanding of the implications of these laws, we consider a number of simple one-dimensional flows. From these applications we will see how we can arrive at reasonable predictions of flow behavior by making simplifying assumptions.

1. Flow Properties and Characteristics

In studying the flow of fluids, we encounter a wide variety of distinct ways in which the flows may be characterized. Many times, the terminology used is of an either/or nature; that is, a flow is either steady or unsteady, laminar or turbulent, uniform or nonuniform. In many applications the boundaries between the various classifications can be imprecise, and when we do use these broad categories they may apply only to portions of the flow region, rather than to the entire flow region. As in most things in engineering, simplified models are always the starting point for the analysis of a problem.

Many of the flow quantities in which we are interested (e.g., velocity, acceleration, pressure) are *field quantities*. By that we mean they depend on the location, or position, of a point in space and time. Some of our terms used to describe a flow have to do with how various field quantities depend on these coordinates. For example, a flow is said to be a *steady flow* when its flow properties at a particular point in space do not vary with time. If the flow properties at a point *do* vary with time, the flow is said to be an *unsteady flow*. Note that even in a steady flow, as a fluid particle moves from point to point in space its flow properties such as velocity, pressure, density, and the like may change. By steady flow we mean only that if we observe the flow at any given point, the properties of this flow at that point do not vary with time.

The spatial counterpart of steady/unsteady is uniform/nonuniform. A *uniform flow* is one for which the flow is the same at every point in the flow space at a given point in time. If at a given time the flow quantities take on different values at different points, the flow is said to be a *nonuniform flow*. A uniform flow may, or may not, vary with time. Flows that are everywhere uniform are rare. More often we encounter flows whose properties do not vary across an area, such as a conduit cross section. In such cases, we say that the flow is uniform across the given area.

An important flow classification is whether a flow is laminar or turbulent. In *laminar flows*, flow particles move smoothly along well-defined, relatively simple, paths, or in layers (laminas), without mixing. *Turbulent flows* on the other hand have pronounced random, chaotic characteristics with much particle mixing, and are best defined in terms of their statistical properties such as averages and their deviations. Generally, we can expect that very slow flows are laminar. As the speed of a flow is increased, most flows become unstable and change character. This change frequently occurs abruptly and intermittently, until eventually a developed turbulent state occurs. In the process, the transition from laminar flow to turbulent flow can either first go through several progressively more complicated laminar flows, or it can alternate in time between laminar and turbulent flow states. Some of the parameters that describe the point where this transition occurs for a given flow will be discussed in Chapters 6 and 7.

Many times, in certain regions of a flow, the viscous stresses play a very minor part in governing the flow. In such a case, the fluid in that region is sometimes said to be an *ideal fluid*. Such "ideal" fluids, of course, do not exist, and it is the *flow* that is negligibly influenced by viscosity. A better nomenclature is to refer to these situations as regions of *frictionless flow*, or *inviscid flow*, since the flow is the same as if the viscosity of the fluid were zero in that region. It is important to keep the distinction between fluid properties and flow properties clear, since not doing so can lead to unnecessary confusion.

We will deal largely with *one-dimensional* flows; that is, the flow properties will be considered to vary with only one of the spatial coordinates, usually the one in the direction of flow. The simplification in dealing with one space dimension in contrast with two or three turns out to be enormous, and still leads to many results of fundamental engineering usefulness. A thorough understanding of one-dimensional flows is a necessary prerequisite to the understanding of flows in a greater number of spatial dimensions.

In order to visualize what is happening in a given flow, we make use of several line concepts. The simplest of these is the *path line*, which is the path, or trajectory, of a given fluid particle. It tells us the travel history of any one fluid particle over a given time period. We can imagine a conceptual experiment where one particle of our fluid is made luminous, and the flow takes place in an apparatus with transparent walls and in a dark room. A time exposure taken by a camera would then show the path line of that one particular particle.

A *streamline* is a somewhat more abstract quantity, although it is a concept used much more frequently than a path line. A streamline is a line drawn in the flow at a particular instant of time such that the velocity vector of a fluid particle at any point on the streamline is tangent to the streamline at that point. If (v_x, v_y, v_z) are the components of the velocity vector at a given point and we look at two points separated by a distance with components (dx, dy, dz), then the requirement that the velocity be tangent to the streamline gives

$$\frac{dy}{dx} = \frac{v_y}{v_x}, \qquad \frac{dz}{dy} = \frac{v_z}{v_y}, \qquad \frac{dz}{dx} = \frac{v_z}{v_x},$$

or

$$\frac{dx}{v_x} = \frac{dy}{v_y} = \frac{dz}{v_z}. \tag{3.1.1}$$

Integration of equations (3.1.1) at a given instant of time gives the equation of the streamline. If a flow is steady, the streamlines and path lines coincide. If a flow is unsteady, they may, but usually do not, coincide, depending on the nature of the unsteadiness.

The notion of streamlines can be extended to collections of streamlines as well. A *stream surface* is a surface in space at a given time, constructed so that the velocity vector is tangent to that surface at any given point. It is thus made up of many streamlines adjacent to one another. The intersection of two stream surfaces is then a streamline. A stream surface that closes to form a conduit is called a *stream tube*. There is no flow across a stream tube; thus a solid surface such as a pipe is an example of a stream tube.

Another important line is the *streak line*. It is the locus of all particles that have passed through a given point in space during a time interval. It can be obtained in the laboratory by injecting particles or a dye into a transparent liquid, or smoke into a gas. If the flow is laminar, the dye or smoke trail remains coherent and represents the streak line. If, additionally, the flow is steady and laminar, the streak line is also a path line and a streamline.

Example 3.1.1. Stream- and path-lines for rigid-body rotation

For rigid-body rotation it was shown in Chapter 2 that the speed of a fluid particle is of magnitude $r\Omega$ and directed tangent to a circle of radius r. In Cartesian coordinates this gives the Eulerian description of the velocity components as

$$v_x = -y\Omega, \qquad v_y = x\Omega \qquad \text{with } r = \sqrt{x^2 + y^2}.$$

Find the equations for the streamlines and path lines.

Given: Velocity components.

Assumptions: Steady flow.

Solution: According to (3.1.1) the equation for the streamlines is

$$dx/v_x = dy/v_y = dz/v_z.$$

Since for this flow v_z is zero, dz must be zero. Thus the streamlines all lie parallel to the xy plane.

Putting the given velocity components into (3.1.1) gives

$$dx/(-y\Omega) = dy/(x\Omega),$$

or, upon canceling Ω from both sides,

$$x\, dx = -y\, dy.$$

Integration of this expression gives

$$0.5x^2 = -0.5y^2 + \text{constant of integration.}$$

For the streamline passing through a point (x_0, y_0, z_0), evaluation of the constant of integration, rearrangement, and multiplication by 2 gives

$$x^2 + y^2 = x_0^2 + y_0^2,$$

proving that the streamlines are circles.

Path lines would be arcs of circles, starting at some given point x_0, y_0 at time t_0, and reaching the point x, y at time t. The relation between these Lagrangian variables is

$$x = x_0 + \sqrt{x_0^2 + y_0^2}\ \cos \Omega\, (t - t_0), \qquad y = y_0 + \sqrt{x_0^2 + y_0^2}\ \sin \Omega\, (t - t_0).$$

To find the velocity from these Lagrangian variables requires only differentiation with respect to time, namely,

$$v_x = \partial x / \partial t = -\sqrt{x_0^2 + y_0^2}\ \sin \Omega\, (t - t_0), \qquad v_y = \partial y / \partial t = \sqrt{x_0^2 + y_0^2}\ \cos \Omega\, (t - t_0).$$

Example 3.1.2. Stream- and path lines for steady shear flow

Viscous flow due to a pressure gradient between parallel plates spaced a distance b apart has the velocity components

$$v_x(y) = 6U(by - y^2)/b^2, \qquad v_y = v_z = 0,$$

where U is the average velocity. The lower plate is at $y = 0$, the upper at $y = b$. Find the stream- and path lines.

Sought: Equations of the stream- and path lines.

Given: $v_x(y) = 6U(by - y^2)/b^2$, $v_y = v_z = 0$.

Assumptions: Steady shearing flow. Constant fluid properties.

Solution: From (3.2.1) since v_y and v_z are both zero and v_x does not depend on either x or z, the streamlines are straight lines parallel to the xz plane. Thus the streamlines and streak lines are given by

y = constant.

Since this flow is steady, the path lines and streamlines coincide. The path and streamlines are straight lines. The position of a fluid particle initially at (x_0, y_0, z_0) is given by the velocity times the elapsed time; thus the path lines are given by

$$x = x_0 + \Delta t v_x(y) = x_0 + \Delta t 6U(by - y^2)/b^2,$$

$$y = y_0,$$

$$z = z_0.$$

Example 3.1.3. Particle position for a vortex flow

A velocity field is given in cylindrical polar coordinates by $\mathbf{v} = (v_r, v_\theta, v_z) = (0, r\Omega, 0)$. For a fluid particle initially at (b, Θ_0, z_0), find the position as a function of time.

Sought: Find particle position as a function of initial position and time.
Given: $\mathbf{v} = (v_r, v_\theta, v_z) = (0, r\Omega, 0)$. The initial position of a fluid particle is (b, Θ_0, z_0).
Assumptions: Steady flow. Constant fluid properties.
Solution: Since only the tangential (θ) velocity component is nonzero, the streamlines are concentric circles. Since the velocity is independent of the angle theta, the fluid particle will travel around a circle of radius b at a constant speed $b\Omega$. The position at any time t (measured from the initial position) is then given by

$$(r, \Theta, z) = (b, \Omega t + \Theta_0, z_0).$$

Example 3.1.4. Streamlines for inviscid flow past a circular cylinder

A two-dimensional steady inviscid flow is given by

$$\mathbf{v} = (v_x, v_y, v_z)$$

$$= U[1 - a^2(x^2 - y^2)/(x^2 + y^2)^2, -2xya^2/(x^2 + y^2)^2, 0].$$

Find the equations of the streamlines.

Sought: Equations of the streamlines.
Given: $\mathbf{v} = (v_x, v_y, v_z) = U[1 - a^2(x^2 - y^2)/(x^2 + y^2)^2, -2xya^2/(x^2 + y^2)^2, 0]$.
Assumptions: Steady flow, constant fluid properties.
Solution: Rewrite equation (3.1.1) in the form $0 = -v_y\,dx + v_x\,dy$ and integrate from (x_0, y_0, z) to (x, y, z) by integrating first from (x_0, y_0, z) to (x, y_0, z), and then from (x, y_0, z) to (x, y, z). The result is

$$0 = \int_{x_0}^{x} \left[\frac{2xya^2U}{(x^2+y^2)^2} \right]_{y=y_0} dx + \int_{y_0}^{y} \left[\frac{2xya^2U}{(x^2+y^2)^2} \right]_{x=x_0} dy$$

$$+ \int_{x}^{x} U \left[1 - a^2 \frac{(x^2 - y^2)}{(x^2 + y^2)^2} \right]_{y=y_0} dx + \int_{y_0}^{y} U \left[1 - a^2 \frac{(x^2 - y^2)}{(x^2 + y^2)^2} \right]_{x=x_0} dy$$

$$= Uy_0 a^2 \left(\frac{-1}{x^2 + y_0^2} + \frac{1}{x_0^2 + y_0^2} \right) + 0 + 0 + \int_{y_0}^{y} U \left[1 + \frac{a^2}{x^2 + y^2} - \frac{2a^2 x^2}{(x^2 + y^2)^2} \right] dy$$

$$= Uy_0 a^2 \left(\frac{-1}{x^2 + y_0^2} + \frac{1}{x_0^2 + y^2} \right) + U(y - y_0) + \int_{y_0}^{y} U \left(\frac{a^4}{x^2 + y^2} \right) dy - 2a^2 x^2 U \left(\frac{y}{2x^2(x^2 + y^2)} \right)$$

$$- \frac{2a^4 x^2}{2x^2} \int_{y_0}^{y} U \left(\frac{1}{x^2 + y^2} \right) dy$$

$$= U \left(y - y_0 + \frac{y_0 a^2}{x_0^2 + y_0^2} - \frac{y a^2}{x^2 + y^2} \right).$$

Dividing by U and rearranging, we have

$$y[1 - a^2/(x^2 + y^2)] = y_0[1 - a^2/(x_0^2 + y_0^2)] = \text{constant through any point } (x_0, y_0).$$

Thus the streamlines are given by the equation $y[1 - a^2/(x^2 + y^2)] = \text{constant}$. In particular we note as special cases that $y = 0$ and $x^2 + y^2 = a^2$ are both streamlines.

An important concept unique to fluid mechanics is that of **vorticity**. It can be thought of as an analog of rigid-body rotation, but in a nonrigid medium. If we could suddenly freeze a very small portion of a fluid, it would spin with an angular velocity that would be the local vorticity. Vorticity is a vector; thus we can have vortex surfaces and vortex tubes defined analogously to stream surfaces and stream tubes.

An interesting visualization of vorticity is seen in the series of photographs in Figure 3.1. The fluid is initially at rest. A piston in the bottom of the conically shaped nozzle is given a sharp blow, and a quantity of dyed fluid is emitted from the nozzle. As it passes the nozzle exit a vortex ring is formed, similar to the smoke rings formed in days when people smoked. As the vortex ring rises it spreads out and develops a richer structure. Readers who have seen photographs of nuclear bomb blasts will note the strong similarity in the vortex structures. Sir Geoffrey Taylor in England was able to obtain a good estimate of the bomb energy from these photos years before this data was unclassified.

Examples of vorticity at the edge of a jet are shown in Figures 3.2 and 3.3. (See the insert between pages 128 and 129 for full-color photographs of Figures 3.2 and 3.3.) The Reynolds number in the first of these figures is less than in the second. The main body of the jet is seen to be sinuous, and the vortices at the jet edge are periodic in structure. Typically the two rows of vortices are staggered, and form what has been named a

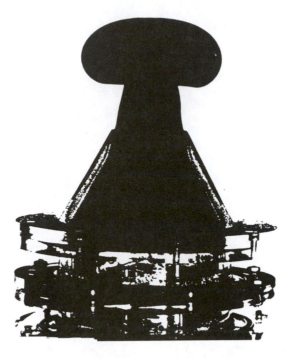

Figure 3.1. Formation of a vortex ring. A cylinder below the nozzle is a sharp blow to create the vortex. Stages in the movement of the mushroom vortex: 1. Initial formation. (Photos courtesy Professor Werner Dahm, The University of Michigan. For more details see the *Journal of Fluid Mechanics*, vol. **205**, page 1.)

Kármán vortex street, after Theodore von Kármán,[1] who first successfully analyzed the stability of rows of vortices. In Figure 3.3 at the higher Reynolds number, the vortex structure is much stronger and richer in detail. In this case the vortices have dominated the jet and there does not appear to be a coherent central core.

Vortices are the principal feature of turbulent flow. Our present understanding of turbulent flow is that for around 90% of the time at a given cross section in the flow no new turbulent structure is created. However, in the rest of the time a vortex cylinder forms near the wall that becomes deformed into a hairpin shape and rises into the main portion of the flow. These turbulent bursts sustain and refresh the turbulence.

You have undoubtedly used the term *wake* at one time or another to refer to the flow pattern behind boats, cars, and trucks. Wakes are caused on bodies when the flow separates and reverse flow regions are established. Kármán vortex streets are typical

---

[1]Kármán found that if the vortices traveled in the x direction and the x separation was denoted by a, the y separation by b, then $b/a = 0.281$.

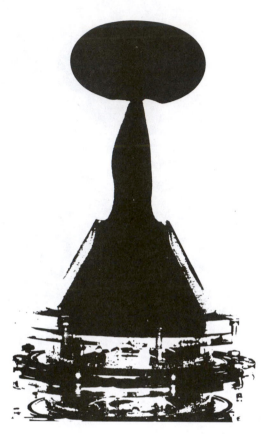

Figure 3.1. (*Continued.*) 2. Start of vortex break-off

of wakes. They can cause damage to structures, such as the famous collapse of the Tacoma Narrows Bridge in 1940. In that case the frequency of the vortex shedding was near that of natural frequencies of the bridge structure, and so wind energy was converted to vibration energy. Power lines will also shed vortices in a cross wind, and can oscillate slowly up and down until supports at the crossbar break.[2] On the annoying side, when automobile windows are opened while traveling at medium to high speeds, strong low-frequency pressure pulses can be felt on the ear drum. Wakes created by side-view mirrors and even the indentation of windows can have a surprising influence on the overall drag on a vehicle.

On the beneficial side, most wind musical instruments use vortices to provide their sounds. Whistling is the production of vortices of proper strength and frequency.

---

[2]You may have noticed blocks of concrete suspended along power lines, spaced about a quarter of the distance away from a pole. These blocks detune the frequency of the power line so that much higher wind speeds are needed for oscillation.

Figure 3.1. (*Continued.*) 3. Roll-up of vortex bottom.

Figure 3.1. (*Continued.*) 4. Progression of roll-up.

Figure 3.1. (*Continued.*) 5. Later development of vortex.

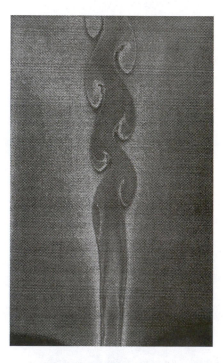

Figure 3.2. Laminar jet at a value of the Reynolds number about 200. (Photo courtesy Professor Mory Gharib and M. Beizaie, University of California, San Diego. This jet is simulated using a "Soap film tunnel," *Physics of Fluids*, vol. **31**, page 2389, 1988.)

Figure 3.3. Jet at a high value of the Reynolds number. Note vortex development at edge of jet. (Photo courtesy Professor Mory Gharib and P. Derango, University of California, San Diego. This jet is simulated using a "Soap film tunnel," *Physics of Fluids*, vol. **31**, page 2389, 1988.)

Blowing across the mouthpiece of a flute—or even a soft drink bottle—provides the oscillations that are amplified by the volume of the flute body or bottle. Organ pipes sound in a similar manner, with the length of the pipe providing the basic resonant frequency and the shape of the pipe determining the strength of the higher harmonics. The air blows across a sharp edge to provide the vortices. The ancient Greeks found that wind blowing through trees provided pleasant sounds. By tying stringed harps and bottles to trees, they produced what they termed aeolian tones. The practice has long continued in Mediterranean countries, where bottles are tied to the arms of windmills for a similar effect.

You can perform your own experiments with vortices the next time you have a cup of coffee or any other dark-colored beverage. Brushing the tip of a spoon across the surface of the liquid produces a pair of vortices that move toward the cup side, where they spread and suddenly disappear. Because of the size of the cup you will not produce a vortex street. You could do so if you have access to a channel in your school laboratory or a low bridge over a slow-moving river. In this case you can suspend a cylinder on a string partially submerged in the stream. When the shedding frequency equals the period of the pendulum, you will see the cylinder oscillate in a plane perpendicular to the flow.

2. Acceleration and the Material Derivative

In the study of the dynamics of a single particle, or even of a system made up of a finite number of particles, it is customary to consider the position vector of the particle to be the fundamental describing quantity. The time derivative of the particle position is the particle velocity, and the second time derivative of the position is the particle acceleration. This approach, while conceptually simple in particle mechanics, is almost always much too detailed for fluid mechanics. This is because there are so many particles in a fluid (an unbounded number). Also, the history of any individual particle as given by its position vector as a function of time usually contains much more information than we are interested in, or can afford to keep track of. Only in the case of some very special problems involving free surfaces or transport of particles or sediment is this approach useful in fluid mechanics. This individual particle description is termed a ***Lagrangian description***.

The vast majority of flow problems are most easily dealt with using what is called the ***Eulerian description***. In this approach, basic quantities such as pressure and velocity are given as functions of time and position at a point in space, and not for an individual fluid particle. The Eulerian approach can be realized in the laboratory where a measuring instrument inserted at a given position in the flow is allowed to record data as a function of time.

Calculating acceleration from an Eulerian description of the velocity is slightly more involved than when a Lagrangian velocity description is used. Consider, for example, the rigid-body rotation we studied in the previous chapter. We know a fluid particle travels in a circular path with constant velocity, and thus must experience a centripetal acceleration whose magnitude is velocity squared divided by the path radius. If, however, we look at only one point in the flow, all we see is a velocity constant in time, and we have no way of knowing whether the particle travels on a straight line, circular path, or whatever.

The resolution of this is through the definition of the term acceleration, which is the time rate of change of the velocity **of a particular fluid particle**. We also use the term "following a particle" to have the same meaning. Since velocity is a vector, in taking time derivatives changes in both magnitude (speed) and direction must be accounted for in determining acceleration.

A particle that at time t was at position $\mathbf{r}$ will at time $t + \Delta t$ be at position $\mathbf{r} + \mathbf{v}\,\Delta t$. (See Figure 3.4.) Since acceleration is given by

$$\mathbf{a}(\mathbf{r}, t) = \operatorname*{limit}_{\Delta t \to 0} \Delta \mathbf{v}/\Delta t,$$

where $\Delta \mathbf{v}$ refers to the change in the velocity of a specific fluid particle over time Δt, using an Eulerian description a more detailed form of the above is

$$\mathbf{a}(\mathbf{r}, t) = \lim_{\Delta t \to 0} \left[\frac{\mathbf{v}(\mathbf{r} + \mathbf{v}\Delta t,\ t + \Delta t) - \mathbf{v}(\mathbf{r}, t)}{\Delta t} \right]$$

$$= \lim_{\Delta t \to 0} \left[\frac{\mathbf{v}(x + v_x\Delta t,\ y + v_y\Delta t,\ z + v_z\Delta t,\ t + \Delta t) - \mathbf{v}(x, y, z, t)}{\Delta t} \right].$$

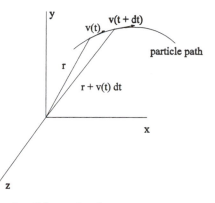

Figure 3.4. Determination of particle acceleration.

Taking the limit as Δt goes to zero requires that we look at the changes induced in each component of the particle position, as well as the change locally with respect to time. Thus the limit is

$$\mathbf{a}(\mathbf{r},\, t) = \frac{\partial \mathbf{v}}{\partial t} + v_x \frac{\partial \mathbf{v}}{\partial x} + v_y \frac{\partial \mathbf{v}}{\partial y} + v_z \frac{\partial \mathbf{v}}{\partial z}. \tag{3.2.1}$$

(Notice that in computing the acceleration in the Eulerian description, we have "borrowed" the Lagrangian description for a brief, infinitesimal time interval.) The first term in equation (3.2.1) is called the ***temporal***, or ***local*, *acceleration***, and vanishes if the flow is steady. The remaining terms are called either the ***convective acceleration*** terms or the ***advective acceleration*** terms. Both names imply the concept of conveyance, or carrying along with, to account for direction changes. They are the chief source of nonlinear effects in flows, and introduce the greatest difficulties and complexities encountered in understanding the nature of fluid flows.

The del operator

$$\nabla = \mathbf{i}\frac{\partial}{\partial x} + \mathbf{j}\frac{\partial}{\partial y} + \mathbf{k}\frac{\partial}{\partial z}$$

is a convenient tool for expressing the convective acceleration terms. With its use the writing of the convective terms can be shortened by replacing them with

$$\mathbf{v} \cdot \nabla = v_x \frac{\partial}{\partial x} + v_y \frac{\partial}{\partial y} + v_z \frac{\partial}{\partial z}, \tag{3.2.2}$$

the right-hand side being the form of the operator in Cartesian coordinates. A vector dot product is by definition the magnitude of the two vectors times the cosine of the angle between them. Thus the operator given in (3.2.2) is equivalent to

$$\mathbf{v} \cdot \nabla = |\mathbf{v}|\frac{\partial}{\partial s}, \tag{3.2.3}$$

where s is the coordinate in the direction of the velocity vector, e.g., tangent to the streamline. This again emphasizes that we are measuring the change in velocity in the direction in which the fluid particle actually travels.

We will henceforth use the notation

$$\mathbf{a} = \frac{D\mathbf{v}}{Dt}, \tag{3.2.4}$$

where

$$\frac{D}{Dt} \equiv \frac{\partial}{\partial t} + \mathbf{v} \cdot \boldsymbol{\nabla} \tag{3.2.5}$$

is called either the ***material derivative*** or the ***substantial derivative***. Its name comes from the fact that we are taking a time derivative while following a specific fluid particle in the fluid (material). It will be seen to be useful not only in computing acceleration, but also changes of other flow quantities described in an Eulerian manner (e.g., changes in mass density).

An alternative and equivalent form for the convected acceleration is given by

$$(\mathbf{v} \cdot \boldsymbol{\nabla})\mathbf{v} = \boldsymbol{\nabla}\left(\frac{|\mathbf{v}|^2}{2}\right) - \mathbf{v} \times (\boldsymbol{\nabla} \times \mathbf{v}). \tag{3.2.6}$$

You can verify this result either through the use of vector identities, or more easily by writing out both sides of equation (3.2.6) in component form and comparing them.

Note from (3.2.1) that if we want to experimentally measure the acceleration at a point in our fluid, we would accomplish this by measuring the velocity at the selected point plus also at three adjacent points, taking all measurements simultaneously. We then at a given time would need to perform the space differentiation needed to compute the convective acceleration terms. The space derivatives will of course be difference approximations, and we must keep our four points close together to have the necessary accuracy.

Obviously this is a very difficult, tedious, and costly process. For most flows it would be impossible to carry this process out with any degree of accuracy, and it is in fact seldom if ever done. Usually acceleration is arrived at by indirect means, wherein other quantities are measured and the acceleration is then inferred from the linear momentum equation, discussed later in this chapter.

Example 3.2.1. Acceleration for rigid-body rotation

For rigid-body rotation it was shown that the speed of a fluid particle is of magnitude $r\Omega$ and directed tangent to a circle of radius r. In Cartesian coordinates this gives the velocity components

$$v_x = -y\Omega, \quad v_y = x\Omega \quad \text{with } r = \sqrt{x^2 + y^2}.$$

Find the acceleration.

Sought: Acceleration as a function of radius.

Given: $v_x = -y\Omega$, $v_y = x\Omega$, $v_z = 0$.

Assumptions: Rigid-body rotation with constant angular velocity. Constant density, steady flow.

Solution: From (3.2.1), since $v_z = 0$ and since v_x does not depend on x or z and v_y does not depend on y or z, the acceleration is then

$$a_x = v_y \partial v_x/\partial y = -x\Omega^2, \qquad a_y = v_x \partial v_y/\partial x = -y\Omega^2, \qquad a_z = 0.$$

The magnitude of the acceleration is then

$$|a| = \sqrt{a_x^2 + a_y^2 + a_z^2} = r\Omega^2 = v^2/r$$

with

$$v^2 = v_x^2 + v_y^2 = (r\Omega)^2.$$

Here v is the magnitude of the velocity vector, and the acceleration is seen to be of centripetal type, directed radially inward toward the center of the circle.

Example 3.2.2. Acceleration for a steady shearing flow

Viscous flow due to a pressure gradient between parallel plates spaced a distance b apart has the velocity components

$$v_x(y) = 6U(by - y^2)/b^2, \qquad v_y = v_z = 0,$$

where U is the average velocity. The lower plate is at $y = 0$, the upper at $y = b$. Find the acceleration.

Sought: Acceleration as a function of y.

Given: $v_x(y) = 6U(by - y^2)/b^2$, $v_y = v_z = 0$.

Assumptions: Steady shearing flow. Constant fluid properties.

Solution: From (3.2.1) since v_y and v_z are both zero and v_x does not depend on either x or z, the acceleration is zero.

---

3. Control Volume and Control Surface Concepts

In our study of hydrostatics we used a free-body diagram to isolate a body of fluid and thereby define the internal and external forces that act upon it. In this section we will develop an alternative concept that is better suited for analyzing the mechanics of flowing fluids.

In several of the sections of Chapters 1 and 2 we used the concepts of "system" and "free-body diagrams." A *system* is defined as a quantity or volume of mass chosen for analysis. Anything external to the system is called the *surroundings*. The *system boundary* divides the system from the surroundings. In general, the system can be chosen so that the size and shape of the volume of the system may change, and the system volume may move in space. If the system always consists of the same fluid particles, and the mass of the system thus does not change, we refer to this as a ***closed system***. If on the other hand our system is defined so that fluid can enter and leave the system, we have an ***open system***.

Instead of the free-body diagrams used in our study of static flows, ***control volumes*** are used to study flows in motion. A control volume is a volume of fixed size and shape, fixed in a position in space. All external surface and body forces are shown along with masses, momentum, and energies entering and leaving the control volume. A "generic" control volume is shown in Figure 3.5[3] with a simpler control volume consisting of a single inlet region and a single outflow region is shown in Figure 3.6. Mass can enter and/or leave the control volume through several regions. (As an example, Figure 3.5 has two inlet regions and three outflow regions.) The control volume is thus an example of an open system. The surface bounding the control volume will be called the control surface. It separates the control volume from the surroundings, and is always a closed surface. We will use the abbreviations CV and CS to designate control volume and control surface, respectively.[4]

Different fluid particles will occupy a control volume as time changes. While the choice of the control volume is arbitrary, in a given problem there are probably only one or two choices that best aid in analyzing a given problem. Selecting this "best" control volume is an engineering decision to be made by you, the engineer. By following the examples later in this chapter, you will see some of the decisions that are involved in arriving at a suitable control volume.

By dividing the universe into what is inside the boundaries of the control volume and what is outside, the control surface effectively defines which forces are internal forces and which are external forces in our system. The portion of the universe external to the control volume acts on the contents internal to the control volume in two ways: by direct contact at the surface (e.g., surface stresses, entering mass, momentum, and energy) and by remote action throughout the interior volume of our control volume (e.g., body forces due to gravity as well as possibly electric and magnetic fields). Thus in analyzing changes of any property within our control volume, we must be sure to include both surface and volume effects.

The importance of the control volume analysis point of view is that, since the choice of the control volume is up to the engineer, the engineer can make the choice in a manner that utilizes all information known about the problem and that also brings out information about those quantities to be found. In many problems, no detailed knowledge is needed about what happens in the interior of the control volume, only about what happens on the control surface. Since many flow problems are so complex that detailed knowledge of the flow is frequently beyond our capabilities of analysis and computation, control volume analysis thus gives us results that could be found in no other way.

---

[3] It is possible, and sometimes convenient, to introduce control volumes that can move in space, and even change in size and/or shape. Except for the development of a control volume moving with a constant velocity we will not consider these elaborations.

[4] It has been said that the control volume concept is an example that illustrates how engineers think differently about problems than do scientists, and also points out the difference in the goals of engineering and science.

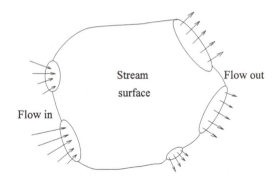

Figure 3.5. A general control volume.

The basic physical laws that we will be applying to our study of fluid mechanics are the conservation of mass (matter), Newton's expressions for the rate of change of linear and angular momentum, and the balance of energy. These laws are simply an accounting of what is happening to the quantity in question, listing what comes out of the control volume, what goes into the control volume, and accounting for any difference. They can be stated as

$$\text{Rate at which} \begin{bmatrix} \text{mass} \\ \text{momentum} \\ \text{energy} \end{bmatrix} \text{accumulates within the control volume}$$

$$+ \text{rate at which} \begin{bmatrix} \text{mass} \\ \text{momentum} \\ \text{energy} \end{bmatrix} \text{enters the control volume}$$

$$- \text{rate at which} \begin{bmatrix} \text{mass} \\ \text{momentum} \\ \text{energy} \end{bmatrix} \text{leaves the control volume}$$

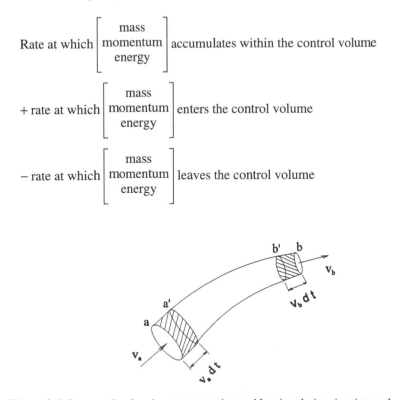

Figure 3.6. Stream tube showing mass entering and leaving during time interval dt.

$$= \begin{bmatrix} 0 \\ \text{net force acting on control volume} \\ \text{rate of heat addition} + \text{rate of work done on control volume} \end{bmatrix}.$$

We next consider these laws individually.

4. Conservation of Mass—the Continuity Equation

Of the basic laws that govern the flow of fluids, conservation of mass is perhaps the most important one and also the one most frequently used. It states that since matter is neither created nor destroyed within our control volume, the net mass must remain constant. In rate form, it translates into the statement that the rate at which mass accumulates inside the control volume plus the rate at which it enters through the control surface must be balanced by the rate at which it leaves the control surface, or in equation form:

Rate at which mass accumulates within the control volume

+ the rate at which mass enters the control volume

− the rate at which mass leaves the control volume

= 0.

Looking at the control volume shown in Figure 3.6 we see that the mass of fluid entering control surface a during a time Δt is the mass density of the fluid times the volume of the fluid that enters. The length of the entering volume of fluid is $v_n \Delta t$, where v_n is the velocity component normal to the area. Thus the entering mass is

$$\Delta m = \rho (A v_n \Delta t).$$

The rate at which mass enters, $\dot{m}$, is the limit as Δm is divided by Δt and then Δt taken to approach zero. With time differentiation indicated by a superposed dot, this gives

$$\dot{m} = \rho A v_n.$$

In equation form for the control volume shown in Figure 3.6, this gives

$$0 = d/dt \underset{\text{cv}}{\iiint} \rho \, dV + \underset{\text{cs in}}{\iint} \rho v_a \, dA - \underset{\text{cs out}}{\iint} \rho v_{bj} \, dA. \tag{3.4.1}$$

Here v_a and v_b are the velocities normal to the inlet and exit areas of the control volumes, respectively. They are positive if v_a enters the control volume and v_b exits the control volume. If we have M inlet areas and N outflow areas (e.g., Figure 3.7 has $M = 1$, $N = 2$), then

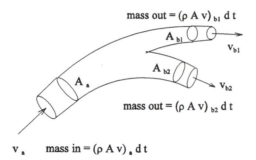

Figure 3.7. Control volume with two exiting areas.

$$0 = d/dt\iiint\limits_{cv} \rho\, dV + \sum_{i=1}^{M} \iint\limits_{cs\ in\ i} \rho v_{ai}\, dA - \sum_{j=1}^{N} \iint\limits_{cs\ out\ j} \rho v_{bj}\, dA. \qquad (3.4.2)$$

We could also use vectors to write this in the form

$$0 = d/dt\iiint\limits_{cv} \rho\, dV + \iint\limits_{cs} \rho\mathbf{v} \cdot d\mathbf{A}$$

$$= d/dt \iiint\limits_{cv} \rho\, dV + \iint\limits_{cs} \rho|v|\mathbf{n} \cdot d\mathbf{A}, \qquad (3.4.3)$$

where n is the unit outward normal from the control volume as seen in Figure 3.8.

In our studies in this book we will deal almost exclusively with steady flows. For steady flows the first (volume integral) terms in equations (3.4.1) to (3.4.3) vanish, and we are left with the statement that the net rate at which mass crosses our control surface must be zero. If the mass density is constant, the control surface integrals can be written as

$$\rho v_{ave} A = \iint\limits_{cs} \rho\mathbf{v} \cdot d\mathbf{A},$$

where v_{ave} is the average of the ***normal*** component for the velocity over the area. Then the above continuity equation can be approximated by

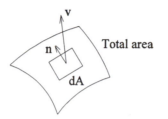

Figure 3.8. Flow through an area.

$$\dot{m}_{\text{in}} = \dot{m}_{\text{out}}, \tag{3.4.4a}$$

or equivalently,

$$(\rho v A)_{\text{in}} = (\rho v A)_{\text{out}}, \tag{3.4.4b}$$

if there is only one inlet region and one outlet region, and

$$\sum_{i=1}^{M} \dot{m}_{\text{in } i} = \sum_{j=1}^{N} \dot{m}_{\text{out } j}, \tag{3.4.5a}$$

or equivalently,

$$\sum_{i=1}^{M} (\rho v A)_{\text{in } i} = \sum_{j=1}^{N} (\rho v A)_{\text{out } j}, \tag{3.4.5b}$$

if there is more than one inlet or outlet region such as in Figures 3.5 and 3.7. As a reminder, in (3.4.3), (3.4.4), and (3.4.5) v represents the velocity component normal to an area A averaged over that area.

In Chapter 1 we defined an incompressible fluid as being one in which the bulk modulus is large enough and the pressure changes small enough that there is no change of density with pressure anywhere in the fluid. From that definition it follows that the mass density would everywhere have the same constant value. For many flows, including most flows of liquids and moderately fast flows of gases, small variations in mass density have little effect on the flow, and thus we speak of such flows as being incompressible flows to a high degree of approximation. However, there are situations where mass density is variable, and yet for practical purposes the flow is still incompressible. An example is a freshwater river discharging into a saltwater ocean. Near the river outlet, the saltwater will be at the bottom with freshwater on top, giving a density stratification. Compressibility effects are no more important here than if we were dealing with either the freshwater or saltwater alone. Thus we use as a better definition of incompressibility the following: a flow is said to be an incompressible flow if as we follow a particular fluid particle the density of that particle does not change. From our definition of the material derivative D/Dt, this can be expressed in equation form by

$$\frac{D\rho}{Dt} = 0. \tag{3.4.6}$$

The case of constant mass density is a special case of (3.4.6).

From the above argument it is seen that the mass discharge (mass per unit time, either slugs/s or kg/s), the rate at which mass flows through a given area, is given by

$$\dot{m} = \iint_{\text{area}} \rho \mathbf{v} \cdot d\mathbf{A} \tag{3.4.6a}$$

in integral form, and

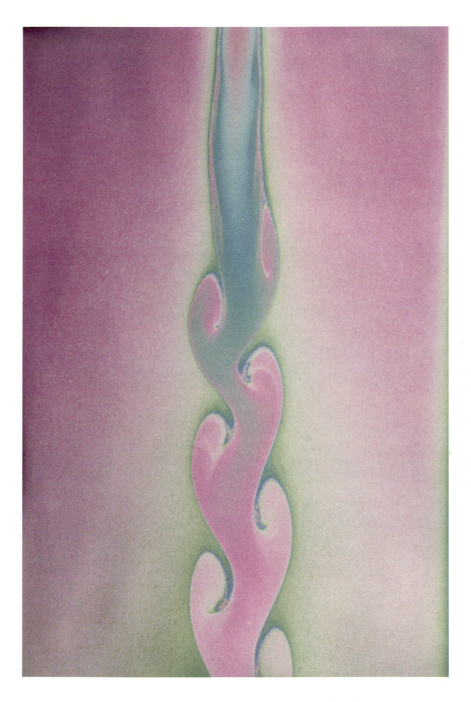

Figure 3.2. Laminar jet at a value of the Reynolds number about 200. (Photo courtesy Professor Mory Gharib and M. Beizaie, University of California, San Diego. This jet is simulated using a "Soap film tunnel," *Physics of Fluids,* vol. **31**, page 2389, 1988.)

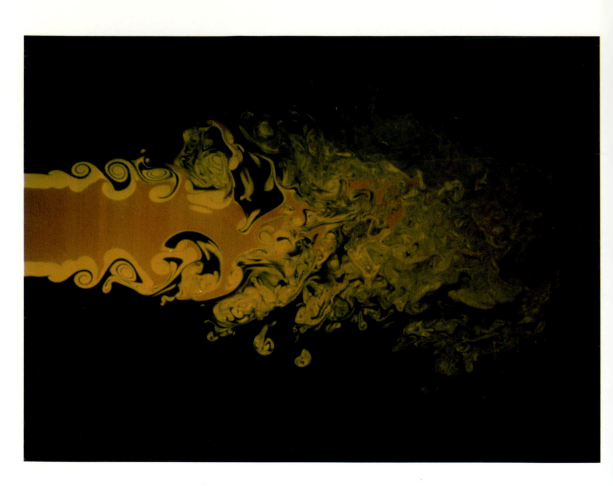

Figure 3.3. Jet at a high value of the Reynolds number. Note vortex development at edge of jet. (Photo courtesy Professor Mory Gharib and P. Derango, University of California, San Diego. This jet is simulated using a "Soap film tunnel," *Physics of Fluids,* vol. **31**, page 2389, 1988.)

$$\dot{m} = \rho v_n \, dA \tag{3.4.6b}$$

in terms of average quantities. We often have a need to give a quantitative measure of the volume rate at which flow is taking place through an area such as the cross section of a conduit or stream tube. We do this by finding the volumetric discharge (volume per unit time, either ft^3/s or m^3/s), often just called discharge, given by

$$Q = \iint_{\text{area}} \mathbf{v} \cdot d\mathbf{A} \tag{3.4.7a}$$

in integral form, or

$$Q = v_n A \tag{3.4.7b}$$

in terms of average quantities.

Example 3.4.1. Conservation of mass with one inlet, one outlet

Water enters a chamber through an inlet having an area of $0.1 \, m^2$ and leaves through an outlet with area $0.5 \, m^2$. The inlet normal velocity is 7 m/s. What is the velocity component normal to the exit area?

Sought: Normal velocity component of the exiting fluid.
Given: Area $A_{in} = 0.1 \, m^2$, area $A_{out} = 0.5 \, m^2$, $v_{in} = 7$ m/s.
Assumptions: Since the fluid is water and the flow is not fast, the flow can be taken to be of constant density.
Solution: Since there is only a single inlet and single outlet, the continuity equation reduces to

$$(vA)_{in} = (vA)_{out},$$

or

$$7 \times 0.1 = v_{out} \times 0.5.$$

Thus $v_{out} = 1.4$ m/s. The volumetric flow rate is $Q = 7 \times 0.1 = 0.7 \, m^3/s$.

Example 3.4.2. Conservation of mass with two inlets, two outlets

Fluid flows through a chamber having two inlets and two outlets. At the inlets, the flow rates are 0.9 and 1.3 ft^3/s. The outlets have flow rates of 0.5 and 1.5 ft^3/s. Is conservation of mass satisfied?

Sought: Answer whether conservation of mass is or is not satisfied.
Given: $Q_{in\,1} = 0.9 \, ft^3/s$, $Q_{in\,2} = 1.3 \, ft^3/s$, $Q_{out\,1} = 0.5 \, ft^3/s$, $Q_{out\,2} = 1.5 \, ft^3/s$.
Assumptions: The density of the fluid is constant. We can work with average quantities.
Solution: The rate of flow into the chamber is

$$Q_{in\,1} + Q_{in\,2} = 0.9 + 1.3 = 2.2 \, ft^3/s,$$

and the rate of flow out of the chamber is

$$Q_{out\,1} + Q_{out\,2} = 0.5 + 1.5 = 2.0 \text{ ft}^3/\text{s}.$$

Thus fluid is coming into the chamber an amount 0.2 ft^3/s faster than it is leaving. The only way that conservation of mass can be satisfied is if the volume inside the chamber is increasing at a rate of 0.2 ft^3/s, or if there is a mass density difference.

---

5. Newton's Law—the Linear Momentum Equation

Our second basic law is due to Newton. It states that the sum of all external forces acting on the control volume is equal to the time rate of change of linear momentum. In equation form this is

Rate at which momentum accumulates within the control volume

+ rate at which momentum enters the control volume

– rate at which momentum leaves the control volume

= net sum of forces acting on the control volume.

$$(3.5.1)$$

The momentum inside our control volume is the mass times the velocity, or

$$d/dt \iiint_{cv} \rho \mathbf{v} \, dV.$$

The rates at which momentum enters and leaves the control volume are the velocity times the rate at which mass enters and leaves, or

$$\text{Rate at which momentum enters the control volume} = \sum_{i=1}^{M} \iint_{cs\ in\ i} \rho v v_{ai} \, dA,$$

$$\text{Rate at which momentum leaves the control volume} = \sum_{j=1}^{N} \iint_{cs\ out\ j} \rho v v_{bj} \, dA.$$

For the control volume shown in Figure 3.9, putting the above together with (3.5.1) gives

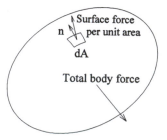

Figure 3.9. Surface and body forces acting on a control volume.

$$d/dt \iiint_{cv} \rho \mathbf{v} \, dV + \sum_{i=1}^{M} \iint_{cs \, in \, i} \rho v v_{ai} \, dA - \sum_{j=1}^{N} \iint_{cs \, out \, j} \rho v v_{bj} \, dA = \sum \mathbf{F} \qquad (3.5.2)$$

in an integral formulation. The force $\mathbf{F}$ consists of surface forces due to contact with the fluid and other surfaces outside the control volume, and body forces within the control volume. For steady flows the first term on the left-hand side of (3.5.2) vanishes.

As was the case for the continuity equation, for steady constant density flows with the flow uniform over each inlet and outlet area, equation (3.5.2) simplifies considerably. Using the same notation for average velocities used in developing (3.4.4b), we simplify (3.5.2) to

$$\sum \mathbf{F} = \sum_{i=1}^{M} {}_{out} (\rho v A)_i \mathbf{v}_i - \sum_{j=1}^{N} {}_{in} (\rho v A)_j \mathbf{v}_j = \sum_{i=1}^{M} {}_{out} \dot{m}_i \mathbf{v}_i - \sum_{j=1}^{N} {}_{in} \dot{m}_j \mathbf{v}_j. \qquad (3.5.3)$$

Note that in the above, $\mathbf{v}$ refers to the velocity (magnitude and direction), while v is the speed (the magnitude of $\mathbf{v}$).

If there is only one inlet and one outlet, by mass continuity $\dot{m}_{in} = \dot{m}_{out} = \dot{m}$, allowing (3.5.3) to be further reduced to

$$\sum \mathbf{F} = \dot{m}(\mathbf{v}_{out} - \mathbf{v}_{in}). \qquad (3.5.4)$$

Example 3.5.1. Conservation of momentum with one inlet, one outlet
A chamber has water entering horizontally with a velocity of 80 m/s through an inlet with area 0.1 m^2, and leaving through an opening of area 0.15 m^2. The exiting flow makes an angle of 30° with the entering flow. What force is needed to hold the chamber in place?

Sought: Force needed to hold the chamber in place.
Given: The fluid is water with $\rho = 1{,}000$ kg/m^3, $v_{in} = 80$ m/s, $A_{in} = 0.1$ m^2, angle$_{in} = 0$, $A_{out} = 0.15$ m^2, angle$_{out} = 30°$.
Assumptions: For water at the velocities involved, the density can be considered to be constant.
Solution: By the continuity equation, the flow rate through the chamber is constant, so that $Q_{in} = Q_{out} = Q = 80 \times 0.1 = 8$ m^3/s $= 0.15 \, v_{out}$. Thus $v_{out} = 8/0.15 = 53.3$ m/s. Taking the x axis in the direction of the inlet jet, we have

$$F_x = \rho Q(v_{out} \cos 30° - v_{in}) = 1{,}000 \times 8 \, (53.3 \times 0.866 - 80) = -593.8 \text{ kN,}$$

the minus sign indicating that this force component is directed opposite to the incoming velocity. In the perpendicular direction, we have

$$F_y = \rho Q(v_{out} \sin 30° - 0) = 1{,}000 \times 8 \times 53.3 \times 0.5 = 213.3 \text{ kN.}$$

These force components are the force components needed to hold the chamber in place, provided that the incoming and exiting flows are at the same gage pressure.

Example 3.5.2. Conservation of momentum with two inlets, two outlets

Water flows through a chamber with two inlets and two outlets. The following conditions hold:

	Area (ft^2)	Speed (ft/s)	Angle with horizontal (degrees)
Inlet #1	0.3	50	0
Inlet #2	0.15	40	45
Outlet #1	0.2	70	30
Outlet #2	0.25	?	−30

All areas are measured perpendicular to their respective areas. What force is needed to hold the chamber in place?

Sought: Force needed to hold chamber in place.

Given: Values for area and velocity as given in the table above. $\rho = 1.94$ slugs/ft^3.

Assumptions: Since water at low speeds is involved, its density can be considered to be constant.

Solution: We can solve for the unknown velocity out of the control volume by using the continuity equation, since all the other flows into and out of the control volume are known. Since the density is constant, the net discharge into the control volume must equal the net discharge out of the control volume. Thus

$$Q_{in\ 1} + Q_{in\ 2} = 0.3 \times 50 + 0.15 \times 40 = Q_{out\ 1} + Q_{out\ 2} = 0.2 \times 70 + 0.25 \times v_{2\ out}.$$

Solving this for $v_{2\ out}$ we have

$$v_{2\ out} = (15 + 6 - 14)/0.25 = 28 \text{ ft/s}.$$

To find the forces, we note that since all the velocity components are now known, we can compute the rates of change of momentum in the various directions. Then from our momentum equation,

$$F_x = 1.94 \times \{14 \times (70 \cos 30°) + 3.5 \times [28 \cos (−30°)] − 15 \times 50 − 6 \times (40 \cos 45°)\}$$

$$= 26.9 \text{ lb},$$

$$F_y = 1.94 \times \{14 \times (70 \sin 30°) + 3.5 \times [28 \sin (−30°)] − 0 − 6 \times (40 \sin 45°)\}$$

$$= 526.3 \text{ lb}.$$

As in the previous example, these force components are the force components needed to hold the chamber in place, provided that the incoming and exiting flows are at gage pressure.

The order in which we used our equations in this problem was standard. Typically we use the continuity equation to find unknown velocity components, and the momentum equation to find unknown forces.

6. Balance of Energy Equation

The third basic law that we will use is the time derivative form of the first law of thermodynamics. In its usual rate form for an open system, this is a balance of power terms, stating that

Rate at which energy accumulates within the control volume

+ rate at which energy enters the control volume

− rate at which energy leaves the control volume

= rate of heat addition to the control volume

+ rate of work done on the control volume.

This equation is often referred to as the ***energy equation***, although, since it deals with the rate of energy, ***power equation*** would be a more appropriate name.

The energy is made up of three distinct forms: kinetic energy due to macroscopic motion of the flow; potential energy due to gravity forces; and internal, or thermal, energy, which is due to microscopic motion of the fluid molecules and/or intermolecular forces.[5] Thus we write for the energy per unit mass (***specific energy***)

$$e = \mathbf{v} \cdot \mathbf{v}/2 + gh + u,$$

where h in the potential energy term is a distance measured opposite from that in which gravity acts—that is, if gravity acts downward, h is measured upward. The datum from which h is measured is arbitrary, as long as we do not change the datum during our solution of a problem. This is because we always are interested only in changes in h, and so the position of the datum from which h is measured will disappear from our calculations.

In the above expression for e, u is taken as the ***specific internal energy***. This is a function of the thermodynamic state, and will generally depend on two independent thermodynamic variables, such as pressure and temperature. For liquids or slow-moving gases, changes in internal energy generally are small and can be ignored compared to changes in the other terms in e. For high-speed gas flows, these changes are important and must be included. This will be discussed further in Chapter 9.

For our energy equation, following the procedures used in developing (3.4.2) and (3.5.2), we can write the rate of change of energy as

Rate at which energy accumulates within the control volume

+ rate at which energy enters the control volume

− rate at which energy leaves the control volume

$$= d/dt \iiint_{cv} \rho e \, dV + \sum_{i=1}^{M} \iint_{cs \text{ in } i} \rho e v_{a_i} \, dA - \sum_{j=1}^{N} \iint_{cs \text{ out } j} \rho e v_{b_j} \, dA. \tag{3.6.1}$$

[5]Other possible energy forms not considered here are chemical energy, caused by the arrangement of atoms making up the molecules, and nuclear energy, caused by the cohesive forces that hold the protons and neutrons together at the nuclei of the atoms.

Figure 3.10 is a one-dimensional visualization of these energy terms. Again the velocity components v_a and v_b are those measured normal to the surface.

We will not be going into much detail in dealing with the heat term, except for compressible flows. For now we designate it by dQ_T/dt. In considering the work terms we will write out the pressure work terms explicitly, and lump the viscous rate-of-work terms into one catchall term, dW_{vis}/dt, since these are again more difficult to deal with on a beginning level. Viscous terms extract energy from our system, usually converting this energy to internal energy or heat; hence the introduction of the negative sign with this term.

In many situations we have a pump, turbine, or other mechanical device within our system that produces or extracts mechanical work. We include these devices in a lumped term called ***shaft work***. The rate at which these devices do work is represented by dW_{sh}/dt. Shaft work is done by a mechanical shaft-driven device used for energy transmission. The shaft allows power to enter or leave the system. If work is done *on* the system by a pump or other mechanical device that adds energy to the system, shaft work is a positive quantity acting to increase the rate of change of energy. If work is done *by* the system such as by a turbine, shaft work is a negative quantity and the flow energy is decreased as the flow proceeds through the control volume.

Pressure work is accounted for as follows. Rate of work, or power, is defined as the dot product of force times velocity. For a pressure force acting on an area element dA the force is $-p\, dA$, since the area element has an outward drawn normal and the pressure force acts inward. Dotting this with the velocity and integrating it over the exterior surface of the control volume gives

$$-\iint_{cs} p\mathbf{v} \cdot d\mathbf{A}.$$

The combined rate of heat transfer and the rate of work done on the system is

$$\frac{dQ_T}{dt} + \frac{dW}{dt} = \frac{dQ_T}{dt} - \iint_{cs} p\mathbf{v} \cdot d\mathbf{A} - \frac{dW_{vis}}{dt} + \frac{dW_{sh}}{dt}. \tag{3.6.2}$$

Putting (3.6.1) and (3.6.2) together, our complete energy equation becomes

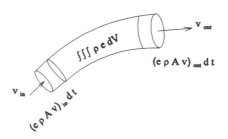

Figure 3.10. Stream-tube control volume showing energy entering, leaving, and changing internally.

$$\frac{d}{dt} \iiint_{cv} \rho \left(\frac{|\mathbf{v}|^2}{2} + gh + u \right) dV + \sum_{i=1}^{M} \iint_{cs\ in\ i} \rho \left(\frac{|\mathbf{v}|^2}{2} + gh + u \right) v_{a_i}\, dA$$

$$- \sum_{j=1}^{N} \iint_{cs\ out\ j} \rho \left(\frac{|\mathbf{v}|^2}{2} + gh + u \right) v_{b_i}\, dA = \frac{dQ_T}{dt} - \iint_{cs} p\mathbf{v} \cdot d\mathbf{A} - \frac{dW_{vis}}{dt} + \frac{dW_{sh}}{dt}. \qquad (3.6.3)$$

For steady flows, the derivative of the volume integral on the left-hand side of the equation is zero.

When the density is constant and the flow is uniform over the inlet and outlet areas of our control volume, we can simplify (3.6.3) as we did for our continuity and momentum equations. Then (3.6.3) becomes

$$\sum_{i=1}^{M} {}_{out}\ \dot{m} \left(\frac{|\mathbf{v}|^2}{2} + gh + u \right) - \sum_{j=1}^{N} {}_{in}\ \dot{m} \left(\frac{|\mathbf{v}|^2}{2} + gh + u \right)$$

$$= \frac{dQ_T}{dt} - \frac{dW_{vis}}{dt} + \frac{dW_{sh}}{dt} + \sum_{j=1}^{N} {}_{in}\ pvA - \sum_{i=1}^{M} {}_{out}\ pvA. \qquad (3.6.4)$$

It is convenient to write the pressure work terms pvA in the form $p(\rho vA/\rho)$, which equals $p\dot{m}/\rho$. [The combination $p\dot{m}/\rho = pAv$ is called the *flow work*, since it is a force (pA) times a velocity.] We can then bring the pressure work term in equation (3.6.4) over to the left-hand side to obtain

$$\sum_{i=1}^{M} {}_{out}\ \dot{m} \left(\frac{p}{\rho} + \frac{|\mathbf{v}|^2}{2} + gh + u \right) - \sum_{j=1}^{N} {}_{in}\ \dot{m} \left(\frac{p}{\rho} + \frac{|\mathbf{v}|^2}{2} + gh + u \right) = \frac{dQ_T}{dt} - \frac{dW_{vis}}{dt} + \frac{dW_{sh}}{dt}.$$
$$(3.6.5)$$

The simple form of the right-hand side of (3.6.5) with the mass rate of flow factored out of the terms suggests that, particularly for steady flow problems with only a few places where flow enters and leaves the control volume, it would be convenient if we could write the right-hand side of the equation in a similar manner. This is accomplished by lumping the portion of the shaft work, viscous work, internal energy, and heat terms that represent energy gain as

$$h_G g \sum_{out} \dot{m}$$

and the terms which represent energy loss as

$$-h_L g \sum_{out} \dot{m},$$

where h_G and h_L have the dimensions of length and are called the ***energy head gain*** and ***energy head loss***, respectively. They represent the rate at which energy per unit weight rate of flow is gained or lost. Then we have

$$(h_G - h_L)g \sum_{i=1}^{M} {}_{\text{out}} \dot{m} = \sum_{i=1}^{M} {}_{\text{out}} \dot{m}\left(\frac{p}{\rho} + \frac{|\mathbf{v}|^2}{2} + gh + u\right) - \sum_{j=1}^{N} {}_{\text{in}} \dot{m}\left(\frac{p}{\rho} + \frac{|\mathbf{v}|^2}{2} + gh + u\right).$$

(3.6.6)

If there is only one inlet and one outlet to our control volume as in Figure 3.10, the mass summations disappear, and $\dot{m}$ can be divided out of (3.6.6). The result is

$$(h_G - h_L)g = \left(\frac{p}{\rho} + \frac{|\mathbf{v}|^2}{2} + gh + u\right)_{\text{out}} - \left(\frac{p}{\rho} + \frac{|\mathbf{v}|^2}{2} + gh + u\right)_{\text{in}}.$$

(3.6.7)

The collection of terms in parentheses, $p/\rho + \mathbf{v} \cdot \mathbf{v}/2 + gh + u$, is called the **Bernoulli terms**.

Example 3.6.1. Conservation of energy—determination of flow direction

At point A, water has a velocity of 10 m/s and the pressure is 100 kPa. Point B is 20 m higher than A, the velocity there is 6 m/s, and the pressure is 130 kPa. Is the flow from A to B, or B to A?

Sought: Direction of flow.

Given: $v_A = 10$ m/s, $p_A = 200{,}000$ Pa, $v_B = 6$ m/s, $p_B = 130{,}000$ Pa, $h_B = h_A + 20$ m, $\rho_A = \rho_B = 1{,}000$ kg/m^3.

Assumptions: Since the fluid is water at a low speed, the flow can be considered to be of constant density. Continuity is satisfied. There is no source of energy between A and B.

Solution: Evaluating the right-hand side of equation (3.6.5), we have

$$\dot{m}[(p/\rho + \mathbf{v} \cdot \mathbf{v}/2 + gh)_A - (p/\rho + \mathbf{v} \cdot \mathbf{v}/2 + gh)_B]$$

$$= \dot{m}[(100{,}000/1{,}000 + 100/2 + 0) - (130{,}000/1{,}000 + 36/2 + 20)]$$

$$= -18 \text{ m.}$$

Thus if the flow was from A to B, the rate of energy gain would be negative, indicating an energy loss. Since no mention was made of a pump or any other energy source between A and B, we would expect an energy loss to occur in the direction of flow. Thus the flow is from A to B.

Example 3.6.2. Conservation of energy—determination of pressure

A wye connection diverts 40% of the water entering at A to B, and the remainder to C. The velocities at A, B, and C are 20, 15, and 12 ft/s, respectively, and the pressures at A and B are 12 and 9 psi. The wye lies in a horizontal plane. Assuming no energy losses, what is the pressure at C?

Sought: Pressure at point C.

Given: $v_A = 20$ ft/s, $v_B = 15$ ft/s, $v_C = 12$ ft/s, $p_A = 12$ psi, $p_B = 9$ psi, $\rho = 1.94$ slugs/ft^3.
Assumptions: No elevation change, no energy losses or gains, constant density because the fluid is water at low speeds.
Solution: Using equation (3.6.5), we get

$$0 = 0.4\dot{m}_A(9 \times 144/1.94 + (225/2) + 0.6\dot{m}_A(p_C/1.94 + 144/2)$$

$$- \dot{m}_A(12 \times 144/1.94 + 400/2),$$

and upon solving for p_C we find

$$p_C = (1/0.6)[-0.4 \times 9 \times 144 + 12 \times 144 + 1.94(-0.4 \times 112.5 - 0.6 \times 72 + 200)]$$

$$= 23{,}377.1 \text{ psf} = 16.5 \text{ psi.}$$

---

7. The Entropy Inequality

There is still another relation that is useful to us, the entropy inequality, also called the "second law of thermodynamics." It differs from the three balance equations just discussed in that it is an inequality, rather than an equation, and thus is not a conservation law. It places restrictions on both mechanical processes and constitutive equations. It tells us, for instance, that a fluid passing through a shock wave has a higher Mach number upstream of the shock wave than it does downstream. Similarly, for a gravity wave known as a "hydraulic jump," it tells us that the surface elevation upstream of the jump is lower than it is downstream of the jump. The prime use we will make of the second law is this indication of the directionality of a process when multiple solutions of our other equations are possible.

The second law states that there is a limit to the rate at which heat can be converted to energy without doing mechanical work. This can be worded in many ways. One of the more useful definitions is one that introduces the concept of entropy. It states that, whenever a real flow or process is carried out, the integral representing the change in entropy during a thermodynamic process,

$$\Delta S = \int \left(\frac{dQ_T}{T}\right)_{rev} \geq 0,$$

must be greater than or equal to zero, and never negative. Here T is the absolute temperature and the subscript rev indicates that the integration is carried out not necessarily for the actual process the flow undergoes, but rather for a reversible process with the same end states. S is termed the *entropy*, and is a function of the endpoints of the path of integration, but not the actual path itself. Thus the absolute temperature T plays the role of transforming the inexact differential dQ_T into the exact differential $dS = dQ_T/T$. Again, at times the specific entropy S, the entropy per unit mass, will be more useful in computations.

Just as terms like force, pressure, stress, strain, and the like have been used outside of the engineering and physics field (and not always with as precise definitions), so too has entropy. Among the more precise usages, communications theory has adopted the term to indicate the measure of uncertainty in a signal after it has been transmitted. The famed "Heisenberg uncertainty principle" of nuclear physics can be interpreted in terms of an "entropy." Beyond this, the term has been included into everyday vocabulary to mean increasing confusion, agreeing well with the spirit of our technical definition.

8. Applications

The four laws we have developed are sufficient to deal with a great many important engineering problems. To aid in understanding how to apply the laws we have just developed, we will first consider a number of simple but still useful cases. To expand our range of applications further, we will then extend these laws to include moving control volumes and the conservation of angular momentum.

An important point for the student to observe in these problems is how the various laws are used. In general, the continuity equation is used for determining velocity components, the momentum equation for determining force components, and the energy equation for determining energy gains or losses. In some applications, however, these roles are interchanged. If forces are known, for instance, the momentum equation can be used for determining velocities; if energy gains and losses are known, the energy equation can be used for determination of the velocity.

a. Applications for flow measurement

Venturi meter

Consider a control volume (which is also a stream tube) as in Figure 3.11 having a circular cross section but with variable diameter. We will take the flow to have constant density, and the velocity to be uniform across any cross section. (This last assumption is valid primarily for turbulent flows, as will be seen in more detail in Chapter 7. For slower laminar flows, a correction must be applied to our final result, as in Chapter 6.) Then the continuity condition gives

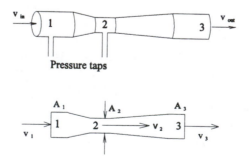

Figure 3.11. Venturi meter and its control volume.

$$v_1 A_1 = v_2 A_2 = v_3 A_3, \tag{3.8.1}$$

which upon solving for v_2 and v_3 in terms of v_1 gives

$$v_3 = v_1, \qquad v_2 = v_1 A_1/A_2. \tag{3.8.2}$$

The place of minimum area, 2 in the figure, is called the ***throat*** of the venturi.

Generally, in converging flows such as between points 1 and 2 in the venturi, there is very little energy loss. If the venturi meter is properly designed—that is, if the diverging section angle is small so that the stream lines do not separate from the wall—there will also be very little energy loss between 2 and 3 as well. The energy equation then gives, for a horizontal meter and neglecting internal energy changes and viscous losses,

$$\frac{p_1}{\rho} + \frac{v_1^2}{2} = \frac{p_2}{\rho} + \frac{v_2^2}{2} = \frac{p_3}{\rho} + \frac{v_3^2}{2}.$$

After the use of (3.8.2) this simplifies to

$$p_1 = p_3, \tag{3.8.3a}$$

$$\left. \frac{p_1 - p_2}{\rho} = \frac{v_2^2 - v_1^2}{2} = \frac{v_1^2}{2}\left(\frac{A_1^2}{A_2^2} - 1\right) \right\} \tag{3.8.3b}$$

and thus we can solve for the velocity v_1, obtaining the expression

$$v_1 = \sqrt{\frac{2(p_1 - p_2)}{\rho\left(\dfrac{A_1^2}{A_2^2} - 1\right)}}. \tag{3.8.4}$$

The pressure difference $p_1 - p_2$ can be measured by a manometer or other pressure transducer. Thus equation (3.8.4) gives a convenient way to measure the flow rate $A_1 v_1$ inside a conduit.

Note from (3.8.3) that the pressure at the throat of the venturi can be reduced much below ambient conditions. Thus the venturi meter can also be used as an aspirator, or suction pump, by attaching a tube to the side near the throat and using this reduced pressure to draw in fluid at a slow rate. These devices are frequently seen in chemistry laboratories and dentist's offices. They are used on airplanes to provide reference pressures for instruments. They also are the means by which some paint sprayers work, they can be used for mixing dry chemicals with their solvents, and they form the main body of an engine carburetor.

Example 3.8.1. Venturi meter
A venturi meter has air flowing with an incoming velocity of 30 m/s. The throat area is 0.6 times the entrance area. What is the pressure difference between the entrance and the throat?

Sought: Pressure difference between entrance and throat of venturi meter.

Given: $v_{in} = 30$ m/s, $A_{throat} = 0.6 \times A_{entrance}$, $\rho = 1.26$ kg/m^3.

Assumptions: Because the velocity is low, density and internal energy can be assumed to be constant. Assume that the velocities given are parallel to the venturi meter axis. If the meter is properly designed so that area changes take place slowly and the meter entrance is perpendicular to the incoming flow, very little energy will be lost as the air flows through the meter. Since these meters are usually not very long and the air density is small, potential energy changes can be neglected.

Solution: While we could use equation (3.8.3b) to solve this problem, it is instructive to go back to our basic equations. According to the continuity equation,

$$v_{in} A_{in} = 30 \times A_{in} = v_{throat} A_{throat} = v_{throat} \times 0.6 \times A_{in}.$$

Dividing by A_{in} gives

$$v_{throat} = 30/0.6 = 50 \text{ m/s}.$$

In the energy equation, it follows from our assumptions that we need consider only the pressure work terms and the kinetic energy changes. Thus the energy equation gives

$$\dot{m}(p_{entrance}/\rho + v_{in}^2/2) = \dot{m}(p_{throat}/\rho + v_{throat}^2)/2.$$

Dividing first by the mass rate of flow and then solving for the pressure difference, we are left with

$$p_{entrance} - p_{throat} = \rho(v_{throat}^2 - v_{in}^2)/2 = 1.26 \times (2{,}500 - 900)/2 = 1{,}008 \text{ Pa} = 1.008 \text{ kPa}$$

Orifice meter

The venturi meter requires the fabrication of relatively complicated pieces of hardware. A mechanically simpler device that gives much the same measurement capabilities consists of a flat plate pierced by a tapered hole and placed across a conduit (Figure 3.12). The flow passing through the hole experiences a contraction, as inertia causes a fluid particle near the wall to continue toward the centerline, and then to turn in the

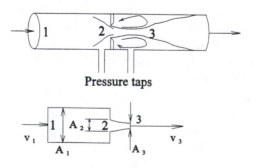

Figure 3.12. Orifice flow meter and its control volume.

opposite direction to resume being parallel to the wall. Thus the flow becomes jetlike a little downstream from the plate.

Point 3 is the location of minimum area of the jet (called the **vena contracta**) where the flow is parallel to the wall. Writing the energy and continuity equations between points 1 and 3 (omitting $\dot{m}$ since it is the same at all sections) gives

$$p_1/\rho + v_1^2/2 = p_3/\rho + v_3^2/2 \tag{3.8.5}$$

and

$$v_1 A_1 = v_3 A_3. \tag{3.8.6}$$

Combining these and solving for v_1, we have

$$v_1 = \sqrt{\frac{2(p_1 - p_3)}{\rho\left(\dfrac{A_1^2}{A_3^2} - 1\right)}}. \tag{3.8.7}$$

The separated region downstream of the plate forms an eddy moving slowly compared to the remainder of the flow. This slow motion is what makes the constant pressure assumption in the eddy a good one. By placing pressure taps near the plate to take advantage of this, we measure the pressure difference $p_1 - p_3$ across the orifice plate.

The question of the determination of the area A_3 is more complicated. Here we have to rely on experiments. The **contraction coefficient** C_c, defined as $C_c = A_3/A_2$, is a function of the area ratio A_2/A_1 and is given in Table 3.1.

A curve fit to this table, accurate with an error of 4% or less, is

$$C_c = 0.602 + 0.370\,(A_2/A_1)^2.$$

Table 3.1.   Coefficients of contraction for the vena contracta

$\dfrac{A_2}{A_1}$	0.0	0.1	0.2	0.3	0.4	0.5	0.6	0.7	0.8	0.9	1.0
C_c	0.611	0.624	0.632	0.643	0.659	0.681	0.712	0.755	0.813	0.920	1.0

Example 3.8.2. Orifice meter
An orifice plate in a 3-in-diameter pipe has a 2-in-diameter hole. Water is flowing, and a pressure difference of 25 in of mercury is recorded across the orifice plate. What is the volumetric flow rate?

Sought: Volumetric flow through an orifice meter
Given: $A_1 = 1^2 \pi$ in^2, $A_2 = (1.5)^2 \pi$ in^2, $A_2/A_1 = (1/1.5)^2 = 0.667$, $\rho = 62.4/32.17 = 1.94$ slugs/ft^3, $\Delta p = (25/12) \times 62.4 \times 13.6 = 1{,}768$ psf.
Assumptions: Constant density flow since the fluid is water at low velocities. No change in internal or potential energies. No energy is lost between points 1 and 3.

Solution: From Table 3.1, the coefficient of contraction for an area ratio of 0.444 is 0.741; thus $A_3/A_1 = (A_3/A_2)(A_2/A_1) = 0.766 \times (0.667)^2 = 0.340$.

In solving this problem we will again start with our basic equations rather than using equations (3.8.6) and (3.8.7). We use the continuity equation to find v_3 in terms of v_1; thus

$$v_1 A_1 = v_3 A_3 = v_3 \times 0.340 \times A_1,$$

and so

$$v_3 = (1/0.340) \times v_1 = 2.941 \times v_1.$$

The energy equation under the assumptions listed is

$$\dot{m}(p_1/\rho + v_1^2/2) = \dot{m}(p_3/\rho + v_3^2/2) = \dot{m}[p_3/\rho + (2.941v_1)^2/2].$$

Dividing by the mass rate of flow and solving for v_1, we have

$$v_1^2 (2.941^2 - 1)/2 = (p_1 - p_3)/\rho,$$

or

$$v_1 = \sqrt{2(p_1 - p_3)/\rho(2.941^2 - 1)}$$

$$= \sqrt{2 \times 1768/1.94 \times (2.941^2 - 1)} = 22.96 \text{ ft/s},$$

$$Q = v_1 A_1 = 22.96 \times (\pi/144) = 0.501 \text{ ft}^3/\text{s}.$$

Note that energy is lost after point 3, where the stream expands to again fill the tube.

Borda mouthpiece

Another flow example that exhibits a vena contracta is the Borda mouthpiece shown in Figure 3.13. This time the throat area is calculable in a simple manner. If a short tube about one diameter long is inserted into the side of a reservoir, the pressure distribution on the reservoir sides is virtually hydrostatic and the pressure forces can all be easily computed. As a result the momentum equation can be applied in the horizontal direction. From Figure 3.13, summation of the pressure forces in the horizontal direction gives

$$\Sigma F = \rho g H A_0 = \rho Q v_{\text{jet}}. \tag{3.8.8}$$

Momentum entering into the sides of the control volume will not contribute to the horizontal forces. Momentum into the left end of the control volume will be negligible because of low velocity there.

If it is assumed that there is no energy loss, writing the energy equation between the top surface and the jet leaving the mouthpiece (at atmospheric pressure) gives

$$v_{\text{jet}}^2/2 = gH. \tag{3.8.9}$$

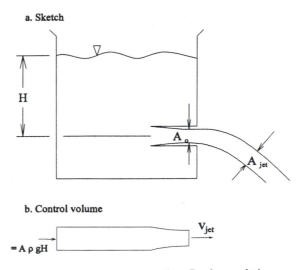

Figure 3.13. Defining sketch and control volume for a Borda mouthpiece.

The discharge is given by $Q = v_{jet} A_{jet}$. Inserting this into (3.3.8) and eliminating v_{jet} by (3.8.9) gives

$$A_{jet} = A_0/2. \tag{3.8.10}$$

Example 3.8.3. Borda mouthpiece

A Borda mouthpiece with a diameter of 20 mm is inserted in an open tank 300 mm down from the free surface. What is the flow rate of water from the tank?

Sought: Flow rate through a Borda mouthpiece.
Given: $D_{tube} = 0.01$ m, $H = 0.3$ m, $g = 9.8$ m/s^2.
Assumptions: Incompressible flow, constant density, no energy losses, uniform velocity at the vena contracta.
Solution: From (3.8.9) and (3.8.10),

$$v = \sqrt{2gH} = \sqrt{2 \times 9.8 \times 0.3} = 2.425 \text{ m/s,}$$

$$A_{jet} = (\pi \times 0.012/4)/2 = 0.392 \times 10^{-4} \text{ m}^2,$$

$$Q = v_{jet} A_{jet} = 0.952 \times 10^{-4} \text{ m}^3/\text{s.}$$

---

b. Fans, propellers, windmills, and wind turbines

The idea of using rotating blades to utilize wind power for mechanical work goes back many centuries. The advent of engines allowed the process to be inverted, so that propellers could be used for propulsion. By applying power to such fans, a thrust could be produced for driving marine vessels, pumping air, and driving planes through the skies. While detailed study of these devices can become quite involved, our elementary theory can provide fundamental information that is necessary to the designer.

The theory presented here was originally developed by W. J. R. Rankine in 1865, and improved by R. E. Froude in 1889. Both were interested in ship propulsion and published their work in the *British Transactions of the Institute of Naval Architecture*. With the advent of the airplane, interest in propeller theory shifted to the aerodynamicists. A. Betz improved the theory in 1920, and since then many more investigators added to our knowledge. The book by Durand (mentioned in the references at the end of this chapter) gives a more detailed account of much of the early work. Modern research in propeller design is now primarily of a computational nature.

We will first consider the "pumping" case, where energy is provided to the propeller to provide a thrust force on the air. This theory, due to Rankine and Froude, is often called the ***actuator-disk theory***, since it replaces the propeller by a disk across which the pressure, but not the fluid velocity, changes abruptly. We will consider the propeller to be stationary. The moving propeller case can be easily determined from this by considerations given in a latter section of this chapter.

The control volume in Figure 3.14 contains all the fluid that passes through the propeller and whose motion has been affected by the propeller. The side boundaries of this control volume are called the ***slipstream***. We will assume that the propeller is sufficiently thin that areas on either side of the propeller, A_2 and A_3, are the same. By continuity, since we are considering the mass density to be constant, the velocities V_2 and V_3 will also be equal. We write, then,

$$A = A_2 = A_3, \qquad V = V_2 = V_3. \tag{3.8.11}$$

If we assume that at all boundaries of this control volume the pressure is the same, then the only net force acting on this control volume is that due to the thrust provided by the propeller, F_{thrust}. Applying the momentum equation to this control volume, we have

$$F_{\text{thrust}} = \rho Q (V_4 - V_1), \tag{3.8.12}$$

where $Q = VA$.

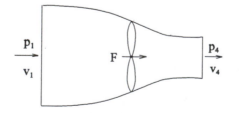

Figure 3.14. Control volume for flow through a fan. The force F is the force that the fan exerts on the fluid.

The thrust of the propeller results in an increase in kinetic energy of the fluid. Breaking our control volume up into the three parts shown in Figure 3.15, we can write three more equations:

Energy equation for control volume a: $p_1 + \rho V_1^2/2 = p_2 + \rho V^2/2,$ (3.8.13)

Momentum equation for control volume b: $F_{\text{thrust}} = A(p_3 - p_2),$ (3.8.14)

Energy equation for control volume c: $p_3 + \rho V^2/2 = p_4 + \rho V_4^2/2,$ (3.8.15)

where we have assumed no energy loss or gain between stations 1 and 2, and 3 and 4. Also, because of our assumption of no area change from 2 to 3, there is no momentum change between those stations.

From (3.8.13) and (3.8.15) we have

$$p_2 = p_1 + \rho V_1^2/2 - \rho V^2/2 \qquad \text{and} \qquad p_3 = p_4 + \rho V_4^2/2 - \rho V^2/2.$$

Combining this with (3.8.14) and (3.8.12), we have

$$F_{\text{thrust}} = A(p_2 - p_3) = A[(p_1 + \rho V_1^2/2 - \rho V^2/2) - (p_4 + \rho V_4^2/2 - \rho V^2/2)]$$

$$= \rho A(V_1^2 - V_4^2)/2 = \rho AV(V_4 - V_1), \tag{3.8.16}$$

where we have used the fact that, since the pressure is the same all around our control volume, $p_1 = p_4$. From (3.8.16) we see that

$$V = (V_4 + V_1)/2; \tag{3.8.17}$$

that is, the fluid velocity through the propeller is the average of the incoming and exiting velocities.

An important result that follows from the above is the question of efficiencies. For a propeller (ship or airplane) moving with a velocity V_1 through a fluid otherwise at rest, the useful power is

$$P_{\text{useful}} = V_1 F = V_1 \, \rho AV \, (V_4 - V_1) = 2\rho AV_1 V(V - V_1),$$

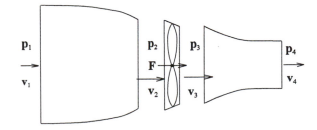

Figure 3.15. Control volume for flow through a fan. The control volume of Figure 3.14 has been broken into three parts.

where we have used (3.8.17) to eliminate V_4. Differentiating this with respect to V to find the maximum power for a given V_1, we find that the maximum is reached when $V = V_1/2$, when $V_4 = 0$. This maximum power is

$$P_{\text{useful maximum}} = -\rho A V_1^3/2. \tag{3.8.18a}$$

The available energy for $V_4 = 0$ is

$$P_{\text{available}} = -\rho A V_1^3/4. \tag{3.8.18b}$$

Thus the efficiency at maximum power is

$$\eta = P_{\text{useful maximum}}/P_{\text{available}} = (-\rho A V_1^3/2)/(-\rho A V_1^3/4) = 50\%. \tag{3.8.18c}$$

This is the ideal efficiency at maximum power.

If the propeller is stationary as in a wind turbine, the control volumes are the same, but with all velocity and force arrows pointing in the opposite directions. From the above analysis the useful power is now

$$P_{\text{useful}} = -F_{\text{thrust}} \, V = 2\rho A V^2 \, (V - V_4),$$

where we have used (3.8.16) and (3.8.17) to eliminate V_1. Differentiating this with respect to V while holding the exiting velocity V_4 constant, we find that the maximum useful power is obtained when $V = 4V_4/3$. Thus the maximum power obtainable from a fan or wind turbine is

$$P_{\text{useful maximum}} = 8\rho A V_4^3/27. \tag{3.8.19a}$$

Since the maximum available energy is the incoming kinetic energy

$$P_{\text{available}} = \rho A V_4^3/2, \tag{3.8.19b}$$

the coefficient of performance (the counterpart of the efficiency for energy-extracting devices) is

$$C_{\text{performance}} = P_{\text{useful maximum}}/P_{\text{available}} = (8\rho A V_4^3/27)/(\rho A V_4^3/2)$$
$$= 16/27 = 59.3\%. \tag{3.8.19c}$$

The above analysis is idealized, in that it assumes that the fluid that has passed through the propeller has no swirl component of velocity. This is generally not the case. The swirl velocity will be present because of the rotation of the propeller, and represents a component of kinetic energy that we have not included in our analysis. The above analysis does, however, give us an upper limit on the powers associated with such devices, and gives a starting point on obtaining design specifications.

Example 3.8.4. Efficiency of a propeller

An airplane traveling at 180 mph in still air at 20°C discharges 16,000 ft³/s of air through its 8-ft-diameter propeller. Determine the pressure change across the propeller, the thrust force, theoretical power required, and the theoretical efficiency.

Sought: Pressure change across propeller, thrust force, theoretical power requirement, and theoretical efficiency.
Given: $D_{propeller} = 8$ ft, $A = 4^2\pi = 50.27$ ft^2, $Q = 16,000$ ft^3/s, $V_1 = 180$ mph $= 264$ ft/s. From Appendix B, $\rho = 1.2$ kg/m$^3 = 1.2/515.3788 = 0.00233$ slug/ft^3.
Assumptions: Incompressible flow, constant density, no energy losses, no swirl velocity.
Solution: From continuity,

$V = Q/A = 16,000/50.27 = 318.3$ ft/s.

From (3.8.17), $V_4 = 2V - V_1 = 2 \times 318.3 - 264 = 372.6$ ft/s.

From (3.8.16), the pressure change across the propeller and the thrust force are given by

$p_3 - p_2 = \rho V(V_4 - V_1) = 0.00233 \times 318.3 \times (372.6 - 264) = 80.5$ psf,

$F_{thrust} = A(p_3 - p_2) = 50.27 \times 80.5 = 4,046.7$ lb.

The theoretical power requirement is

$P_{useful} = F_{thrust} V_1 = 4046.7 \times 264 = 1,068,338$ lb-ft/s $= 1,942$ hp.

The power in is

$P_{in} = \rho Q(V_4^2 - V_1^2)/2 = 0.00233 \times 16,000 \times (372.62 - 2642)/2$

$\quad = 1,288,672$ lb-ft/s $= 2,343$ hp.

The efficiency is then

$\eta = 1,942/2,343 = 83\%$.

Example 3.8.5. Efficiency of a wind-driven power generator
On a given day the 125-ft-diameter ERDA/NASA wind power generator at Sandusky, Ohio, experiences an 18-mph wind. The wind velocity behind the propeller is measured at 11 mph. What is the pressure difference across the propeller, the thrust force, power available, and the theoretical coefficient of performance?

Sought: Pressure change across propeller, thrust force, power available, and theoretical coefficient of performance.
Given: $D_{propeller} = 125$ ft, $A = 62.5^2 \pi = 12,271.8$ ft^2, $V_1 = 18$ mph $= 26.4$ ft/s, $V_4 = 11$ mph $= 16.1$ ft/s. From Appendix B, $\rho = 1.2$ kg/m$^3 = 1.2/515.3788 = 0.00233$ slug/ft^3.
Assumptions: Incompressible flow, constant density, no energy losses, no swirl velocity.
Solution: From (3.8.17),

$V = (V_4 + V_1)/2 = (26.4 + 16.1)/2 = 21.25$ ft/s.

From (3.8.16), the pressure change across the propeller and the thrust force are given by

$p_3 - p_2 = \rho V(V_4 - V_1) = 0.00233 \times 21.25 \times (16.1 - 26.4) = 0.510$ psf,

$F_{thrust} = -A(p_3 - p_2) = 12,271.8 \times 0.510 = 6,258.6$ lb.

The theoretical power requirement is

$$P_{\text{useful}} = F_{\text{thrust}}V = 6{,}258.6 \times 21.25 = 132{,}995.6 \text{ lb-ft/s} = 241.8 \text{ hp.}$$

The power available is

$$P_{\text{available}} = \rho A V_4^3/2 = 0.00233 \times 12{,}271.8 \times 16.13/2$$

$$= 59{,}663.9 \text{ lb-ft/s} = 108.5 \text{ hp.}$$

The coefficient of performance is then

$$C_{\text{performance}} = 108.5/241.8 = 0.449.$$

c. Forces on vanes

Forces on stationary vanes

Vanes are used in many forms of turbomachinery to convert flow momentum to force or moment in turbines, and vice versa, in pumps. Usually the fluid is introduced in the form of a jet that is deflected by the vane, the jet thereby having its momentum changed. Since vanes are usually small in size, changes in pressure due to gravity effects can usually be neglected compared to the other forces. Unless the vane is enclosed, the jet free-surface is at constant pressure. Therefore the energy equation with losses neglected tells us that the fluid speed remains constant along the jet. (Remember that speed is the *magnitude* of the velocity. However, the velocity of a fluid particle does change direction in passing along the vane, and so there is still a momentum change.) In combination with the continuity equation, this tells us that the area of the jet remains constant.

We will consider the flow past a vane whose control volume is shown in Figure 3.16. Applying the momentum equation gives

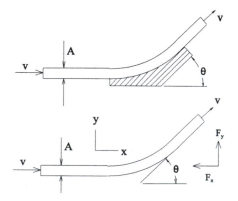

Figure 3.16. Defining sketch and control volume for forces on a single vane.

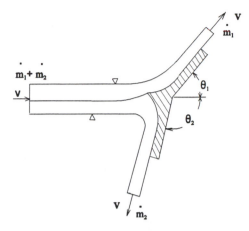

Figure 3.17. Sketch of a splitter vane.

$$-F_x = \dot{m}(v \cos \theta - v), \tag{3.8.20}$$

$$F_y = \dot{m}(v \sin \theta - 0) \tag{3.8.21}$$

for the force components, since the velocity of the jet as it leaves the control volume has x and y components $v \cos \theta$ and $v \sin \theta$. The x component of the force is thus maximum when the turning angle θ is 180°.

Occasionally a vane is used to split a jet into two jets as in Figure 3.17. The analysis of the splitter vane can be carried out either by considering the incoming jet to be made up of two jets and applying (3.8.20) and (3.8.21) to each of them, or by dealing with both jets simultaneously. For the control volume shown in Figure 3.18, application of the continuity and momentum equations gives

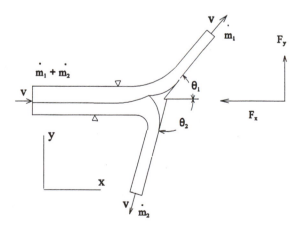

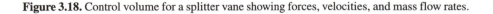

Figure 3.18. Control volume for a splitter vane showing forces, velocities, and mass flow rates.

$$\dot{m} = \dot{m}_1 + \dot{m}_2 \tag{3.8.22}$$

$$-Fx = \dot{m}_1 v \cos \theta_1 + \dot{m}_2 v \cos \theta_2 - \dot{m}v, \tag{3.8.23}$$

$$F_y = \dot{m}_1 v \sin \theta_1 - \dot{m}_2 v \sin \theta_2. \tag{3.8.24}$$

Example 3.8.6. Stationary vane

A water jet with a speed of 100 m/s and a cross-sectional area of 5 cm^2 strikes an unenclosed stationary vane and is deflected upward through an angle of 60°. What force is exerted on the vane?

Sought: The force on a stationary vane.
Given: $v = 100$ m/s, $A = 5 \times 10^{-4}$ m^2, $\rho = 1{,}000$ kg/m^3, angle$_{in} = 0°$, angle$_{out} = 60°$.
Assumptions: The density is constant. No energy is lost. The effects of gravity forces are neglected. The vane is not enclosed.
Solution: The mass flow rate is

$$\dot{m} = 1{,}000 \times 100 \times 5 \times 10^{-4} = 50 \text{ kg/s}.$$

The pressure will remain constant on the free surface of the jet since the vane is not enclosed, and so from the energy equation in the translating coordinate system the fluid speed is everywhere 100 m/s. By equations (3.8.22) and (3.8.22),

$$-F_x = 50(100 \cos 60° - 100) = -2{,}500 \text{ N},$$

$$F_y = 50 \times 100 \sin 60° = 4{,}330 \text{ N}.$$

Thus the force that the vane exerts on the fluid has a component of 2,500 N acting to the left, and a component of 4,330 N acting upward. The force that the fluid exerts on the vane is of equal magnitude, but opposite in direction.

   The force exerted on the vane comes primarily from pressure forces. Even though the free surface of the flow is at constant pressure, the fact that the streamlines are curved means that the pressure at the vane surface will be different from this value.

Example 3.8.7. Stationary vane

Suppose that in the previous example (3.8.6) 40% of the flow leaves the vane at 60°, and the remainder is turned downward and through an angle of 135°. What is the force on the vane?

Sought: The force on a stationary wave.
Given: $v = 100$ m/s, $A_{in} = 5 \times 10^{-4}$ m^2, $A_{out\ 1} = 2 \times 10^{-4}$ m^2, angle$_{out\ 1} = 60°$, angle $_{out\ 2} = -135°$, $\rho = 1{,}000$ kg/m^3.
Assumptions: The density is constant. The effects of gravity are negligible. No energy losses. The vane is not enclosed.

Solution: Now $\dot{m} = 50$ kg/s, $\dot{m}_1 = 20$ kg/s, and by continuity $\dot{m}_2 = 30$ kg/s. By (3.8.23) and (3.8.24),

$$-F_x = 20 \times 50 \cos 60° + 30 \times 50 \cos 135° - 50 \times 100 = -5,560 \text{ N},$$

$$F_y = 20 \times 50 \sin 60° - 30 \times 50 \sin 135° = -194 \text{ N}.$$

Thus the vane exerts a force on the fluid of 5,560 N to the left, and 194 N downward. The force that the fluid exerts on the vane is of equal magnitude and opposite direction.

d. Miscellaneous applications

Siphons

Siphons are useful as a simple and inexpensive form of pump for liquids. They can range in size from small—laboratory devices, flush toilets—to very large. Examples of the latter are several dams in Arizona that have siphons up to 10,000 ft long, inner diameters of 22 ft, and flow rates up to 3,000 ft³/s. A typical configuration of a siphon is seen in Figure 3.19. If the area of the container is large compared to the siphon area, then the kinetic energy at point 1 can be neglected compared to the other kinetic energies. Writing the energy equation between point 1 and points 2 and 3, we find

$$0 = p_2/\rho + v_2^2/2 + h_2 g = v_3^2/2 - h_3 g. \tag{3.8.25}$$

Continuity tells us that $v_2 = v_3 = v$, say. Thus, from the last part of (3.8.25) the velocity of the fluid leaving the siphon is

$$v = \sqrt{2gh_3}, \tag{3.8.26}$$

and from the first part of (3.8.25) the pressure at the highest point (2) is

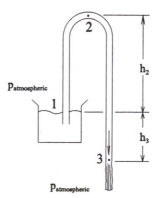

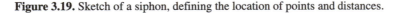

Figure 3.19. Sketch of a siphon, defining the location of points and distances.

$$p_2 = -\rho g(h_2 + h_3).\tag{3.8.27}$$

The discharge of the siphon then can be increased by increasing h_3. If, however, h_3 is made too large the pressure p_2 can reach vapor pressure, and the siphon will cease to function for values of h_3 equal or greater than this value of h_3.

Example 3.8.8. Siphon

A 100-mm-diameter tube is used as a siphon to drain 20°C water from an open tank. The siphon exit extends 2 m below the surface of the tank. What is the volumetric flow rate? How high above the water surface can the top of the siphon be without cavitation occurring?

Sought: Maximum height that a water-containing siphon can be above the free surface.
Given: $\rho = 1,000$ kg/m³, $A = \pi \times (0.005)^2 = 78.54 \times 10^{-4}$ m².
Assumptions: Vapor pressure is approximately zero. No losses. The density is constant.
Solution: From equation (3.8.26) we have

$$v = \sqrt{2 \times 9.8 \times 2} = 6.26 \text{ m/s},$$

$$Q = vA = 6.26 \times 78.54 \times 10^{-4} = 0.0492 \text{ m}^3/\text{s}.$$

From equation (3.8.27), since p_2 will be the vapor pressure, which according to Figure 1.5 is approximately zero absolute pressure, then (since we are working with gage pressures) p_2 is approximately minus the atmospheric pressure. Thus $p_2 = -101$ kPa. Since $v_2 = v_3$, equation (3.8.10) becomes

$$h_2 = -h_1 + p_2/\rho g = -2 + 101,000/1,000 \times 9.8 = 8.3 \text{ m}.$$

Sudden expansions and contractions

When the cross-sectional area of a conduit suddenly changes, there is generally a region of complicated flow near the change that results in an energy loss. Consider first the sudden expansion shown in Figure 3.20. Because of the inertia of the flow, Newton's third law says that the fluid cannot suddenly expand to fill the greater area, but must do so gradually. The fluid trapped between the walls and the expanding flow is slow-moving compared to the expanding flow, and so the pressure at any section can be considered to be constant over that section. Writing the continuity and momentum equations for the control volume shown gives

$$v_1 A_1 = v_2 A_2,\tag{3.8.28}$$

$$(p_1 - p_2)\, A_2 = \rho v_2 A_2 (v_2 - v_1).\tag{3.8.29}$$

The energy equation gives for the head loss

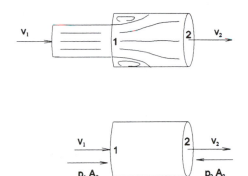

Figure 3.20. Defining sketch and control volume for a sudden expansion.

$$gh_L = \frac{p_1}{\rho} + \frac{|v_1|^2}{2} - \frac{p_2}{\rho} - \frac{|v_2|^2}{2} = \frac{p_1 - p_2}{\rho} + \frac{|v_1|^2 - |v_2|^2}{2} = \frac{|v_1|^2}{2}\left(1 - \frac{A_1}{A_2}\right)^2$$

$$= v_1^2(1 - A_1/A_2)^2/2. \tag{3.8.30}$$

Equation (3.8.30) shows that the head loss is proportional to the kinetic energy, which in fact is true for all the problems considered in this chapter. The power loss is thus proportional to the cube of the speed. We will see later (Chapter 7) that this implies that the flows are turbulent.

A sudden area contraction in a closed conduit (Figure 3.21) is analyzed in a similar manner. In the contracting region (point 1 to point 2), there is very little energy lost. The majority of energy is lost in the expansion from point 2 to point 3. Using equation (3.8.30), solving for the head loss gives

$$gh_L = \frac{v_2^2}{2}\left(1 - \frac{A_2}{A_3}\right)^2. \tag{3.8.31}$$

The contracted area A_2 is a function of the area ratio A_1/A_3, as was the case for the orifice plate. Letting $A_2 = C_c A_3$, (3.8.31) becomes

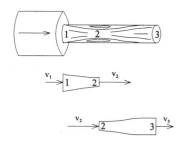

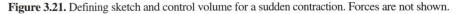

Figure 3.21. Defining sketch and control volume for a sudden contraction. Forces are not shown.

$$gh_L = \frac{v_2^2}{2}(1 - C_c)^2 = \frac{v_3^2}{2}\left(\frac{1}{C_c} - 1\right)^2. \tag{3.8.32}$$

The coefficient of contraction in (3.8.32) is exactly that given in Table 3.1 for the orifice plate.

Example 3.8.9. Sudden expansion

A pipe abruptly changes diameter from 1 to 2 in. The velocity at the 1-in-diameter pipe inlet is 10 ft/s. What is the head loss?

Sought: Head loss caused by a sudden expansion.
Given: $A_2/A_1 = 2^2 = 4$, $v = 10$ ft/s.
Assumptions: Parallel and uniform flow before the expansion and far downstream from the contraction. Constant density.
Solution: From (3.8.30),

$$gh_L = 10^2(1 - 1/4)/2 = 28.125 \text{ ft}^2/\text{s}^2;$$

thus $h_L = 28.125/32.17 = 0.874$ ft.

Example 3.8.10. Sudden contraction

The flow in the pipe in the previous example is now going in the opposite direction. What is the head loss?

Sought: Head loss caused by a sudden contraction.
Given: $A_2/A_1 = 0.25$, $v = 10$ ft/s.
Assumptions: Constant density. Parallel and uniform flow before the contraction and far downstream from the contraction.
Solution: From Table 3.1, since $A_2/A_1 = 0.25$, by interpolation $C_c = 0.6375$. From (3.8.32),

$$gh_L = 10^2(1 - 0.6375)^2/2 = 6.57 \text{ ft}^2/\text{s}^2;$$

thus $h_L = 0.204$ ft.

Nozzles and reducing elbows

In the vane problems we considered previously, we saw that changing the direction of the jet momentum resulted in a net force being exerted on the vane. An extension of this is the reducing elbow, where the entire fluid is contained inside a rigid wall so that the pressure, and thereby the speed, can change.

Consider the control volume shown in Figure 3.22. The elbow exerts force components F_x and F_y on the fluid, through a combination of pressure and viscous stresses. The continuity and momentum equations give

$$Q = v_1 A_1 = v_2 A_2, \tag{3.8.33}$$

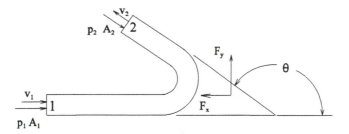

Figure 3.22. Control volume for a reducing elbow.

$$-F_x + p_1A_1 - p_2A_2 \cos\theta = \dot{m}(v_2\cos\theta - v_1),\tag{3.8.34}$$

$$F_y - p_2A_2 \sin\theta = \dot{m}v_2\sin\theta.\tag{3.8.35}$$

To complete the determination of the force components, we thus need to know the pressures.

A simple example of this is a nozzle discharging into the atmosphere (see Figure 3.23). Since the nozzle is horizontal, the angle θ is zero. There is no vertical force, so F_y must vanish. Since the nozzle is discharging into the atmosphere, p_2 is zero (gage pressure), and from (3.8.34) and (3.8.33),

$$F_x = p_1A_1 - \dot{m}(A_1/A_2 - 1)v_1.\tag{3.8.36}$$

If we assume no energy loss, the pressure is given by the energy equation

$$p_1/\rho + v_1^2/2 = v_2^2/2,\tag{3.8.37}$$

so that

$$F_x = \dot{m}v_1(A_1/A_2 - 1)^2/2 = \rho A_1(v_2 - v_1)^2/2.\tag{3.8.38}$$

The force of the nozzle on the fluid thus acts to the left. In turn, the force of the fluid on the nozzle acts to the right. Thus the hose or pipe attached to the nozzle would be in tension as a result of the flow.

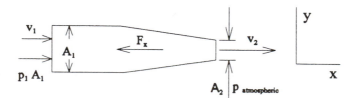

Figure 3.23. Control volume for a nozzle.

Example 3.8.11. Reducing elbow

A horizontal reducing elbow has a diameter change from 6 to 2 in, and turns the flow through an angle of 45°. Oil (SG = 0.8) is flowing at a speed of 5 ft/s, and the pressure at the 6-in-diameter end of the nozzle is 30 psi. What force is exerted on the nozzle, neglecting any losses?

Sought: Force on a reducing nozzle.

Given: $A_1/A_2 = (6/2)^2 = 9$, $v_1 = 5$ ft/s, $p_1 = 30$ psi $= 30 \times 144 = 4{,}320$ psf, $\rho = 1.94 \times 0.8 = 1.552$ slugs/ft^3.

Assumptions: Constant density. Uniform flow entering and leaving the nozzle. No energy loss.

Solution: Using the continuity and energy equations,

$$v_2 = v_1 A_1/A_2 = 5 \times 9 = 45 \text{ ft/s,}$$

$$p_1/\rho + v_1^2/2 = p_2/\rho + v_2^2/2.$$

Therefore

$$p_2 = p_1 + \rho(v_1^2 - v_2^2)/2 = 30 \times 144 + 1.552 \times (5^2 - 45^2)/2 = 2{,}314 \text{ psf} = 16.1 \text{ psi.}$$

From the definition of mass flow rate,

$$\dot{m} = 5 \times \pi \times (3/144)^2 \times 1.94 \times 0.8 = 0.01058 \text{ slug/s.}$$

From the momentum equations in the x and y directions,

$$-F_x + 30 \times \pi \times 32 - 16.1 \times \pi \times 12 \cos 45° = -0.01058 \times (45 \cos 45° - 5);$$

thus $F_x = 812.2$ lb.

$$F_y - 16.1 \times \pi \times 12 \sin 45° = 0.01058 \times 45 \sin 45°;$$

thus $F_y = 36.1$ lb.

The force exerted on the oil is thus 812.2 lb to the left and 36.1 lb upward. The force exerted on the elbow is of the same magnitude, but in the opposite direction. In this problem the momentum terms play a small role because of the relatively large pressures.

---

9. Unsteady Flows and Translating Control Volumes

For some problems, a moving control volume can be a more convenient analytical tool than a stationary one. For turbomachinery, for example, a control volume fixed to the vane or rotor may be convenient. In this section we consider the case of a control volume moving with a constant linear velocity, to see how this affects our basic equations and our method of analysis.

In situations involving moving objects, the problem we are considering often appears much simpler if we choose a control volume moving with the device and a coordinate system *xyz* fixed to the device we are investigating. We denote the velocity of the origin of this moving control by $\mathbf{v}_0$ and its acceleration by $\mathbf{a}_0$. If $\mathbf{r}_0$ is the position vector of a fluid point P with respect to our moving reference frame (Figure 3.24), then the velocity of that point is

$$\mathbf{v} = \mathbf{v}_0 + d\mathbf{r}_0/dt = \mathbf{v}_0 + \mathbf{v}_r, \tag{3.9.1}$$

where $\mathbf{v}_r$ is the velocity as seen in the moving coordinate system. The acceleration is found by differentiating this velocity expression with respect to time, with the result that

$$\mathbf{a} = \mathbf{a}_0 + \mathbf{a}_r \tag{3.9.2}$$

where $\mathbf{a}_r$ is the fluid acceleration as seen in our moving system. Use of (3.9.1) and (3.9.2) in the continuity, momentum, and energy equations, along with the mass rates as measured in the moving system, gives the needed results. Rather than inserting these results into our equations, it is more useful and provides better understanding to illustrate their use by example.

In this section we will consider only those control volumes whose motions are translations at constant velocity. Thus $\mathbf{a}_0$ is zero. A reference frame that is moving with a constant velocity is an inertial frame, e.g., one in which Newton's law holds with no additional terms. Thus little complication will be added to our analysis.

a. Unsteady flows

Moving vanes

Suppose the vane is moving in the *x* direction with a constant speed *u* as in Figure 3.25a. The analysis is made considerably simpler by analyzing the problem in a coordinate system moving with the vane, as in Figure 3.25b. The jet inlet velocity that the vane sees is $v - u$, and the mass flow rate seen by the vane is

$$\dot{m}_r = \rho A(v - u) = \dot{m}(v - u)/v, \tag{3.9.3}$$

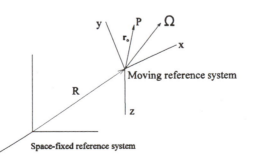

Figure 3.24. Defining sketch for a moving reference frame.

where $\dot{m} = \rho A v$ is the mass flow rate the vane would see if it were stationary. In our moving reference frame, the jet leaves tangent to the vane, and at the vane angle θ. Applying our momentum analysis in the moving coordinate system, we find

$$-F_x = \dot{m}_r[(v - u)\cos\theta - (v - u)], \tag{3.9.4}$$

$$F_y = \dot{m}_r(v - u)\sin\theta. \tag{3.9.5}$$

To complete the analysis we must find the absolute velocity of the jet as it leaves the vane. In going from the fixed reference frame to the moving reference frame, it was required to subtract the horizontal velocity component u from all quantities. In going back to the fixed frame we must add u in the same manner. This is shown in Figure 3.25c. The speed of the jet leaving the vane is thus

$$v_a = \sqrt{[(v - u)\cos\theta + u]^2 + [(v - u)\sin\theta]^2}, \tag{3.9.6}$$

and the angle it makes with the x axis is

$$\xi = \tan^{-1}\{(v - u)\sin\theta / [(v - u)\cos\theta + u]\}. \tag{3.9.7}$$

The above analysis is for a single moving vane. When there is a series of vanes, as in turbomachinery, the analysis is the same except that all mass flow rate from the incoming jet is assumed to strike one or another of the vanes. Then equation (3.9.4) and (3.9.5) can be used with $\dot{m}_r$ replaced by $\dot{m}$.

Example 3.9.1. A single moving vane

A vane with a turning angle of 30° moves to the right with a speed of 40 ft/s. A water jet with a speed of 100 ft/s and a diameter of 0.1 ft strikes the vane. Find the force that the jet exerts on the vane and the absolute velocity of the jet as it exits the vane.

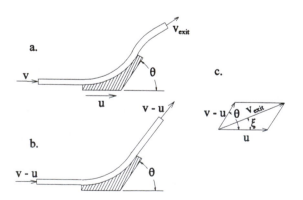

Figure 3.25. Moving vane. (a) Space-fixed reference frame. (b) Moving reference frame attached to vane. (c) Exiting absolute velocity.

Sought: Force exerted on a single moving vane by a jet.

Given: $v_{\text{in fluid}} = 100$ ft/s, $v_{\text{vane}} = 40$ ft/s, angle $_{\text{in}} = 0°$, angle $_{\text{out}} = 30°$, $A = \pi \times (0.05)^2 = 78.54 \times 10^{-4}$ ft^2, $\rho = 1.94$ slugs/ft^3, single vane.

Assumptions: Constant density. No losses. Unenclosed vane.

Solution: By (3.9.3),

$$\dot{m}_r = 1.94 \times 100 \times 78.54 \times 10^{-4}(100 - 40)/100 = 0.914 \text{ slug/s.}$$

From the momentum equations (3.9.4) and (3.9.5) and in a reference frame moving with the vane,

$$-F_x = 0.914 \times (100 - 40) \times (\cos 30° - 1) = -7.35 \text{ lb,}$$

$$F_y = 0.9145 \times (100 - 40) \times \sin 30° = 27.4 \text{ lb.}$$

These are the components of the force that the vane exerts on the jet. The force of the jet on the vane is thus 7.35 lb to the left, and 27.4 lb downward. By (3.9.6) and (3.9.7),

$$v_a = \sqrt{(60 \cos 30° + 40)^2 + (60 \sin 30°)^2} = 96.7 \text{ ft/s,}$$

$$\xi = \tan^{-1}[60 \sin 30°/(60 \cos 30° + 40)] = 18.1°.$$

Example 3.9.2. Multiple moving vanes

Suppose that in the previous example the single moving vane is replaced by multiple moving vanes, such as might occur in turbomachinery. How does the answer change?

Sought: Force exerted on multiple moving vanes by a jet.

Given: $v_{\text{in fluid}} = 100$ ft/s, $v_{\text{vane}} = 40$ ft/s, angle $_{\text{in}} = 0°$, angle $_{\text{out}} = 30°$, $A = \pi \times (0.05)^2 = 78.54 \times 10^{-4}$ ft^2, $\rho = 1.94$ slugs/ft^3, multiple vanes.

Assumptions: Constant density. No losses. Unenclosed vanes.

Solution: The mass flow rate is now

$$\dot{m}_r = 1.94 \times 100 \times 78.54 \times 10^{-4} = 1.52 \text{ slugs/s.}$$

From the momentum equations (3.9.4) and (3.9.5),

$$-F_x = 1.52 \times (100 - 40) \times (\cos 30° - 1) = -12.22 \text{ lb,}$$

$$F_y = 1.52 \times (100 - 40) \times \sin 30° = 45.6 \text{ lb.}$$

These again are the components of the force that the vane exerts on the jet. The remaining answers are unchanged.

---

b. Approximately unsteady flows

There are a number of flows that are technically unsteady but for which the effect of the unsteadiness is small enough to be of minor importance. Consider for instance the draining of a container (Figure 3.26), where the drain hole is small compared to the container cross-sectional area. Writing the continuity and energy equations between points 0 and 1, we find

$$A_1 v_1 = A_0 v_0 = -A_0 \, dh/dt \tag{3.9.8}$$

and

$$hg = v_1^2/2. \tag{3.9.9}$$

In writing these equations we have neglected the kinetic energy term at the top surface as being small compared to the kinetic energy of the draining jet. Similarly we have neglected the distance of the vena contracta below the container as being small compared to h. We cannot, however, forget about v_0 in the continuity equation, since this would lead to v_1 being zero.

Eliminating v_1, we have the equation

$$dh/dt = -(A_1/A_0) \sqrt{2gh}. \tag{3.9.10}$$

This equation can be integrated by separating variables to put it in the form

$$dh/\sqrt{h} = -(A_1/A_0) \sqrt{2g} \, dt,$$

which after integration gives

$$2h^{1/2} = -(A_1/A_0) \times \sqrt{2g}t + C, \quad \text{where } C \text{ is a constant of integration.}$$

If we say that the height is h_0 at $t = 0$, then

$$2h_0^{1/2} = 0 + C.$$

Solving for C we see that h is given for any time by

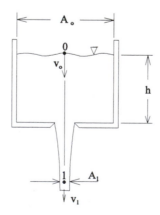

Figure 3.26. Defining sketch for a draining container.

$$h = (\sqrt{h_0} - A_1\sqrt{2g}\ t/2A_0)2. \tag{3.9.11}$$

The container is thus empty ($h = 0$) when

$$t = 2(A_0/A_1)\sqrt{h_0/2g}. \tag{3.9.12}$$

Example 3.9.3. Emptying tank

A circular tank 4 m in diameter is filled to a depth of 3 m. How long does it take to empty the tank through a 7-cm-diameter hole in the tank bottom?

Sought: Time needed to empty a tank.
Given: $A_{\text{hole}} = \pi \times (0.035)^2 = 38.48 \times 10^{-4}$ m^2, $V_{\text{initial}} = \pi \times 3 \times 2^2 = 37.70$ m^3.
Assumptions: Almost steady flow. No energy losses. Constant density.
Solution: From (3.9.18), since $h_0 = 3$ and $A_0/A_1 = (4/0.07)^2 = 3{,}265$, the time needed to empty the tank is

$$t = 2 \times 3{,}265\ \sqrt{3/2 \times 9.8} = 2{,}555\,\text{s} = 42.58\ \text{min}.$$

Rockets

An example of a control volume with variable mass is the rocket. Since part of its mass is used as fuel, from conservation of mass, the mass inside the rocket at any time will be

$$m(t) = m_0 - m_e(t), \tag{3.9.13}$$

where m_0 is the initial mass (rocket plus fuel) and $m_e(t)$ is the mass of the fuel expended since the rocket was launched. We take the expelled mass to exit from the rocket at a constant velocity u, measured relative to the rocket. The forces on the rocket will be a combination of drag and gravity. If v is the absolute speed of the rocket, then taking a control volume fixed to the rocket (Figure 3.27), summing forces in the direction in which the rocket is traveling, and applying the momentum equation gives

$$-R = m\ dv/dt - \dot{m}_e u, \tag{3.9.14}$$

where R is the net retarding force (drag plus gravity) in the direction opposed to v and $\dot{m}_e u$ is the thrust force.

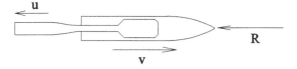

Figure 3.27. Defining sketch for a rocket.

As a special case example, suppose that $R = mg + C_D v^2$, a combination of gravity and air resistance, and that $\dot{m}_e$ is a constant β. Then (3.9.14) becomes

$$(m_0 - \beta t) \, dv/dt + C_D v^2 = \beta u - (m_0 - \beta t)g. \tag{3.9.15}$$

Generally, a numerical solution of equation (3.9.15) is necessary, using numerical integration procedures such as are given in Appendix C. However, if air drag is negligible so that C_D can be neglected, the variables in equation (3.9.15) can be separated in the form

$$dv = [\beta u - (m_0 - \beta t)g] \, dt/(m_0 - \beta t).$$

Integration of this equation with respect to time gives

$$v = v_0 - gt + u \ln[m_0/(m_0 - \beta t)], \tag{3.9.16}$$

where we have chosen v_0 to be the velocity at $t = 0$. For this case of air drag neglected, we see that βu must be greater than the initial weight $m_0 g$ in order to have an initial acceleration that is positive.

Example 3.9.4. Rocket

A toy rocket weighing 0.05 lb is filled with 0.4 lb of pressurized water as a fuel. The rocket is fired vertically, with no initial velocity. If all of the water is expelled in 1.5 s, at a relative speed of 60 ft/s, what is the maximum speed the rocket can attain?

Sought: Maximum speed attainable by this rocket.
Given: $v_0 = 0$, $m_0 = (0.05 + 0.4)/32.17 = 0.0140$ slug.
Assumptions: No air drag. Constant density.
Solution: From the data,

$$\beta = 0.4/1.5 \times 32.17 = 0.00829 \text{ slug/s},$$

$$u = 60 \text{ ft/s}.$$

Putting these values into (3.9.16), we have

$$v = -32.17t + 60 \times \ln [0.014/(0.014 - 0.00829t)].$$

The maximum speed is obtained at the end of 1.5 s, when all the fuel has been expelled. (The equation for v holds up only to that time, i.e., for $0 \le t \le 1.5$.) At time 1.5 s,

$$v = -32.17 \times 1.5 + 60 \times \ln (0.45/0.05) = 83.6 \text{ ft/s}.$$

---

10. Conservation of Moment-of-Momentum and Rotating Control Volumes

a. Moment-of-momentum equations for stationary control volumes

For turbomachinery and other flow problems involving rotating bodies, it is frequently more convenient to deal with torques and moment of momentum (or angular momentum) rather than forces and linear momentum. Development of the necessary equations is carried out in a fashion similar to the derivation of the linear momentum equation, with the moment of the forces and momentums being taken first for a small control volume of mass $\rho \, dV$, and then summed over the finite control volume. The result is

$$\Sigma \, \mathbf{moments} = \Sigma \, \mathbf{r} \times \mathbf{F} = d/dt \iiint_{cv} \mathbf{r} \times \mathbf{v} \, \rho \, dV + \iint_{cs} \mathbf{r} \times \mathbf{v} \, \rho \mathbf{v} \cdot d\mathbf{A}. \qquad (3.10.1)$$

Here $\mathbf{r}$ is the position vector for a point in the fluid contained in the control volume ($\mathbf{r}$ is used for both $\mathbf{F}$ and $\mathbf{v}$) measured from the reference point about which moments are being taken. The first integral on the right-hand side of (3.10.1) vanishes for steady flows. The second integral, the integral over the control volume, represents the moment of momentum of the flow entering and exiting the control volume.

b. Moment-of-momentum equations for rotating control volumes

We will restrict our development of moving moment-of-momentum equations to the case where the reference frame is rotating but not translating, since this is the case most often encountered. Let the control volume rotate at an angular rate $\mathbf{\Omega}$. If $\mathbf{r}_0$ is the position vector of a fluid point measured with respect to the origin of our rotating reference frame (Figure 3.24), then the velocity of that point is

$$\mathbf{v} = d\mathbf{r}_0/dt = \mathbf{v_r} + \mathbf{\Omega} \times \mathbf{r}_0, \qquad (3.10.2)$$

where $\mathbf{v_r}$ is the velocity as seen in the moving coordinate system and $\mathbf{\Omega} \times \mathbf{r}_0$ is the correction due to the rotation of the coordinate system. The acceleration is found by time differentiating this once more, with the result that

$$\mathbf{a} = \mathbf{a_r} + \dot{\mathbf{\Omega}} \times \mathbf{r}_0 + \mathbf{\Omega} \times (\mathbf{\Omega} \times \mathbf{r}_0) + 2\mathbf{\Omega} \times \mathbf{v_r}. \qquad (3.10.3)$$

where $\mathbf{a_r}$ is the fluid acceleration as seen in our moving system. The third term on the right-hand side of this equation is the centripetal acceleration needed to correct for the rotation of the coordinate system, and the last term is the Coriolis acceleration, which is familiar to you if you have taken a course in elementary dynamics. It represents the correction to the acceleration necessary when the particle we are considering is moving perpendicular to the angular velocity of rotation.

The moment-of-momentum equation then is

$$\Sigma \, \mathbf{moments} = \Sigma \, \mathbf{r} \times \mathbf{F} = d/dt \iiint_{cv} \mathbf{r} \times \mathbf{v_r} \rho \, dV + \iint_{cs} \mathbf{r} \times \mathbf{v_r} \rho \mathbf{v_r} \cdot d\mathbf{A}$$

$$+ \iiint_{cv} \{\mathbf{r} \times (\dot{\boldsymbol{\Omega}} \times \mathbf{r}) + \mathbf{r} \times [2\boldsymbol{\Omega} \times \mathbf{v_r} + \boldsymbol{\Omega} \times (\boldsymbol{\Omega} \times \mathbf{r})]\} \rho \, dV, \qquad (3.10.4)$$

where the first and second terms on the right-hand side correspond to the moment-of-momentum terms in (3.10.10) and the additional terms are corrections for the moving reference frame. The second correction term is the Coriolis acceleration and the last is the centrifugal acceleration. Note that for steady flows the first and third terms on the right-hand side vanish.

Use of equations (3.10.1) and (3.10.4) will be shown by the following examples.

Lawn sprinklers

A common type of lawn sprinkler consists of N arms free to rotate, with jets at the ends (Figure 3.28). Note that if the working fluid was steam, this would be the classical steam turbine of Hero. The discharge flow rate in each arm provided to the sprinkler is Q, and the arms have an inside area A_a. The jet at the end of the tip has an area A_j. The velocity at which the water passes through the jets when the arms are stationary is $v_j = Q/A_j$. The flow speed in each arm is $v_a = Q/A_a$.

Referring to the coordinate system in Figure 3.28, we write

$$\boldsymbol{\Omega} = \Omega \mathbf{k}, \qquad \mathbf{T} = \Sigma \, \mathbf{moments},$$

$$\mathbf{r} = R\mathbf{i} + a(\mathbf{i} \cos \theta - \mathbf{j} \sin \theta)\cos \beta + \mathbf{k} \sin \beta] \text{ on the tip of the arm,}$$

$$\boldsymbol{\Omega} \times \mathbf{r} = \Omega[R\mathbf{j} + a(\mathbf{j} \cos \theta + \mathbf{i} \sin \theta) \cos \beta] \text{ on the tip of the arm,}$$

$$\mathbf{v} = \boldsymbol{\Omega} \times \mathbf{r} + v_j[(\mathbf{i} \cos \theta - \mathbf{j} \sin \theta) \cos \beta + \mathbf{k} \sin \beta] \text{ on the tip of the arm.}$$

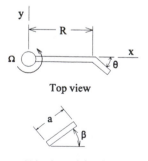

Top view

Side view of tip of arm

Figure 3.28. Defining sketch for one arm of a lawn sprinkler.

$\mathbf{r} \times \mathbf{v} = -a\,\Omega\,\sin\beta\,(R + a\cos\theta\cos\beta)\mathbf{i} + (a^2\Omega\sin\theta\sin\beta\cos\beta - v_j R\sin\beta)\mathbf{j}$

$\qquad + [\Omega(R^2 + 2aR\cos\theta\cos\beta + a^2\cos^2\beta) - v_jR\sin\theta\cos\beta]\mathbf{k}.$

Note that the control surface integrals in (3.10.4) are all evaluated at the arm tip, while the control integral terms must be integrated throughout the volume. We will consider only the case where Ω is a constant angular velocity; thus

$$d/dt \iiint_{\text{cv}} \mathbf{r} \times \mathbf{v}_r\,\rho\,dV = 0.$$

For the control surface integral,

$$\mathbf{T} = \iint_{\text{cs}} \mathbf{r} \times \mathbf{v}\rho\mathbf{v} \cdot d\mathbf{A} = \sum_{N\,\text{arms}} \rho Q_a\{-a\,\Omega\,\sin\beta(R + a\cos\theta\cos\beta)\mathbf{i}$$

$\qquad + (a^2\,\Omega\,\sin\theta\sin\beta\cos\beta - v_jR\sin\beta)\mathbf{j}$

$\qquad + [\Omega\,(R^2 + 2aR\cos\theta\cos\beta + a^2\cos^2\beta) - v_jR\sin\theta\cos\beta]\mathbf{k}\}. \qquad (3.10.5)$

The terms in (3.10.5) that lie in the x-y plane (terms multiplying $\mathbf{i}$ and $\mathbf{j}$) generally would balance out if there were more than the arm, and the arms were placed symmetrically. The moment-of-momentum term in the z direction (the term multiplied by $\mathbf{k}$) is, in the absence of a driving torque, balanced by to friction and wind resistance. Denoting this torque by $-T_0$, we have

$$-T_0 = \sum_{N\,\text{arms}} \rho Q_a\{\Omega[R^2 + (2aR\cos\theta + a^2\cos\beta)\cos\beta] - v_jR\sin\theta\cos\beta\}. \qquad (3.10.6)$$

When the arms are all the same the angular velocity then can be solved for, giving

$$\Omega = \frac{v_jR\sin\theta\cos\beta - T_0/N\,\rho Q_a}{R^2 + a\,(2R\cos\theta + a\cos\beta)\cos\beta}. \qquad (3.10.6a)$$

The flow rate, friction, and geometry thus determine the angular speed at which the sprinkler spins.

The first term on the right-hand side of the above expression for Ω represents the change of moment of momentum of the fluid leaving the sprinkler that drives the sprinkler rotation, and the second represents the slowing down of the sprinkler due to mechanical friction.

Further examples of flows with angular momentum are given in Chapter 10.

Example 3.10.1. Lawn sprinkler
A flow rate of 10 gpm is applied to a three-arm lawn sprinkler with arms 1.2 ft long. The inner diameter of the arm is 3/8 in, and the jet hole diameter is 1/8 in. The end of the arm is bent upward 20°, and backward 60°. At what angular rate does the sprinkler rotate in the absence of friction?

Sought: Angular speed of a three-arm lawn sprinkler.

Given: $Q = 1$ gpm $= 1/60 \times 7.49 = 2.225 \times 10^{-3}$ ft³/s, $Q_a = Q/3 = 7.417 \times 10^{-4}$ ft³/s, $R = 1.2$ ft, $a \approx 0$, $\theta = 60°$, $\beta = 20°$, $A_a = \pi(3/16 \times 12)^2 = 7.670 \times 10^{-4}$ ft², $A_j = \pi \times (1/16 \times 12)^2 = 8.522 \times 10^{-5}$ ft².

Assumptions: No friction. Incompressible flow.

Solution: The velocity of the jet leaving the arm is

$$v_j = Q_a/A_{\text{jet}} = 7.417 \times 10^{-4}/8.522 \times 10^{-5} = 8.703 \text{ ft/s}.$$

The velocity of the flow within the arm is

$$v_a = v_j/9 = 0.9670 \text{ ft/s}.$$

For the case where $a = 0$ and $T_0 = 0$, (3.10.6) reduces to

$$\Omega = v_j \sin \theta \cos \beta/R, \text{ or}$$

$$\Omega = 8.703 \times \sin 60° \times \cos 20°/1.2 = 5.902 \text{ rad/s} = 0.9393 \text{ rev/s} = 56.36 \text{ rpm}.$$

Atmospheric and oceanic flows

In many flows of engineering interest, flow occurs in the direction of the negative pressure gradient. For example, for flow in a pipe the pressure is higher at the inlet end of the pipe and lower at the outlet end, and the flow goes from the high-pressure point to the low-pressure point. A familiar exception to this correlation between flow direction and pressure gradient is the weather map you see in the newspaper or on television. These maps show isobars (lines of constant pressure), and the air flow is along the isobar. The reason for this is the earth's rotation. To see this, we consider a plane attached tangent to the surface of the earth at the point of interest with the x axis pointing east, the y axis pointing north, and the z axis pointing upward perpendicular from the surface of the earth. The component of the earth's rotation of importance to this coordinate system is

$$\boldsymbol{\Omega} = \Omega_0 \sin \beta \ \mathbf{k},$$

where Ω_0 is the earth's rotational speed (1 revolution per day) and β is the latitude at which our coordinate system touches the earth. The latitude is zero at the earth's equator and $\pm 90°$ at the north and south poles. The "flat earth" approximation we are using is referred to by meteorologists as the beta-plane approximation.

Since Ω_0 and β are constants, expression (3.10.3) simplifies to

$$\mathbf{a} = \mathbf{a_r} + \boldsymbol{\Omega} \times (\boldsymbol{\Omega} \times \mathbf{r_0}) + 2\boldsymbol{\Omega} \times \mathbf{v_r}. \tag{3.10.7}$$

This acceleration times the mass density is approximately equal to the negative of the pressure gradient in the upper atmosphere, where viscous forces are generally small. The second term in (3.10.7), the centrifugal acceleration, is independent of the air velocity and hence adds an air-velocity independent term to the pressure gradient, much as we saw in Chapter 2 when discussing rotating flows. The last term becomes

$$2\mathbf{\Omega} \times \mathbf{v_r} = 2\Omega_0 \sin\beta\, \mathbf{k} \times \mathbf{v_r} = 2\Omega_0 \sin\beta\,(-v_y\,\mathbf{i} + v_x\,\mathbf{j}). \qquad (3.10.8)$$

For steady flows $\mathbf{a_r}$ will be zero. The term $\mathbf{\Omega} \times (\mathbf{\Omega} \times \mathbf{r_0})$ in (3.10.7) gives a pressure gradient typical of rigid-body rotation. Since $\Omega = \Omega_{earth}$ appears squared, this term is small. Thus the acceleration term that has to do with air or ocean circulation is the Coriolis term $2\mathbf{\Omega} \times \mathbf{v_r}$. Since the pressure gradient is proportional to this acceleration, $\mathbf{v_r}$ must be perpendicular to the pressure gradient. Thus it follows that $\mathbf{v_r}$ is tangent to the isobars.

The magnitude of $2\Omega_0 \sin\beta$ is small. For example, at a latitude of $45°$ it is about 1×10^{-4} rad/s. However, when as in the atmosphere and oceans we deal with length scales of the order of hundreds of miles, these are the terms that control the flow, since they are greater in magnitude than the other acceleration terms.

Frequently terms such as (3.10.8) are used to argue that the flow down a drain in a sink will go in one direction in the northern hemisphere and the opposite direction in the southern hemisphere. While this is theoretically logical, unfortunately in practice it is not often true. As you can easily demonstrate in a wash basin or kitchen sink, you can make the rotation go either way just by filling it from the hot or cold faucet, or by initially imparting a swirl with your hand. The starting condition will completely mask the very tiny effect of the Coriolis term. If you were to construct a much more symmetric basin than the kitchen sink and wait 24 hours or more so that the filling conditions have died out by viscous dissipation, you might find an even more interesting effect. Provided that you open the drain very carefully so as not to disturb the fluid, if you were to continually drip a colored marker on the fluid near the rim of the basin you would first see the markers move radially in toward the drain. As the depth in the container gets thinner and thinner you would see the markers start spiraling in toward the drain, first in one direction, then the other, this pattern continuing until all the liquid had drained.

The reason for this result is that the Coriolis terms are so small that an extremely accurate measurement is needed to detect them. However, when the depth of the water is of the order of

$$\sqrt{\nu/\Omega_0 \sin\beta},$$

viscous effects start to become important and in fact dominate the pressure gradient terms. If you could measure the flow when the depth is of this order, you would find that the velocity vector spirals with the depth, pointing in all of the horizontal directions. This result is termed the ***Ekman spiral***, named after V. W. Ekman, the oceanographer who discovered it in 1902. The thin layer where viscous effects are important is called the ***Ekman layer***. Ekman's analysis is likely the first study of a boundary layer in fluid mechanics.[6] Other examples of geophysical boundary layers

---

[6]A simpler experiment showing the Ekman-layer effect can be performed with a cup of tea containing a moderate number of tea leaves. Stirring the tea in a circular manner causes the tea leaves to travel in approximately circular paths. As we found for the rotating flows studied in Chapter 2, the centripetal acceleration is balanced by a radial pressure gradient. However, at the bottom of the cup the no-slip condition means that the centripetal acceleration has to vanish, but the pressure gradient will still exist, balanced by viscous forces. Thus as the tea leaves settle out, the pressure gradient $2\rho r\Omega^2$ causes the tea leaves to collect at the center of the cup bottom.

are the jet stream in the atmosphere and the Gulf Stream and Japanese Current in the Atlantic and Pacific Oceans.

11. Path Coordinates—the Euler and Bernoulli Equations

In the beginning of this chapter we derived the continuity, momentum, and energy equations for finite-size control volumes. In applying them, we found that it was possible to obtain many results in regions where the streamlines are either parallel or nearly parallel. When this is not true, the one-dimensional equations are much less useful. To find details of the flow in such regions, it is necessary to analyze the flow through a control volume of infinitesimal size, so that we can work with equations valid at a point, rather than over a finite area. We will derive one such equation here, to see some of the restrictions that our finite-size control volumes impose, as well as to extend our capability for understanding more complicated flows.

Consider the control volume shown in Figure 3.29, of length ds and constant cross-sectional area dA. The control volume is oriented so as to be along the path line of a fluid particle entering the center of the area dA at the instant our analysis takes place. If v is the speed of the fluid particle at the point where it enters our control volume, then according to the development in Section 3.2, equation (3.2.5) would give the fluid acceleration along the path line as

$$\partial v/\partial t + v\, \partial v/\partial s. \tag{3.11.1}$$

The net force due to pressure along the path line will be

$$-\partial p/\partial s\; ds\; dA, \tag{3.11.2}$$

since the force on the face that the fluid particle enters is $p\, dA$, and where it leaves is $[p + (\partial p/\partial s)ds]\, dA$.

We have allowed the pressure to vary along the path line, and expanded the pressure variation along the streamline in a Taylor series in the distance along the streamline.

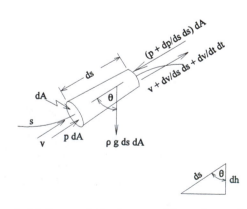

Figure 3.29. Defining sketch for a control volume used for path coordinates.

It is sufficient to consider only the terms of first order in the distance, since higher order terms will drop out. Then the gravity force along the path line will be

$$-\rho g \cos \theta \, ds \, dA. \tag{3.11.3}$$

Summing forces along the path line, and neglecting viscous forces, we have

$$-\partial p/\partial s \, ds \, dA - \rho g \cos \theta \, ds \, dA = \rho \, ds \, dA(\partial v/\partial t + v \, \partial v/\partial s).$$

Since the volume $ds \, dA$ appears in every term, we can divide our equation by it, obtaining

$$-\partial p/\partial s - \rho g \cos \theta = \rho \, (\partial v/\partial t + v \, \partial v/\partial s). \tag{3.11.4}$$

Normal to the path line, the acceleration is a centripetal term of the form

$$v^2/R, \tag{3.11.5}$$

where R is the local radius of curvature of the path line. The net pressure force will be

$$-(\partial p/\partial n) \, ds \, dA, \tag{3.11.6}$$

where n is measured perpendicular to the path line along the radius of curvature. Summing forces in this direction, again neglecting viscous forces, after dividing by the volume, we have

$$-\partial p/\partial n - \rho g \sin \theta = -\rho v^2/R. \tag{3.11.7}$$

Equations (3.11.4) and (3.11.7) are called the Euler equations in path coordinates, in the directions along a path line and perpendicular to the path line. They show us that where streamlines are curved, the pressure will increase due to the centripetal acceleration as well as due to the hydrostatic terms. We have already seen this in Section 3.5.

For steady flows, the path line becomes a streamline, and equation (3.11.4) can be integrated along this streamline. If h is in the direction opposed to gravity, then we have

$$\cos \theta = dh/ds$$

as seen from Figure 3.19. Then equation (3.11.4) can be rewritten as

$$-\partial p/\partial s - \rho g \, dh/ds = \rho v \, \partial v/\partial s = \rho \, \partial(v^2/2)/\partial s,$$

where the product rule of calculus has been used on the acceleration term. Integrating with respect to s gives

$$p + \rho g h + \rho v^2/2 = \text{constant along any given streamline.} \tag{3.13.8}$$

This is Bernoulli's equation, an integrated form of the Euler equation. The terms are formally like the Bernoulli terms in the energy equation, but the two equations are distinct and come from distinctly different physical laws—Newton's law for the

Bernoulli equation and the first law of thermodynamics for the energy equation. The first law contains the Bernoulli terms above, plus other terms that are not all of mechanical origin and that do not appear in Newton's law. The fact that under certain situations the Bernoulli equation and the first law reduce to the same equation is a statement only that the two laws are not in conflict, and does not imply any commonality of origin between these two laws.

a. Application—pitot tubes

A standard device for measuring velocity is the pitot tube, shown in Figure 3.30. It is a tube of circular cross section, usually bent in an L shape. One end has a rounded nose, with a hole at the center. This hole is connected to an inner tube, which runs through the pitot tube body and is connected to one end of a manometer or other type of pressure transducer. If the pitot tube is aligned parallel to the flow, by symmetry the flow would split around the nose, and the hole would be a stagnation point, or point of zero velocity.

Along the side of the tube, 10 or so diameters from the nose, there is another hole, slot, or ring of holes. This opening reads the static ambient pressure, and connects to the inside of the body of the tube. It is connected to the other side of the pressure measuring device. From the energy equation, since the velocity v_1 at the stagnation point is zero,

$$p_1/\rho = p_2/\rho + v_2^2/2 \tag{3.11.9}$$

or

$$v_2 = \sqrt{2(p_1 - p_2)/\rho}. \tag{3.11.10}$$

Equation (3.11.10) is much like equation (3.8.4), but without any geometric factors.

The pitot tube is used in conduits to measure flow rates, and is frequently used for heating and air-conditioning measurements. It is used to measure external as well as internal flows, and is a favorite device for measuring airspeeds on planes. Many variations exist for special purposes. If ambient pressure can be determined by other

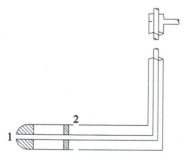

Figure 3.30. Cross section of a pitot tube.

means, the side holes are left off, leading to a simpler device called an impact tube. (Miniature impact tubes can be made from small tubes or hypodermic needles for use in the laboratory.) When the direction of the flow is not known a priori, more holes may be added to the front nose to make available further pressure readings which allow determination of the flow direction.

Example 3.11.1. Pitot tube

A pitot tube measuring in air records a pressure difference of 10 in of water. What velocity is measured by the pitot tube?

Sought: Velocity measured by a pitot tube.

Given: $\Delta p = 10$ in of water = $(10/12) \times 62.4 = 52$ psf, $\rho = 0.00244$ slug/m^3, equation (3.11.10).

Assumptions: Constant density. No energy losses.

Solution: From equation (3.11.10),

$$v = \sqrt{2 \times 52/0.00244} = 206.5 \text{ ft/s}.$$

This velocity is just marginal for our incompressibility assumption, as we will see in Chapter 9.

Suggestions for Further Reading

Curle, N., and H. J. Davies, *Modern Fluid Dynamics*, vol. **1**, *Incompressible Flow*, Van Nostrand, Princeton, NJ, 1968.

Durand, W. F., *Aerodynamic Theory*, vol. **IV**, Dover, New York, 1963. (First published by J. Springer in 1935.)

McConnell, A. J., *Applications of Tensor Analysis*, Dover, New York, 1957. (First published in 1931.)

Prandtl, L., *Essentials of Fluid Dynamics*, Hafner, New York, 1952. (Translated from the German edition.)

Tritton, D. J., *Physical Fluid Dynamics*, Van Nostrand Reinhold, New York, 1977.

Problems for Chapter 3

Acceleration

3.1. A fluid particle moves on a circular path of radius 3 ft. Its only acceleration is a centripetal one, of magnitude 5 m/s^2. What is the speed of the particle? Express the velocity in terms of x and y components, with origin at the circle center.

3.2. For the velocity $\mathbf{V} = 8x\mathbf{i} + (-8y + 10)\mathbf{j}$, find the equations of the streamlines. Find also the equation of the path line that passes through the point having coordinates (1, 2, 3).

3.3. A velocity field is given by $\mathbf{V} = [5 + 14y + (1 + t)x]\mathbf{i} - [3 + 9x^2 + (1 + t)y]\mathbf{j}$. What is the acceleration associated with this flow at any point and any time? What is the acceleration at the origin at $t = 0$?

Mass

3.4. Benzene (SG = 0.88) flows through a pipe of diameter 50 mm at a mean velocity of 2.3 m/s. Find the volumetric discharge, the mass rate of flow, and the weight rate of flow.

3.5. Water (SG = 1) flows through a tube of conical cross section. At a point where the diameter is 2 in, the velocity is 4 ft/s. Find the volumetric discharge, the mass flow rate, and the velocity at the section where the diameter is 0.5 in.

3.6. Glycerine (SG = 1.26) and water (SG = 1.0) are being mixed to a ratio of 2 parts water by volume to 7 parts glycerine. Assuming perfect mixing, what will be the mass density of the mixture with a 3.0 ft³/s flow rate?

3.7. For flow through a conical nozzle the axial velocity component is given approximately by $v_x = V/(1 - x/L)^2$, where V is the speed at the nozzle entrance $x = 0$ and L is the distance between the entrance and the cone apex. Compute the velocity and acceleration at the two points $x = 0$ and $x = L/2$. Use $V = 3$ m/s and $L = 2$ m.

Momentum

3.8. A 2-in-diameter jet of water (SG = 1) with a discharge of 0.5 ft³/s is aimed to hit the inside of a large stationary container at an angle of 45° with respect to the horizontal. What force does the jet exert on the container?

3.9. Calculate the resultant force on the end plate in the figure. The mass density is 2 slugs/ft³ and the discharge is 3 ft³/s.

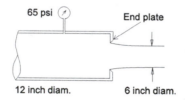

Problem 3.9

3.10. For the flow situation shown in the figure, with a fluid density of 2 slugs/ft³, what is the volumetric discharge Q, necessary to ensure that continuity is satisfied?

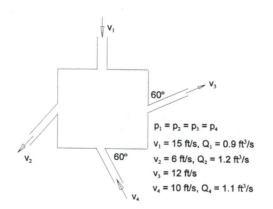

Problem 3.10

3.11. A two-dimensional wedge is placed in a water tunnel with a 4-ft-square test section. The upstream velocity is a uniform 15 ft/s. Downstream of the wedge the velocity is measured approximately in the distribution shown in the figure. What is the velocity V_2? If the measured drag on the wedge is 2,000 lb total, what is $p_{upstream} - p_{downstream}$, assuming that the pressure is distributed uniformly at each location?

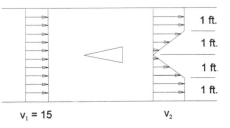

$$v_1 = 15 \qquad\qquad V_2$$

Problem 3.11

3.12. During the early 1940s the New York Central Railroad experimented with picking up water for the engine boilers while running at high speeds. It eventually achieved a design that could pick up water at the rate of 7,113 gal/hour from a trough below the rails while traveling at 80 mph. The scoop entrance was 8 in high and 13 13/16 in wide, and extended 7 1/2 in below the rail line. What vertical force did this exert on the train?

Energy

3.13. In the system shown in the figure, with water flowing, the pump supplies enough energy to the flow so that the exit velocity is 20 ft/s. Assuming no losses in any part of the system, how much energy is supplied to the fluid by the pump? Express your answer both as a head gain and as lb-ft/s. What are the pressures at points 2 and 3? What force is exerted on the nozzle? Is the final length of straight pipe in tension or compression?

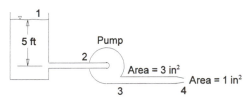

Problem 3.13

3.14. It is proposed to lift a circular cylinder 4 ft in diameter by directing a jet of air down on the body as shown in the figure. This jet would reduce the pressure on the top of the cylinder, so that surrounding atmospheric pressure lifts the body. If the circular cylinder weighs 100 lb, what mass discharge of air ($\gamma = 0.078$ lb/ft^3) is necessary to

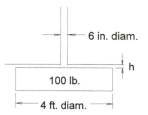

Problem 3.14

produce the required lift force? Assume that air pressure is constant at the region where the jet changes direction. The spacing h is 0.1 in.

3.15. Calculate the thrust (both magnitude and direction) on the turbine in the figure, which extracts 5.0 hp from the flow. The water enters at a pressure of 50 psi.

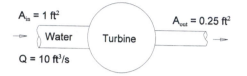

Problem 3.15

3.16. Compute the resultant force on the elbow shown in the figure due to momentum changes. The head loss is $3V^2/2g$. The fluid is water at 68°F. $Q = 15$ ft³/s.

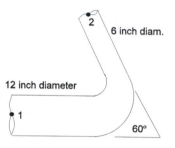

Problem 3.16

3.17. Neglecting any losses, for the configuration shown in the figure, find the discharge and the dynamic force of the fluid on the container. Neglect the weight of the fluid.

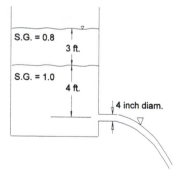

Problem 3.17

3.18. Water enters the box shown in the figure at A and B, and leaves at C. The flow speeds at A and B are 15 m/s and 35 m/s, respectively, and the areas there are 0.05 and 0.01 m². The area at C is 0.03 m². What is the velocity at C? What are the x and y components of force on the box? At what rate is energy gained or lost as the fluid flows through the box?

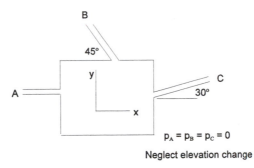

PA = PB = PC = 0

Neglect elevation change

Problem 3.18

3.19. A gradual expansion (no energy loss) in a pipeline changes the diameter from 6 in to 12 in. The upstream pressure is 45 psi, with 2 ft³/s of water flowing. Determine the magnitude and direction of the fluid force on the expanding section.

3.20. With water flowing in the box shown in the figure, what force must be exerted on the box to hold it in place? Assume all inlet and outlet pressures to be zero gage. How much power is lost or gained in the box, assuming that the flow is in the horizontal plane? Express your answer both in ft-lb/s and in head loss.

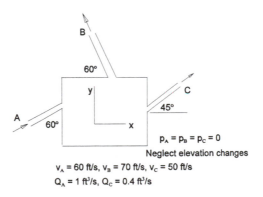

$p_A = p_B = p_C = 0$

Neglect elevation changes

$v_A = 60$ ft/s, $v_B = 70$ ft/s, $v_C = 50$ ft/s

$Q_A = 1$ ft³/s, $Q_C = 0.4$ ft³/s

Problem 3.20

3.21. Find the force exerted on the discharge pipe in the figure by the discharged water. Neglect energy losses due to the entrance region. The jet area is 1 in².

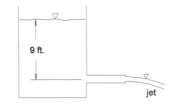

9 ft.

jet

Problem 3.21

3.22. A fan exerts a net force F on the air passing through its blades. For the fan shown in the figure, if the incoming velocity is 10 ft/s, determine F and also the rate at which energy must be supplied to the fan.

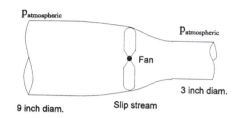

9 inch diam.   Slip stream

Problem 3.22

3.23. A discharge of 0.4 ft³/s is flowing in the 2.5-in-diameter pipe in the figure. If the flow of air between the circular plates of diameter 20 in discharges into the atmosphere, what is the pressure 5 in from the center of the plates?

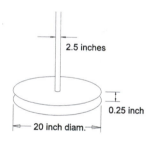

Problem 3.23

3.24. A pump and reservoir in the basement provide water to a swimming pool on the upper level of a 50-ft-high hotel. The pump provides a head gain of 70 ft to the fluid, and the head losses are given by $8V^2/2g$, with all losses occurring after the pump. What is the fluid velocity?

3.25. A 3-in-diameter 180° bend carries a liquid ($\rho = 2$ slugs/ft³) at 20 ft/s. If the entire flow is at zero gage pressure, what is the force on the bend? If the flow were at a gage pressure of 10 psi, what is the force on the bend?

Venturi meter

3.26. A venturi meter has an inlet diameter of 50 mm and a throat diameter of 30 mm. If standard air flows through the meter at 50 m/s, what is the difference in elevation of the 2 legs of the manometer? $SG_{\text{manometer fluid}} = 0.8$.

3.27. What is the flow rate for water in the venturi meter shown in the figure, with the manometer reading 6 in as shown?

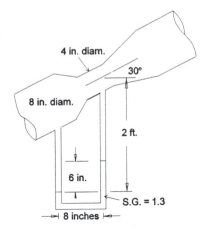

Problem 3.27

3.28. Find the velocity and pressure in the middle of the pipe upstream of the venturi meter in the figure, assuming no losses. The fluid is water at 68°F.

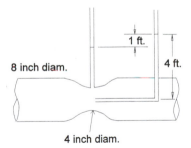

Problem 3.28

3.29. At what manometer reading h in the figure will the discharge be 0.314 ft^3/s? The fluid is water at 68°F. What force does the fluid exert on the nozzle? Neglect viscous effects.

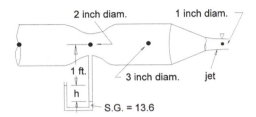

Problem 3.29

Orifice meter

3.30. A flat-plate orifice with a 1-in-diameter hole is in a 3-in-inner-diameter pipe with water flowing as shown in the figure. The mercury-water manometer reading is a 4-in difference in the levels in each leg of the manometer. What are the volume and mass rates of flow?

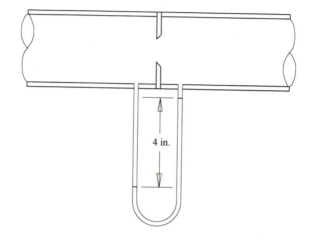

4 in.

Problem 3.30

3.31. Repeat the previous problem when air is flowing in the pipe. The mercury-air manometer again reads a 4-in difference.

Borda mouthpiece

3.32. A Borda mouthpiece with diameter 25 mm is inserted into the side of a large tank of water 5 m below the free surface. The gas above the water is pressurized to 10,000 Pa. Find the velocity and discharge of the exiting jet.

Vanes

3.33. Find the force components on the vane in the figure. The fluid is water at 68°F, $Q_0 = 2$ ft^3/s, and $V_0 = 150$ ft/s.

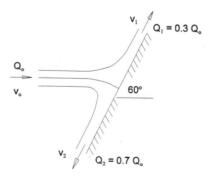

V_1

$Q_1 = 0.3\,Q_0$

Q_0

V_0

60°

V_2

$Q_2 = 0.7\,Q_0$

Problem 3.33

3.34. What force components are necessary to hold the vane in the figure stationary given that $Q_0 = 3$ ft³/s and $A_0 = 0.015$ ft²? The fluid is water at 68°F.

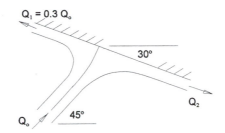

Problem 3.34

3.35. A jet impinges on a long flat plate inclined at an angle of 30° with the horizontal. Assuming that the viscous forces are negligible so that all forces act perpendicular to the plate, find the division of the discharge and the total force on the plate in terms of ρ, Q, and v.

3.36. The nozzle shown in the figure discharges water ($\rho = 1.935$ slugs/ft) at a rate of 2 ft³/s and a velocity of 100 ft/s. What are the components of force exerted on the vane by the fluid?

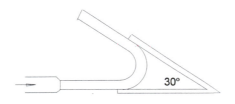

Problem 3.36

3.37. For a discharge of 40 ft³/s of water at a speed of 100 ft/s, compute the force on the stationary vane in the figure.

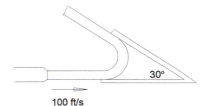

100 ft/s

Problem 3.37

3.38. The fixed vane shown in the figure divides the water jet so that 1 ft³/s goes in each direction. If the incoming velocity is 50 ft/s, what force components are necessary to hold the vane in place?

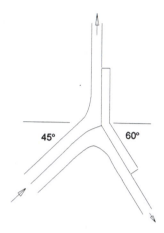

45°   60°

Problem 3.38

Siphons

3.39. A container of oil and water (see the figure) is being drained by a combination of a tube in the side and a siphon. What are the discharges for each, using the following for head losses: $h_{L\,\text{siphon}} = 3V^2/2g$, $h_{L\,\text{tube}} = V^2/2g$?

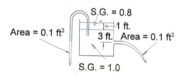

S.G. = 0.8

Area = 0.1 ft²   1 ft.
                        3 ft.   Area = 0.1 ft²

S.G. = 1.0

Problem 3.39

3.40. In the siphon shown in the figure, there is a head loss proportional to the kinetic energy head given by head loss = $3V^2/2g$. All the loss takes place within the siphon. The fluid has a specific gravity of 1.7, and the air pressure in the tank is 100 kPa above atmospheric. The cross-sectional area of the siphon is 0.02 m². What is the volumetric discharge from the siphon?

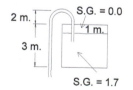

2 m.   S.G. = 0.0
              1 m.
3 m.

S.G. = 1.7

Problem 3.40

3.41. Assuming no energy loss, compute the following quantities for the siphon shown in the figure: discharge; velocity and pressure at point 2; pressure at point 3; velocity at point 4.

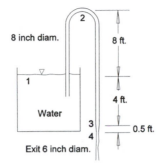

Problem 3.41

3.42. Find the pressure at points A and B in the figure with the valve closed. If the valve is opened, find the pressure and velocity at point A. The head loss is given by $h_L = 6V^2/2g$. Half of the head loss occurs between the free surface and point A.

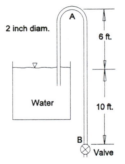

Problem 3.42

3.43. A siphon of 0.1-ft inside diameter is used to empty the container of water shown in the figure. The container is rectangular, with a 3-ft by 4-ft cross section. When the container is filled to a depth of 3 ft, (i.e., when $L = 3$ ft), what is the volume discharged per second?

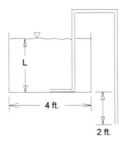

Problem 3.43

3.44. The bell siphon in the figure serves to ensure that a reservoir stays below a set maximum depth. When the water level inside the bell reaches the central lip of the discharge nozzle, the siphon will lower the reservoir level until the free surface reaches the lip of the outer bell. For the bell shown, find the discharge once siphoning begins. Assume that the siphon runs full of water, that there are no losses, and that the pressure at the siphon exit (3.5 m diameter) is approximately hydrostatic. The throat is 1.5 m below the free surface.

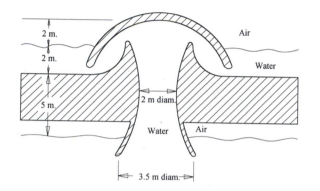

Problem 3.44

Expansion/contraction

3.45. A sudden enlargement in a pipeline from an area A_1 to an area $A_2 = 2A_1$, followed by a sudden contraction from the area A_2 back to an area A_1 both occur in the same pipeline in which there is steady flow. Compute the head loss in the sudden expansion and the sudden contraction separately. Which change in section produces the greater head loss? By what percent?

3.46. A 3-in-diameter pipe abruptly changes to a 1-in diameter. After 10 ft it changes back to the larger diameter. The velocity in the 1-in pipe is 9 ft/s. What is the rate of energy loss in the abrupt contraction? In the abrupt expansion? Compute the ratio of the two rates.

Nozzles, reducing elbows

3.47. In the system shown in the figure, what would be the gage pressure at point 2 if $p_1 = 10$ psig and there is no energy loss going from point 1 to point 2? What is the force the fluid exerts on the contraction? If the pressure at point 2 were measured and found to be 10% lower than your previous answer, at what rate is energy being lost in the system? $v_1 = 6$ ft/s.

Problem 3.47

3.48. The discharge of the 68°F water at point A in the figure is 1 ft^3/s and the pressure at A is 4 psi. Find the velocity given and pressure at B and C, given $Q_B = 0.8\ Q_A$ and $p_A = 2$ lb/ft^2. Assume outflow occurs at both B and C. What force (give x and y components) is needed to hold the tee in place?

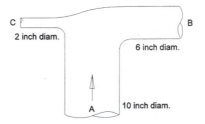

Problem 3.48

3.49. Water at 68°F flows at a rate of 20 ft/s through an 18-in-diameter pipeline. A horizontal 90° bend reduces the diameter from 18 to 12 in following the bend. If the pressure before the bend is 10 psi, and losses can be neglected, what is the pressure in the 12-in region of the pipe? What force components are needed to hold the bend in place?

3.50. Water at 68°F is flowing in the pipe elbow shown in the figure at a rate of 15 ft^3/s from point 1 to point 2. What are the velocities at points 1 and 2? If the pressure at point 1 is 2 psi, what is the pressure at point 2? Assume no losses occur. What force does the fluid exert on the elbow?

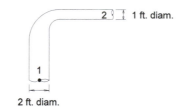

Problem 3.50

3.51. An expanding elbow turns a flow of 1.0 ft^3/s of water at 68°F through 90°. The inlet pressure (point A in the figure) is 2.5 psi, the outlet pressure (point B) is 1 psi, and the diameter increases from 4 to 12 in in a 4-ft elevation change. What are the head loss and the rate of power loss?

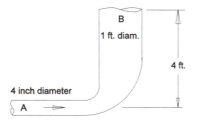

Problem 3.51

3.52. Calculate the reading (both direction and magnitude) of the gasoline-mercury manometer in the figure. The fluid is gasoline with a specific gravity of 0.78. Assume no energy losses.

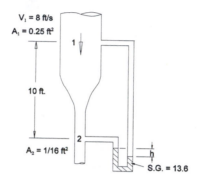

Problem 3.52

3.53. Water is flowing in the closed tee section shown in the figure. Neglecting losses and taking the tee to lie in a horizontal plane, what is the discharge at C? What are the pressures at B and C? What are the force components that the fluid exerts on the tee?

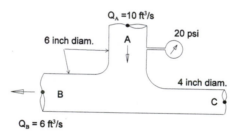

Problem 3.53

3.54. Given the following: losses from point A to point B in the figure are $0.5V_a^2/2g$; the cross section at any point is circular; the fluid is water at 68°F; the discharge is 20 ft^3/s; the pressure at A is 500 psf. What is the pressure at B? What force (in terms of x and y components) is necessary to hold the bend in place? Neglect the weight of the tee in computing the force.

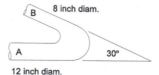

Problem 3.54

3.55. The pressure at A in the figure is 1.0 psig. Water at 68°F flows at a rate of 2.0 ft³/s, and the manometer fluid has a specific gravity of 1.6. What is the pressure at B? Is the flow from A to B, or B to A? What force components must be applied to hold this bend in place? Neglect elevation changes in computing p_B.

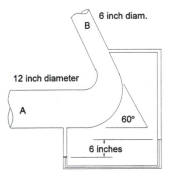

Problem 3.55

3.56. The reducing bend in the figure has a pressure at the inlet of 2×10^5 Pa and a flow rate of 0.5 m³/s. The fluid is oil with a specific gravity of 0.8. Assuming no losses, what is the velocity at the inlet? What are the velocity and pressure at the outlet? What are the horizontal and vertical force components exerted by the fluid on the bend? Neglect the weight of the oil.

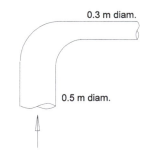

Problem 3.56

3.57. Water (SG = 1) flows through a tee in a pipeline shown in the figure. The flow rate before the tee (point 1) is 12 ft³/s, and 30% of this goes into branch 3. The pressure before the tee is 80 lb/ft². Assuming there are no losses in the tee, find the pressures at 2 and 3. What are the horizontal and vertical components of the force needed to hold the tee in place? The area inside each leg of the tee is 1.2 ft².

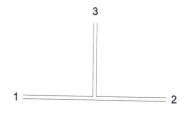

Problem 3.57

Unsteady flows

3.58. A jet plane traveling at 300 mph takes air into its engine at the rate of 3 slugs/s and discharges it at a speed of 1,800 ft/s relative to the plane. Neglecting the mass of any fuel added to the discharged air, what is the thrust if the air is discharged with no change in air direction? What would be the force exerted on the plane if this air were diverted vertically downward (relative to the plane) by a 90° nozzle at the same relative speed?

3.59. A small boat is propelled by taking in water at inlets in the front of the hull, passing it through a pump, and expelling the water through an 8-in-diameter pipe at the rear of the boat. If the boat travels at 10 mph and the drag force is 300 lb, what must be the exiting velocity of the jet as measured with respect to the boat?

Moving vanes

3.60. A water jet has a speed of 25 m/s and an area of 0.01 m^2. It strikes the vane shown in the figure, which moves to the left at a speed of 10 m/s. The flow splits so that 40% goes to the upper part of the vane, and 60% to the lower part of the vane. Show a sketch giving the known speeds and directions as seen by a stationary observer. Show a sketch giving the known speeds and directions as seen by an observer moving with the vane. What are the x and y components of the force exerted on the vane? Indicate whether the force component acts to the right or left, up or down. What is the absolute velocity of the jet leaving the vane at its upper edge? What is the absolute velocity of the jet leaving the vane at its lower edge?

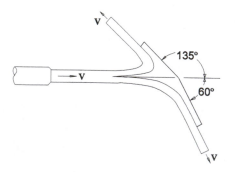

Problem 3.60

3.61. For the vane shown in the figure, compute the x and y force components on the vane. The nozzle discharges 20 ft^3/s of water at 68°F with a velocity of 50 ft/s. Do your computations for the following conditions: (a) the vane is stationary; (b) the

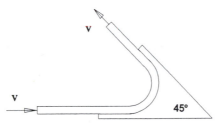

Problem 3.61

      vane is moving to the left with a velocity of 25 ft/s. For this case, at what angle does the water leave the vane in an absolute reference frame?

3.62. A vane moves to the left with a velocity of 50 ft/s. A nozzle discharges 3 ft³/s of water with a velocity of 100 ft/s. The vane angle is 150°. (a) What is the force on the vane? Give your answer in terms of horizontal and vertical components. (b) What is the absolute speed (magnitude of the velocity) and angle at which the fluid leaves the vane? (c) If instead of a single vane, there was a series of many vanes, what would be your answer to part (a)?

3.63. A nozzle discharges 120 lb/s of 68°F water horizontally at a speed of 20 ft/s. This strikes a vertical plate moving away from the nozzle at a speed of 5 ft/s. Draw a diagram of the velocities in a frame of reference in which the flow is steady. What is the force on the vane and the absolute velocity of the fluid leaving the vane?

3.64. Compute the force components on the vane shown in the figure. The fluid is water at 68°F, and the discharge is 10 ft³/s at a velocity of 50 ft/s. Give your answer for the following three cases: (a) the vane is stationary; (b) the vane moves to the left with a velocity of 40 ft/s; (c) the vane moves to the right with a velocity of 40 ft/s.

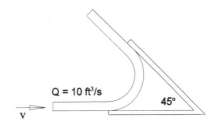

Problem 3.64

3.65. At what speed (give x, y components) does the water (SG = 1) leave the moving vane in the figure? What is the force on the vane?

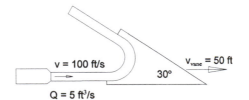

Problem 3.65

3.66. A vane is moving to the left at a speed of 10 ft/s into a jet of water having a velocity of 50 ft/s and an area of 0.1 ft^2. For the geometry shown in the figure, make a sketch showing the velocities an observer moving with the vane would see. Provide numbers and angles. What is the force on the vane (x and y components)? At what angle does the absolute velocity leave the vane?

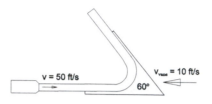

Problem 3.66

3.67. What thrust force is exerted on the car by the jet of water (SG = 1) in the figure as the car passes under the jet? The car moves at a speed of 10 ft/s away from the jet.

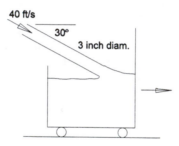

Problem 3.67

3.68. The nozzle shown in the figure emits 10 ft^3/s of water (SG = 1) at a velocity of 50 ft/s. (a) If the vane is stationary and the water leaves the vane at an angle of 30°, what is the force which the fluid exerts on the vane? (b) If the vane is moving away from the nozzle at a speed of 20 fps, and the water leaves the vane at an angle of 30° when viewed by a viewer moving with the vane, what is the force exerted by the fluid on the vane, and what is the angle of the fluid leaving the vane as seen by a stationary observer?

Problem 3.68

3.69. The power produced by a jet of water striking a moveable vane is equal to the force exerted on the vane times the velocity of the vane. What should be the ratio of vane velocity to jet velocity to produce maximum power for a given Q and vane angle?

3.70. A single vane is moving with a speed of 50 ft/s away from a nozzle discharging water (SG = 1) at a rate of 3 ft^3/s. The nozzle splits the flow, dividing it so that 70% of the flow is going up and 30% is going down, as shown in the figure. What are the components of the net force of the fluid on the vane? What is the absolute velocity of the fluid as it leaves the vane at the top? Give your answer in component form. By what factor are the force components changed if there is a series of moving vanes instead of just one?

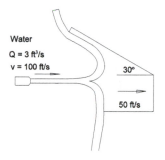

Water
Q = 3 ft³/s
v = 100 ft/s
30°
50 ft/s

Problem 3.70

Draining

3.71. A tank of volume 1 ft^3 is evacuated to essentially zero pressure. It is then connected by means of a 2-in-diameter tube to a source of air with fixed density and pressure ρ_{source} and p_{source}. If the velocity at which the air enters the tank is proportional to the square root of the difference in pressure (i.e., $v = C\sqrt{p_{source} - p}$, where p is the pressure in the tank), and the density is proportional to the pressure ($\rho = Kp$), find how long it takes to bring the pressure in the tank up to the level p_{source} in terms of the parameters ρ_{source}, p_{source}, K, and C.

3.72. The water surface in a canal lock 40 ft wide by 200 ft long stands 10 ft above that of the river below it. The lock level is to be reduced to river level by opening submerged gates having discharge coefficients of 0.40. ($Q = C_Q A\sqrt{2gh}$.) How much gate area must be provided if the water levels are to be equalized in 5 min?

3.73. A tank of water (SG = 1) 5 ft in diameter and 6 ft in height discharges through a 1-in-diameter hole in the bottom into the atmosphere. How long will it take to empty the tank?

Rockets

3.74. When mounted in a test stand, a rocket engine is found to develop a thrust force of 10^6 lb. The exhaust gases were found to exit the engine at a velocity of 11,000 ft/s. What is the mass rate of flow out of the rocket engine? If this engine is mounted in a rocket that has a weight of 145,000 lb (this includes rocket, engine, and initial fuel), what is the acceleration at launch?

3.75. Find the velocity of the rocket in the previous question after 30,000 lb of fuel has burned. The fuel mass was consumed at a constant rate.

Angular momentum

3.76. A nonsymmetrical lawn sprinkler is designed with one arm 15 cm long and the other arm 5 cm long. The internal cross-sectional area of each arm is 0.005 m². The bearing is frictionless. A flow of 0.1 m³ is supplied, and it divides equally between the two arms. The velocity of the fluid leaving each jet is 15 m/s measured relative to the arm. At what angular velocity does the sprinkler turn if the system is frictionless? Does the sprinkler rotate in a clockwise or counterclockwise direction as viewed from above? The jets are in the plane of the arms and perpendicular to the arms, $a = 1.5$ cm.

3.77. Find the torque T required to prevent the lawn sprinkler in the figure from rotating if the total discharge is 18 gpm. Each nozzle is inclined 10° above the horizontal and has an opening of 1/4 in diameter.

Problem 3.77

Euler equation

3.78. Along a streamline in a steady flow, the speed varies according to $V = 10 - 3s + 2s^2$ ft/s, where s is the distance along the streamline. If the mass density is 1.94 slugs/s, find the acceleration and pressure gradient along the streamline.

Pitot tube

3.79. A pitot tube immersed in water gives the manometer reading shown in the figure. What is the velocity at which the fluid is flowing?

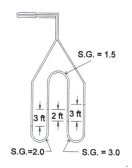

Problem 3.79

3.80. The impact tube shown in the figure reads stagnation pressure. If it is known that the pressure in the horizontal pipe is 1 psi when the impact tube is absent, what is the velocity in the pipe, assuming no losses? What is the pressure read by the impact tube? Neglect any interference effect that the vertical leg of the impact tube may have on the flow.

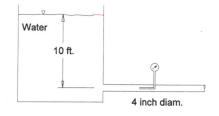

Problem 3.80

3.81. The pitot tube shown in the figure is used to measure the velocity of a high-speed air jet from a nozzle. Given that the manometer fluid is kerosene with a specific gravity of 0.81, the ratio of specific heats for air is $k = 1.4$, and the ambient temperature and static pressure of the air are 50°F and 12 psia ($\gamma = 0.078$ lb/ft^3), what is the velocity of the stream when the deflection of the manometer fluid is 30 in? In which leg of the manometer is the fluid highest?

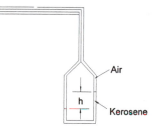

Problem 3.81

3.82. Given that for the figure $V = 15$ ft/s, and that the fluid is air, what is the velocity at B? What is the pressure at B? What is the reading of gage C, assuming that it reads stagnation pressure? What is the force on the elbow?

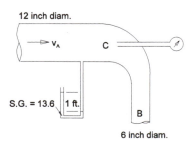

Problem 3.82

3.83. A pitot tube is used to measure velocities in a 30-cm-diameter duct. The duct area is first divided into five rings of radii 3, 6, 9, 12, and 15 cm, and then measurements are made at distances of 0, 4.5, 7.5, 10.5, and 13.5 cm from the duct centerline. The discharge is found according to $Q = \Sigma A_i V_i$ summed over the five regions. Find Q for the following data:

r (cm)	0	4.5	7.5	10.5	13.5
v (m/s)	7.1	6.4	5.2	3.6	1.3

chapter 4

Differential Analysis

Chapter Overview and Goals

In the previous chapter, we considered global forms of the basic equations governing fluid mechanics. In many cases, these are sufficient for obtaining information of engineering interest. However, there are important occasions where a full three-dimensional local description of the flow is necessary to the engineer. A local description means that it is presented in the form of differential equations valid at an arbitrary point in the fluid. In this chapter, we derive the basic equations governing flows whose pressure, velocity, and density can vary in three space directions, and also possibly in time.

After the derivation of the continuity equation, the use of stream functions to describe the velocity field is introduced. This is followed by a derivation of the Euler equations in three dimensions. The concepts of vorticity and circulation are then presented. The use of velocity potentials and Bernoulli's equation for irrotational flows is developed. The basic velocity potentials that allow flow fields to be constructed for all irrotational flows are given, along with some illustrative examples of their use.

Our study of three-dimensional fluid mechanics is completed with the inclusion of viscous effects. We consider the general description of the rates of deformation, the quantity that relates flow kinematic to stress. Three-dimensional stress components are then defined, and Newton's law of viscosity is used to relate stress to rate of deformation. Combining these with the momentum equation, we have the Navier-Stokes equations. Its uses are discussed along with some simple applications.

1. The Local Continuity Equation

In order to derive our local equations, which hold true at any point in our fluid, we locate a general point in the fluid and draw an infinitesimal box around it. This box is our control volume. For derivation purposes, it is easiest to work in Cartesian coordinates; the box will then have dimensions dx by dy by dz, as shown in Figure 4.1. Positive sign convention for the velocity components are given.

We represent the density and velocity components at the center of the box by ρ, v_x, v_y, and v_z, respectfully. Since these quantities vary throughout the fluid, they will

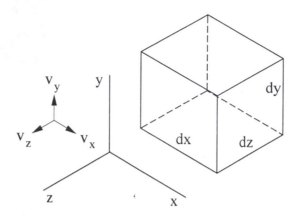

Figure 4.1. Three-dimensional differential control volume showing positive directions of velocity.

vary also throughout our box. We represent this variation by considering the quantities at any point as the zero- and first-order terms in a Taylor series. Thus density for instance would vary within the box according to

$$\rho(x, y, z, t) = \rho(x_0, y_0, z_0, t) + \frac{\partial \rho}{\partial x}(x - x_0) + \frac{\partial \rho}{\partial y}(y - y_0) + \frac{\partial \rho}{\partial z}(z - z_0)$$

+ higher order terms.

Here the subscript 0 is the center of the box, and (x, y, z) is any point in the box. Derivatives are evaluated at the box center.

In order to obtain our local continuity equation, we will sum mass flow rates and find that the lowest order terms are all multiplied by the box volume. Dividing by the volume and then letting the volume shrink to zero eliminates all higher order terms, so we need not consider them.

A figure showing the variation of all components needed in three dimensions to our analysis soon becomes very complicated. In Figure 4.2 we show the simplified case for two dimensions. Table 4.1 summarizes the mass rates of flow in three dimensions for the six faces of our box. Recall from the previous chapter that the mass rate of flow is given by

$$\dot{m} = \rho v_{\text{normal}} A,$$

where A is the surface area and v_{normal} is the velocity component normal to that area. Thus, for example, an area that is facing in the x direction would have a normal velocity component of v_x.

Using the sign convention for velocities as being positive in the direction of positive coordinates, we see that all flows through areas having normals facing in the plus coordinate direction are exiting the box control volume, and all those from areas with

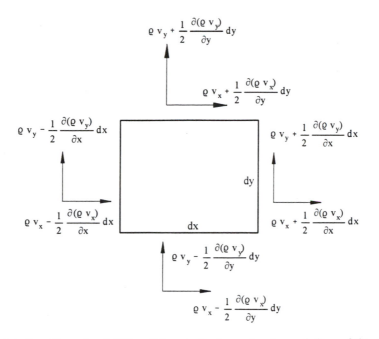

Figure 4.2. Two-dimensional differential control volume showing variations of the velocity components on the faces of the control volume.

Table 4.1.   Mass flow components and summation in three dimensions

Face normal pointing in	Mass rate of flow in x direction	Mass rate of flow in y direction	Mass rate of flow in z direction
+x direction	$\left[\rho v_x + \dfrac{\partial(\rho v_x)}{\partial x}\dfrac{dx}{2}\right] dy\, dz$		
+y direction		$\left[\rho v_y + \dfrac{\partial(\rho v_y)}{\partial y}\dfrac{dy}{2}\right] dx\, dz$	
+z direction			$\left[\rho v_z + \dfrac{\partial(\rho v_z)}{\partial z}\dfrac{dz}{2}\right] dx\, dy$
−x direction	$\left[\rho v_x - \dfrac{\partial(\rho v_x)}{\partial x}\dfrac{dx}{2}\right] dy\, dz$		
−y direction		$\left[\rho v_y - \dfrac{\partial(\rho v_y)}{\partial y}\dfrac{dy}{2}\right] dx\, dz$	
−z direction			$\left[\rho v_z - \dfrac{\partial(\rho v_z)}{\partial z}\dfrac{dz}{2}\right] dx\, dy$
Net sum in all directions	$\dfrac{\partial(\rho v_x)}{\partial x} dx\, dy\, dz + \dfrac{\partial(\rho v_y)}{\partial y} dx\, dy\, dz + \dfrac{\partial(\rho v_z)}{\partial z} dx\, dy\, dz$		

normals facing in the negative coordinate direction are entering the box. An algebraic summation of flows over the six faces of the box gives a mass rate of

$$\frac{\partial(\rho v_x)}{\partial x} \, dx \, dy \, dz + \frac{\partial(\rho v_y)}{\partial y} \, dx \, dy \, dz + \frac{\partial(\rho v_z)}{\partial z} \, dx \, dy \, dz$$

exiting our control volume.

To ensure that mass is conserved, this must be balanced by a corresponding rate of decrease of mass inside our control volume. Consequently, to have mass conserved we must have

$$\frac{\partial(\rho v_x)}{\partial x} \, dx \, dy \, dz + \frac{\partial(\rho v_y)}{\partial y} \, dx \, dy \, dz + \frac{\partial(\rho v_z)}{\partial z} \, dx \, dy \, dz = -\frac{\partial \rho}{\partial t} \, dx \, dy \, dz,$$

where the minus sign represents the fact that the mass is decreasing, and the volume dx by dy by dz is constant and hence can be brought out of the time derivative.

Since the volume multiplies every term in our equation, it can be divided out. Letting the volume shrink to zero in order to eliminate the higher order terms, we are left with

$$\frac{\partial(\rho v_x)}{\partial x} + \frac{\partial(\rho v_y)}{\partial y} + \frac{\partial(\rho v_z)}{\partial z} = -\frac{\partial \rho}{\partial t}$$

or, after putting all terms on the same side of the equation,

$$\frac{\partial \rho}{\partial t} + \frac{\partial(\rho v_x)}{\partial x} + \frac{\partial(\rho v_y)}{\partial y} + \frac{\partial(\rho v_z)}{\partial z} = 0. \tag{4.1.1}$$

This is the local (i.e., differential equation) form of the continuity equation.

To better understand the meaning of this equation, we expand the derivatives using the product rule of calculus to obtain

$$\frac{\partial \rho}{\partial t} + v_x \frac{\partial \rho}{\partial x} + \rho \frac{\partial v_x}{\partial x} + v_y \frac{\partial \rho}{\partial y} + \rho \frac{\partial v_y}{\partial y} + v_z \frac{\partial \rho}{\partial z} + \rho \frac{\partial v_z}{\partial z} = 0,$$

which after a little rearranging becomes

$$\frac{\partial \rho}{\partial t} + v_x \frac{\partial \rho}{\partial x} + v_y \frac{\partial \rho}{\partial y} + v_z \frac{\partial \rho}{\partial z} + \rho \frac{\partial v_x}{\partial x} + \rho \frac{\partial v_y}{\partial y} + \rho \frac{\partial v_z}{\partial z} = 0.$$

We note that the first four terms are the material derivative of the density, $D\rho/Dt$.

Therefore we can rewrite our continuity equation as

$$\frac{D\rho}{Dt} + \rho \left(\frac{\partial v_x}{\partial x} + \frac{\partial v_y}{\partial y} + \frac{\partial v_z}{\partial z} \right) = 0. \tag{4.1.2}$$

The $D\rho/Dt$ term represents the time rate of change of mass density within the control volume. The terms $\partial v_x/\partial x + \partial v_y/\partial y + \partial v_z/\partial z$ represent the time rate of change of fluid volume per unit fluid volume. Thus we could also think of the continuity equation as telling us that as we follow a given mass of fluid,

$$\frac{D(\text{mass})}{Dt} = \frac{D\,(\rho \text{ times volume})}{Dt} = 0.$$

Those readers familiar with vector calculus will notice that the volume rate terms are the gradient of the velocity, and can be written as

$$\frac{\partial v_x}{\partial x} + \frac{\partial v_y}{\partial y} + \frac{\partial v_z}{\partial z} = \text{grad } \mathbf{v} = \nabla \cdot \mathbf{v}.$$

Recall that for incompressible flows $D\rho/Dt$ vanishes. Consequently, for incompressible flows the continuity equation reduces to

$$\frac{\partial v_x}{\partial x} + \frac{\partial v_y}{\partial y} + \frac{\partial v_z}{\partial z} = \text{grad } \mathbf{v} = \nabla \cdot \mathbf{v} = 0. \tag{4.1.3}$$

As a final reminder, we note that the continuity equation is a single scalar equation.

2. The Stream Function

The continuity equation imposes a restriction on the velocity components. It can be thought of as relating one of the velocity components to the other two (and of course mass density). It is not possible to directly integrate the continuity equation for this velocity component. Rather, this is handled most easily through the use of an intermediary scalar function called a ***stream function***. The stream function is closely related to our earlier concept of a streamline.

We will here study stream functions principally in connection with incompressible flows; that is, flows where the density of individual fluid particles does not change as the particle moves in the flow. In that case, we saw above that the continuity equation reduces to

$$\partial v_x/\partial x + \partial v_y/\partial y + \partial v_z/\partial z = 0. \tag{4.1.3}$$

Because the equations relating stream functions to velocity differ in two and three dimensions, the two cases will be considered separately.

a. Two-dimensional flows—Lagrange's stream function

For two-dimensional flows, (4.1.3) reduces to

$$\frac{\partial v_x}{\partial x} + \frac{\partial v_y}{\partial y} = 0, \tag{4.2.1}$$

indicating that one of the velocity components can be expressed in terms of the other. If we were to do this directly, we would find relations such as

$$v_y = -\int \frac{\partial v_x}{dx} \, dy.$$

This cannot be integrated directly in any straightforward form. Instead, it is more convenient to introduce a scalar function ψ, called *Lagrange's stream function*, that allows the integration to be carried out explicitly. We let

$$v_x = \partial \psi / \partial y. \tag{4.2.2}$$

Then (4.2.1) becomes

$$\partial v_y / \partial y + \partial^2 \psi / \partial x \, \partial y = 0,$$

which can be integrated with respect to y to give

$$v_y = -\partial \psi / \partial x. \tag{4.2.3}$$

Therefore expressing the two velocity components in terms of (4.2.2) and (4.2.3) guarantees that continuity is satisfied for an incompressible flow.

Expressions (4.2.2) and (4.2.3) are useful when Cartesian coordinates are being used. Sometimes, however, it is more convenient to use cylindrical polar coordinates, where the continuity equation for incompressible flows is of the form

$$\nabla \cdot \mathbf{v} = \frac{\partial v_r}{\partial r} + \frac{v_r}{r} + \frac{1}{r} \frac{\partial v_\theta}{\partial \theta} + \frac{\partial v_z}{\partial z} = 0.$$

For two-dimensional flows, the last term $\partial v_z / \partial z$ vanishes. Therefore when cylindrical polar coordinates are used,[1] instead of (4.2.2) and (4.2.3), the suitable expressions for the velocity components are

$$v_r = \frac{\partial \psi}{r \partial \theta} \tag{4.2.4}$$

and

$$v_\theta = -\frac{\partial \psi}{\partial r}. \tag{4.2.5}$$

Note from either (4.2.4) or (4.5.2) that the dimension of ψ is length squared per unit time.

A physical interpretation of the nature of ψ can be found by looking at the discharge through a curve C as in Figure 4.3. The discharge through the curve C per unit distance into the paper is

---

[1]Cylindrical polar coordinates are defined so that $x = r \cos \theta$, $y = r \sin \theta$, $z = z$. Inversely, $r = (x^2 + y^2)^{1/2}$, $\theta = \arctan y/x$.

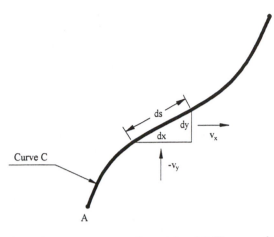

Figure 4.3. Curve drawn between two streamlines at A and B. The normal velocity components are showing for an infinitesimal length.

$$Q = \int_A^B (v_x\, dy - v_y\, dx). \tag{4.2.6}$$

When the velocity components are replaced by their expressions in terms of ψ, (4.2.6) becomes

$$Q = \int_A^B \left(\frac{\partial \psi}{\partial y}\, dy + \frac{\partial \psi}{\partial x}\, dx \right) = \int_A^B d\psi = \psi_B - \psi_A. \tag{4.2.7}$$

Thus the discharge through the curve C is equal to the difference of the value of ψ at the endpoints of C.

Since the equation of a streamline was given in Chapter 3 as

$$\frac{dx}{v_x} = \frac{dy}{v_y},$$

use of (4.2.2) and (4.2.3) reduces this to

$$\frac{dx}{\dfrac{\partial \psi}{\partial y}} = \frac{dy}{-\dfrac{\partial \psi}{\partial x}},$$

or, upon rearrangement,

$$0 = \frac{\partial \psi}{\partial x}\, dx + \frac{\partial \psi}{\partial y}\, dy = d\psi. \tag{4.2.8}$$

Equation (4.2.8) states that $d\psi$ vanishes along a streamline, implying that on a streamline ψ is constant. This is the motivation for the name stream function for ψ.

Example 4.2.1. Lagrange stream function

Find the stream function associated with the two-dimensional incompressible flow

$$v_r = U(1 - a^2/r^2) \cos \theta,$$

$$v_\theta = -U(1 + a^2/r^2) \sin \theta.$$

Given: The velocity components for a two-dimensional incompressible flow.
Solution: Since $v_r = \partial\psi/r\partial\theta$, by integration of v_r with respect to θ we find

$$\psi = \int U(1 - a^2/r^2) \cos \theta \, dr = U(r - a^2/r) \sin \theta + f(r).$$

The "constant of integration" f here possibly depends on r since we integrated a partial derivative that had been taken with respect to the angle theta. Differentiating this ψ we found with respect to r, we have

$$\partial\psi/\partial r = -v_\theta = U(1 + a^2/r^2) \sin \theta + df(r)/dr.$$

Comparing the two expressions for v_θ, we see that df/dr must vanish; thus f must be a constant. We can set this constant to any value convenient for us without affecting the velocity components. Here for simplicity we set it to zero. This gives

$$\psi = U(r - a^2/r) \sin \theta.$$

Example 4.2.2. Path lines

Find the path lines for the flow of Example 4.2.1, and equations for the position of a fluid particle along the path line as a function of time.

Given: The velocity components and stream function for a two-dimensional incompressible steady flow.
Solution: For steady flows, path lines and streamlines coincide. Therefore on a path line, ψ is constant. Using the stream function from the previous example we have

$$\psi = U(r - a^2/r) \sin \theta = \text{constant} = B \text{ (say)}.$$

The equations for a path line in cylindrical polar coordinates are

$$dt = dr/v_r = r \, d\theta/v_\theta.$$

Since on the path line $\sin \theta = B/(r - a^2/r)$,

$$v_r = U(1 - a^2/r^2)\sqrt{1 - \sin^2 \theta} = \frac{U}{r^2}\sqrt{(r^2 - a^2) - (Br)^2}.$$

If for convenience we introduce new constants

$$F^2 = a^2 + B^2/2 + B\sqrt{a^2 + B^2/4}, \qquad G^2 = a^2 + B^2/2 - B\sqrt{a^2 + B^2/4}, \qquad k = F/G,$$

and a new variable $s = r/F$. Then, since $a^4 = (FG)^2$ and $2a^2 + B^2 = F^2 + G^2$,

$$v_r = (U/F^2s^2)\sqrt{F^4s^4 - (F^2 + G^2)F^2s^2 + F^2G^2} = (UG/Fs^2)\sqrt{(1-s)(1-k^2s^2)}.$$

Consequently we have

$$dt = \frac{F^2s^2}{UG\sqrt{(1-s)(1-k^2s^2)}}\, ds.$$

Upon integration, this gives

$$t - t_0 = (F^2/UG)\int_{r_0/F}^{r/F} s^2\, ds/\sqrt{(1-s)(1-k^2s^2)},$$

where r_0 is the value of r at t_0.

The integral is related to what are called "elliptic integrals." Its values can be found tabulated in many handbooks, or by numerical integration. Once r is found as a function of t on a pathline (albeit in an inverse manner, since we have t as a function of r), the angle is found from

$$\theta = \sin^{-1}[B/(r - a^2/r)].$$

From this, we see that the constant B can also be interpreted as

$$B = (r_0 - a^2/r_0)\sin\theta_0.$$

---

b. Three-dimensional flows

For three dimensions, (4.1.3) states that there is one relationship between the three velocity components; hence the velocity can be expressed in terms of two scalar functions. There are at least two ways of accomplishing this. The one that retains the interpretation of stream function as introduced in two dimensions is

$$\mathbf{v} = \nabla\psi \times \nabla\xi, \tag{4.2.9}$$

where ψ and ξ are each constant on stream surfaces. The intersection of these stream surfaces is the streamline. Equations (4.2.2) and (4.2.3) are in fact a special case of this result with $\xi = z$.

While (4.2.9) preserves the advantage of having the stream functions constant on a streamline, it has the disadvantage that $\mathbf{v}$ is a nonlinear function of ψ and ξ. Introducing further nonlinearities into an already highly nonlinear problem is not usually helpful!

Another possibility is to write

$$\mathbf{v} = \nabla \times \mathbf{A}, \tag{4.2.10}$$

which guarantees $\nabla \cdot \mathbf{v} = 0$ for any vector $\mathbf{A}$. For the two-dimensional case, $\mathbf{A} = (0, 0, \psi)$ corresponds to our Lagrange stream function. In three dimensions, since only two scalars are needed, one component of $\mathbf{A}$ can be arbitrarily set to zero. (Some thought must be used in doing this. Obviously in the two-dimensional case, difficulty would be encountered if one of the components of **A** we set to zero is the z component.) The form (4.2.10), while guaranteeing satisfaction of continuity, has not been much used, since the appropriate boundary conditions to be imposed on **A** can be awkward.

A particular three-dimensional case in which a stream function is useful is that of axisymmetric flow. Taking the z axis as the axis of symmetry, we can use either spherical polar coordinates,

$$R = \sqrt{x^2 + y^2 + z^2}, \qquad \beta = \cos^{-1}(z/R), \qquad \theta = \tan^{-1}(y/x) \tag{4.2.11}$$

(see Figure 4.4), or cylindrical polar coordinates with

$$r = \sqrt{x^2 + y^2}, \qquad \theta = \tan^{-1}(y/x), \qquad z = z \tag{4.2.12}$$

(see Figure 4.5). By axisymmetry we mean that the flow appears the same in any $\theta =$ constant plane, and the velocity component normal to that plane is zero. (There could in fact be a swirl velocity component v_θ without changing anything we have said. It would not be related to ψ.) Since any plane given by θ equal to a constant is therefore a stream surface, we can use (4.2.9) with $\xi = \theta$, giving

$$v_R = \frac{1}{R^2 \sin\beta} \frac{\partial\psi}{\partial\beta}, \qquad v_\beta = -\frac{1}{R\sin\beta} \frac{\partial\psi}{\partial R} \tag{4.2.13}$$

in spherical coordinates with $\psi = \psi(R, \beta)$, or

$$v_r = -\frac{1}{r} \frac{\partial\psi}{\partial z}, \qquad v_z = \frac{1}{r} \frac{\partial\psi}{\partial r} \tag{4.2.14}$$

in cylindrical coordinates with $\psi = \psi(r, z)$. This stream function in either coordinate system is called the ***Stokes stream function***. Note that the dimensions of the Stokes stream function is length cubed per unit time, differing from Lagrange's stream function by a length dimension.

The volumetric discharge through an annular region is given in terms of the Stokes stream function by

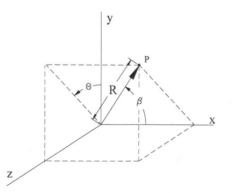

Figure 4.4. Spherical polar coordinates.

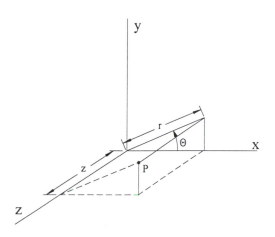

Figure 4.5. Cylindrical polar coordinates.

$$Q = \int_A^B 2\pi r(v_z\, dr - v_r\, dz) = \int_A^B 2\pi r\left[\frac{1}{r}\frac{\partial\psi}{\partial r}\, dr - \frac{(-1)}{r}\frac{\partial\psi}{\partial z}\, dz\right]$$

$$= 2\pi \int_A^B \left(\frac{\partial\psi}{\partial r}\, dr + \frac{\partial\psi}{\partial z}\, dz\right) = 2\pi \int_A^B d\psi = 2\pi(\psi_B - \psi_A). \qquad (4.2.15)$$

The principal differences between (4.2.7) and (4.2.15) are the factor 2π and the dimensions of Q and of ψ.

Example 4.2.3.   Stokes stream function

A flow field in cylindrical polar coordinates is given by

$$v_r = -1.5Ua^3rz/(r^2 + z^2)^{5/2}, \qquad v_\theta = 0.5Ua^3(r^2 - 2z^2)/(r^2 + z^2)^{5/2}.$$

Find the Stokes stream function for this flow.

Given: The velocity components for an axisymmetric incompressible flow.

Solution: Since $v_r = -\partial\psi/r\,\partial z$, integration of v_r with respect to z gives

$$\psi = \int \frac{-1.5Ua^3rz}{(r^2 + z^2)^{5/2}}\, r\, dz = -\frac{0.5Ua^3r^2}{(r^2 + z^2)^{3/2}} + f(r).$$

Differentiating this with respect to r, we have

$$\frac{\partial\psi}{\partial r} = rv_z = \frac{0.5Ua^3r(r^2 - 2z^2)}{(r^2 + z^2)^{5/2}} + \frac{df(r)}{dr}.$$

Comparing this with the previous expression for v_z, we see that $df(r)/dr = rU$, and so $f(r) = 0.5Ur^2$. Thus

$$\psi = -\frac{0.5Ua^3r^2}{(r^2+z^2)^{3/2}} + 0.5Ur^2 = 0.5Ur^2\left[1 - \frac{a^3}{(r^2+z^2)^{3/2}}\right].$$

The above derivations were all for incompressible flow, whether the flow be steady or unsteady. The derivations can easily be extended to steady compressible flow by recognizing that for these flows, since the continuity equation can be written as $\nabla \times (\rho\mathbf{v}) = 0$, the previous stream functions can be used simply by replacing $\mathbf{v}$ by $\rho\mathbf{v}$.

A further extension to unsteady compressible flows is possible by regarding time as a fourth dimension, and using the extended four-dimensional vector $\mathbf{V} = (\rho v_x, \rho v_y, \rho v_z, \rho)$ together with the augmented del operator $\nabla_{\mathbf{a}} = (\partial/\partial x, \partial/\partial y, \partial/\partial z, \partial/\partial t)$. The continuity equation (4.1.1) can then be written as

$$\nabla_{\mathbf{a}} \cdot \mathbf{V} = 0. \tag{4.2.16}$$

The previous results can be extended to this general case in a straightforward manner.

3. Equations Governing Inviscid Flows

We next apply Newton's momentum equation to our control volume. We first consider the case where the only external forces are due to pressure and gravity. In computing momentum transfer through surfaces and forces acting on surfaces, we expand the velocity, density, and pressure in Taylor series as we did in deriving the continuity equation (4.1.1).

Change in momentum will occur internally in our control volume and through the surface of our control volume. The internal momentum change will be simply the time derivative of the momentum in our volume, or to third-order terms:

$$\frac{\partial}{\partial t}(\rho\mathbf{v})\,dx\,dy\,dz.$$

The momentum transfer through the surface of our control volume are shown in Table 4.2, and for the pressure forces acting on the control volume in Table 4.3. The two-dimensional control volume showing pressure variations is given in Figure 4.6.

Summing both the internal and surface momentum changes and equating these to the pressure and gravity forces, we have the following results in each of the three directions:

$$x\text{ direction: } \left[\frac{\partial}{\partial t}(\rho v_x) + v_x\frac{\partial(\rho v_x)}{\partial x} + \rho v_x\frac{\partial v_x}{\partial x} + v_x\frac{\partial(\rho v_y)}{\partial y} + \rho v_y\frac{\partial v_x}{\partial y} + v_x\frac{\partial(\rho v_z)}{\partial z} + \rho v_z\frac{\partial v_x}{\partial z}\right]dx\,dy\,dz$$

$$= -\frac{\partial p}{\partial x}dx\,dy\,dz + \rho g_x\,dx\,dy\,dz, \tag{4.3.1a}$$

Table 4.2. Momentum flow components and summation in three dimensions

Face normal pointing in	Momentum rate of flow in x direction	Momentum rate of flow in y direction	Momentum rate of flow in z direction
+x direction	$\left[\rho v_x + \dfrac{\partial(\rho v_x)}{\partial x}\dfrac{dx}{2}\right]\left(v_x + \dfrac{\partial v_x}{\partial x}\dfrac{dx}{2}\right)dy\,dz$	$\left[\rho v_x + \dfrac{\partial(\rho v_x)}{\partial x}\dfrac{dx}{2}\right]\left(v_y + \dfrac{\partial v_y}{\partial x}\dfrac{dx}{2}\right)dy\,dz$	$\left[\rho v_x + \dfrac{\partial(\rho v_x)}{\partial x}\dfrac{dx}{2}\right]\left(v_z + \dfrac{\partial v_z}{\partial x}\dfrac{dx}{2}\right)dy\,dz$
+y direction	$\left[\rho v_y + \dfrac{\partial(\rho v_y)}{\partial y}\dfrac{dy}{2}\right]\left(v_x + \dfrac{\partial v_x}{\partial y}\dfrac{dy}{2}\right)dx\,dz$	$\left[\rho v_y + \dfrac{\partial(\rho v_y)}{\partial y}\dfrac{dy}{2}\right]\left(v_y + \dfrac{\partial v_y}{\partial y}\dfrac{dy}{2}\right)dx\,dz$	$\left[\rho v_y + \dfrac{\partial(\rho v_y)}{\partial y}\dfrac{dy}{2}\right]\left(v_z + \dfrac{\partial v_z}{\partial y}\dfrac{dy}{2}\right)dx\,dz$
+z direction	$\left[\rho v_z + \dfrac{\partial(\rho v_z)}{\partial z}\dfrac{dz}{2}\right]\left(v_x + \dfrac{\partial v_x}{\partial z}\dfrac{dz}{2}\right)dx\,dy$	$\left[\rho v_z + \dfrac{\partial(\rho v_z)}{\partial z}\dfrac{dz}{2}\right]\left(v_y + \dfrac{\partial v_y}{\partial z}\dfrac{dz}{2}\right)dx\,dy$	$\left[\rho v_z + \dfrac{\partial(\rho v_z)}{\partial z}\dfrac{dz}{2}\right]\left(v_z + \dfrac{\partial v_z}{\partial z}\dfrac{dz}{2}\right)dx\,dy$
−x direction	$\left[\rho v_x - \dfrac{\partial(\rho v_x)}{\partial x}\dfrac{dx}{2}\right]\left(v_x - \dfrac{\partial v_x}{\partial x}\dfrac{dx}{2}\right)dy\,dz$	$\left[\rho v_x - \dfrac{\partial(\rho v_x)}{\partial x}\dfrac{dx}{2}\right]\left(v_y - \dfrac{\partial v_y}{\partial x}\dfrac{dx}{2}\right)dy\,dz$	$\left[\rho v_x - \dfrac{\partial(\rho v_x)}{\partial x}\dfrac{dx}{2}\right]\left(v_z - \dfrac{\partial v_z}{\partial x}\dfrac{dx}{2}\right)dy\,dz$
−y direction	$\left[\rho v_y - \dfrac{\partial(\rho v_y)}{\partial y}\dfrac{dy}{2}\right]\left(v_x - \dfrac{\partial v_x}{\partial y}\dfrac{dy}{2}\right)dx\,dz$	$\left[\rho v_y - \dfrac{\partial(\rho v_y)}{\partial y}\dfrac{dy}{2}\right]\left(v_y - \dfrac{\partial v_y}{\partial y}\dfrac{dy}{2}\right)dx\,dz$	$\left[\rho v_y - \dfrac{\partial(\rho v_y)}{\partial y}\dfrac{dy}{2}\right]\left(v_z - \dfrac{\partial v_z}{\partial y}\dfrac{dy}{2}\right)dx\,dz$
−z direction	$\left[\rho v_z - \dfrac{\partial(\rho v_z)}{\partial z}\dfrac{dz}{2}\right]\left(v_x - \dfrac{\partial v_x}{\partial z}\dfrac{dz}{2}\right)dx\,dy$	$\left[\rho v_z - \dfrac{\partial(\rho v_z)}{\partial z}\dfrac{dz}{2}\right]\left(v_y - \dfrac{\partial v_y}{\partial z}\dfrac{dz}{2}\right)dx\,dy$	$\left[\rho v_z - \dfrac{\partial(\rho v_z)}{\partial z}\dfrac{dz}{2}\right]\left(v_z - \dfrac{\partial v_z}{\partial z}\dfrac{dz}{2}\right)dx\,dy$

(continued)

Table 4.2. (*Continued.*)

	Net sum in various directions to third-order terms
Momentum rate of flow in x direction	$\left\{\left[v_x\dfrac{\partial(\rho v_x)}{\partial x}\right]dx + \rho v_x\dfrac{\partial v_x}{\partial x}dx\right\}dy\,dz + \left\{\left[v_x\dfrac{\partial(\rho v_y)}{\partial y}\right]dy + \rho v_y\dfrac{\partial v_x}{\partial y}dy\right\}dx\,dz + \left\{\left[v_x\dfrac{\partial(\rho v_z)}{\partial z}\right]dz + \rho v_z\dfrac{\partial v_x}{\partial z}dz\right\}dx\,dy$
Momentum rate of flow in y direction	$\left\{\left[v_y\dfrac{\partial(\rho v_x)}{\partial x}\right]dx + \rho v_x\dfrac{\partial v_y}{\partial x}dx\right\}dy\,dz + \left\{\left[v_y\dfrac{\partial(\rho v_y)}{\partial y}\right]dy + \rho v_y\dfrac{\partial v_y}{\partial y}dy\right\}dx\,dz + \left\{\left[v_y\dfrac{\partial(\rho v_z)}{\partial z}\right]dz + \rho v_z\dfrac{\partial v_y}{\partial z}dz\right\}dx\,dy$
Momentum rate of flow in z direction	$\left\{\left[v_z\dfrac{\partial(\rho v_x)}{\partial x}\right]dx + \rho v_x\dfrac{\partial v_z}{\partial x}dx\right\}dy\,dz + \left\{\left[v_z\dfrac{\partial(\rho v_y)}{\partial y}\right]dy + \rho v_y\dfrac{\partial v_z}{\partial y}dy\right\}dx\,dz + \left\{\left[v_z\dfrac{\partial(\rho v_z)}{\partial z}\right]dz + \rho v_z\dfrac{\partial v_z}{\partial z}dz\right\}dx\,dy$

Table 4.3. Pressure force components and summation in three dimensions

Face normal pointing in	Pressure force in x direction	Pressure force in y direction	Pressure force in z direction
+x direction	$-\left(p+\dfrac{\partial p}{\partial x}\dfrac{dx}{2}\right)dy\,dz$		
+y direction		$-\left(p+\dfrac{\partial p}{\partial y}\dfrac{dy}{2}\right)dx\,dz$	
+z direction			$-\left(p+\dfrac{\partial p}{\partial z}\dfrac{dz}{2}\right)dx\,dy$
−x direction	$-\left(p-\dfrac{\partial p}{\partial x}\dfrac{dx}{2}\right)dy\,dz$		
−y direction		$-\left(p-\dfrac{\partial p}{\partial y}\dfrac{dy}{2}\right)dx\,dz$	
−z direction			$-\left(p-\dfrac{\partial p}{\partial z}\dfrac{dz}{2}\right)dx\,dy$
Net force sum in each direction	$-\dfrac{\partial p}{\partial x}dx\,dy\,dz$	$-\dfrac{\partial p}{\partial y}dx\,dy\,dz$	$-\dfrac{\partial p}{\partial z}dx\,dy\,dz$

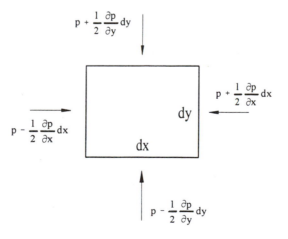

Figure 4.6. Two-dimensional differential control volume showing variations of the pressure on the faces of the control volume.

y direction:
$$\left[\frac{\partial}{\partial t}(\rho v_y) + v_y\frac{\partial(\rho v_x)}{\partial x} + \rho v_x\frac{\partial v_y}{\partial x} + v_y\frac{\partial(\rho v_y)}{\partial y} + \rho v_y\frac{\partial v_y}{\partial y} + v_y\frac{\partial(\rho v_z)}{\partial z} + \rho v_z\frac{\partial v_y}{\partial z}\right]dx\,dy\,dz$$

$$= -\frac{\partial p}{\partial y}dx\,dy\,dz + \rho g_y\,dx\,dy\,dz, \tag{4.3.1b}$$

z direction:
$$\left[\frac{\partial}{\partial t}(\rho v_z) + v_z\frac{\partial(\rho v_x)}{\partial x} + \rho v_x\frac{\partial v_z}{\partial x} + v_z\frac{\partial(\rho v_y)}{\partial y} + \rho v_y\frac{\partial v_z}{\partial y} + v_z\frac{\partial(\rho v_z)}{\partial z} + \rho v_z\frac{\partial v_z}{\partial z}\right]dx\,dy\,dz$$

$$= -\frac{\partial p}{\partial z}dx\,dy\,dz + \rho g_z\,dx\,dy\,dz. \tag{4.3.1c}$$

These have been summed to include up to third-order terms. All terms of order lower than three have canceled. Here the gravity force is represented by its components $\mathbf{g} = (g_x, g_y, g_z)$, where the magnitude of $\mathbf{g}$ is 32.17 or 9.80, depending on whether the units used are British or SI.

Since each term is multiplied by the volume, this can be divided out of each equation. The result is, after rearrangement of the momentum terms,

x direction:
$$v_x\frac{\partial\rho}{\partial t} + \rho\frac{\partial v_x}{\partial t} + v_x\frac{\partial(\rho v_x)}{\partial x} + \rho v_x\frac{\partial v_x}{\partial x} + v_x\frac{\partial(\rho v_y)}{\partial y} + \rho v_y\frac{\partial v_x}{\partial y} + v_x\frac{\partial(\rho v_z)}{\partial z} + \rho v_z\frac{\partial v_x}{\partial z} = -\frac{\partial p}{\partial x} + \rho g_x,$$
$$\tag{4.3.2a}$$

y direction:
$$v_y\frac{\partial\rho}{\partial t} + \rho\frac{\partial v_y}{\partial t} + v_y\frac{\partial(\rho v_x)}{\partial x} + \rho v_x\frac{\partial v_y}{\partial x} + v_y\frac{\partial(\rho v_y)}{\partial y} + \rho v_y\frac{\partial v_y}{\partial y} + v_y\frac{\partial(\rho v_z)}{\partial z} + \rho v_z\frac{\partial v_y}{\partial z} = -\frac{\partial p}{\partial y} + \rho g_y,$$
$$\tag{4.3.2b}$$

z direction:
$$v_z\frac{\partial\rho}{\partial t} + \rho\frac{\partial v_z}{\partial t} + v_z\frac{\partial(\rho v_x)}{\partial z} + \rho v_x\frac{\partial v_z}{\partial x} + v_z\frac{\partial(\rho v_y)}{\partial y} + \rho v_y\frac{\partial v_z}{\partial y} + v_z\frac{\partial(\rho v_z)}{\partial z} + \rho v_z\frac{\partial v_z}{\partial z} = -\frac{\partial p}{\partial z} = \rho g_z.$$
$$\tag{4.3.2c}$$

The first, third, fifth, and seventh terms on the left-hand side of each of these three equations sum to zero by virtue of the continuity equation (4.1.1). Looking at the remaining terms on the left side, we see that they are the mass density times the components of the acceleration. Hence we can rewrite these equations as

$$\rho\frac{Dv_x}{Dt} = -\frac{\partial p}{\partial x} + \rho g_x, \tag{4.3.3a}$$

$$\rho\frac{Dv_y}{Dt} = -\frac{\partial p}{\partial y} + \rho g_y, \tag{4.3.3b}$$

$$\rho \frac{Dv_z}{Dt} = -\frac{\partial p}{\partial z} + \rho g_z, \tag{4.3.3c}$$

or, in vector notation,

$$\rho \frac{D\mathbf{v}}{Dt} = -\mathbf{\nabla}p + \rho\mathbf{g}. \tag{4.3.4}$$

This form of the momentum equation is called the Euler equation. It is the three-dimensional Cartesian coordinate form of what we developed in Section 12 of Chapter 3 along a streamline. It is to be solved, together with the continuity equation, for the velocity vector and the pressure. If mass density is not constant, an additional relation is necessary.

These equations are to be solved along with the boundary condition that the relative velocity component normal to a solid wall be zero.

While the Euler equations do not include any effects due to viscosity, they still have many important practical applications, including flows around bodies, computation of lift forces and pressure drag forces, and many cases of compressible flows.

4. Vorticity and Circulation

Any motion of a small region of a fluid can be thought of as a combination of translation, rotation, and deformation. Translation is described by velocity of a point. Deformation is described in Section 6. Here we consider the rotation of a fluid element.

In rigid-body mechanics, the concept of angular rotation is an extremely important one, and rather intuitive. Along with translational velocity, it is one of the basic descriptors of the motion. In fluid mechanics we can introduce a similar concept in the following manner.

Consider the two-dimensional picture shown in Figure 4.7. At time t we have selected three neighboring points ABC, lined up (for our convenience) so that they make an angle of 90°. At a later time $t + dt$, these points will have moved to A', B', and C', as shown. The fluid that initially was at point A, a distance dx to the right of B, will have velocity components that, using Taylor series expansions to the first order, are given by $[v_x + (\partial v_x/\partial x)\,dx]$ and $[v_y + (\partial v_y/\partial x)\,dx]$. The Taylor expansion distance is dx, since A is a distance dx to the right of B. Thus the fluid initially at point A will move a distance $[v_x + (\partial v_x/\partial x)\,dx]\,dt$ to the right of A, and a distance $[v_y + (\partial v_y/\partial x)dx]\,dt$ above A. The fluid initially occupying point B has moved to point B', which is $v_x\,dt$ to the right of B and $v_y\,dt$ above B.

The line $A'B'$ is not horizontal as was line AB, but as seen from the figure is rotated an amount $d\theta_1$, where

$$d\theta_1 \cong \tan^{-1} d\theta_1 \cong \frac{(\partial v_y/\partial x)\,dx\,dt}{dx[1 + (\partial v_x/\partial x)\,dt]} \cong \frac{\partial v_y}{\partial x}\,dt. \tag{4.4.1}$$

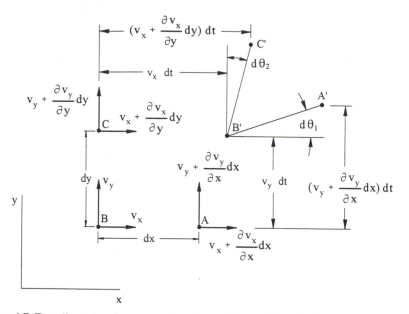

Figure 4.7. Two-dimensional geometry showing variations of the velocity components used in determining vorticity.

The final form is obtained after canceling the dx and neglecting the higher order terms in the denominator. The rate of rotation of the line connecting the fluid particles initially at points A and B is found by dividing (4.4.1) by dt, giving

$$\dot{\theta}_1 = d\theta_1/dt = \partial v_y/\partial x. \tag{4.4.2}$$

A similar argument for the fluid initially at point C shows that it has moved to C', which is to the right of C an amount $[v_x + (\partial v_x/\partial y)dy]\, dt$, and above C an amount $[v_y + (\partial v_y/\partial y)\, dy]\, dt$. (Since C is above B, in the Taylor series expansion we consider changes only in the y direction.) Hence a line connecting the fluid particles initially at B and C has rotated an amount

$$d\theta_2 \cong \tan^{-1} d\theta_2 = \frac{(\partial v_x/\partial y)\, dy\, dt}{dy[1 + (\partial v_y/\partial y)\, dt]} \cong (\partial v_x/\partial y)\, dt, \tag{4.4.3}$$

with a rate of rotation found by dividing both sides of (4.4.3) by dt to be

$$\dot{\theta}_2 = d\theta_2/dt = \partial v_x/\partial y. \tag{4.4.4}$$

From (4.4.2) and (4.4.4) we see that the angular velocity of a line depends on the orientation of that line. To develop our analog of angular velocity, we want a definition that is independent of direction at a point, and depends only on local conditions at the point itself. In considering the transformation of ABC into $A'B'C'$, we see that two things have happened: the angle has changed, or deformed, by an amount $d\theta_1 + d\theta_2$, and the bisector of the angle ABC has rotated an amount $0.5(d\theta_1 - d\theta_2)$. Considering only this rotation, we see that the rate of rotation of the bisector is

$0.5[\partial v_y/\partial x - \partial v_x/\partial y].$ (4.4.5)

Writing the curl of **v** in Cartesian coordinates, we have

$$\text{curl } \mathbf{v} = \nabla \times \mathbf{v} = (\partial v_z/\partial y - \partial v_y/\partial z)\mathbf{i} + (\partial v_x/\partial z - \partial v_z/\partial x)\mathbf{j} + (\partial v_y/\partial x - \partial v_x/\partial y)\mathbf{k}.$$

We see by comparison that (4.4.5) is one-half the z component of the curl of **v**.

If we were to consider similar neighboring points in the yz and xz planes, we would find that similar arguments would yield the x and y components of one-half the curl of the velocity. We therefore define the ***vorticity vector*** as being the curl of the velocity, that is,

$$\boldsymbol{\omega} = \text{curl } \mathbf{v} = \nabla \times \mathbf{v},$$ (4.4.6)

and note that the vorticity is twice the local angular rotation of the fluid. (Omitting the one-half from the definition saves some unimportant arithmetic, and does not obscure the physical interpretation of the concept.) Since vorticity is a vector, it is independent of the coordinate frame used. Our definition agrees with the usual "right-hand-rule" sign convention of angular rotation.

Flows with vorticity are said to be ***rotational flows***; flows without vorticity are said to be ***irrotational flows***. Note for later use that, from a well-known vector identity, it follows that

$$\text{div } \boldsymbol{\omega} = \nabla \cdot (\nabla \times \mathbf{v}) = 0.$$ (4.4.7)

Example 4.4.1. Rigid-body rotation
Find the vorticity associated with the velocity field $\mathbf{v} = (-y\Omega, x\Omega, 0)$.

Given: The velocity components for a two-dimensional incompressible steady flow in rigid-body rotation.
Solution: This is the case of rigid-body rotation studied in Chapter 2, with the speed being given by Ω times the distance of the point from the origin, and the streamlines being concentric circles. Taking the curl of the velocity, we have

$$\boldsymbol{\omega} = \nabla \times (-y\Omega\mathbf{i} + x\Omega\mathbf{j}) = 2\Omega\mathbf{k}.$$

Since the vorticity is a constant, this shows that the flow everywhere has the same vorticity.

Example 4.4.2. Vortex motion
Find the vorticity associated with the velocity field $\mathbf{v} = [-yG/(x^2 + y^2), xG/(x^2 + y^2), 0]$.

Given: The velocity field associated with a two-dimensional incompressible steady flow.
Solution: This is a velocity field again with streamlines that are concentric circles, but with the speed now being proportional to the reciprocal distance from the origin. This flow is called a ***line vortex***. Taking the curl of this velocity, we have

$$\boldsymbol{\omega} = \nabla \times [-yG\mathbf{i}/(x^2 + y^2) + xG\mathbf{j}/(x^2 + y^2)] = 0.$$

We see that the vorticity is zero except perhaps at the origin, where the derivatives become infinite, and a more careful examination must be made.

The above two examples help to point out what vorticity is, and also what it is not. In both flows the streamlines are concentric circles, and a fluid particle travels around the origin of the coordinate system. In the rigid-body rotation case, particles on two neighboring streamlines radius travel at slightly different velocities, the particle on the streamline with the greater radius having the greater velocity. A line connecting the two particles on different streamlines will travel around the origin as shown in Figure 4.8.

Considering a similar pair of neighboring streamlines in the line vortex case, the outer streamline has a slower velocity, and the outer particle will lag behind the inner one (Figure 4.9). A line connecting the two particles in the limit of zero distance will always point in the same direction, much as the needle of a compass would. It is this rotation that vorticity deals with, and not the rotation of a point about some arbitrary reference point such as the origin.

Since vorticity is a vector, many of the concepts we encountered with velocity and stream functions can be carried over. Thus we can define *vortex lines* as being lines instantaneously tangent to the vorticity vector, satisfying the equations

$$dx/\omega_x = dy/\omega_y = dz/\omega_z. \tag{4.4.8}$$

Vortex sheets are surfaces of vortex lines lying side by side. *Vortex tubes* are closed vortex sheets.

Analogous to the concept of volume flow through an area, $\iint_S \mathbf{v} \cdot d\mathbf{A}$, we can define the vorticity flow through an area, termed *circulation*, as

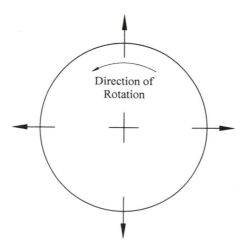

Figure 4.8. Circular path followed by a particle in rotational flow. Arrows show how a very small floating object would orient itself.

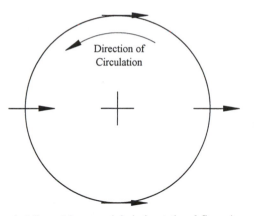

Figure 4.9. Circular path followed by a particle in irrotational flow. Arrows show how a very small floating object would orient itself.

$$\text{Circulation} = \Gamma \equiv \oint_C \mathbf{v} \cdot d\mathbf{s} = \iint_S \boldsymbol{\omega} \cdot d\mathbf{A}. \tag{4.4.9}$$

The relation between the line and surface integral forms following from Stokes theorem, where C is a closed path bounding the area S.

Example 4.4.3.   Circulation for a rigid rotation

Find the circulation through the square with corners at (+1, +1, 0), (−1, +1, 0), (−1, −1, 0), (+1, −1, 0) for rigid-body rotation flow shown in Figure 4.8.

Given: The velocity field associated with a two-dimensional incompressible rotational flow is $(-y\Omega, x\Omega, 0)$.

Solution: Starting first with the line integral form of the definition in (4.4.9), we see that

$$\Gamma = \oint_C \mathbf{v} \cdot d\mathbf{s} = \int_{+1}^{-1} v_x\big|_{y=1}\, dx + \int_{+1}^{-1} v_y\big|_{x=-1}\, dy + \int_{-1}^{+1} v_x\big|_{y=-1}\, dx + \int_{-1}^{+1} v_y\big|_{x=1}\, dy$$

$$= \int_{+1}^{-1} -\Omega\, dx + \int_{+1}^{-1} -\Omega dy + \int_{-1}^{+1} \Omega\, dx + \int_{-1}^{+1} \Omega\, dy = -\Omega(-2) - \Omega(-2) + \Omega(2) + \Omega(2) = 8\Omega.$$

Of course, we could have obtained this result much faster by using the area integral form of the definition (4.4.9),

$$\Gamma = \iint_S \boldsymbol{\omega} \cdot d\mathbf{A}.$$

Since the integrand is constant (2Ω) and the area is 4, the result for the circulation follows from a simple arithmetic multiplication.

Example 4.4.4. Circulation for a vortex motion

Using the square given in Example 4.4.3, find the circulation for the vortex flow given in Example 4.4.2.

Given: The velocity field for a two-dimensional incompressible flow that is irrotational everywhere except at the origin of the coordinate system is

$$\mathbf{v} = [-yG/(x^2 + y^2), \, xG/(x^2 + y^2), \, 0].$$

Solution: In this case, we cannot easily use the area form of the definition, since the vorticity is not defined at the origin. The line integral form of (4.3.9) gives us

$$\Gamma \equiv \oint_C \mathbf{v} \cdot ds = \int_{+1}^{-1} [-G/(x^2 + 1)] \, dx + \int_{+1}^{-1} [-G/(1 + y^2)] \, dy$$

$$+ \int_{-1}^{+1} [G/(x^2 + 1)] \, dx + \int_{-1}^{+1} [G/(1 + y^2)] \, dy$$

$$= -G(-\pi/4 - \pi/4) - G(-\pi/4 - \pi/4) + G(\pi/4 + \pi) + G(\pi/4 + \pi/4) = 2\pi G.$$

If we take any path that does not include the origin, we could in fact use the area form of the definition, and we would conclude that the circulation about that path was zero. Any path containing the origin would have circulation $2\pi G$. Therefore we say that the vorticity at the origin is infinite for the line vortex, being infinite in such a manner that the infinite vorticity times the zero area gives a finite, nonzero, value for the circulation.

Differential equations governing the change of vorticity can be formed from the Euler equations. When we divide (4.3.1) by the mass density and then take the curl, the result after some manipulation and use of the continuity equation is

$$\frac{D\omega}{Dt} = (\omega \cdot \nabla)\mathbf{v} - \left(\nabla \frac{1}{\rho}\right) \times \nabla \mathbf{p}. \tag{4.4.10}$$

The right-hand side of (4.4.10) tells us that as we follow a fluid particle, there are two mechanisms by which its vorticity can change. The first term, $(\omega \cdot \nabla)\mathbf{v}$, is vorticity change due to *vortex-line stretching*. The operator $\omega \cdot \nabla$ is the magnitude of the vorticity times the derivative in the direction of the vortex line. Consequently if the velocity vector changes along the vortex line (thus "stretching" the vortex line), there will be a contribution to the change of vorticity. The second term says that unless the pressure gradient and the density gradient are not aligned so that they are parallel to one another, the local vorticity will be changed.

Some further insights into the action of vorticity can be gained by considering Helmholtz's first theorem, which states that the circulation taken over any cross-sectional area of a vortex tube is constant. The proof is simple. Use a vortex tube segment with ends S_1 and S_2, and side area S_0. By one of Green's theorems,

$$\iint_{S_0} \boldsymbol{\omega} \cdot d\mathbf{A} + \iint_{S_1} \boldsymbol{\omega} \cdot d\mathbf{A} + \iint_{S_2} \boldsymbol{\omega} \cdot d\mathbf{A} = \iiint_V \text{div } \boldsymbol{\omega} \, dV = 0 \text{ by (4.4.7)}.$$

On S_0, $\boldsymbol{\omega}$ is normal to the area; hence that integral vanishes. Thus

$$\iint_{S_2} \boldsymbol{\omega} \cdot d\mathbf{A} = -\iint_{S_1} \boldsymbol{\omega} \cdot d\mathbf{A},$$

which with the proper interpretation of signs of the outward normals proves the theorem.

Notice that in proving Helmholtz's theorem, we did not use any dynamical information. Only kinematics and definitions were used. Therefore the theorem is also true if viscosity is taken into account.

A very important corollary of this theorem is that vortex lines can neither originate nor terminate in the interior of a flow. Either they are closed curves or they originate at the boundaries.

Another useful theorem has to do with the rate of change of circulation. Starting with the line integral definition and noting that $(D/Dt)\, d\mathbf{s} = d\mathbf{v}$, where $d\mathbf{v}$ is the change in $\mathbf{v}$ over a distance $d\mathbf{s}$ of C, then

$$D\Gamma/Dt = (D/Dt) \oint_C \mathbf{v} \cdot d\mathbf{s} = \oint_C \{(D\mathbf{v}/Dt) \cdot d\mathbf{s} + \mathbf{v} \cdot d\mathbf{v}\} = \oint_C [D\mathbf{v}/Dt + 0.5\, \nabla|\mathbf{v}|^2] \cdot d\mathbf{s}.$$

The second term in the last integral is an exact differential; hence its integration around a closed path is zero. Therefore our result for the rate of change of circulation reduces to

$$\frac{D\Gamma}{Dt} = \oint_C \frac{D\mathbf{v}}{Dt} \cdot d\mathbf{s}, \tag{4.4.11a}$$

or, with the help of the Euler equations, for inviscid flows we have

$$\frac{D\Gamma}{Dt} = -\oint_C \left(\frac{1}{\rho}\right) \nabla p \cdot d\mathbf{s}. \tag{4.4.11b}$$

Here the body force terms vanish because they are an exact differential, so an integration around a closed path gives zero. Therefore, as we follow a curve C drawn in the flow as it is carried along with a flow, the circulation associated with the curve can change only if $(1/\rho)\nabla p \, d\mathbf{s}$ is not an exact differential.

5. Irrotational Flows and the Velocity Potential

From the circulation theorem (4.4.11b), we see that if mass density is constant and viscous effects can be neglected, the integrand on the right-hand side is an exact differential, and so the integral vanishes. Consequently, for a flow with no upstream circulation, as the flow moves downstream it must continue to be vorticity-free, or *irrotational*. Of course, viscosity effects at a boundary will introduce vorticity, but at high Reynolds numbers this vorticity will be convected downstream, and chiefly confined to the boundary layer and wake.

By the definition of irrotational flows,

$$\boldsymbol{\omega} = \nabla \times \mathbf{v} = 0. \tag{4.5.1}$$

This suggests that two of the velocity components can be solved for in terms of the third component, or alternatively that, as in the case of the continuity equation, scalar functions can be introduced which have the effect of accomplishing this. An easier approach is to realize that since for irrotational flows

$$\Gamma = \oint_C \mathbf{v} \cdot d\mathbf{s} = 0 \text{ for any } C,$$

it follows that the integrand $\mathbf{v} \cdot d\mathbf{s}$ must be an exact differential, and therefore for an irrotational flow field the velocity $\mathbf{v}$ must be expressible as the gradient of a scalar. This allows us to write

$$\mathbf{v} = \nabla \phi, \tag{4.5.2}$$

where ϕ is called the *velocity potential*. For any velocity field written as the gradient of a scalar as in (4.5.3), we are guaranteed that for any scalar function ϕ we chose, $\mathbf{v}$ will automatically be an irrotational velocity field.

The introduction of a velocity potential guarantees irrotationality, but we still must require that the flow field satisfy our basic dynamical equations. We will consider here only incompressible flows. Then continuity requires that the divergence of the velocity field vanish. Therefore the continuity equation for an irrotational incompressible flow is

$$0 = \nabla \cdot \mathbf{v} = \nabla^2 \phi. \tag{4.5.3}$$

This is the equation we will use to determine ϕ for a given flow situation.

What then of the dynamics of the flow? Our flow field at this point seems to be completely determined from irrotationality and continuity, yet we have not considered Euler's equation. Recall from (4.3.1) that Euler's equation is

$$\rho D\mathbf{v}/Dt = -\nabla p + \rho \mathbf{g}. \tag{4.3.1}$$

From (3.2.6), since $\nabla \times \mathbf{v} = 0$, this can be written in the form

$$\rho(\partial \mathbf{v}/\partial t + \nabla |\mathbf{v}|^2/2) = -\nabla p + \rho \mathbf{g}.$$

For irrotational flows $\mathbf{v} = \nabla \phi$, and $\mathbf{g}$ can be written as $\mathbf{g} = -g\nabla h$, where h is the elevation of the point in the direction in which gravity acts. Euler's equation can then be rearranged after dividing by ρ in the form

$$\nabla(\partial \phi/\partial t + |\mathbf{v}|^2/2 + gh + p/\rho) = 0,$$

which upon integration gives

$$\frac{\partial \phi}{\partial t} + \frac{|\mathbf{v}|^2}{2} + gh + \frac{p}{\rho} = f(t), \tag{4.5.4}$$

where $f(t)$ is a constant, or at most a function of time, and is determined from either conditions at a reference point or far upstream.

Equation (4.5.4) is the ***Bernoulli equation*** for irrotational flows. Note that it differs from that for rotational flows found in Section 12 of Chapter 3 in that we do not have to require the flow to be steady, and that the integration does not have to be performed along any special path such as a streamline.

For most incompressible flows, then, the velocity field is found using only the conditions of irrotationality (usually by the introduction of the velocity potential ϕ) and the continuity equation in the form (4.5.3), along with the imposition of conditions on the normal velocity at boundaries. Pressure is found independently from ϕ by the Bernoulli equation (4.5.4). Note that all mathematical nonlinearities appear only in the Bernoulli equation. (For interface problems, however, further nonlinearities can be introduced by boundary conditions.) The linearity of equation (4.5.3) allows superposition of velocity fields. The nonlinearity of (4.5.4) means that pressure fields may not be superposed in a linear manner.

Note that for irrotational flows (4.5.1) can be written as

$$\Gamma = \oint_C \nabla\phi \cdot d\mathbf{s} = \phi_2 - \phi_1 = \Delta\phi, \tag{4.5.5}$$

where ϕ_1 and ϕ_2 are the values of the velocity potential at the start and end points of the traverse of closed path C. Thus, if ϕ is a single-valued function (i.e., if we go around a closed loop and the value of ϕ has not changed), since the curve C is closed, Γ will be zero.

It is, however, possible that ϕ can be multivalued if there exist points or isolated regions where ϕ is either singular or not uniquely defined. A line vortex, used in Example 4.4.2, is one such nonuniquely defined function. For such functions the circulation will usually be different from zero.

Since we will be looking at some methods for solving Laplace's equation, you may wonder whether the solution you obtain for a given flow is unique. That is, if you and your neighbor both solve the same problem, but use different methods, will you end up with the same velocity field? The answer is yes, provided that you both stay with the same set of rules (and of course both do your work accurately). The irrotational flow field around a body is unique for a given set of boundary conditions, provided we also specify the circulation, and do not allow cavities to develop that do not contain fluid. We must always remember that our primary interest is in finding a flow field that models to some degree of accuracy a real physical phenomenon. Since vorticity, and hence circulation, is present in the boundary layer, we may need to include it in our model to give a realistic model of the flow field. Cavities might be a reasonable model for wake flows. As long as we prescribe what the circulation is, as well as rules about whether or not cavities are present, our flow field will be unique.

It is possible that the methods we use to solve Laplace's equation will introduce a "mathematical" flow inside the body as well. Therefore in that region the flow is not unique. Any flows our methods generate inside bodies lie outside our domain of interest, and are artifices of our mathematics with no physical meaning. Their presence might, however, be convenient to us, and even at times useful in our thinking.

a. Intersection of velocity potential lines and streamlines

The lines of constant velocity potential and of constant stream function will intersect one another. We next investigate to see at what angles these intersections take place. Looking at the relationship between ϕ and ψ lines, we see that

$$d\phi = \frac{\partial\phi}{\partial x}\, dx + \frac{\partial\phi}{\partial y}\, dy = v_x\, dx + v_y\, dy,$$

and

$$d\psi = \frac{\partial\psi}{\partial x}\, dx + \frac{\partial\psi}{\partial y}\, dy = -v_y\, dx + v_x\, dy.$$

Consequently, locally the slope of the $\phi = \text{constant}$ line is

$$\left[\frac{dy}{dx}\right]_\phi = -\frac{v_x}{v_y},$$

and the slope of the $\psi = \text{constant}$ line is

$$\left[\frac{dy}{dx}\right]_\psi = \frac{v_y}{v_x}.$$

Multiplying the two slopes gives

$$\left[\frac{dy}{dx}\right]_\phi \cdot \left[\frac{dy}{dx}\right]_\psi = -1,$$

and we conclude that ϕ and ψ lines are orthogonal to one another, except at places where the velocity is zero (stagnation points) or infinite (singularities).

In either of these cases we must investigate the situation by considering the second-order terms to see what occurs. For example, at a stagnation point a Taylor series expansion gives, to second-order terms in dx and dy,

$$d\phi = \frac{\partial\phi}{\partial x}\, dx + \frac{\partial\phi}{\partial y}\, dy + \frac{\partial^2\phi}{\partial x^2}\frac{(dx)^2}{2} + \frac{\partial^2\phi}{\partial x\,\partial y}\, dx\, dy + \frac{\partial^2\phi}{\partial y^2}\frac{(dy)^2}{2}.$$

Since the first derivatives vanish at the stagnation point, on a line of constant ϕ passing through a stagnation point this becomes

$$d\phi = 0 = \frac{\partial^2\phi}{\partial x^2}\frac{(dx)^2}{2} + \frac{\partial^2\phi}{\partial x\,\partial y}\, dx\, dy + \frac{\partial^2\phi}{\partial y^2}\frac{(dy)^2}{2}$$

or

$$0 = \left[\frac{\partial^2\phi}{\partial x^2} + 2\frac{\partial^2\phi}{\partial x\,\partial y}\frac{dy}{dx} + \frac{\partial^2\phi}{\partial y^2}\frac{(dy)^2}{(dx)}\right]\frac{(dx)^2}{2}.$$

Solving this quadratic equation for the slope at the stagnation point, we have

$$\left[\frac{dy}{dx}\right]_{\phi} = \left[-\frac{\partial^2\phi}{\partial x\,\partial y} \pm \sqrt{\left(\frac{\partial^2\phi}{\partial x\,\partial y}\right)^2 - \frac{\partial^2\phi}{\partial x^2}\frac{\partial^2\phi}{\partial y^2}}\right] \Bigg/ \frac{\partial^2\phi}{\partial y^2}.$$

A similar expression can be found for ψ by the same process. We see that there will be two values for the slope at the stagnation point and hence the ϕ line divides, or bifurcates. (Note that the term underneath the square root sign must always be positive, since by Laplace's equation

$$\frac{\partial^2\phi}{\partial x^2} = -\frac{\partial^2\phi}{\partial y^2},$$

and so this term is the sum of two squares.) For details of the angle between ϕ and ψ, individual examples must be considered. Since the Laplace equation is a great averager of things (for instance, it can be shown that the value of ϕ at the center of a circle is the average of all the values it takes on the circle), the constant ϕ lines can be expected to fall midway between the constant ψ lines.

b. Simple two-dimensional irrotational flows

We consider several basic simple flows that are the building blocks of potential flow theory, from which all other potential flows can be constructed. We will adapt the practice of first writing the velocity potential, then examining it to see its nature. The basic flows that we study have their counterparts in electrostatics and electromagnetics (point charges, dipoles), beam deflection theory (concentrated loads), and many other branches of engineering physics, and are special cases of what are termed *Green's functions*.

In presenting the various basic flows, we will first write the velocity potential and then find the stream function by integrating the equations

$$v_x = \frac{\partial\phi}{\partial x} = \frac{\partial\psi}{\partial y} \tag{4.5.6a}$$

and

$$v_y = \frac{\partial\phi}{\partial y} = -\frac{\partial\psi}{\partial x}. \tag{4.5.6b}$$

Notice that

$$\nabla^2\phi = 0 \tag{4.5.7}$$

is the incompressible continuity equation for irrotational flows, corresponding to (4.1.2), while

$$\nabla^2\psi = 0 \tag{4.5.8}$$

is the irrotationality condition for two-dimensional incompressible flows satisfying continuity, corresponding to (4.4.7).

Uniform stream

A uniform stream is one whose velocity is the same at every point in space. Therefore the velocity components are

$$v_x = U_x = \partial\phi/\partial x, \tag{4.5.9a}$$

$$v_y = U_y = \partial\phi/\partial y. \tag{4.5.9b}$$

Integrating (4.4.9a) and (4.4.9b), we find that

$$\phi_{\text{uniform stream}} = xU_x + yU_y, \tag{4.5.10}$$

where we have arbitrarily set the constant of integration to zero since it does not contribute to the velocity field in any way. Lines of constant ϕ and ψ (both straight lines and mutually orthogonal) are shown in Figure 4.10. The stream function is found from use of (4.5.6) to be

$$\psi_{\text{uniform stream}} = yU_x - xU_y. \tag{4.5.11}$$

Line source or sink (monopole)

The velocity potential

$$\phi = (m/2\pi) \ln |\mathbf{r} - \mathbf{r}_0| = (m/2\pi) \ln \sqrt{(x - x_0)^2 + (y - y_0)^2} \tag{4.5.12}$$

is called a ***line source*** (if m is positive) or a ***line sink*** (if m is negative) of strength m, located at (x_0, y_0). It extends from $-\infty$ to ∞ in the z direction. We note that the velocity components are given by

$$v_x = \partial\phi/\partial x = m/(2\pi |\mathbf{r} - \mathbf{r}_0|) \tag{4.5.13a}$$

and

$$v_y = \partial\phi/\partial y = m/(2\pi |\mathbf{r} - \mathbf{r}_0|). \tag{4.5.13b}$$

Figure 4.10. Streamlines (solid) and equipotential lines (dashed) for a uniform stream.

Differentiating the velocity components, we have

$$\partial^2\phi/\partial x^2 = (m/2\pi)[\,|\mathbf{r} - \mathbf{r}_0|^2 - 2(x - x_0)^2]/|\mathbf{r} - \mathbf{r}_0|^4$$

and

$$\partial^2\phi/\partial y^2 = (m/2\pi)[\,|\mathbf{r} - \mathbf{r}_0|^2 - 2(y - y_0)^2]/|\mathbf{r} - \mathbf{r}_0|^4 = -\partial^2\phi/\partial x^2.$$

Therefore continuity is satisfied everywhere except possibly at the point (x_0, y_0).

To investigate what is happening at that point, we integrate the flow rate about a 2 by 2 square centered at (x_0, y_0) as seen in Figure 4.11. The size of the square is actually arbitrary, since the same result would be obtained for a contour of any size or shape that encloses the source. Then, starting from the definition of discharge, we have

$$Q = \int_{y_0-1}^{y_0+1} (\,v_x|_{x_{0}+1} - v_x|_{x_{0}-1}\,)\,dy + \int_{x_0-1}^{x_0+1} (\,v_y|_{y_{0}+1} - v_y|_{y_{0}=-1})\,dx$$

$$= \int_{y_0-1}^{y_0+1} \frac{m}{2\pi}\left[\frac{1}{1 + (y - y_0)^2} - \frac{-1}{1 + (y - y_0)^2}\right]dy$$

$$+ \int_{x_0-1}^{x_0+1} \frac{m}{2\pi}\left[\frac{1}{1 + (x - x_0)^2} - \frac{-1}{1 + (x - x_0)^2}\right]dx = m.$$

Thus m represents the flow rate per unit length in the z direction being emitted from (x_0, y_0). It is called the **_strength_** of the source. If m is positive, we say that this represents a source; if negative, it is a sink. Lines of constant ϕ (radial lines) and ψ (concentric circles) are shown in Figure 4.12.

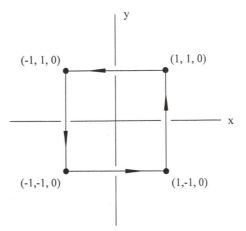

Figure 4.11. Path used in computing circulation.

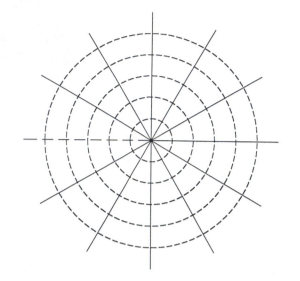

Figure 4.12. Streamlines (solid) and equipotential lines (dashed) for a line source or sink.

The question of irrotationality at (x_0, y_0) should also be considered. If we compute the circulation around a square similar to the one we used above, the result is zero. Therefore there is a concentrated source of mass at (x_0, y_0) but no concentrated source of vorticity there.

The stream function for a source is, from (4.4.6) and (4.4.13),

$$\psi = (m/2\pi) \tan^{-1} (y - y_0)/(x - x_0). \tag{4.5.14}$$

The arctangent is a multivalued function, changing by 2π as we go around a contour enclosing (x_0, y_0). Therefore the change in ψ as we go around the contour is m.

From (4.5.12), we see that constant ϕ lines are concentric circles with centers at (x_0, y_0). From (4.5.14), we see that the constant ψ lines are radial lines emanating from (x_0, y_0). Thus the flow will be along the radial streamlines emanating from (x_0, y_0). To conserve mass, the velocity must decrease inversely with the distance from (x_0, y_0).

Line doublet (dipole)

If we think of a source and sink pair of equal strengths, as they approach one another along a connecting line the net discharge in a small region enclosing the pair is zero. If we let them approach one another in such a manner that their strengths increase inversely as the distance between them, we have

$$[\phi_{\text{source}}(\mathbf{r} + \mathbf{a}) + \phi_{\text{sink}}(\mathbf{r} - \mathbf{a})]/2|\mathbf{a}|,$$

which in the limit as $|\mathbf{a}|$ goes to zero becomes a derivative. We therefore define a *line doublet* as

$$\phi_{\text{doublet}} = \mathbf{B} \cdot \nabla\phi_{\text{source of strength } 2\pi} = [B_x(x - x_0) + B_y(y - y_0)]/|\mathbf{r} - \mathbf{r}_0|^2, \tag{4.5.15}$$

where **B** gives the strength and direction of the doublet. We denote a doublet by a half-filled circle, the filled part representing the "source end" and the unfilled part the "sink end." From its relation to the source, we expect that there is no concentrated discharge or vorticity at (x_0, y_0), which can be verified by taking appropriate integrations around that point. To find what lines of constant ϕ are like, we rearrange (4.4.15) as

$$(x - x_0)^2 + (y - y_0)^2 = [B_x(x - x_0) + B_y(y - y_0)]/\phi.$$

For constant values of ϕ this is the equation of a circle, for if we add

$$(B_x/2\phi)^2 + (B_y/2\phi)^2$$

to each side and rearrange, we have

$$(x - x_0 - B_x/2\phi)^2 + (y - y_0 - B_y/2\phi)^2 = (B_x/2\phi)^2 + (B_y/2\phi)^2 = [|\mathbf{B}|/2\phi]^2,$$

the equation of a circle whose radius is $|\mathbf{B}|/2\phi$. Thus the constant ϕ lines are circles of radius $|\mathbf{B}|/2\phi$ centered at $(x_0 + B_x/2\phi, y_0 + B_y/2\phi)$, which are on a line through (x_0, y_0) in the direction of **B**. Constant ϕ and ψ lines are shown in Figure 4.13. The constant ϕ lines are nested circles centered at $(B_x/2\phi, B_y/2\phi)$ and with radius $[(B_x/2\phi)^2 + (B_y/2\phi)^2]^{1/2}$. Constant ψ lines are a similar family of circles, but rotated 90° with respect to the ϕ circles. They are centered at $(B_y/2\psi, -B_x/2\psi)$ and have radius $[(B_x/2\psi)^2 + (B_y/2\psi)^2]^{1/2}$.

The velocity field associated with the doublet is

$$v_x = \frac{\partial \phi}{\partial x} = \frac{B_x[(y - y_0)^2 - (x - x_0)^2] - 2B_y(x - x_0) \times (y - y_0)}{[(x - x_0)^2 + (y - y_0)^2]^2}$$

$$v_y = \frac{\partial \phi}{\partial y} = \frac{B_y[(x - x_0)^2 - (y - y_0)^2] - 2B_x(x - x_0) \times (y - y_0)}{[(x - x_0)^2 + (y - y_0)^2]^2}.$$

(4.5.16)

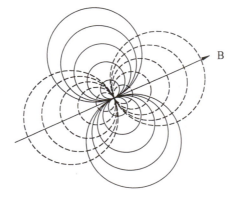

Figure 4.13. Streamlines (solid) and equipotential lines (dashed) for a line doublet.

Along a line in the direction of **B**, the velocity dies out as the square of the distance. Since ϕ_{source} satisfies Laplace's equation, and since

$$\nabla(\nabla^2 \phi_{source}) = \nabla(0) = 0,$$

therefore

$$0 = \mathbf{B} \cdot \nabla(\nabla^2 \phi_{source}) = \mathbf{B} \cdot \nabla^2(\nabla \phi_{source}) = \nabla^2(\mathbf{B} \cdot \nabla \phi_{source}) = \nabla^2 \phi_{doublet}.$$

The velocity potential for a line doublet therefore satisfies Laplace's equation.

From (4.5.6) the stream function for a line doublet is

$$\psi = \frac{B_y(x - x_0) - B_x(y - y_0)}{|\mathbf{r} - \mathbf{r}_0|^2}. \tag{4.5.17}$$

Since we have also labeled the source and doublet as monopole and dipole, you might wonder whether taking even higher derivatives would be useful. The derivative of the dipole is the **quadrapole**, which is mainly of interest in acoustic problems. The dipole is usually sufficient for our use.

Line vortex

The vortex is a "reverse analog" of the source, in that it has concentrated vorticity rather than concentrated discharge, and because the constant ϕ lines are radial lines rather than concentric circles. Its velocity potential is

$$\phi_{vortex} = (\Gamma/2\pi) \tan^{-1} (y - y_0)/(x - x_0), \tag{4.5.18}$$

with velocity components

$$v_x = \frac{\partial \phi}{\partial x} = -\frac{\Gamma(y - y_0)}{2\pi[(x - x_0)^2 + (y - y_0)^2]}$$

$$v_y = \frac{\partial \phi}{\partial y} = \frac{\Gamma(x - x_0)}{2\pi[(x - x_0)^2 + (y - y_0)^2]}. \tag{4.5.19}$$

Thus the velocity decreases inversely with the distance from (x_0, y_0). Constant ϕ (concentric circles) and ψ (radial lines) lines are shown in Figure 4.14.

Since

$$\frac{\partial^2 \phi}{\partial x^2} = \frac{\Gamma(y - y_0)(x - x_0)}{\pi[(x - x_0)^2 + (y - y_0)^2]^2},$$

and

$$\frac{\partial^2 \phi}{\partial y^2} = -\frac{\Gamma(y - y_0)(x - x_0)}{\pi[(x - x_0)^2 + (y - y_0)^2]^2} = -\frac{\partial^2 \phi}{\partial x^2},$$

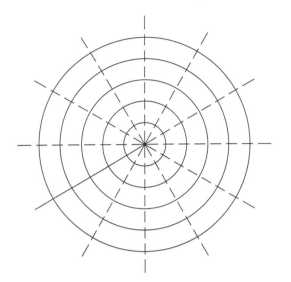

Figure 4.14. Streamlines (solid) and equipotential lines (dashed) for a line vortex.

Laplace's equation is seen to be satisfied. The vortex is counterclockwise if Γ is positive, clockwise if Γ is negative.

Since ϕ_{vortex} is multivalued [in traversing a path around (x_0, y_0), ϕ_{vortex} changes by Γ], we anticipate that there could be circulation associated with vortex flows. Checking this by calculating the circulation around a 2 by 2 square centered at (x_0, y_0) we have

$$\Gamma = \int_{y_0-1}^{y_0+1} v_y|_{x_0+1}\, dy + \int_{x_0+1}^{x_0-1} v_x|_{y_0+1}\, dx + \int_{y_0+1}^{y_0-1} v_y|_{x_0-1}\, dy + \int_{x_0-1}^{x_0+1} v_x|_{y_0-1}\, dx$$

$$= \frac{\Gamma}{2\pi}\left[\int_{y_0-1}^{y_0+1} \frac{dy}{1+(y-y_0)^2} + \int_{x_0+1}^{x_0-1} \frac{-dx}{1+(x-x_0)^2} + \int_{y_0+1}^{y_0-1} \frac{-dy}{1+(y-y_0)^2} + \int_{x_0-1}^{x_0+1} \frac{dx}{1+(x-x_0)^2} \right] = \Gamma,$$

a result that depends only on the fact that the path encircles (x_0, y_0), not on any other details of the path such as shape or size. Any path not enclosing the vortex has zero circulation, as can be verified by Stokes' theorem. Therefore a vortex has concentrated vorticity, but no concentrated mass discharge.

From (4.5.6) the stream function for a vortex is

$$\psi_{\text{vortex}} = -(\Gamma/2\pi) \ln|\mathbf{r} - \mathbf{r}_0|. \tag{4.5.20}$$

The velocity potentials and stream functions for all of these flows plus their three-dimensional counterparts are summarized in Table 4.4.

Table 4.4. Velocity potentials and stream functions for irrotational flows

Flow	Two-dimensional		Three-dimensional	
	ϕ	ψ	ϕ	ψ (axisymmetric)
Uniform stream	$xU_x + yU_y$	$yU_x - xU_y$	$xU_x + yU_y + zU_z$	$0.5U_z r^2$
Source or sink	$\dfrac{m}{2\pi} \ln \sqrt{(x-x_0)^2 + (y-y_0)^2}$	$\dfrac{m}{2\pi} \tan^{-1} \dfrac{(y-y_0)}{(x-x_0)}$	$\dfrac{-m}{4\pi\lvert \mathbf{r} - \mathbf{r}_0 \rvert}$	$\dfrac{-m(z-z_0)}{4\pi\sqrt{r^2 + (z-z_0)^2}}$
Doublet	$\dfrac{B_x(x-x_0) + B_y(y-y_0)}{\lvert \mathbf{r} - \mathbf{r}_0 \rvert^2}$	$\dfrac{B_y(x-x_0) - B_x(y-y_0)}{\lvert \mathbf{r} - \mathbf{r}_0 \rvert^2}$	$\dfrac{\mathbf{B} \cdot (\mathbf{r} - \mathbf{r}_0)}{\lvert \mathbf{r} - \mathbf{r}_0 \rvert^3}$	$\dfrac{-B_z r^2}{[r^2 + (z-z_0)^2]^{3/2}}$
Line vortex	$\dfrac{\Gamma}{2\pi} \tan^{-1}\left(\dfrac{y-y_0}{x-x_0}\right)$	$-\dfrac{\Gamma}{2\pi} \ln\lvert \mathbf{r} - \mathbf{r}_0 \rvert$	None	None

c. Hele-Shaw flows

The four solutions we have just looked at, uniform stream, source/sink, doublet, and vortex, are the fundamental solutions for two-dimensional potential flow. Before considering combinations of these to obtain physically interesting flows, we shall find it useful to see how these basic flows can be produced in a laboratory.

One way of producing flows by means of a Hele-Shaw table, shown in cutaway view in Figure 4.15. It consists of a flat horizontal floor with a trough at one end for introducing a uniform stream and one at the other end for removing it. Holes in the bottom of the floor are connected to bottles that may be raised or lowered. These provide the sources and sinks, the elevation of the bottle controlling their strength. Doublets can be produced by putting a source and sink pair almost together, having the source height above the table being equal to the sink height below. Sometimes a top transparent plate is added, to ensure that the flow is of the same depth everywhere.

Vortices are more difficult to produce in Hele-Shaw flows than are sources or sinks. One way of producing them would be to use a vertical circular rod driven by an electric motor.

Hele-Shaw flows are slow flows, and because the velocity will vary parabolically with the coordinate in the direction normal to the floor, there is inherent to these flows a great deal of vorticity. However, this vorticity is largely parallel to the bed, the vorticity component normal to the floor being virtually zero. Thus Hele-Shaw flows viewed perpendicular to the floor are good models of two-dimensional irrotational flows.

Streamlines can be traced by inserting dye into the flow. A permanent record of these streamlines can be made by photography. Alternatively, one method that has

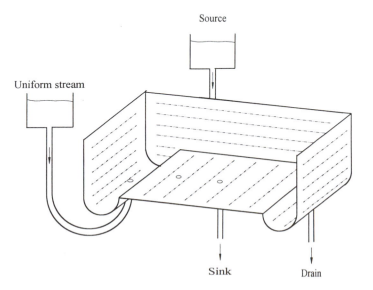

Figure 4.15. Cutaway view of a Hele-Shaw table.

been used is to cast the bed of the table out of a hard plaster, with inlets for the source and sink permanently cast into the plaster. If the bed is painted with a white latex paint, streamlines can be recorded by carefully placing potassium permanganate crystals on the bed. A record of the streamlines remains as dark stains on the paint.

d. Simple three-dimensional irrotational flows

Except for the vortex, all our two-dimensional irrotational flows have three-dimensional counterparts that qualitatively are much like their two-dimensional counterparts. (Examples of a three-dimensional vortex are smoke rings and the mushroom clouds produced by explosions. Mathematical representation of these are much more complicated than for the flows we are considering.) Here we simply list these counterparts. The analysis proceeds as in the two-dimensional case.

Uniform stream

The velocity potential for a uniform stream is

$$\phi_{\text{uniform stream}} = xU_x + yU_y + zU_z = \mathbf{r} \cdot \mathbf{U} \tag{4.5.21}$$

with a velocity field

$$\mathbf{v} = \nabla \phi = \mathbf{U}.$$

Surfaces of constant ϕ are planes perpendicular to $\mathbf{U}$.

When $\mathbf{U}$ has only a component in the z direction, a Stokes stream function can be found in the form

$$\psi_{\text{uniform stream}} = 0.5 U_z R^2 \sin^2 \beta = 0.5 U_z r^2. \tag{4.5.22}$$

Point source or sink (point monopole)

The velocity potential for a point source of strength m at $\mathbf{r}_0$ is

$$\phi_{\text{source}} = -\frac{m}{4\pi |\mathbf{r} - \mathbf{r}_0|}. \tag{4.5.23}$$

Here m is the volume discharge from the source, with continuity satisfied everywhere except at $\mathbf{r}_0$. If m is positive, ϕ represents a source. If negative, it represents a sink. Irrotationality is satisfied everywhere.

Surfaces of constant ϕ are concentric spheres centered at $\mathbf{r}_0$. The velocity is directed along the radius of these spheres, and dies out like the reciprocal of the distance squared to satisfy continuity. The velocity is given by

$$\mathbf{v} = \nabla \phi = \frac{m(\mathbf{r} - \mathbf{r}_0)}{4\pi |\mathbf{r} - \mathbf{r}_0|^3}. \tag{4.5.24}$$

When the source lies on the z axis, a Stokes stream function can be found in the form

$$\psi_{\text{source}} = -\frac{m(z - z_0)}{4\pi\sqrt{r^2 + (z - z_0)^2}}.$$ (4.5.25)

Point doublet (point dipole)

The velocity potential for a doublet can again be found by differentiating the potential for a source, giving

$$\phi_{\text{doublet}} = \mathbf{B} \cdot \nabla\phi_{\text{source of strength } 4\pi} = \frac{\mathbf{B} \cdot (\mathbf{r} - \mathbf{r}_0)}{|\mathbf{r} - \mathbf{r}_0|^3}$$ (4.5.26)

with velocity components

$$\mathbf{v} = \frac{\mathbf{B}}{|\mathbf{r} - \mathbf{r}_0|^3} - \frac{3(\mathbf{r} - \mathbf{r}_0)\mathbf{B} \cdot (\mathbf{r} - \mathbf{r}_0)}{|\mathbf{r} - \mathbf{r}_0|^5}.$$ (4.5.27)

The constant ϕ surfaces are no longer simple geometries like circles and spheres, but are seen from (4.5.6) to be represented by cubic equations.

When the doublet lies on the z axis, and additionally, $\mathbf{B}$ points parallel to the z axis, a Stokes stream function for a doublet can be found in the form

$$\psi_{\text{doublet}} = -\frac{B_z r^2}{[r^2 + (z - z_0)^2]^{3/2}}.$$ (4.5.28)

These results are summarized in Table 4.4.

e. Superposition and the method of images

In a number of simple cases, the solution for flow past a given body shape can be obtained by recognizing an analogy between potential flow and light, since Laplace's equation also governs the passage of light waves. Boundaries can be thought of as mirrors, with images of the fundamental solutions appearing at appropriate points to generate the "mirror" boundary.

Source near a plane wall

Suppose we have a source of strength m a distance b from a plane wall. According to the method of images, the plane wall can be regarded as a mirror. As the source "looks" into the mirror, it sees an ***image source*** of the same strength a distance b behind the mirror. (See Figure 4.16. A mirror interchanges right and left. This does not affect the source, but will affect signs of vortices and doublets.) The potential for the two-dimensional case is, with the x axis acting as the wall,

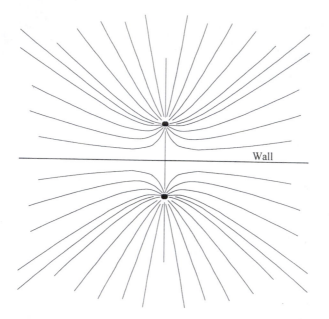

Figure 4.16. Streamlines for a line source near a wall.

$$\phi = (m/2\pi)[\ln \sqrt{x^2 + (y-b)^2} + \ln \sqrt{x^2 + (y+b)^2}],$$ (4.5.29)

with velocity components

$$v_x = \frac{m}{2\pi}\left[\frac{x}{x^2 + (y-b)^2} + \frac{x}{x^2 + (y+b)^2}\right],$$

$$v_y = \frac{m}{2\pi}\left[\frac{y-b}{x^2 + (y-b)^2} + \frac{y+b}{x^2 + (y+b)^2}\right].$$ (4.5.30)

We note that v_y vanishes on $y = 0$, satisfying the necessary boundary condition that the velocity normal to the stationary wall be zero.

This problem is a simple model of physical problems such as water intakes or pollution sources near a straight coastline. Additional sources can be easily added.

Note that if the source is in a corner, both walls act as mirrors, and there are three image sources plus the original source. The third image comes about because the mirrors extend to plus and minus infinity, thus the two images that you might initially think would suffice in turn have their own image. The above procedure can be easily extended to vortices and doublets near a wall, and to three-dimensional examples. It holds also for acoustic sources. Some loudspeaker enclosures are designed to be placed in corners or near walls to increase their apparent power output by the image speakers.

Example 4.4.1. Point source near walls

A point source of strength m is located at the point $(1, 2, 3)$. There is a wall at $x = 0$, and a second wall at $y = 0$. What is the velocity potential for this flow?

Given: A three-dimensional flow is to be made up of a source located at $(1, 2, 3)$ plus image sources needed to generate two walls. From $(4.5.23)$ the given source has a velocity potential

$$\phi_0 = -m/4\pi |\mathbf{r} - \mathbf{r}_0|,$$

where $\mathbf{r}_0 = \mathbf{i} + 2\mathbf{j} + 3\mathbf{k}$.

Solution: To generate the first wall, an image source of strength m at $(-1, 2, 3)$ is needed. This will have a velocity potential

$$\phi_1 = -m/4\pi |\mathbf{r} - \mathbf{r}_1|,$$

where $\mathbf{r}_1 = -\mathbf{i} + 2\mathbf{j} + 3\mathbf{k}$. When the second wall is added, both the original source and the image at $(-1, 2, 3)$ will have images, with a combined velocity potential

$$\phi_2 + \phi_3 = -m/4\pi |\mathbf{r} - \mathbf{r}_2| - m/4\pi |\mathbf{r} - \mathbf{r}_3|,$$

where $\mathbf{r}_2 = \mathbf{i} - 2\mathbf{j} + 3\mathbf{k}$ and $\mathbf{r}_3 = -\mathbf{i} - 2\mathbf{j} + 3\mathbf{k}$.

Thus our velocity potential is the sum of the potentials of four sources, each of strength m, and located at the points $(1, 2, 3)$, $(-1, 2, 3)$, $(1, -2, 3)$, and $(-1, -2, 3)$. The total potential is

$$\phi_{\text{total}} = \phi_0 + \phi_1 + \phi_2 + \phi_3 = -\frac{m}{4\pi}\left[\frac{1}{|\mathbf{r} - \mathbf{r}_0|} + \frac{1}{|\mathbf{r} - \mathbf{r}_1|} + \frac{1}{|\mathbf{r} - \mathbf{r}_2|} + \frac{1}{|\mathbf{r} - \mathbf{r}_3|}\right].$$

Note that if there were a third wall at $z = 0$, eight sources would be needed.

Vortices near walls

An interesting variation of the source near a wall is that of a vortex near a wall. The image will have a circulation in the reverse direction of the original vortex because left and right are interchanged in a reflection. See Figure 4.17. For a wall at $x = 0$ the velocity potential for the original vortex plus its image is

$$\phi = (\Gamma/2\pi)[\tan^{-1}(y - b)/x - \tan^{-1}(y + b)/x] \tag{4.5.31}$$

with velocity components

$$v_x = \frac{\partial\phi}{\partial x} = \frac{\Gamma}{2\pi}\left\{\frac{-(y - b)}{[x^2 + (y - b)^2]} + \frac{y + b}{[x^2 + (y + b)^2]}\right\}, \tag{4.5.32a}$$

$$v_y = \frac{\partial\phi}{\partial y} = \frac{\Gamma}{2\pi}\left\{\frac{x}{[x^2 + (y - b)^2]} + \frac{x}{[x^2 + (y + b)^2]}\right\}. \tag{4.5.32b}$$

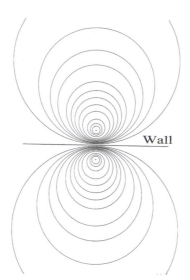

Figure 4.17. Streamlines for a line vortex near a wall.

The velocity at $(0, b)$ due to the image vortex is $(\Gamma/4\pi b, 0)$. This is called the ***induced velocity***. The vortex at $(0, b)$ will tend to travel with the induced velocity, carrying its image with it, since the induced velocity at the image will be the same value.

The stream function to accompany (4.4.30) is

$$\psi = (\Gamma/2\pi)[\ln \sqrt{x^2 + (y-b)^2} - \ln \sqrt{x^2 + (y+b)^2}] = (\Gamma/4\pi) \ln \left[\frac{x^2 + (y-b)^2}{x^2 + (y+b)^2}\right] \quad (4.5.33)$$

Therefore lines of constant ψ are given by

$$x^2 + (y-b)^2 = C[x^2 + (y+b)^2], \quad\quad\quad\quad (4.5.34)$$

where C is a constant related to the value of the stream function by $\psi = (\Gamma/4\pi) \ln C$.

We expand and rearrange, and (4.3.33) becomes

$$x^2 + [y - (1 + C)b/(1 - C)]^2 = 4Cb^2/(1 - C)^2.$$

Thus the streamlines are circles of radius $r = 2b\sqrt{C/(1 - C)}$ centered at $(0, d)$, where $d = b(1 + C)/(1 - C)$.

Since the streamlines in this example are circles, another physical realization of this flow is a single vortex in a cup of radius r, the vortex being a distance b from the center of the cup. The vortex will travel around the cup on a circular path of radius r at a speed of $\Gamma/4\pi b$.

Still another realization of a vortex flow is a vortex pair near a wall, the vortices being of equal but opposite circulation. We take them to be a distance a from the wall, and separated by a distance $2b$. To generate the wall, it is necessary to add a vortex image pair as well. The total velocity potential is then

$$\phi = (\Gamma/2\pi)[\tan^{-1}(y-a)/(x-b) - \tan^{-1}(y-a)/(x+b)$$
$$- \tan^{-1}(y+a)/(x-b) + \tan^{-1}(y+a)/(x+b)], \tag{4.5.35}$$

with stream function

$$\psi = -(\Gamma/2\pi)\{0.5 \times \ln[(x-b)^2 + (y-a)^2] - 0.5 \times \ln[(x+b)^2 + (y-a)^2]$$
$$- 0.5 \times \ln[(x-b)^2 + (y+a)^2] + 0.5 \times \ln[(x+b)^2 + (y+a)^2]\}, \tag{4.5.36}$$

and with velocity components

$$v_x = \frac{\partial \phi}{\partial x} = \frac{\Gamma}{2\pi}\left[\frac{-(y-a)}{(x-b)^2 + (y-a)^2} + \frac{y-a}{(x+b)^2 + (y-a)^2}\right.$$
$$\left. + \frac{y+a}{(x-b)^2 + (y+a)^2} - \frac{y+a}{(x+b)^2 + (y+a)^2}\right],$$

$$v_y = \frac{\partial \phi}{\partial y} = \frac{\Gamma}{2\pi}\left[\frac{x-b}{(x-b)^2 + (y-a)^2} - \frac{x+b}{(x+b)^2 + (y-a)^2}\right.$$
$$\left. - \frac{x-b}{(x-b)^2 + (y+a)^2} + \frac{x+b}{(x+b)^2 + (y+a)^2}\right]. \tag{4.5.37}$$

The induced velocity at (b, a) is

$$\mathbf{V}_{\text{induced}} = \frac{\Gamma}{4\pi}\left[\left(\frac{1}{a} - \frac{a}{a^2 + b^2}\right)\mathbf{i} + \left(\frac{-1}{b} + \frac{b}{a^2 + b^2}\right)\mathbf{j}\right].$$

Therefore the equations of motion for this vortex are

$$\frac{da}{dt} = \frac{\Gamma}{4\pi}\left[\frac{-1}{b} + \frac{b}{a^2 + b^2}\right], \qquad \frac{db}{dt} = \frac{\Gamma}{4\pi}\left[\frac{1}{a} - \frac{a}{a^2 + b^2}\right]. \tag{4.5.38}$$

With appropriate sign changes, similar equations hold for the other three vortices.

The path the vortices travel can be found by rearranging (4.5.38) in the form

$$\frac{\Gamma}{4\pi}\frac{dt}{a^2 + b^2} = \frac{a}{b^2}db = -\frac{b}{a^2}da.$$

The variables a and b can be separated and the resulting equation integrated, giving as the path the vortex travels

$$1/b^2 = 1/b_0^2 + 1/a_0^2 - 1/a^2, \tag{4.5.39}$$

(a_0, b_0) being the initial position of the vortex.

This model of a traveling vortex pair is useful in describing the spreading of the vortex pair left by the wing tips of an aircraft on take-off. In this case the wall represents the ground.

Example 4.4.2. A vortex near a plane wall

A wall is located at $x = 0$. A vortex with circulation 20π m^2/s is placed 1 m above the wall. What is the velocity potential, and at what speed does the vortex move?

Given: The flow is to be made up of a vortex plus the image needed to generate the wall. The vortex will move due to the velocity induced by its image.

Solution: For convenience, take the instantaneous position of the vortex to be $(0, 1, 0)$. Then the velocity potential for the original vortex is $\phi_{orig} = 10 \tan^{-1} (y - 1)/x$. Remember that the circulation of the image vortex is reversed, then from (4.7.3) the velocity potential for the original vortex plus its image is

$\phi_{total} = \phi_{orig} + \phi_{image}$

$$= 10[\tan^{-1} (y - 1)/x - \tan^{-1} (y + 1)/x].$$

Take the gradient of the velocity potential, and the velocity components are

$$v_x = 10\left[\frac{-(y - 1)}{x^2 + (y - 1)^2} + \frac{y + 1}{x^2 + (y + 1)^2}\right], \qquad v_y = 10\left[\frac{x}{x^2 + (y - 1)^2} + \frac{x}{x^2 + (y + 1)^2}\right].$$

The induced velocity at $(0, 1, 0)$ is the velocity at that point due to the image vortex. This gives a velocity at $(0, 1, 0)$ of

$$v_x = 10(2/2^2) = 5 \text{ m/s}, \qquad v_y = 0.$$

The vortex thus moves parallel to the wall at a speed of 5 m/s.

Example 4.4.3. A vortex pair in a cup

A vortex pair is generated in a cup of coffee of radius c by brushing the tip of a spoon lightly across the surface of the coffee. The pair so generated will have opposite circulations. (Perform the experiment to verify this.) If the vortex with positive circulation is at (a, b), and the vortex with negative circulation at $(a, -b)$, verify that the flow with the cup is generated by an image vortex with positive circulation at $[ac^2/(a^2 + b^2), -bc^2/(a^2 + b^2)]$, plus an image vortex with negative circulation at $[ac^2/(a^2 + b^2), bc^2/(a^2 + b^2)]$. These image points are located at what are called the inverse points of our cup.

Given: A pair of opposite-rotating vortices, plus the images needed to generate the cup. Each vortex moves because of the induced velocity generated by the images and the other vortex. The given vortex pair has a stream function

$$\psi = -(\Gamma/2\pi)\{0.5 \ln [(x - a)^2 + (y - b)^2] - 0.5 \ln [(x - a)^2 + (y + b)^2]\}.$$

Solution: The proposed stream function consists of the original stream function plus the stream function due to a pair of vortices at (ak, bk) and $(ak, -bk)$, where $k = c^2/(a^2 + b^2)$. The combined stream function is then

$$\psi = -(\Gamma/2\pi)\{0.5 \ln [(x-a)^2 + (y-b)^2] - 0.5 \ln [(x-a)^2 + (y+b)^2]$$

$$+ 0.5 \ln [(x-ak)^2 + (y-bk)^2] - 0.5 \ln [(x-ak)^2 + (y+bk)^2]\}.$$

As a check of the result, on the circle of radius c, $x = c \cos \theta$ and $y = c \sin \theta$, and so

$$\psi = -(\Gamma/4\pi)\{\ln [(c \cos \theta - a)^2 + (c \sin \theta - b)^2] + \ln [(c \cos \theta + -ak)^2 + (c \sin \theta + bk)^2]$$

$$- \ln [(c \cos \theta - a)^2 + (c \sin \theta + b)^2] - \ln [(c \cos \theta - ak)^2 + (c \sin \theta - bk)^2]\}.$$

But since $\cos^2 \theta + \sin^2 \theta = 1$ and from the definition of k,

$$[(c \cos \theta - ak)^2 + (c \sin \theta - bk)^2] = c^2 - 2k(a \cos \theta + b \sin \theta) + (a^2 + b^2)k^2$$

$$= k[c^2/k - 2(a \cos \theta + b \sin \theta) + (a^2 + b^2)k]$$

$$= k[(a^2 + b^2 - 2(a \cos \theta + b \sin \theta) + c^2]$$

$$= k[(c \cos \theta - a)^2 + (c \sin \theta - b)^2]$$

and

$$[(c \cos \theta - ak)^2 + (c \sin \theta + bk)^2] = c^2 - 2k(a \cos \theta - b \sin \theta) + (a^2 + b^2)k^2$$

$$= k[c^2/k - 2(a \cos \theta - b \sin \theta) + (a^2 + b^2)k]$$

$$= k[(a^2 + b^2 - 2(a \cos \theta - b \sin \theta) + c^2]$$

$$= k[(c \cos \theta - a)^2 + (c \sin \theta + b)^2].$$

Substituting these into the expression of the stream function, we have

$$\psi = -(\Gamma/4\pi)\{\ln [(c \cos \theta - a)^2 + (c \sin \theta - b)^2] + \ln k[(c \cos \theta - a)^2 + (c \sin \theta + b)^2]$$

$$- \ln [(c \cos \theta - a)^2 + (c \sin \theta + b)^2] - \ln k[(c \cos \theta - ak)^2 + (c \sin \theta - bk)^2]\}$$

$$= -(\Gamma/4\pi)\{\ln [(c \cos \theta - a)^2 + (c \sin \theta - b)^2] + \ln [(c \cos \theta - a)^2 + (c \sin \theta + b)^2]$$

$$+ \ln k - \ln [(c \cos \theta - a)^2 + (c \sin \theta + b)^2] - \ln [(c \cos \theta - a)^2 + (c \sin \theta - b)^2] - \ln k\} = 0.$$

Thus the cup of radius c is a streamline.

To find the equations which govern how the vortex at (a, b) moves, the induced velocity components are computed by taking the derivatives of ψ, omitting the term from the vortex at (a, b), and then letting $x = a$, $y = b$. The result is

$$\dot{a} = \frac{\Gamma b}{2\pi}\left[\frac{1}{2b^2} - \frac{(a^2 + b^2)(a^2 + b^2 + c^2)}{a^2(a^2 + b^2 - c^2)^2 + b^2(a^2 + b^2 + c^2)^2} + \frac{1}{a^2 + b^2 - c^2}\right]$$

$$\dot{b} = \frac{\Gamma a}{2\pi}\left[\frac{(a^2 + b^2)(a^2 + b^2 + c^2)}{a^2(a^2 + b^2 - c^2)^2 + b^2(a^2 + b^2 + c^2)^2} - \frac{1}{a^2 + b^2 - c^2}\right].$$

Rankine half-body

A source located at the origin in a uniform stream (Figure 4.18) has the velocity potential and a stream function

$$\phi = xU + (m/2\pi) \ln \sqrt{x^2 + y^2}, \tag{4.5.40}$$

$$\psi = yU + (m/2\pi) \tan^{-1} y/x \tag{4.5.41}$$

in two-dimensional flow, and

$$\phi = zU - m/4\pi\sqrt{r^2 + z^2}, \tag{4.5.42a}$$

$$\psi = 0.5Ur^2 - mz/4\pi\sqrt{r^2 + z^2} \tag{4.5.42b}$$

in three-dimensional flow.

Examining the three-dimensional case in further detail, we find the velocity components to be

$$v_r = \frac{\partial \phi}{\partial r} = \frac{mr}{4\pi(r^2 + z^2)^{3/2}}, \qquad v_z = \frac{\partial \phi}{\partial z} = U + \frac{mz}{4\pi(r^2 + z^2)^{3/2}}. \tag{4.5.43}$$

It is seen from (4.5.43) that there is a stagnation point ($\mathbf{v}_{\text{stagnation point}} = 0$) at the point $r = 0$, $z = -\sqrt{m/4\pi U}$.

On $r = 0$, the stream function takes on values $\psi = -m/4\pi$ for z positive and $\psi = m/4\pi$ for z negative. The streamline $\psi = m/4\pi$ goes from the source to minus infinity. At the stagnation point, however, it bifurcates, and goes along the curve

$$z^2 = (b^2 - 2r^2)^2/4(b^2 - r^2), \tag{4.5.44}$$

with $b^2 = m/\pi U$. This follows by putting $\psi = m/4\pi$ into (4.5.42b) and solving for z. The radius of the body goes from 0 at the stagnation point to $r_\infty = b$ far downstream from the source. This last result can be obtained either from looking for the value of r needed to make z become infinite in (4.5.44), or by realizing that far downstream from the source, the velocity must be U and all the discharge from the source must be contained within the body. In either case the result is that far downstream the body radius is $b + \sqrt{m/\pi U}$.

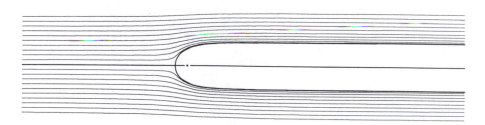

Figure 4.18. Streamlines for a Rankine body made up of a uniform stream and a line source.

This flow could be considered as a model for flow past a pitot tube of a slightly unusual shape. A pitot tube determines velocity by measuring the pressure at the stagnation point and another point far enough down the body so that the speed is essentially U. The difference in pressure between these two points is proportional to U^2. This pressure difference can be found from our analytical results by writing the Bernoulli equation between the stagnation point and infinity.

In the two-dimensional counterpart of this, the velocity is

$$v_x = \frac{\partial \phi}{\partial x} = U + \frac{mx}{2\pi(x^2 + y^2)}, \qquad v_y = \frac{\partial \phi}{\partial y} = \frac{my}{2\pi(x^2 + y^2)},$$

with the stagnation point located at $(-m/2\pi U, 0)$. On $y = 0$, $\psi = 0$ for x positive and $m/2$ for x negative. The streamline $\psi = m/2$ starts at $(-\infty, 0)$, going along the x axis to the source. At the stagnation point the streamline bifurcates, having the shape given by

$$2\pi U y/m = \pi - \tan^{-1} y/x. \tag{4.5.45}$$

The asymptotic half-width of the body is thus $y_\infty = m/2U$.

It is instructive to examine the behavior of the ϕ and ψ lines at the stagnation point. For the two-dimensional case we first expand ψ about the stagnation point in a Taylor series about $(-m/2\pi U, 0)$, giving to second order in $x + m/2\pi U$ and y,

$$\psi = m/2 - (4\pi U^2/m)(x + m/2\pi U)y + \cdots.$$

This tells us that at the bifurcation point, the $\psi = m/2$ line is either along $y = 0$, or perpendicular to it (i.e., locally either on $y = 0$ or on $x = -m/2\pi U$).

Expanding ϕ about $(-m/2\pi U, 0)$, to second order in $x + m/2\pi U$ and y, we have

$$\phi = (m/2\pi)[-1 + \ln (m/2\pi U)] + (2\pi U^2/m)[-(x + m/2\pi U)^2 + y^2] + \cdots.$$

This tells us that $\phi = (m/2\pi)[-1 + \ln (m/2\pi U)]$ along lines with slope ± 45 $[y = \pm(x + m/2\pi U)]$, thereby bisecting the ψ lines as we had earlier shown must happen.

Note that if we wish to model this flow on a Hele-Shaw table, one way of accomplishing this without the necessity of drilling any holes in our table would be to cut out a solid obstacle of the shape given by $\psi = m/2$ and place it on the table aligned with the flow. The flow exterior to the obstacle is the same as if we had drilled a hole and inserted the source.

Rankine oval

The previous Rankine half-body was not closed because there was a net unbalance in mass discharge. By putting an aligned source and sink pair in a uniform stream with the source upstream of a sink, a closed oval shape is obtained. The velocity potential and stream function then become

$$\phi = xU + (m/2\pi)[\ln \sqrt{(x + a)^2 + y^2} - \ln \sqrt{(x - a)^2 + y^2}], \tag{4.5.46a}$$

$$\psi = yU + (m/2\pi)[\tan^{-1} y/(x + a) - \tan^{-1} y/(x - a)] \tag{4.5.46b}$$

for a two-dimensional body, and

$$\phi = zU + (m/4\pi)[1/|\mathbf{r} + a\mathbf{k}| - 1/|\mathbf{r} - a\mathbf{k}|], \tag{4.5.47a}$$

$$\psi = 0.5Ur^2 + (m/4\pi)[(z + a)/|\mathbf{r} + a\mathbf{k}| - (z - a)/|\mathbf{r} - a\mathbf{k}|] \tag{4.5.47b}$$

for a three-dimensional body.

The velocity for the two-dimensional body is

$$v_x = \frac{\partial \phi}{\partial x} = U + \frac{m}{2\pi}\left[\frac{x + a}{(x + a)^2 + y^2} - \frac{x - a}{(x - a)^2 + y^2}\right],$$

$$v_y = \frac{\partial \phi}{\partial y} = \frac{my}{2\pi}\left[\frac{1}{(x + a)^2 + y^2} - \frac{1}{(x - a)^2 + y^2}\right].$$

The stagnation points are therefore at $(\pm\sqrt{a^2 + ma/\pi U}, 0)$.

For the two-dimensional case, the streamline that makes up the body is given by $\psi = 0$, according to (4.5.46b). (Note that along $y = 0$, $\psi = 0$ except in the range $-a < x < a$, where $\psi = -m/2$.) Therefore the equation giving the body shape is

$$0 = yU + (m/2\pi)[\tan^{-1} y/(x + a) - \tan^{-1} y/(x - a)]. \tag{4.5.48}$$

From symmetry, the maximum height of the body will be at $x = 0$. This height is given from (4.5.48) as a solution of the equation

$$y_{max} = (m/2\pi U)(\pi - 2 \tan^{-1} y_{max}/a). \tag{4.5.49}$$

Similar results hold for the three-dimensional case. The parameter governing shape is m/Ua in the two-dimensional case and m/Ua^2 in the three-dimensional case. If m/Ua is large, the body is long and slender. If m/Ua is small, the body is short and rounded.

More complicated Rankine ovals can be formed by putting more sources and sinks in a uniform stream. For the body to close, it is necessary that the sum of the source and sink strengths be zero. This, however, is not a sufficient condition. Notice that in our simple example, if the source and sink were interchanged (just change the sign of m), there will be no closed streamlines about the source and sink pair.

Circular cylinder or sphere in a uniform stream

The Rankine oval is a somewhat unfamiliar geometrical shape, but if one plots its shape for various values of the separation a and the shape parameter m/Ua, it is seen that as the source-sink pair get closer together while the shape parameter is held constant, the oval shape becomes more and more circular. This suggests that in the limit as the source-sink pair becomes a doublet, a circular shape would be achieved. The source portion of the doublet should be facing upstream and the sink portion facing downstream in order to generate a closed stream surface.

The velocity potential and stream function for a uniform stream plus a doublet is

$$\phi = xU + \frac{B_x x}{x^2 + y^2}, \qquad \psi = yU - \frac{B_x y}{x^2 + y^2}, \tag{4.5.50}$$

in two-dimensional flow, and

$$\phi = z\left[U + \frac{B_z}{(r^2 + z^2)^{3/2}}\right], \qquad \psi = r^2\left[0.5U - \frac{B_z}{(r^2 + z^2)^{3/2}}\right], \tag{4.5.51}$$

in three-dimensional flow.

If we let $B_x = Ub^2$ in the two-dimensional case, (4.5.50) shows that $\psi = 0$ both on $y = 0$ and on $x^2 + y^2 = b^2$. Therefore we have flow past a circular cylinder of radius b. Letting $B_z = Ub^3/2$ in three dimensions gives $\psi = 0$ on a sphere of radius b.

How does this relate to the method of images? The interpretation is complicated by having to consider a curved mirror, but the flow can be thought of as the body focusing the uniform stream upstream (a very large distributed source) into the source part of the doublet. The downstream part of the uniform stream (a very large distributed sink) is focused into the sink part of the doublet.

More orderly ways of distributing sources to generate flows about given body shapes are known. For thin bodies such as wings or airplane fuselages, sources are distributed on the centerline, the strength of the source distribution per unit length being proportional to the rate at which the cross-sectional area changes. For more complicated shapes, sources are distributed on the surface of the body. Vortices are included if lift forces are needed, as indicated in the following section.

Lift forces

To compute the force on any body due to inviscid effects, it is necessary to carry out the integration

$$\mathbf{F} = \iint -p\mathbf{n}\, dA,$$

where $\mathbf{n}$ is the unit outward normal on the body surface and the integration is taken over the entire surface of the body. It can be shown that for bodies generated by source-sink distributions, there will never be forces perpendicular to the uniform stream. These forces are the *lift forces* that we would normally expect to find even with the neglect of viscous effects. This absence of lift can be corrected by including vorticity in any model where lift forces are desired. For instance, for the cylinder in the previous example, including a vortex at the center of the cylinder would give the velocity potential and stream function

$$\phi = xU + \frac{B_x x}{x^2 + y^2} + \frac{\Gamma}{2\pi}\tan^{-1}\frac{y}{x}, \qquad \psi = yU - \frac{B_x y}{x^2 + y^2} + \frac{\Gamma}{2\pi}\ln\sqrt{x^2 + y^2}. \tag{4.5.52}$$

It is seen that the stream function is constant on the cylinder $x^2 + y^2 = a^2$; therefore the boundary condition on the body is still satisfied. Evaluation of the pressure force now, however, gives a lift force proportional to $\rho U \Gamma$, called the *Magnus effect* after its discoverer.

Where would this vorticity come from in a physical situation? We could rotate the cylinder, and the effects of viscosity thus provide the tangential velocity that is provided in our mathematical model by the vortex. This has been attempted in ships and experimental airplanes, but it requires an additional power source and is not generally practical. The "effect" of this rotation is instead provided by having a sharp trailing edge for a wing, or by providing a "flap" on a blunter body. This is done to force the velocity on the body to appear the same as in our model and thereby generate the desired force. The relationship between lift force and vorticity is called the *Joukowski theory of lift*.

6. Rates of Deformation

In our study of inviscid flows, we looked at the behavior of three neighboring points *ABC* (Figure 4.7) that were chosen to make up a right angle at an initial time *t*. We saw how the motion of these points through a time interval *dt* described the rotation of a fluid element, and we defined a quantity we called vorticity. At that point in our study, it was not necessary for our study of the motion of inviscid flows to complete the kinematic analysis of the motion of these three points. This additional information is however needed for viscous flows, so we resume our analysis at the point where we left off.

Looking first at changes of length, we see that after a time interval *dt* point *B* has moved a distance $v_x\, dt$ measured along the *x* axis. Point *A*, initially a distance *dx* from *B*, has moved a distance $[v_x + (\partial v_x/\partial x)\, dx]\, dt$ measured along the *x* axis. The rate of change of length along the *x* axis per unit length, which we will denote by d_{xx}, is this change in length divided by the original length, all divided by *dt*, or

$$d_{xx} = \frac{\left(v_x + \dfrac{\partial v_x}{\partial x}\, dx\right) dt - v_x}{dx\, dt} = \frac{\partial v_x}{\partial x}. \tag{4.6.1}$$

A similar analysis along the *y* axis would give the rate of change of length per unit length as measured along the *y* axis, d_{yy}, as

$$d_{yy} = \frac{\left(v_y + \dfrac{\partial v_y}{\partial y}\, dy\right) dt - v_y}{dy\, dt} = \frac{\partial v_y}{\partial y}, \tag{4.6.2}$$

and similarly in the *z* direction,

$$d_{zz} = \frac{\left(v_z + \dfrac{\partial v_z}{\partial z}\, dz\right) dt - v_z}{dz\, dt} = \frac{\partial v_z}{\partial z}. \tag{4.6.3}$$

The d_{xx}, d_{yy}, and d_{zz} are the ***normal rates of deformation***, and can loosely be thought of as rates of normal, or extensional, strain. The term "loosely" is used since the definitions of strain you are probably familiar with from the study of solid mechanics are for infinitesimal strains. In fluid mechanics, strains are always finite, and there are many definitions of finite strain and rates of strain.

Besides changes of length, changes of angles are involved in the deformation. While discussing vorticity earlier in this chapter, we showed from Figure 4.7 that

$$\dot{\theta}_1 = \partial v_y/\partial x \qquad \text{and} \qquad \dot{\theta}_2 = \partial v_x/\partial y,$$

and the difference of the two angular rates made up one of the components of vorticity. The sum of the two angular rates,

$$\dot{\theta}_1 + \dot{\theta}_2 = \partial v_y/\partial x + \partial v_x/\partial y,$$

represents the rate of change of the angle ABC. We let

$$d_{xy} = d_{yx} = \frac{1}{2}(\dot{\theta}_1 + \dot{\theta}_2) = \frac{1}{2}\left(\frac{\partial v_x}{\partial y} + \frac{\partial v_y}{\partial x}\right). \tag{4.6.4}$$

be the ***rate of shear deformation*** as measured in the xy plane. Similarly we define

$$d_{yz} = d_{zy} = \frac{1}{2}\left(\frac{\partial v_y}{\partial z} + \frac{\partial v_z}{\partial y}\right), \tag{4.6.5}$$

$$d_{zx} = d_{xz} = \frac{1}{2}\left(\frac{\partial v_z}{\partial x} + \frac{\partial v_x}{\partial z}\right) \tag{4.6.6}$$

to be the rates of shear deformation as measured in the yx and xz planes, respectively.

The one-half factor in the definition of the rate of deformation components is introduced so that the components transform independent of axis selection. As in our definition of vorticity, what we are primarily interested in is some measure of the deformation. In this case we wish to relate the deformation rate to stress.

It may be helpful to your physical understanding of rate of deformation to look at what is happening from a slightly different viewpoint. Consider any two neighboring fluid particles a distance $d\mathbf{r}$ apart, where the distance $d\mathbf{r}$ changes with time but must remain small since the particles were initially close together. To find the rate at which the particles separate, we differentiate $d\mathbf{r}$, obtaining

$$\frac{D(d\mathbf{r})}{Dt} = d\left(\frac{D\mathbf{r}}{Dt}\right) = d\mathbf{v}, \tag{4.6.7}$$

where $d\mathbf{v}$ is the difference in velocity between the two points as shown in Figure 4.19. Since the magnitude of the distance between the two points, or more conveniently, its square, is $dr^2 = d\mathbf{r} \cdot d\mathbf{r}$, we have

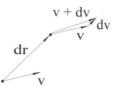

Figure 4.19. Two points separated by a vector distance $d\mathbf{r}$, showing the velocity difference.

$$\frac{D(dr^2)}{Dt} = 2\,d\mathbf{r} \cdot \frac{D(d\mathbf{r})}{Dt} = 2\,d\mathbf{r} \cdot d\mathbf{v} = 2\,dx\left(\frac{\partial v_x}{\partial x}\,dx + \frac{\partial v_x}{\partial y}\,dy + \frac{\partial v_x}{\partial x}\,dz\right)$$

$$+\,2\,dy\left(\frac{\partial v_y}{\partial x}\,dx + \frac{\partial v_y}{\partial y}\,dy + \frac{\partial v_y}{\partial x}\,dz\right) + 2\,dz\left(\frac{\partial v_z}{\partial x}\,dx + \frac{\partial v_z}{\partial y}\,dy + \frac{\partial v_z}{\partial x}\,dz\right)$$

$$=\,2\frac{\partial v_x}{\partial x}\,dx^2 + 2\frac{\partial v_y}{\partial y}\,dy^2 + 2\frac{\partial v_z}{\partial z}\,dz^2$$

$$+\left(\frac{\partial v_x}{\partial y} + \frac{\partial v_y}{\partial x}\right)dx\,dy + \left(\frac{\partial v_x}{\partial z} + \frac{\partial v_z}{\partial x}\right)dx\,dz + \left(\frac{\partial v_y}{\partial z} + \frac{\partial v_z}{\partial y}\right)dy\,dz$$

$$=\,2(d_{xx}\,dx^2 + d_{yy}\,dy^2 + d_z\,dz^2 + 2\,d_{xy}\,dx\,dy + 2\,d_{xz}\,dx\,dz + 2\,d_{yz}\,dy\,dz).$$

$$(4.6.8)$$

Thus after choosing the two points that we wish to describe (i.e., selecting $d\mathbf{r}$), to find the rate at which the distance between the points change, we need to know the local values of the six components of the rate of deformation, that is, $d_{xx}, d_{yy}, d_{zz}, d_{xy}, d_{yz}, d_{zx}$.

Note that

$$\nabla \cdot \mathbf{v} = d_{xx} + d_{yy} + d_{zz}, \qquad (4.6.9)$$

so for incompressible flows the sum of the three normal components of the rate of deformation will always be zero by continuity. Note also that vorticity has no effects on length changes.

Example 4.6.1. Rigid-body rotation
Find the rate of deformation for rigid-body rotation as given by the velocity field $(-y\Omega, x\Omega, 0)$.

Given: Velocity components for a two-dimensional incompressible flow.
Solution: From the definition of rate of deformation, $d_{xx} = d_{yy} = d_{zz} = 0$, $d_{xy} = d_{yz} = d_{xz} = 0$. The absence of rates of deformation confirms that the fluid is behaving as a rotating rigid-body.

Example 4.6.2. Vortex motion
Find the rate of deformation for a line vortex with velocity $\mathbf{v} = [-yG/(x^2 + y^2), xG/(x^2 + y^2), 0]$.

Given: Velocity components for a two-dimensional incompressible flow.
Solution: Again from the definition,

$$d_{xx} = -d_{yy} = 2xyG/(x^2 + y^2)^2, \qquad d_{zz} = 0, \qquad d_{xy} = xyG/(x^2 + y^2)^2, \qquad d_{xz} = d_{yz} = 0.$$

7. Stress

When we treat a material as a continuum, a force must be applied as a quantity distributed over an area. (In analysis, a concentrated force can be a convenient idealization. In a real material, any concentrated force would provide very large changes—in fact infinite changes—both in deformation and in the material.) For our control volume of Figure 4.1 we consider first a force $\Delta\mathbf{F}$ acting on a surface ΔA_x with unit normal pointing in the x direction; thus $\mathbf{n} = \mathbf{i}$. Then we write the stress on this face of our box as

$$\tau^{(x)} = \lim_{\Delta A_x = 0} \Delta\mathbf{F}/\Delta A_x = \tau_{xx}\mathbf{i} + \tau_{xy}\mathbf{j} + \tau_{xz}\mathbf{k}, \tag{4.7.1}$$

where τ_{xx} is the limit of the x component of the force per unit area acting on this face, τ_{xy} is the limit of the y component of the force per unit area acting on this face, and τ_{xz} is the limit of the z component of the force per unit area acting on this face.

Similarly, for normals pointing in the $y(\mathbf{n} = \mathbf{j})$ and $z(\mathbf{n} = \mathbf{k})$ directions we would have

$$\tau^{(y)} = \tau_{yx}\mathbf{i} + \tau_{yy}\mathbf{j} + \tau_{yz}\mathbf{k} \tag{4.7.2}$$

and

$$\tau^{(z)} = \tau_{zx}\mathbf{i} + \tau_{zy}\mathbf{j} + \tau_{zz}\mathbf{k}. \tag{4.7.3}$$

Stress components associated with force components acting in the same direction as their normal (i.e., τ_{xx}, τ_{yy}, τ_{zz}) are referred to as ***normal stresses***. Stress components associated with force components acting perpendicular to their normal (i.e., τ_{yx}, τ_{yz}, τ_{zx}) are referred to as ***shear stresses***. Note that the first subscript on the components tells us the direction in which the area faces, and the second subscript gives us the direction of the force component on that face. Positive sign conventions for viscous stress are given in Figure 4.20.

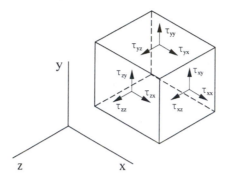

Figure 4.20. Three-dimensional differential control volume showing positive directions of stress. On the three "invisible" faces with normals pointing in negative coordinate directions, the stresses would be reversed from their counterparts on faces with normals pointing in positive coordinate directions.

We next apply Newton's law to a control volume. This follows exactly what we did earlier in this chapter in deriving Euler's equation. We expand the viscous forces in Taylor series about the center of our control volume. These are shown in Figure 4.21 for the two-dimensional case. The various terms are summarized in Table 4.5.

If we add the net forces in the x, y, and z directions to the right-hand sides of equations (4.3.1a), (4.3.1b), and (4.3.1c), respectively, the result after dividing by the volume is

x direction:
$$\frac{\partial(\rho v_x)}{\partial t} + v_x \frac{\partial(\rho v_x)}{\partial x} + \rho v_x \frac{\partial v_x}{\partial x} + v_x \frac{\partial(\rho v_y)}{\partial y} + \rho v_y \frac{\partial v_x}{\partial y}$$

$$+ v_x \frac{\partial(\rho v_z)}{\partial z} + \rho v_z \frac{\partial v_x}{\partial z} = -\frac{\partial p}{\partial x} + \rho g_x + \frac{\partial \tau_{xx}}{\partial x} + \frac{\partial \tau_{yx}}{\partial y} + \frac{\partial \tau_{zx}}{\partial z}, \quad (4.7.4a)$$

y direction:
$$\frac{\partial(\rho v_y)}{\partial t} + v_y \frac{\partial(\rho v_x)}{\partial x} + \rho v_x \frac{\partial v_y}{\partial x} + v_y \frac{\partial(\rho v_y)}{\partial y} + \rho v_y \frac{\partial v_y}{\partial y}$$

$$+ v_y \frac{\partial(\rho v_z)}{\partial z} + \rho v_z \frac{\partial v_y}{\partial z} = -\frac{\partial p}{\partial y} + \rho g_y + \frac{\partial \tau_{xy}}{\partial x} + \frac{\partial \tau_{yy}}{\partial y} + \frac{\partial \tau_{zy}}{\partial z}, \quad (4.7.4b)$$

z direction:
$$\frac{\partial(\rho v_z)}{\partial t} + v_z \frac{\partial(\rho v_x)}{\partial x} + \rho v_x \frac{\partial v_z}{\partial x} + v_z \frac{\partial(\rho v_y)}{\partial y} + \rho v_y \frac{\partial v_z}{\partial y}$$

$$+ v_z \frac{\partial(\rho v_z)}{\partial z} + \rho v_z \frac{\partial v_z}{\partial z} = -\frac{\partial p}{\partial z} + \rho g_z + \frac{\partial \tau_{xz}}{\partial x} + \frac{\partial \tau_{yz}}{\partial y} + \frac{\partial \tau_{zz}}{\partial z}, \quad (4.7.4c)$$

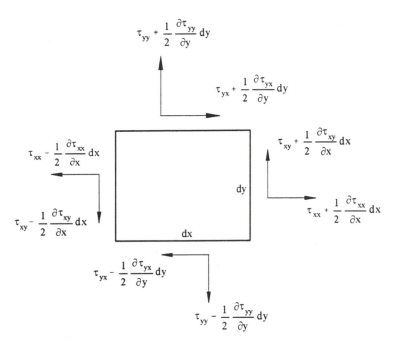

Figure 4.21. Two-dimensional differential control volume showing variations of the stress components on the faces of the control volume.

or in vector form,

$$\rho \frac{Dv}{Dt} = \rho \mathbf{g} + \nabla p + \mathbf{i} \left(\frac{\partial \tau_{xx}}{\partial x} + \frac{\partial \tau_{yx}}{\partial y} + \frac{\partial \tau_{zx}}{\partial z} \right)$$

$$+ \mathbf{j} \left(\frac{\partial \tau_{xy}}{\partial x} + \frac{\partial \tau_{yy}}{\partial y} + \frac{\partial \tau_{zy}}{\partial z} \right) + \mathbf{k} \left(\frac{\partial \tau_{xz}}{\partial x} + \frac{\partial \tau_{yz}}{\partial y} + \frac{\partial \tau_{zz}}{\partial z} \right). \tag{4.7.5}$$

Moments can be balanced in the same manner as forces. Using our control volume and taking moments about any corner of the box, we find the moments of the viscous forces to be of third order in the control volume dimensions, while the moments of the pressure and gravity forces and of the moment of momentum are of fourth order. The result tells us that

$$\tau_{xy} = \tau_{yx}, \quad \tau_{xz} = \tau_{zx}, \quad \tau_{yz} = \tau_{zy}. \tag{4.7.6}$$

Therefore the stress is symmetric in the subscripts.

Table 4.5. Viscous forces and their summation

Face normal pointing in	Force in x direction	Force in y direction	Force in z direction
$+x$ direction	$\left(\tau_{xx}+\dfrac{\partial\tau_{xx}}{\partial x}\dfrac{dx}{2}\right)dy\,dz$	$\left(\tau_{xy}+\dfrac{\partial\tau_{xy}}{\partial x}\dfrac{dx}{2}\right)dy\,dz$	$\left(\tau_{xz}+\dfrac{\partial\tau_{xz}}{\partial x}\dfrac{dx}{2}\right)dy\,dz$
$+y$ direction	$\left(\tau_{yx}+\dfrac{\partial\tau_{yx}}{\partial y}\dfrac{dy}{2}\right)dx\,dz$	$\left(\tau_{yy}+\dfrac{\partial\tau_{yy}}{\partial y}\dfrac{dy}{2}\right)dx\,dz$	$\left(\tau_{yz}+\dfrac{\partial\tau_{yz}}{\partial y}\dfrac{dy}{2}\right)dx\,dz$
$+z$ direction	$\left(\tau_{zx}+\dfrac{\partial\tau_{zx}}{\partial z}\dfrac{dz}{2}\right)dx\,dy$	$\left(\tau_{zy}+\dfrac{\partial\tau_{zy}}{\partial z}\dfrac{dz}{2}\right)dx\,dy$	$\left(\tau_{zz}+\dfrac{\partial\tau_{zz}}{\partial z}\dfrac{dz}{2}\right)dx\,dy$
$-x$ direction	$\left(\tau_{xx}-\dfrac{\partial\tau_{xx}}{\partial x}\dfrac{dx}{2}\right)dy\,dz$	$\left(\tau_{xy}-\dfrac{\partial\tau_{xy}}{\partial x}\dfrac{dx}{2}\right)dy\,dz$	$\left(\tau_{xz}-\dfrac{\partial\tau_{xz}}{\partial x}\dfrac{dx}{2}\right)dy\,dz$
$-y$ direction	$\left(\tau_{yx}-\dfrac{\partial\tau_{yx}}{\partial y}\dfrac{dy}{2}\right)dx\,dz$	$\left(\tau_{yy}-\dfrac{\partial\tau_{yy}}{\partial y}\dfrac{dy}{2}\right)dx\,dz$	$\left(\tau_{yz}-\dfrac{\partial\tau_{yz}}{\partial y}\dfrac{dy}{2}\right)dx\,dz$
$-z$ direction	$\left(\tau_{zx}-\dfrac{\partial\tau_{zx}}{\partial z}\dfrac{dx}{2}\right)dx\,dy$	$\left(\tau_{zy}-\dfrac{\partial\tau_{zy}}{\partial z}\dfrac{dx}{2}\right)dx\,dy$	$\left(\tau_{zz}-\dfrac{\partial\tau_{zz}}{\partial z}\dfrac{dx}{2}\right)dx\,dy$

Net force in x direction	$\dfrac{\partial\tau_{xx}}{\partial x}dx\,dy\,dz+\dfrac{\partial\tau_{yx}}{\partial y}dx\,dy\,dz+\dfrac{\partial\tau_{zx}}{\partial z}dx\,dy\,dz$
Net force in y direction	$\dfrac{\partial\tau_{xy}}{\partial x}dx\,dy\,dz+\dfrac{\partial\tau_{yy}}{\partial y}dx\,dy\,dz+\dfrac{\partial\tau_{zy}}{\partial z}dx\,dy\,dz$
Net force in z direction	$\dfrac{\partial\tau_{xz}}{\partial x}dx\,dy\,dz+\dfrac{\partial\tau_{yz}}{\partial y}dx\,dy\,dz+\dfrac{\partial\tau_{zz}}{\partial z}dx\,dy\,dz$

8. Constitutive Relations

The physical laws so far presented, while of general validity, still allow very little to be said in the way of definite statements concerning the behavior of any substance, as can be ascertained by noting that at this point we have many more unknowns then we have equations. The "missing" relations are those that describe how a material is made up, or constituted, and relate stress to the geometric and thermodynamic variables. Hooke's law and the state equations of an ideal gas are two familiar examples of *constitutive equations*. While in a few cases constitutive relations can be derived from statistical mechanics considerations using special mathematical models for the molecular structure, the usual procedure is to decide, based on experiments, which quantities must go into the constitutive equation, and then formulate from these a set of equations that agree with fundamental ideas such as invariance with respect to the observer and the like.

There is much inventiveness in going from the first step to the second, and much attention has been given to the subject in recent decades. The mental process of generating a description of a particular fluid involves a continuous interchange between theory and practice. Once a constitutive model is proposed, mathematical predictions can be made that then hopefully can be compared with experiment. Such a procedure can show a model to be wrong, but cannot guarantee that it will always be correct, since many models can predict the same velocity field for the very simple flows used in viscometry and rheogoniometry. As an example, if a fluid is contained between two large plane sheets placed parallel to each other, and the sheets are allowed to move in their planes at different velocities, many constitutive models will predict a fluid velocity varying linearly with the distance from one plate and giving the same shear stress. The various models usually, however, predict quite different normal stresses.

The familiar model presented by Newton and elaborated on by Navier and Stokes has withstood many of these tests for fluids of relatively simple molecular structure, and holds for many of the fluids that one normally encounters. This model is not valid for fluids such as polymers, suspensions, or many of the fluids encountered in the kitchen such as cake batter, catsup, and the like. A good rule of thumb is that, if the molecular weight of the fluid is less than a half million or so, and if the distance between molecules is not too great (as in rarefied gases), a Newtonian model is likely to be valid.

We will not delve deeper into a rigorous justification of a particular constitutive equation here. We simply put down the minimum requirements that we expect of our constitutive law, and give a partial justification of the results.

Considering a fluid of simple molecular structure such as water or air, experience and many experiments suggest the following:

1. Stress will depend explicitly on only pressure and the rate of deformation. Temperature can enter implicitly through coefficients such as viscosity.
2. When the rate of deformation is identically zero, all shear stresses vanish, and the normal stresses are each equal to the negative of the pressure.
3. The fluid is isotropic. That is, the material properties of a fluid at a point are the same in all directions.
4. The stress must depend on rate of deformation in a linear manner, according to the original concepts of Newton.

The most general constitutive relation satisfying all of the above requirements is

$$\tau_{xx} = \mu^1 \nabla \cdot \mathbf{v} + 2\mu d_{xx}, \qquad \tau_{yy} = \mu^1 \nabla \cdot \mathbf{v} + 2\mu d_{yy}, \qquad \tau_{zz} = \mu^1 \nabla \cdot \mathbf{v} + 2\mu d_{zz},$$

$$\tau_{xy} = 2\mu d_{xy}, \qquad \tau_{xz} = 2\mu d_{xz}, \qquad \tau_{yz} = 2\mu d_{yz}, \qquad\qquad (4.8.1)$$

where we have used the abbreviation

$$d_{xx} + d_{yy} + d_{zz} = \frac{\partial v_x}{\partial x} + \frac{\partial v_y}{\partial y} + \frac{\partial v_z}{\partial z} = \nabla \cdot \mathbf{v},$$

Here μ is the **viscosity** and μ^1 is the **second viscosity coefficient**. Both of these viscosities can depend on temperature and even pressure. The fluid described by (4.8.1) is called a **Newtonian fluid**, although occasionally the term "Navier-Stokes fluid" is also used.

Up until now we have deliberately left pressure undefined, and the reader no doubt has assumed all along an implicit understanding of what we meant by the symbol p. We must note, however, that people mean many different things by the term pressure. For instance, in elementary thermodynamics texts, the term pressure is commonly used for the negative of mean normal stress. Summing our constitutive equations gives

Mean normal stress $\equiv 1/3$(sum of the total normal stress components)

$$= -p + \frac{\tau_{xx} + \tau_{yy} + \tau_{zz}}{3}.$$

From (4.8.1), however, we see that

$$\frac{\tau_{xx} + \tau_{yy} + \tau_{zz}}{3} = \left(\mu^1 + \frac{2}{3}\mu\right)\nabla \cdot \mathbf{v}. \tag{4.8.2}$$

The coefficient $\mu^1 + (2/3)\mu$ is called the **bulk viscosity**, or **volume viscosity**, since it represents the amount of normal stress change needed to get a unit specific volume rate change. If we are to have the mean normal stress equal to the negative of pressure, it must be that the bulk viscosity must be zero. Stokes at one time suggested that this might in general be true. Since for most flows the term $[\mu^1 + (2/3)\mu]\nabla \cdot \mathbf{v}$ is numerically much smaller than p, this assumption is widely used. Later, however, Stokes wrote that he never put much faith in this relation.

Usually one thinks of p as being the thermodynamic pressure, given by an equation of state (e.g., $p = \rho RT$ for an ideal gas). In such a case, the bulk modulus is not necessarily zero [the second law of thermodynamics can be used to show that $\mu^1 + (2/3)\mu \geq 0$], and so the thermodynamic pressure generally must differ from the mean normal stress. Values of μ^1 for various fluids have been determined experimentally in flows involving very-high-speed sound waves,[2] but the data are still quite sparse. Statistical mechanics tells us that for a monatomic gas, the bulk viscosity is zero. In any case, in flows where both the dilatation and the bulk viscosity tend to be very small compared to pressure, the effects of the bulk viscosity can usually be neglected.

Some elaboration of the four postulates for determining our constitutive equation is in order. We have said that the only kinematic quantity appearing in the stress is the rate of deformation. What about strain or vorticity?

The type of fluid we are considering is completely without a sense of history. (One class of fluids with a sense of history of their straining is the **viscoelastic fluids**. For these fluids strain and strain rates appear explicitly in their constitutive equations.) A Newtonian fluid is aware only of the present. It cannot remember the past, even the immediate past, and hence strain cannot enter the model. While such a model predicts most of the basic features of flows, it does have its disturbing aspects, such as infinite speed of propagation of information. For most flows, however, it seems a reasonable assumption.

[2]See, for example, Lieberman, *Phys. Rev.*, vol. **75**, 1955.

Should vorticity enter into the constitutive equation? What has been called the "principle of material objectivity," or "principle of isotropy of space," or "material frame indifference," among other things, states that all observers, regardless of their frame of reference (inertial or otherwise), must observe the same material behavior. Therefore an observer stationed on a rotating platform, say, sees the same fluid behavior as an observer standing on the floor of the laboratory. As we have seen in the last chapter, vorticity is not satisfactory in this regard, in that it is sensitive to rigid rotations. (If the reader finds the idea of material frame indifference unsettling, read page 6 of the book by Truesdell in the references at the end of the chapter. Truesdell was one of the first proponents of this concept, but apparently had many doubts on the same question initially.) There have been many constitutive equations postulated that violate this principle (both intentionally and unintentionally). Present work in constitutive equations tends to obey the principle religiously, although doubts are still sometimes expressed.

Our second assumption, that when the rate of deformation is zero the stresses reduce to the pressure, is simply a reaffirmation of the principles of hydrostatics, and is a basic law used for practically all materials.

The isotropy of a fluid is a realistic assumption for a fluid of simple molecular structure. If we had in mind materials made up of small rods, ellipsoids, or complicated molecular chains, all of which have directional properties, other models would be called for, and this constraint would have to be relaxed.

The linearity assumption can be justified only by experiment. The remarkable thing is that it quite often works! If we relax this point, but retain all the other assumptions, the effect is to add only one term to the right side of (4.8.1), and to note that the various viscosities can depend on invariant combinations of the rate of strain as well as on the thermodynamic variables. While this adds to the mathematical generality, no fluids are presently known to behave according to this more complicated law.

We have already remarked that a state equation is also a necessary addition to the constitutive description of our fluids. Examples frequently used are incompressibility ($D\rho/Dt = 0$) and the ideal gas law ($p = \rho RT$). Additionally information on the heat flux and internal energy must be added to the list. Familiar laws are Fourier's law of heat conduction, where the heat flux is proportional to the negative of the temperature gradient, or

$$\mathbf{q} = -k\,\nabla T, \tag{4.8.3}$$

and the ideal gas relation

$$U = U(T). \tag{4.8.4}$$

The latter is frequently simplified further by the assumption of linearity, so that

$$U = U_0 + c_\rho(T - T_0), \tag{4.8.5}$$

c_ρ being the specific heat at constant density or volume.

Our knowledge of the constitutive behavior of non-Newtonian fluids is unfortunately much more sparse than our knowledge of Newtonian fluids. Particularly with the ever-increasing use of plastics in our modern society, the ability to predict the behavior of such fluids is of great economic importance in manufacturing processes. Unfortunately, while many theoretical models have been put forth over the last century, the situation in general is far from satisfactory. In principle, from a few simple experiments the parameters in a given constitutive model can be found. Then predictions of other flow geometries can be made from this model, and compared with further experiments. The result more often than not is that the predictions may be valid for only a very few simple flows whose nature is closely related to the flows from which the parameters in the constitutive model were determined. There are thus many gaps in our fundamental understanding of these fluids.

9. Equations for Newtonian Fluids

Substitution of equation (4.8.1) into equation (4.7.5) gives the result

$$\rho \frac{Dv_x}{Dt} = -\frac{\partial}{\partial x}(p - \mu^1 \nabla \cdot \mathbf{v}) + \rho g_x + \frac{\partial}{\partial x}\left[\mu\left(\frac{\partial v_x}{\partial x} + \frac{\partial v_x}{\partial x}\right)\right] + \frac{\partial}{\partial y}\left[\mu\left(\frac{\partial v_x}{\partial y} + \frac{\partial v_y}{\partial x}\right)\right]$$
$$+ \frac{\partial}{\partial z}\left[\mu\left(\frac{\partial v_x}{\partial z} + \frac{\partial v_z}{\partial x}\right)\right],$$

$$\rho \frac{Dv_y}{Dt} = -\frac{\partial}{\partial y}(p - \mu^1 \nabla \cdot \mathbf{v}) + \rho g_y + \frac{\partial}{\partial x}\left[\mu\left(\frac{\partial v_y}{\partial x} + \frac{\partial v_x}{\partial y}\right)\right] + \frac{\partial}{\partial y}\left[\mu\left(\frac{\partial v_y}{\partial y} + \frac{\partial v_y}{\partial y}\right)\right]$$
$$+ \frac{\partial}{\partial z}\left[\mu\left(\frac{\partial v_y}{\partial z} + \frac{\partial v_z}{\partial y}\right)\right],$$

$$\rho \frac{Dv_z}{Dt} = -\frac{\partial}{\partial z}(p - \mu^1 \nabla \cdot \mathbf{v}) + \rho g_z + \frac{\partial}{\partial x}\left[\mu\left(\frac{\partial v_z}{\partial x} + \frac{\partial v_x}{\partial z}\right)\right] + \frac{\partial}{\partial y}\left[\mu\left(\frac{\partial v_z}{\partial y} + \frac{\partial v_y}{\partial z}\right)\right]$$
$$+ \frac{\partial}{\partial z}\left[\mu\left(\frac{\partial v_z}{\partial z} + \frac{\partial v_z}{\partial z}\right)\right]. \tag{4.9.1}$$

When ρ and μ are constant, and for incompressible flows, this simplifies greatly with the help of the continuity condition $\nabla \cdot \mathbf{v} = 0$ to the components

$$\rho \frac{Dv_x}{Dt} = -\frac{\partial p}{\partial x} + \rho g_x + \mu \nabla^2 v_x,$$

$$\rho \frac{Dv_y}{Dt} = -\frac{\partial p}{\partial y} + \rho g_y + \mu \nabla^2 v_y,$$

$$\rho \frac{Dv_z}{Dt} = -\frac{\partial p}{\partial z} + \rho g_z + \mu \nabla^2 v_z, \tag{4.9.2}$$

where

$$\nabla^2 = \partial^2/\partial x^2 + \partial^2/\partial y^2 + \partial^2/\partial z^2$$

is the Laplace operator.

Equation (4.9.2a) can be written in vector notation as

$$\rho \frac{D\mathbf{v}}{Dt} = -\nabla p + \rho \mathbf{g} + \mu \nabla^2 \mathbf{v}. \tag{4.9.3}$$

Either form (4.9.2) or (4.9.3) is referred to as the ***Navier-Stokes equations***.

In Cartesian coordinates with $\mathbf{v} = (u, v, w)$, (4.9.2) becomes

$$\rho \left[\frac{\partial u}{\partial t} + u\frac{\partial u}{\partial x} + v\frac{\partial u}{\partial y} + w\frac{\partial u}{\partial z} \right] = -\frac{\partial p}{\partial x} + \rho g_x + \mu \nabla^2 u,$$

$$\rho \left[\frac{\partial v}{\partial t} + u\frac{\partial v}{\partial x} + v\frac{\partial v}{\partial y} + w\frac{\partial v}{\partial z} \right] = -\frac{\partial p}{\partial y} + \rho g_y + \mu \nabla^2 v,$$

$$\rho \left[\frac{\partial w}{\partial t} + u\frac{\partial w}{\partial w} + v\frac{\partial w}{\partial y} + w\frac{\partial w}{\partial z} \right] = -\frac{\partial p}{\partial z} + \rho g_z + \mu \nabla^2 w. \tag{4.9.4}$$

This last form (4.9.4) is the form that we will use when solving problems in a Cartesian coordinate system. The Navier-Stokes equations in other coordinate systems are given in Appendix C.

Another equation that is derived from the Navier-Stokes equation and that is of some use is the ***vorticity equation***, obtained by taking the curl of (4.9.3). Again we consider only the case where ρ and μ are constants. As we did for the Euler equations, we first divide (4.9.3) by ρ. After a good deal of rearrangement we are left with

$$\frac{D\boldsymbol{\omega}}{Dt} = (\boldsymbol{\omega} \cdot \nabla)\mathbf{v} + \nabla \times \mathbf{g} + \nu \nabla^2 \boldsymbol{\omega}, \tag{4.9.5}$$

where ν is the kinematic viscosity. The first two terms on the right-hand side are familiar from our study of inviscid flows, and represent the change in the vorticity vector as we follow the flow. The last term is typical of a diffusion process, and represents the diffusion of vorticity by viscosity.

10. Boundary Conditions

In order to obtain a solution of the preceding equations that suits a particular problem, it is necessary to add conditions that need to be satisfied on the boundaries of the region of interest. The conditions that are most commonly encountered are the following:

1. The fluid velocity component normal to an impenetrable boundary is always equal to the normal velocity of the boundary. If $\mathbf{n}$ is the unit normal to the boundary, then

$$\mathbf{n} \cdot (\mathbf{v}_{fluid} - \mathbf{v}_{boundary}) = 0 \tag{4.10.1}$$

on the boundary. If this condition were not true, fluid would pass through the boundary. This condition must hold true even in the case of vanishing viscosity ("inviscid flows").

    If the boundary is moving, as in the case of a flow with a free surface or moving body, then, with $F(\mathbf{x}, t) = 0$ as the equation of the bounding surface, (4.10.1) is satisfied if

$$\frac{DF}{Dt} = 0 \text{ on the surface } F = 0. \tag{4.10.2}$$

This condition is necessary to establish that $F = 0$ is a material surface, i.e., a surface moving with the fluid and that always contains the same fluid particles.

2. Stress must be continuous everywhere within the fluid. If stress were not continuous, an infinitesimal layer of fluid with an infinitesimal mass would be acted upon by a finite force, giving rise to infinite acceleration of that layer.

    At interfaces where fluid properties such as density are discontinuous, there can, however, be a discontinuity in stress. This stress difference is related to the surface tension. Write the stress in the direction normal to the interface as $\tau^{(\mathbf{n})}$ and denote the surface tension by σ (a force per unit length). By summing forces on an area of the interface, if the surface tensile force acts outwardly along the edge of S, in a direction locally tangent to both S and C, the result is

$$\Delta\tau^{(\mathbf{n})} = \tau^{(\mathbf{n})}_{lower\ fluid} - \tau^{(\mathbf{n})}_{upper\ fluid} = -\nabla\sigma + \mathbf{n}\sigma(1/R_1 + 1/R_2), \tag{4.10.3}$$

where $\mathbf{n}$ is the unit normal directed into the upper fluid. Here R_1 and R_2 are the principal radii of curvature of the interface. By taking components of this equation in directions locally normal and tangential to the surface, we can conveniently split equation (4.10.3) into

$$\mathbf{n} \cdot \Delta\tau^{(\mathbf{n})} = \sigma(1/R_1 + 1/R_2), \tag{4.10.4}$$

$$\mathbf{t} \cdot \Delta\tau^{(\mathbf{n})} = -\mathbf{t} \cdot \nabla\sigma, \tag{4.10.5}$$

where $\mathbf{n}$ is the unit normal and $\mathbf{t}$ is a unit tangent to the surface. In words, if surface tension is present, the difference in normal stress is proportional to the local surface curvature. If gradients in the surface tension can exist, shear stress discontinuities can also be present across an interface.

3. Velocity must be continuous everywhere. That is, in the interior of a fluid, there can be no discrete changes in $\mathbf{v}$. If there were such changes, it would give rise to discontinuous deformation gradients and, from the constitutive equations, discontinuous stresses.

The velocity of most fluids at a solid boundary must have the same velocity tangential to the boundary as the boundary itself. This is the "no-slip" condition, which has been observed over and over experimentally. The molecular forces required to peel away fluid from a boundary are quite large, due to molecular attraction of dissimilar molecules. The only exceptions observed to this are in extreme cases of rarefied gas flow, when the continuum concept is no longer completely valid.[3] We consider next several solutions of the Navier-Stokes equations.

11. Some Solutions to the Navier-Stokes Equations When Convective Acceleration Is Absent

For special flows where the velocity gradients are perpendicular to the velocity, the convective acceleration terms vanish. The results are then independent of the Reynolds number. We will consider two important cases of unsteady flow that were first solved by Stokes.

Both of the problems considered by Stokes have a velocity of the form $\mathbf{v} = (u(y, t), 0, 0)$ and an infinite plate located at $y = 0$. For these flows, since the flow is due solely to the motion of the plate and no other forcing is imposed, the pressure gradient is absent, and the x momentum equation becomes

$$\frac{\partial u}{\partial t} = \nu \frac{\partial^2 u}{\partial y^2}. \tag{4.11.1}$$

a. Stokes' first problem—impulsive motion of a plate

Stokes' first problem considers the case where at initial time the plate is suddenly caused to move in the x direction with a velocity U. Considering the various parameters in the problem, we see that u must depend only on y, t, U, and ν. Thus we have

$$u = u(y, t, U, \nu),$$

which in dimensionless form can be written as

$$u/U = f(y/\sqrt{\nu t}, \, Uy/\nu). \tag{4.11.2}$$

Since the momentum equation (4.11.1) in this case is linear, and since U appears only in the boundary condition, we expect that U will appear only as a multiplying factor, and so we can reduce the dimensionless form (4.11.2) to

$$u/U = f(y/\sqrt{\nu t}). \tag{4.11.3}$$

To verify this, we note that

---

[3]Those interested in the history of this once-controversial condition should read the note at the end of *Modern Developments in Fluid Dynamics*, given in the references at the end of this chapter, for an interesting account.

$$\frac{\partial u}{\partial t} = -(Uy/2t\sqrt{\nu t})f', \qquad \frac{\partial u}{\partial y} = (U/\sqrt{\nu t})f',$$

and

$$\frac{\partial^2 u}{\partial y^2} = (U/\nu t)f'',$$

where a prime denotes differentiation with respect to the dimensionless variable $\eta = y/\sqrt{\nu t}$.

Putting the above into equation (4.11.1), the result is

$$-(Uy/2t\sqrt{\nu t}]f' = (U/t)f''$$

or

$$f'' + 0.5\eta f' = 0. \tag{4.11.4}$$

The boundary conditions on f are the following. Because of the no-slip condition, $u = U$ on the flat plate, so that $f(0) = 1$. At very large values of y, u must go to zero since the velocity is transported away from the plate only by viscous diffusion and it will take a very large amount of time for the effect to be noticed away from the plate. Therefore $f(\infty) = 0$.

Equation (4.11.4) can be integrated once to give

$$f' = A \exp(-\eta^2/4),$$

and again to give

$$f' = A\int_0^\eta \exp(-\zeta^2/4)\, d\zeta + B,$$

where A and B are constants of integration. Since $f(0) = 1$, $B = 1$. Since $f(\infty) = 0$,

$$A\int_0^\infty \exp(-\zeta^2/4)\, d\zeta + 1 = 0,$$

determining A. The integral appearing in this expression has the value $\sqrt{\pi}$; thus $A = -1\sqrt{\pi}$ and

$$f = 1 - \int_0^\eta \exp(-\zeta^2/4)\, d\zeta/\sqrt{\pi}. \tag{4.11.5}$$

The integral appearing in (4.11.5) is a form of the well-known error function, which can be found tabulated in many sets of tables. Typical values for the solution are seen in Table 4.6 and in Figure 4.22. The combination $1 - f$ is the case where the plate is at rest and the outer fluid is impulsively started in motion.

Table 4.6. The error function

η	$f = u/U$	$1 - f$	η	$f = u/U$	$1 - f$	η	$f = u/U$	$1 - f$
0.0	1.0000	0.0000	1.3	0.3580	0.6420	2.6	0.0660	0.9340
0.1	0.9436	0.0564	1.4	0.3222	0.6778	2.8	0.0477	0.9523
0.2	0.8875	0.1125	1.5	0.2888	0.7112	3.0	0.0339	0.9661
0.3	0.832	0.1680	1.6	0.2579	0.7421	3.2	0.0236	0.9764
0.4	0.7773	0.2227	1.7	0.2293	0.7707	3.4	0.0162	0.9838
0.5	0.7237	0.2763	1.8	0.2031	0.7969	3.6	0.0109	0.9891
0.6	0.6714	0.3286	1.9	0.1791	0.8209	3.8	0.0072	0.9928
0.7	0.6206	0.3794	2.0	0.1573	0.8427	4.0	0.0047	0.9953
0.8	0.5716	0.4284	2.1	0.1376	0.8624	4.2	0.0030	0.9970
0.9	0.5245	0.4755	2.2	0.1198	0.8802	4.4	0.0018	0.9982
1.0	0.4795	0.5205	2.3	0.1039	0.8961	4.6	0.0011	0.9989
1.1	0.4367	0.5633	2.4	0.09	0.9103	4.8	0.0007	0.9993
1.2	0.3961	0.6039	2.5	0.077	0.9229	5.0	0.0004	0.9996

The variable η that appears in this solution is called a ***similarity variable***, and f is said to be a ***similarity solution***. By similarity here, we mean that at two different times t_1 and t_2 we have geometric similarity in our flow (one velocity profile is merely an enlargement of the other) at locations y_1 and y_2, providing that

$$y_1/\sqrt{\nu t_1} = y_2/\sqrt{\nu t_2}.$$

This can be stated in an equivalent way by saying that we have similarity at the points where

$$\eta_1 = \eta_2. \tag{4.11.6}$$

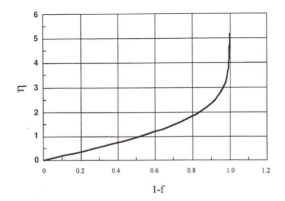

Figure 4.22. Plot of dimensionless velocity $1 - f$ versus η for Stokes' first problem.

Similarity solutions play an important role in fluid mechanics. A similarity solution reduces the Navier-Stokes partial differential equations to ordinary differential equations, which are much easier to solve, and the solution is much easier to understand physically. The existence of a similarity solution depends on simplifying conditions such as the domain having infinite extent, which allows the number of dimensionless parameters to be small. Thus the similarity solutions are very special solutions, and exist only in a relatively few flow cases. We will see more of similarity solutions in the next section.

This Stokes' problem is a reasonably good approximation of the flow on a flat plate if we consider Ut to be a "distance" from the leading edge of the plate. To see how "deep" the viscous effects penetrate, we note that for $\eta = 3$, the fluid velocity is reduced to less than 4% of the plate velocity. Consequently we can consider this to be approximately the outer edge of the effects of viscosity. This thickness is termed the **boundary layer thickness**. In terms of distance from the leading edge of the plate, we have

$$\text{Boundary layer thickness} = 3\sqrt{\nu t} = 3\sqrt{\nu x/U} = 3x/\sqrt{Ux/\nu}. \tag{4.11.7}$$

We will consider this problem again in Chapter 6, and find it to be in general agreement with those results.

b. Stokes' second problem—oscillation of a plate

The second problem considered by Stokes is one where an infinite plate oscillates back and forth with a circular frequency Ω, the motion having taken place sufficiently long that the starting conditions can be ignored. Then $u(0, t) = U \cos \Omega t$.

If we were to try to solve this problem using separation of variables, the "traditional" method of solving partial differential equations, we would find phase differences in time between the boundary velocity and the fluid velocities, and we would need both the sine and cosine of Ωt. The result would be that we would end up having to solve two coupled second-order equations. An easier approach is to let

$$u(0, t) = Ue^{i\Omega t},$$

where i is the imaginary number square root of -1, and

$$e^{i\Omega t} = \cos (\Omega t) + i \sin (\Omega t).$$

Since the real part of $e^{i\Omega t}$ is $\cos \Omega t$, our solution is found by taking only the real part of u at the very end of our analysis.

Proceeding in this manner, we let

$$u = f(y)e^{i\Omega t} \tag{4.11.8}$$

and substitute into equation (4.7.8). The result is

$$i\Omega f = \nu f'', \tag{4.11.9}$$

with the solution

$$f = Ae^{-(1+i)y/a} + Be^{(1+i)y/a},$$ (4.11.10)

where $a \equiv \sqrt{12\nu/\Omega}$ is a characteristic length for the problem. The parameter a tells us how far the viscous effects penetrate from the boundary into the fluid.

Because of the condition at $y = 0$, we have

$$A + B = U.$$

If we consider the only solid boundary to be at $y = 0$, then u must vanish as y becomes large. Since the term that B multiplies grows exponentially as y increases, then we must have $B = 0$ to keep u finite. Then it follows that $A = U$, and (4.7.8) and (4.7.10) combine to give

$$u(y, t) = Ue^{-(1+i)y/a + i\Omega t},$$

or upon taking the real part,

$$u(y, t) = Ue^{-y/a} \cos (\Omega t - y/a).$$ (4.11.11)

A plot of these results is shown in Figure 4.23.

If there were instead a solid boundary at $y = b$, then $u(b, t) = 0$ and

$$0 = Ae^{-(1+i)b/a} + Be^{(1+i)b/a}.$$

Solving for A and B yields

$$f = U \left(\frac{e^{(1+i)(b-y)/a} - e^{-(1+i)(b-y)/a}}{e^{(1+i)b/a} - e^{-(1+i)b/a}} \right).$$ (4.11.12)

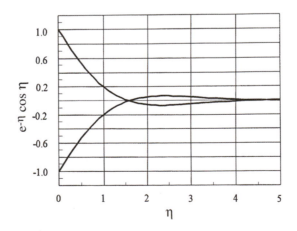

Figure 4.23. Plot of dimensionless velocity extremes versus η for Stokes' second problem.

Note that if a is small, the limit of (4.11.12) is

$$f \approx U e^{-(1+i)y/a},$$

which says that the fluid motion is confined to a thin region near the wall with thickness of order a. In fact (4.11.12) shows us that when $b \geq 2a$, and since $e^{-2} \approx 0.1$, for practical purposes the upper plate can be thought of as being at infinity.

The results of Stokes' second problem are important to the study of pulsatile flows. We have few analytical tools to deal with these flows, and Stokes' solution at least indicates some of the primary effects that can occur.

Example 4.11.1. Work done in an oscillating flow

An acoustic pressure $p_0 \sin (\Omega x/c) \sin (\Omega t)$ is set up in a gas that has a speed of sound c. This in turn induces a velocity field $U_0 \cos (\Omega x/c) \cos (\Omega t)$, where $U_0 = p_0/\rho c$. No useful work is produced by this since the pressure and velocity are 90° out of phase. Show that by introducing a flat plate into the flow and parallel to the velocity field, useful work is produced.

Given: The velocity and pressure far from the plate.

Assumptions: The plate is long enough that edge effects can be ignored. The viscous terms in the Navier-Stokes equations that involve second derivatives with respect to x can be neglected compared to the terms involving second derivatives with respect to y. The justification of this is the usual boundary layer assumption.

Solution: The equation that will govern the flow and that uses the above assumption on the viscous terms is

$$\frac{\rho \partial u}{\partial t} = -\frac{\partial p}{\partial x} + \mu \frac{\partial^2 u}{\partial y^2}.$$

This is the counterpart of equation (4.11.1), which however also includes the pressure gradient term. As suggested by Stokes' second solution and the far field, take

$$u(x, y, t) = U_0[1 + f(y)] \cos (\Omega x/c)e^{i\Omega t},$$

$$p(x, t) = -ip_0 \sin (\Omega x/c)e^{i\Omega t},$$

the $-i$ in front of the pressure taking care of the time phase. Again we will take real values after solving for f.

Substituting these forms into the governing equation we have, after canceling out the common x and t factors that

$$i\Omega \rho U_0 (1 + f) = -\Omega p_0/c + \mu U_0 f''.$$

By virtue of the relation between U_0 and p_0, the constant terms cancel. Thus we are left with

$$i\Omega f = \nu f'',$$

which is the same as (4.11.9). Again we take the solution that decays as y increases. Thus $f = Ae^{-(1+i)y/a}$, where a is defined as in Stokes' second problem.

Applying the boundary condition, we have $A = -1$. Thus

$$u(x, y, t) = U_0(1 - e^{-(1+i)y/a}) \cos (\Omega x/c)e^{i\Omega t},$$

which satisfies all boundary conditions. Taking the real part, we have for the final form of the velocity

$$u(x, y, t) = U_0\{[1 - e^{-y/a} \cos (y/a)] \cos (\Omega t) - e^{-y/a} \sin (y/a) \sin (\Omega t)\} \cos (\Omega x/c).$$

The useful local work LW performed over a period $T = 2\pi/\Omega$ is

$$\text{LW} = (1/T) \int_0^T pu \, dt.$$

Putting our expressions for pressure and velocity, we have

$$\text{LW} = (1/T)\int_0^T p_0 \sin (x\Omega/c) \sin (\Omega t)U_0\{[1 - e^{-y/a} \cos (y/a)] \cos (\Omega t)$$

$$- e^{-y/a} \sin (y/a) \sin (\Omega t)\} \cos (\Omega x/c) \, dt.$$

The time terms that involve the product of the sine and cosine will average to zero over the interval, while the average of the sine squared terms is $T/2$. Thus we have

$$\text{LW} = -0.5e^{-y/a} \sin (y/a) \cos (\Omega x/c).$$

Note that the term independent of y, which represents the work done in the far field, dropped out of LW.

The total work TW per unit width and length is

$$\text{TW} = \int_0^\infty \text{LW} \, dy = -0.5p_0U_0 \cos (\Omega x/c)\int_0^\infty e^{-y/a} \sin (y/a) \, dy = -0.25p_0U_0 \cos (\Omega x/c).$$

Thus useful work is done due to the phase shift introduced in the velocity field by the no-slip condition and the presence of the plate.

The results here are a much simplified version of a device known as the "acoustic refrigerator."

Suggestions for Further Reading

Batchelor, G. K., *An Introduction to Fluid Mechanics*, Cambridge University Press, New York, 1967.

Blasius, H., "Grenzschichtenin Flussigkeiten mit kleiner Reibung," *Z. Math. Phys*, vol. **56**, pages 1–37, 1908. (English translation: NACA TM 1256.)

Goldstein, S., editor, *Modern Developments in Fluid Dynamics,* vols. **I** and **II**, Oxford University Press, London; reprinted by Dover Publications, New York, 1965.

Hinze, J. O., *Turbulence*, McGraw-Hill, New York, 1959.

Lamb, H., *Hydrodynamics*, 6th edition, Macmillan, New York; reprinted by Dover Publications, New York, 1945.

Landau, L. D., and Lifshitz, E. M., *Fluid Mechanics*, vol. 6 of *Course of Theoretical Physics*, Pergamon Press, London, 1959.

Milne-Thomson, L. M., *Theoretical Hydrodynamics*, 3rd edition, Macmillan, New York, 1955.

Moran, J., *An Introduction to Theoretical and Computational Aerodynamics*, Wiley, New York, 1984.

Prandtl, L., *Essentials of Fluid Mechanics*, Hafner, New York, 1952.

Rayleigh, Baron (John William Strutt), *The Theory of Sound*, vols. 1 and 2, Macmillan, London, 1878; reprinted by Dover, New York, 1945.

Reynolds, O., "On the dynamical theory of incompressible viscous fluids and the determination of the criterion," *Phil. Trans. Roy. Soc. London*, vol. **186**, pages 123–164, 1895.

Rosenhead, L., editor, *Laminar Boundary Layers*, Oxford University Press, London, 1963.

Rouse, H., editor, *Advanced Mechanics of Fluids*, Wiley, New York, 1959.

Schlichting, H., *Boundary-Layer Theory*, 7th edition, McGraw-Hill, New York, 1979.

Stoker, J. J., *Water Waves*, Interscience, New York, 1957.

Stokes, G. G., "On the effect of the internal friction of fluids on the motion of pendulums," *Cambr. Phil. Trans. IX*, vol. **8**, 1851.

Tennekes, H., and Lumley, J. L., *A First Course in Turbulence*, MIT Press, Cambridge, 1972.

Truesdell, C., *Six Lectures on Modern Natural Philosophy*, Springer-Verlag, New York, 1966.

van Dyke, M., *Perturbation Methods in Fluid Mechanics*, Parabolic Press, Stanford, 1964.

van Dyke, M. *An Album of Fluid Motion*, Parabolic Press, Stanford, 1982.

Yih, C.-S., *Fluid Mechanics*, West River Press, Ann Arbor, 1977.

Problems for Chapter 4

Continuity and stream functions

4.1. Indicate which of the following velocity components represent flows of incompressible flows:

(a) $v_x = x + y + z,$ $v_y = x - y + z,$ $v_z = x + y + 3.$

(b) $v_x = xy,$ $v_y = yz,$ $v_z = -z(x + y) - z^2.$

(c) $v_x = xyzt,$ $v_y = y(x + z),$ $v_z = z^2(xt^2 - yt)/2.$

4.2. Find the stream functions for the following flows:

(a) $v_x = -y\Omega,$ $v_y = x\Omega.$

(b) $v_x = -y\Omega/(x^2 + y^2),$ $v_y = x\Omega/(x^2 + y^2).$

(c) $v_x = xA/(x^2 + y^2),$ $v_y = yA/(x^2 + y^2).$

(d) $v_x = (y^2 - x^2)A/(x^2 + y^2)^2,$ $v_y = -2xyA/(x^2 + y^2)^2.$

4.3. For the incompressible flows below, two velocity components are given. Find the most general third velocity component such that continuity is satisfied for an incompressible flow. Are your answers unique?

(a) $v_x = x^2 + y^2 + z^2$, $v_y = -xy - yz - xz$, $v_z = ?$

(b) $v_x = \ln{(y^2 + z^2)}$, $v_y = \sin{(x^2 + z^2)}$, $v_z = ?$

(c) $v_x = ?$, $v_y = \dfrac{y}{(x^2 + y^2 + z^2)^{3/2}}$, $v_z = \dfrac{z}{(x^2 + y^2 + z^2)^{3/2}}$.

(d) $v_x = a^2x^2 + by^2 - (a + b)z^2$, $v_y = 2bxy - cyz$, $v_z = ?$

4.4. (a) Find the stream functions for the flow given below:

$v_x = x^2 - 2xy \cos{y^2}$, $v_y = -2xy + \sin{y^2}$, $v_z = 0$.

(b) What is the discharge per unit width between the points $(0, 0, 0)$ and $(3, 2, 0)$?

4.5. The stream function $\psi = 5R^2 \sin^2{\beta}$ represents an axially symmetric flow in three dimensions.

(a) Sketch the streamlines $\psi = 10$, $\psi = 20$, $\psi = 30$.

(b) Find the discharges between the plane containing the z axis and the point $R = 10$, $\beta = \pi/6$, and the plane containing the z axis and the point $R = 1$, $\beta = \pi/4$.

4.6. Determine v_y and ψ for the two-dimensional incompressible flow with $v_x = (k \cos{\theta})/r$. Take $v_\theta = 0$ when $\theta = 0$.

4.7. Determine the stream function for the two-dimensional incompressible flow given in cylindrical polar coordinates by

$v_r = -U(a^2/r^2 - 1) \cos{\theta}$, $v_\theta = -U(a^2/r^2 + 1) \sin{\theta}$.

4.8. A narrow stream discharges fluid at a rate of m m³/s/depth into a river of width w flowing at a speed U. If the river and stream are the same depth, and it is assumed that the two fluids do not mix, (a) what is the speed of the river downstream from the stream? (b) what width of the river is occupied by the discharge from the stream?

4.9. The radial velocity component of a two-dimensional incompressible flow is given by

$v_r = -1.5Ur^{3/2} \cos{(1.5\theta)}$.

(a) Find the v_θ that satisfies continuity and $v_\theta = 0$ when $\theta = 0$.

(b) Find the equation for ψ such that $\psi = 0$ for $\theta = 2\pi/3$.

4.10. (a) Write the stream function for a two-dimensional Rankine body made up of a single line source at the origin plus a uniform stream parallel to the x axis. The uniform stream is 3 ft/s, and the width of the body far downstream is 2 ft.

(b) Write the equation for the body shape.

(c) If the mass density is 1.94 slugs/ft³, find the pressure difference between the pressure at the stagnation point and the pressure on the body far downstream.

(d) Locate the point of minimum pressure on the body.

4.11. Write the stream function for an axisymmetric Rankine body made up of a source-sink pair and a uniform stream. The uniform stream flows at a speed of 2 ft/s and is parallel to the principal axis of the body. The body is 3 ft long and 2 ft in maximum diameter. Find the source and sink strengths and their location.

4.12. The velocity potential for a uniform stream flowing in the z direction with a speed U past a stationary sphere of radius a is

$$\phi = zU \left[1 + \frac{a^3}{2(r^2 + z^2)^{3/2}} \right].$$

A sphere of radius a moving in an otherwise quiescent fluid with velocity $U(t)$ in the z direction is obtained by subtracting the uniform stream from the above, namely,

$$\phi = \frac{zU}{2} \frac{a^3}{(r^2 + z^2)^{3/2}}.$$

Note: This is valid at the instant that the sphere is at the origin.

(a) Check that the boundary condition for the above is satisfied by computing the r and z velocity components on $r^2 + z^2 = a^2$.

(b) Compute the pressure on the surface of the sphere, including the unsteady term.

(c) Integrate the pressure times the unit normal over the surface of the sphere to find the force on the sphere. Which term in the Bernoulli equations did this force come from?

(d) In part (c) you should have found that the force is proportional to dU/dt. What is the constant of proportionality? This constant is called the "added mass of the sphere."

4.13. For steady two-dimensional incompressible flow past a body, it is found that far upstream of the body the velocity is given by

$$v_x = U + Ay, \qquad v_y = 0.$$

If the flow can be considered to be inviscid, derive the equation that the stream function must satisfy at all points in the flow.

Vorticity and velocity potential

4.14. For the velocity components given in Problem 4.1, find which flows are irrotational. For the irrotational flows, find the velocity potential.

4.15. Which of the following velocity potentials satisfy the incompressible continuity equation?

(a) $\phi = x + y + z$. (d) $\phi = zx^2 - y^2 - z^2$.

(b) $\phi = x + xy - xyz$. (e) $\phi = \sin(x + y + z)$.

(c) $\phi = x^2 + y^2 + z^2$. (f) $\phi = \ln(x)$

4.16. Show that if ϕ_1 and ϕ_2 both individually satisfy $\nabla^2 \phi = 0$, then $\phi_1 + \phi_2$, $C\phi_1$, and $C + \phi_1$ are also solutions of Laplace's equation. C depends on time, but not on x, y, or z.

4.17. If you are given an arbitrary function of x, y, z, and t and told to assume that it is a velocity potential,

(a) does it automatically represent an irrotational flow?

(b) does it automatically satisfy the continuity equation?

(c) does it automatically satisfy the Euler equations?

(d) does it automatically satisfy the Navier-Stokes equations?

Give reasons for your answers.

4.18. If you are given an arbitrary function of x, y, z, and t and are told to assume that it is a stream function,

(a) does it automatically represent an irrotational flow?

(b) does it automatically satisfy the continuity equation?

(c) does it automatically satisfy the Euler equations?

(d) does it automatically satisfy the Navier-Stokes equations?

Give reasons for your answers.

4.19. A vortex of strength Γ is placed at $(b, 0)$ inside a circular cup of radius a $(a > b)$ centered at the origin.

(a) Find the image vortex needed to generate the cup wall.

(b) At what velocity will the vortex move?

4.20. A solid sphere of density ρ_s is released from rest in a fluid with density ρ_f. Find the pressure on the sphere from Bernoulli's equation. By integrating this pressure distribution over the sphere surface, find the initial acceleration of the sphere? Note that this answer is valid initially even in a viscous fluid, since no viscous stresses exist at the first instant after release.

Rates of deformation

4.21. Find the components of the rate of deformation for the following flows:

(a) $v_x = -y\Omega, \qquad v_y = x\Omega.$

(b) $v_x = -y\Omega/(x^2 + y^2), \qquad v_y = x\Omega/(x^2 + y^2).$

(c) $v_x = xA/(x^2 + y^2), \qquad v_y = yA/(x^2 + y^2).$

(d) $v_x = (y^2 - x^2)A/(x^2 + y^2)^2, \qquad v_y = -2xyA/(x^2 + y^2)^2.$

Navier-Stokes equations

4.22. Fluid is contained between two closely spaced parallel plates a distance b apart. The lower plate is stationary, while the upper plate is suddenly made to rotate with an angular velocity Ω from rest. Assume that only the theta velocity component is present, that it depends linearly on r, and that the streamlines are circles.

(a) What is the steady-state velocity profile?

(b) What is the steady-state pressure distribution?

(c) Find the unsteady velocity distribution.

4.23. A uniform stream flows with a velocity U. A small solid particle having the same density as the fluid is suspended in the fluid. Is the velocity of the particle greater than, equal to, or less than U? Justify your answer.

4.24. A viscous fluid of viscosity μ_1 and density ρ_1 occupies the space $0 \le y \le h$. A fluid of viscosity μ_2 and density ρ_2 occupies the space $-h \le y \le 0$. Find the velocity profile due to a constant pressure $\partial p/\partial x$. The velocity and the stresses must be continuous at the interface $y = 0$.

4.25. Two vertically superposed viscous fluids flow in a channel. A fluid of viscosity μ and density ρ occupies the space $0 \le y \le h$. A fluid with viscosity $1.5\,\mu$ and density $1.25\,\rho$ occupies the space $-2h \le y \le 0$. Find the velocity profile such that the net mass discharge is zero. The pressure gradient $\partial p/\partial x$ is constant and must be the same in both fluids. The velocity and the stresses must be continuous at the interface $y = 0$.

4.26. Flow between two horizontal plates a distance d apart is caused by an oscillating pressure gradient of the form

$\partial p/\partial x = Ke^{i\Omega t}$, where $i = \sqrt{-1}$.

Find the steady-state velocity profile after all startup effects have died out.

4.27. A line vortex is put into a viscous fluid.

(a) Show that a similarity solution exists for the unsteady flow in the form

$\omega = (\Gamma/2\pi\nu t)f(\eta)$ with $\eta = r/\sqrt{\nu t}$.

(b) Solve the equation for f.

(c) Find the theta velocity component v_θ.

The conditions to be satisfied are that

$$\Gamma = 2\pi \int_0^\infty \omega r \, dr, \qquad \omega(r, 0) = 0 \text{ except at } r = 0.$$

The equation to be satisfied is

$$\frac{\partial \omega}{\partial t} = \nu \, \nabla^2 \omega, \text{ with } \omega = \frac{\partial v_\theta}{\partial r} + \frac{v_\theta}{r}, \qquad v = v_\theta.$$

4.28. (a) Starting with the velocity components

$$v_r = -ar, \qquad v_\theta = (K/r)f(r), \qquad v_z = 2az,$$

by substitution into the Navier-Stokes equations, find the equation that the function f must satisfy.

(b) Solve for f such that $f(\infty) = 1$.

4.29. (a) Show by deriving the ordinary differential equation that the x component Navier-Stokes equation has a similarity solution of the form $\psi = Ux\sqrt{\nu/c}f(\eta)$, where $\eta = y\sqrt{c/\nu}$ and $U = -c/x$. The pressure gradient is given by $\partial p/\partial x = \rho c^2/x^3$. Neglect the $\partial^2 v_x/\partial x^2$ term in the Navier-Stokes equation. Note that on $y = 0$, $v_x = v_y = 0$.

(b) Solve the equation for f such that v_x approaches U as η gets large. Note that the order of the equation can be reduced by first multiplying the equation by f'' and then integrating to obtain a second-order equation. This can be integrated once more to give f', which will involve the square of the hyperbolic tangent.

4.30. Two parallel plane circular disks of radius R are separated by a small distance. The space between the disks is occupied by a fluid. The top disk approaches the bottom stationary disk at a constant velocity U. Find the velocity components between the disks assuming the Navier-Stokes equations can be approximated as follows:

$$0 = -\frac{\partial p}{\partial r} + \mu\frac{\partial^2 v_r}{\partial z^2} \quad (r \text{ Navier–Stokes}), \qquad 0 = \frac{\partial p}{\partial z} \quad (z \text{ Navier–Stokes}),$$

$$\frac{\partial v_r}{\partial r} + \frac{v_r}{r} + \frac{\partial v_z}{\partial z} = 0 \quad (\text{continuity})$$

with $p = p(r)$. Note that global continuity requires that

$$\pi r^2 U = 2\pi r \int_0^h v_r \, dz.$$

chapter 5

Dimensional Analysis

Chapter Overview and Goals

Because the fundamental behavior of flows can change drastically as flow parameters are varied, it is customary to describe flows in terms of dimensionless parameters. For instance, when we say a flow is "fast," a relative term, we mean that the flow speeds are large compared to some other quantity that has the dimensions of velocity. Another way to state this is to say that in equations such as our momentum and energy equations, the word "fast" is a comparison of the relative magnitudes of two terms in one or the other of these equations. Since all the terms in a given equation must have the same dimensions, a comparison of two terms gives a dimensionless ratio.

In this chapter we will work with Buckingham's pi theorem. This gives the number of dimensionless parameters that can be formed for a particular flow. We will learn how to form dimensionless parameters, study some common dimensionless parameters that recur often in fluids problems, and see how dimensionless parameters can be used to guide and interpret scale-model studies and flow simulations.

Dimensional analysis can be used in any field of engineering. It is particularly important in analyzing experimental data, since it can lend order and insight to our understanding.

1. Introduction

The subjects of dimensional analysis and dimensionless parameters are of fundamental importance to all branches of engineering and applied physics. They provide a means of consolidating experimental, analytical, and computational results into a compact form, and are an aid in designing both experiments and techniques for obtaining analytical results. They play a central role in fluid mechanics because of the large number of parameters that are frequently involved in a problem, and also because of the highly complex behavior of fluid flows. In our presentation of dimensional analysis we will restrict our attention primarily to applications in fluid flows, and will deal solely with the dimensions of force (or mass), length, and time. By adding the dimension of temperature, we could include the subjects of heat transfer and thermodynamics. Similarly, by adding the unit of electrical and/or magnetic charge, the subjects of electricity and magnetism could be included. The references at the end of the chapter give examples in some of these other fields.

2. Buckingham's Pi Theorem

A fundamental but often unstated bit of mathematics that is important to understand from the beginning is that the units of each and every term in an equation must be the same. Each term may be made up of a number of different quantities, but the combination that appears as a term in the equation must have the same units as every other term, and must appear as the product or quotient of the quantities. Also, when we use trigonometric, logarithmic, exponential, or any similar function that you might be familiar with, we must have that function operating on a dimensionless quantity. Taking the sine or logarithm of units such as feet or newtons is *not* a legitimate operation.

The fundamental tool of dimensional analysis is a theorem credited to E. Buckingham that tells us what is possible in a given study. It states that in any physical problem where there are "q" quantities (e.g., velocity, pressure, discharge) involving "d" basic dimensions (e.g., mass, length, time) needed to describe the problem, these quantities can be rearranged into *at most* $(q - d)$ independent dimensionless parameters. These dimensionless parameters are in the form of products of powers of the q quantities. The dimensions needed to describe most of the incompressible fluid mechanics problems are force (or mass), length, and time, and so "d" will usually be 3. In the case of compressible flows it is also necessary to include temperature, so "d" is usually 4 for these flows.

Usually the words "at most $q - d$" in the theorem can be replaced by "exactly $q - d$." In a very few problems, usually those where the number of quantities q is 2 or 3, it may happen that these dimensions always appear in special combinations, so that in fact dimensions can be combined in such a manner so that d is less than 3, giving the need for the "at most" qualification.

Incidentally, the word "pi" in the name of the theorem has nothing to do with the number 3.14159.... Buckingham denoted his dimensionless parameters as $\Pi_1, \Pi_2, \ldots$; hence the name.

The proof of the theorem is simple, and also provides a means of finding the dimensionless parameters. Suppose that we have the q quantities $Q_1, Q_2, \ldots, Q_q$ with d basic dimensions $D_1, D_2, \ldots, D_d$. Then as stated earlier, each dimensionless parameter must be in the form of a product such as

$$\Pi = Q_1^{P_1} Q_2^{P_2} \cdots Q_q^{P_q} = \sum_{j=1}^{q} Q_j^{P_j} \tag{5.2.1}$$

We next take the dimensions of both sides of equation (5.2.1). We will use square brackets [ ] to mean "the dimension of" whatever is inside of them. Since Π is dimensionless, and since each of the Q's must be products of powers of the basic dimensions, taking the dimensions of both sides of (5.2.1) gives

$$[\Pi] = \sum_{i=1}^{d} D_i^0 = \sum_{j=1^q} [Q_j]^{p_j} = \sum_{j=1}^{q} \sum_{i=1}^{d} D_j^{a_i p_j} = \sum_{i=1}^{d} \sum_{j=1}^{q} D_i^{a_i p_j} \qquad (5.2.2)$$

Upon equating powers of D_i on both sides of equation (5.2.2), we find the result to be

$$0 = \sum_{j=1}^{q} a_i p_j, \qquad \text{where } i = 1, 2, \ldots, d. \qquad (5.2.3)$$

Thus (5.2.3) represents d equations in terms of the q unknown powers p. If we regard $q - d$ of the powers as being "known" (the Q quantities associated with these $q - d$ powers are called the **repeating variables**, and the remaining d powers are called the **nonrepeating variables**), then we have d equations in d unknowns, which can be solved providing that the determinant of the coefficients of the d unknowns is neither zero nor infinite. The possibility of a zero or infinite determinant is what provides the "at most" in the theorem. If this occurs we must go back and review our dimensions to see whether they can be combined into a reduced set.

After solving these equations, we can group our Q's together in terms of the $q - d$ "known" powers. Since for any choice of these powers Π is dimensionless, we have thus found $q - d$ dimensionless parameters. The $q - d$ powers can be assigned arbitrarily at our convenience, provided we don't make them zero.

In case you did not follow all the algebraic steps in the above proof, we will go through examples where we exactly follow the above steps, using numbers instead of symbols. This should clarify your understanding of the theorem.

Like most mathematical theorems, Buckingham's pi theorem tells us how many independent dimensionless parameters can be formed, but does not tell us how to form them. We will look at three methods of forming them: a formal algebraic method (that follows the steps in our proof); inspection; and finally, knowledge of the pertinent forces and the standard parameters that represent them. While the formal method always works, it involves the solution of a set of algebraic equations, a task that is frequently error-prone and time-consuming. Once experience with dimensional parameters has been gained, the other methods are faster and should be used whenever possible.

Notice also that the theorem states how many "independent" dimensionless parameters can be found. Two students working with the same q quantities could end up with two different sets of parameters, say,

$\Pi_1, \Pi_2, \ldots, \Pi_{q-d}$

and

$P_1, P_2, \ldots, P_{q-d}.$

While the two sets may appear to be different, if they were correctly arrived at with no algebraic errors, they are not in fact independent. The second set must be obtainable from the first by products of powers of the Π's, such that

$$P_i = \Pi_1^{a_1}\Pi_2^{a_2} \ldots \Pi_{q-d}^{a_{(q-d)}}. \tag{5.2.4}$$

This is important, since we may find in doing a problem that part way through our work the results would somehow be "better"—that is, in terms of familiar dimensionless parameters—if we had used a different set of dimensionless parameters. Equation (5.2.4) suggests how we can change to a new set without the need for starting all over again.

A final comment on the theorem is that it assumes we know what quantities are important in a given problem. Deciding what these quantities are is in fact the greatest difficulty in using dimensional analysis. The starting set of dimensional quantities comes from our physical understanding of the problem. The final result can be no better than this. If we put in too many quantities, the result may be made unduly complicated, physical understanding of the problem may be obscured, and unnecessary work and expense may be involved in carrying out whatever we had set out to do. If we put in too few quantities, we will generally find that experiments, if done over a sufficiently broad range of parameters, will show little correlation with our parameters. Sometimes we are fortunate and dimensional analysis will tell us that somehow we have erred in our understanding of a problem. In any case, a clear physical understanding of the forces that govern the flow in question is of prime importance.

3. Introductory Example

To understand the power of dimensional analysis, let us look at a problem you are already familiar with from an elementary physics or mechanics course. Suppose that we want to determine experimentally the period of a pendulum without resorting to analysis and Newton's laws. To simplify matters we will suppose the pendulum to be a concentrated mass on the end of a string, and that we start the pendulum by moving the string through some angle and then release it. We will neglect air resistance. After some thought we might decide that the period of oscillation depended on the string length h, the concentrated mass m, the starting angle β, and the acceleration of gravity g. (Gravity is the main driving force, represented here by the gravitational acceleration g. We would expect the period to be different on earth from that on the moon.) If τ is the period of oscillation, then our decision could be put in the functional form

$$\tau = f(h, m, \beta, g). \tag{5.3.1}$$

Assuming that we have included every important term, equation (5.3.1) tells us that in order to provide a good experimental solution we should obtain many lengths of string, many masses, and do the experiments for many starting angles and on planets with many different gravities. We can then plot our results in a five-dimensional plot, and thereby provide a thorough knowledge of the simple pendulum. Obviously the number of experiments to be carried out, and their time and expense, will be large!

To avoid unnecessary work and expense and at the same time improve our understanding of the physical side of the problem, we first realize that in equation (5.3.1) no matter how complicated the function may be, each term in the equation must have the same dimension. To make sure that our function in (5.3.1) obeys this law, we will recast the entire equation in dimensionless form, guaranteeing compliance with this principle.

Using Buckingham's theorem we note that there are five quantities (τ, h, m, β, g, so $q = 5$), involving three basic dimensions mass, length, and time (M, L, T, so $d = 3$). Thus, at most two ($q - d = 5 - 3$) independent dimensionless parameters can be formed. We will find these by inspection. To aid in this, first write down the five quantities along with their dimensions. Using square brackets to indicate the dimension of whatever is inside the brackets, and M, L, T to denote the dimensions mass, length, and time, we have

$$[\tau] = T, \qquad [h] = L, \qquad [m] = M, \qquad [\beta] = 1, \qquad [g] = L/T^2.$$

The "one" for the dimension of β indicates that it is already a dimensionless parameter; that is, in terms of our three basic dimensions M, L, T,

$$[\beta] = ML^0T^0. \tag{5.3.2}$$

Thus half of our task of finding the dimensionless parameters is already completed, for we have found that $\Pi_1 = \beta$!

We note that only one of our quantities (m) has the unit of mass. Therefore, one of two things is true: either we have included an extraneous quantity m that cannot be put into a dimensionless parameter and thus must be rejected, or we have neglected other quantities that should have been included and that would involve the dimension of mass. Dimensional analysis cannot tell us which of these possibilities is correct. We rely here on our physical understanding of the problem, and have confidence that the first situation is the correct one.

Thus our list is down to four quantities in two dimensions, with still two independent dimensionless parameters. One of them we already know. We find the second by our method above, writing

$$\Pi_1 = L^0T^0 = [\tau]^{p_1} [h]^{p_2} [g]^{p_3} = T^{p_1} L^{p_2} (L/T^2)^{p_3}$$

$$= L^{p_2 + p_3} T^{p_1 - 2p_3}$$

Equating powers on the left and right sides of the equation gives

$$0 = p_2 + p_3, \qquad 0 = p_1 - 2p_3,$$

or equivalently,

$$p_2 = -p_3, \qquad p_1 = 2p_3.$$

Thus we can write

$$\Pi_2 = (\tau^2 g/h)^{p_3}.$$

The selection of p_3 is ours to make, since any value other than zero is satisfactory. Since we started off by saying we wish to find the period τ, a logical choice for p_3 is one-half, so that τ appears to the first power. Then

$$\Pi_2 = \tau\sqrt{g/h}. \tag{5.3.3}$$

In (5.3.1) we assumed that the period was a function of the other quantities involved with the problem. This translates into saying that there is a relationship between the dimensionless parameters, or

$$\Pi_2 = F(\Pi_1).$$

Putting in the values for the Π's gives

$$\tau\sqrt{g/h} = F(\beta), \tag{5.3.4}$$

where F is a function of the starting angle β, different from f, but still unknown. By comparison of (5.3.1) and (5.3.4), we see that $F = f\sqrt{g/h}$.)

Equation (5.3.4) now tells us that to conduct a complete set of experiments, we need use only one mass and one length of string, and one planet on which to do our experiments. All we have to do is vary the starting angle and make a two-dimensional plot of $\tau\sqrt{g/h}$ versus β. Anyone doing the experiment with a different length of string and a different mass should find their experimental results lie on the same curve—**providing the basic assumption as to which quantities are important holds true for both sets of experiments**.

4. Algebraic Approach for the Formulation of Dimensionless Parameters

When formulating a set of $q - d$ dimensionless parameters from q quantities, we first split these quantities into d nonrepeating variables and $q - d$ repeating variables, as indicated in the proof of the pi theorem. The nonrepeating variables will each appear in only one dimensionless parameter, while the repeating variables will appear in one or more dimensionless parameters. The nonrepeating variables must contain all the dimensions of the problem. The major quantities being sought in a given case usually are chosen as the nonrepeating variables. Often a length, velocity, and density (all associated with the inertia force) are selected as repeating variables. The list of repeating variables must include all physical dimensions present in the problem.

Next, the dimensionless parameters are written as a product of powers of all the q quantities. This is most easily explained by working through an example such as we did in the above. For instance, if the variables are a length l, velocity V, density ρ, and pressure p, then

$$\Pi = l^a V^b \rho^c p^d g^e. \tag{5.4.1}$$

Then we take the dimensions of each side of equation (5.4.1). For most fluid problems, when temperature is not of direct concern, our basic set of units can be taken as either mass, length, and time (M, L, T), or force, length, and time (F, L, T), for by Newton's law the set of units F, M, L, T are related by

$$F = ML/T^2. \tag{5.4.2}$$

In principle, it does not make any difference whether we use F, L, T or M, L, T as the basic unit set, although in any given problem one of the choices may save some algebra. We will arbitrarily use F, L, T in our efforts.[1]

Proceeding with examining the dimensions of equation (5.4.1), since Π is dimensionless, we have

$$[\Pi] = F^0 L^0 T^0 = [l^a V^b \rho^c p^d g^e] = [l]^a [V]^b [\rho]^c [p]^d [g^e]$$

$$= L^a (L/T)^b (FT^2/L^4)^c (F/L^2)^d (L/T^2)^e,$$

or, after collecting terms,

$$F^0 L^0 T^0 = F^{c+d} \, L^{a+b-4c-2d+e} \, T^{-b+2c-2e}. \tag{5.4.3}$$

Since the powers on the left side of this equation must be the same as those on the right, we are led to three equations ($d = 3$) in the five unknowns ($q = 5$) a, b, c, d, e. These equations are

$$0 = c + d,$$

$$0 = a + b - 4c - 2d + e,$$

$$0 = -b + 2c - 2e. \tag{5.4.4}$$

Rewrite these equations, considering the powers associated with the repeating variables to be the unknowns, and those with the repeating variable to be the knowns. For this example, we select l, V, and ρ as the repeating variables, and p and g as the nonrepeating variables. Rearranging (5.4.4) to have the powers associated with p and g (these are d and e) on the right-hand side, we have

$$c = -d,$$

$$a + b - 4c = 2d - e, \tag{5.4.5}$$

$$-b + 2c = 2e.$$

Solving this set, we find

$$a = e, \qquad b = -2d - 2e, \qquad c = -d. \tag{5.4.6}$$

Substituting this result into the expression for our dimensionless parameter (5.4.1), we have upon collecting powers,

$$\Pi = (p/\rho V^2)^d \, (lg/V^2)^e. \tag{5.4.7}$$

[1]We have said that we can "usually" use either set. Again "usually" means "almost always," except in degenerate cases. For example, suppose we tried to form a dimensionless parameter from the three quantities V, ρ, p. A too cursory attempt to use the previous results might lead one to conclude that no dimensionless parameters can be formed, since $3 - 3 = 0$. However, the quantity $p/\rho V^2$ is certainly dimensionless. The reason for the difficulty here is that there are really only two basic dimensions for this set, F/L^2 and L/T. This is the "at most" qualification of the theorem. Again, this situation arises only when the number of quantities q is small.

This expression is dimensionless no matter what values d and e take on; hence $p/\rho V^2$ and lg/V^2 must both be dimensionless. Thus with one solution of a set of algebraic equations we have found all our dimensionless parameters. The nonrepeating variables each appear in the numerator of only one of the parameters, and the repeating variables usually appear in more than one parameter.

Particularly if q is large the solution to the set of algebraic equations may be cumbersome and, unless sufficient care is taken, algebraic errors will be made. (In any case it is a good policy to always check each parameter at the end to see if it is indeed dimensionless.) Knowledge of some commonly used parameters can help to reduce the net value of q by removing a priori known dimensionless parameters, as we did in our pendulum example.

Example 5.4.1. Forming dimensionless parameters—pipe flow
In a problem involving a pipe, the engineer decides that the flow depends on the fluid density and viscosity, the pipe diameter D, and the volumetric discharge Q. Form the maximum number of independent dimensionless parameters possible.

Sought: Dimensionless parameter(s) based on four given quantities.
Given: The dimensional quantities of interest are μ, ρ, D, and Q. Their dimensions are

$$[\mu] = FT/L^2, \qquad [\rho] = FT^2/L^4, \qquad [D] = L, \qquad [Q] = L^3/T.$$

Assumptions: All quantities of importance to the problem have been included in the analysis.
Solution: Since there are four quantities and three dimensions, we expect at most one independent dimensionless parameter. Writing

$$\Pi = \mu^a \rho^b D^c Q^d,$$

and taking dimensions of both sides,

$$[\Pi] = F^0 L^0 T^0 = (FT/L^2)^a (FT^2/L^4)^b L^c (L^3/T)^d = F^{a+b} L^{-2a-4b+c+3d} T^{a+2b-d}.$$

Upon equating powers on both sides of this equation, we have

$$a + b = 0,$$

$$-2a - 4b + c + 3d = 0,$$

$$a + 2b - d = 0.$$

Arbitrarily choosing $Q(d)$ to be the repeating variable, we find upon solving these equations that

$$a = -d, \qquad b = d, \qquad c = -d,$$

and so

$$\Pi = (\rho Q/\mu D)^d.$$

Since d can be chosen arbitrarily (but neither zero or infinite), a choice might be $d = 1$, giving

$$\Pi = \rho Q/\mu D.$$

Example 5.4.2. Forming dimensionless parameters—pipe flow

The engineer in the previous problem suddenly realizes that his pipe is not flowing completely full. Thus there is a free surface, and for this problem, surface tension was also important. How does the problem change?

Sought: Dimensionless parameters based on five given quantities.

Given: The dimensional quantities of interest are μ, ρ, D, Q, and σ. Their dimensions are

$$[\mu] = FT/L^2, \quad [\rho] = FT^2/L^4, \quad [D] = L, \quad [Q] = L^3/T, \quad [\sigma] = F/L.$$

Assumptions: All quantities of importance to the problem have been included in the analysis.

Solution: With five quantities and three dimensions, we now expect at most two independent dimensionless parameters. Proceeding as in the previous example, we have

$$\Pi = \mu^a \rho^b D^c Q^d \sigma^e,$$

and, upon taking dimensions of both sides,

$$[\Pi] = F^0 L^0 T^0 = (FT/L^2)^a \, (FT^2/L^4)^b \, L^c \, (L^3/T)^d \, (F/L)^e$$

$$= F^{a+b+e} \, L^{-2a-4b+c+3d-e} \, T^{a+2b-d}.$$

Upon equating powers on both sides of this equation, we have

$$a + b + e = 0,$$

$$-2a - 4b + c + 3d - e = 0,$$

$$a + 2b - d = 0.$$

Arbitrarily choosing $Q(d)$ and σ (e) to be the repeating variables, we find upon solving these equations that

$$a = -d - 2e, \quad b = d + e, \quad c = -d + e.$$

Thus

$$\Pi = (\rho Q/\mu D)^d \, (\sigma \rho D/\mu^2)^e.$$

Since d and e can be chosen arbitrarily (but neither zero or infinite), we can arbitrarily choose

$$\Pi_1 = \rho Q/\mu D, \quad \Pi_2 = \sigma \rho D/\mu^2.$$

---

5. Interpretation of Dimensionless Parameters as Force Ratios

As has been mentioned, the most difficult problem in dimensional analysis is in formulating the list of important quantities that we need to work with. A guide to this formulation is to think about the problem in terms of the basic forces and energies that are of importance to the particular flow case of interest. In Chapters 2, 3, and 4 we summed these forces and energies with the aid of Newton's law and the energy balance equation. In dimensional analysis, we merely list the terms that are important, and the quantities that are associated with them.

Suppose, as an example, we wish to study the fluid resistance on a submerged submarine in steady forward motion. We are interested in the drag force resulting from the viscous and momentum forces in the fluid due to the flow past the hull. (When we say momentum force, we are really thinking of the convective terms in the acceleration. An alternative name would be inertia force. We will use the two names interchangeably.) Thus, listing the forces and the quantities associated with them,

Drag F_D

Viscous μ, V, l

Momentum $\rho, V, l.$

Thus we have five quantities F_D, V, l, ρ, and μ, with three dimensions, and therefore we can form at most two independent dimensionless parameters. We can do this using the algebraic procedure already presented, or take a short cut with the help of Table 5.1. Since each of the terms in the right-hand column has the dimension of force, then the ratio of any two of them must be dimensionless. That is, we have that

Drag force/momentum force = $F_D/\rho V^2 l^2$ Π_1 (5.5.1)

and

Table 5.1. Representative forces

Force	Representation
Buoyancy	$\Delta\rho g l^3$
Compressibility	$\rho c^2 l^2$
Drag	F_D
Gravity	$\rho g l^3$
Momentum (inertia)	$\rho V^2 l^2$
Lift	F_L
Pressure	$p l^2$ or $\Delta p l^2$
Unsteady	$\rho \omega^2 l^4$
Viscous	$\mu V l$

Momentum force/viscous force $= \rho V l/\mu = \Pi_2$ (5.5.2)

must each be dimensionless. Thus we have found our dimensionless parameters without solving any algebraic equations. We would then conclude that, providing we have included all the forces important to our problem, there must be a relationship between the dimensionless parameters of the form

$\Pi_1 = f(\Pi_2)$,   (5.5.3a)

or equivalently, after substitution for the pi's,

$F_D = \rho V^2 l^2 f(\rho V l/\mu)$.   (5.5.3b)

In the previous example, you might question why the pressure force was not included. The reason is that by Newton's law the drag force is made up of contributions from the sum of the pressure forces due to wake formation behind the body, and the viscous forces acting on the body surface. Thus by including any two of these forces, the third force is also taken care of by Newton's law. Generally, we include pressure force only when we are interested in finding pressure, or pressure difference, directly at a specific point or points.

We have couched this discussion in terms of listing forces, which usually is sufficient for most problems. In some problems, however, moments or energies might also need to be included. If, for instance, we were interested in the ability of the submarine to turn, we would include the turning moment exerted by the rudder, and would include the moment of one of the forces (usually the momentum force) in our list by multiplying it by the length l. If we were interested in thermal effects, heat terms from the energy equation would similarly be added.

Note that in the list in Table 5.1, we have not said which length, which velocity, etc., should be used. Generally, quantities are chosen that are somehow appropriate to the particular problem, or whose significance has been well established over years of experience. For instance, in our submarine problem, V would be the forward speed of the vessel relative to the fluid speed. If we are interested in forward drag, custom would suggest that l be taken as the square root of the projected area, while if we are interested in the turning of the submarine, the length of the rudder, square root of the side area, or some combination of these might be a suitable choice for a length.

6. Summary of Steps Involved in Forming Dimensionless Parameters

The previous discussion leads to an orderly procedure for forming dimensionless parameters. It can be summarized as follows:

1. Write down the parameters that are important to the problem that you are studying. Make your decisions by deciding which forces and/or moments are important to your study, and use Table 5.1 to choose the parameters.

2. Select which of the parameters will be repeating variables. The number of repeating variables equals the number of parameters (q) minus the number of dimensions (d). The repeating variables must include all dimensions involved in the problem.

3. Write the dimensionless parameter as the product of all of the repeating variables raised to a power.

4. Rewrite the expression written in step 3, this time replacing each parameter by the dimensions of that parameter. Generally, you will use one of force, length, and time as the dimensions, or mass, length, and time.

5. Collect the powers of all similar dimensions so that your expression is in the form $F^a L^b T^c$ (or $M^d L^e T^f$). Then for the dimensionless parameter to be dimensionless, you will have a series of algebraic equations of the form $a = 0$, $b = 0$, and $c = 0$ (or $d = 0$, $e = 0$, $f = 0$).

6. Rearrange these equations so that the repeating variables are on the right-hand side and the nonrepeating variables are on the left-hand side. Solve these equations for the nonrepeating variables in terms of the repeating variables.

7. Rewrite the dimensionless parameter from step 3 in terms of the powers found in step 6. Collect the terms on the powers associated with each repeating variable. Each of the terms associated with these powers is a dimensionless parameter in its own right.

8. Inspect your set of dimensionless parameters to see if any of them can be easily converted to standard versions of dimensionless parameters.[2] (See the next section for a discussion of these standard parameters.) You may prefer to invert one or more of your parameters, or raise them to a power to eliminate fractional powers, to put them in a more convenient form.

7. Some Common Dimensionless Parameters

There is a group of dimensionless parameters that appear so often that they have been given names, quite often the names of the persons who either first introduced them or first used them extensively. They can all be thought of as force or moment ratios, and most of them use the momentum force as the reference force.

Reynolds number, Re $= Vl/\nu = \rho Vl/\mu$ (ratio of momentum forces to viscous forces)

The Reynolds number is probably the most used of all the dimensionless parameters in fluid mechanics, since it is the most descriptive parameter for distinguishing the nature of flows of engineering interest. Generally, Re is an important parameter in deciding whether a flow is laminar or turbulent, such as in a pipe, and in determining the drag on a solid surface. Typically at small Reynolds numbers, flows are laminar,

---

[2] You can save some of the algebra involved in the solving of the algebraic equations if you anticipate these standard dimensionless parameters from your list of the forces and moments involved in your problem, and pull the dimensional parameter appropriate to that dimensionless parameter from your list. (Be sure, however, that along the way you include the corresponding dimensionless parameter in your list.)

while at large Reynolds numbers, flows are turbulent. For example, in pipe flows with l being the pipe diameter and V the average velocity, flows with Reynolds numbers less than 2,000 are laminar, while flows with Reynolds numbers greater than 2,000 exhibit various degrees of turbulence.

Froude number, $Fr = V\sqrt{gh}$ (square root of the ratio of momentum forces to gravity forces)

The Froude number is of particular importance in flows that are affected by gravity, such as wave phenomena. In many references the square of our definition is also often called the Froude number. Because of its importance for waves, it plays a major role in determining that part of the drag of a ship due to the waves produced at the surface. Froude numbers can be expected to arise in any problem where free surfaces are present. When the Froude number is very large, gravity forces are usually unimportant.

When there is a density change in a flow, either abrupt or gradual, a ***densimetric Froude number*** (also called a ***Richardson number***) is used, in either the form

$$V/\sqrt{\Delta\rho gh/\rho}$$

or

$$V/\sqrt{(d\rho/dh)gh^2/\rho}.$$

Mach number, $M = V/c$ (square root of the ratio of momentum forces to compressibility forces)

For high-speed flows, many of the flow characteristics are governed by the Mach number. The c appearing in the definition is the local speed of sound. The Mach number is a deciding factor in deciding whether compressibility plays a role in a given problem. As a rule of thumb, if the Mach number is less than 0.25 or so, the flow can be considered to be incompressible, regardless of whether the fluid is a liquid or gas. Flows with Mach numbers less than 1 are termed subsonic flows, and flows with Mach numbers greater than 1 are termed supersonic flows. The proposed National Space Plane is scheduled to fly at speeds up to Mach 23.

Strouhal number, $St = \omega l/V$ or fd/V (ratio of the unsteady momentum force to the convective momentum force)

The ***Strouhal number*** plays a central role in the study of the frequency of oscillations of a flow or body (ω) or the frequency (f) of shedding of vortices in the wake of a body immersed in a steady flow.

Pressure coefficient, $C_p = \Delta p/0.5\rho V^2$ (ratio of pressure forces to one-half the momentum forces)

The ***pressure coefficient*** without the factor of one-half in the denominator is termed the ***Euler number***. Usage has made the term pressure coefficient more common than the Euler number. When $\Delta p = p - p_v$, where p_v is the vapor pressure at local conditions, C_p is termed the **cavitation number**. C_p has already appeared in our study of venturi meters, pitot tubes, and the like.

Drag coefficient, $C_D = F_D / 0.5\rho V^2 l^2$ *(ratio of total drag force to one-half the momentum force)*

The force F_D encompasses all the drag terms, which may include viscous drag, pressure drag, form or profile drag (due to separation in the wake), unsteady drag, wave drag, and so on. The drag coefficient is used extensively in studies relating to the drag on various body shapes.

Moment coefficient, $C_M = M / 0.5\rho V^2 l^3$ *(ratio of a turning moment to the moment of one-half the momentum force)*

The **moment coefficient** is of use in calculating the moments on rudders, airfoils, turbine wheels, and the like.

Lift coefficient, $C_L = F_L \, 0.5\rho V^2 l^2$ *(ratio of lift force to one-half the momentum force)*

The lift force is that force perpendicular to the main flow direction. The **lift coefficient** is used extensively in the design of wings, propellers, and turbine blades.

Weber number, $We = V^2 l \rho / \sigma$ *(ratio of the momentum force to the surface tension force)*

The **Weber number** is of importance in interface problems, such as surface waves or the motion of bubbles.

    This listing of dimensionless parameters is by no means exhaustive, but it does cover the more commonly encountered ones. For a more detailed listing, see the references at the end of the chapter.

8. Examples of the Use of Dimensionless Parameters

The discussion so far has given the general techniques for developing dimensionless parameters. We turn next to several examples to see how they can be applied in a given situation.

Example 5.8.1. Losses in a pipe
When fluid flows in a steady fashion through a horizontal pipe of diameter D and length l, the dominant forces will be the driving force (pressure), the resisting force (wall shear, or viscous force), and the inertia force. Find the governing dimensionless parameters.

Sought: Dimensionless parameters given two quantities and three forces.

Given: The quantities involved are Δp, ρ, V, l, D, k, and μ. The first six quantities come from the geometry and the forces as detailed in Table 5.1; k is another length quantity, the average height of the roughness in the pipe. The dimensions of the seven quantities are

$$[\Delta p] = F/L^2, \quad [\rho] = FT^2/L^4, \quad [V] = L/T, \quad [l] = L, \quad [D] = L,$$

$$[k] = L, \quad [\mu] = FT/L^2.$$

Assumptions: All quantities of importance to the problem have been included in the analysis.

Solution: Since there are seven dimensionless quantities and three dimensions, there can be at most four independent dimensionless parameters. Forming these parameters from knowledge of force ratios, we find

$$C_p, \quad \text{Re}, \quad l/D, \quad \text{and} \quad k/D.$$

If these are the only quantities of importance to the problem, we are thus able to write

$$C_p = C_p(\text{Re}, l/D, k/D). \tag{5.8.1}$$

A little thought can make this result more informative. Since we are dealing with a situation where all quantities are independent of the pipe length except the pressure difference, and since we would expect that doubling the length should double the pressure difference needed to keep the average velocity the same, then

$$C_p = (l/D)\, f(\text{Re}, k/D), \tag{5.8.2}$$

where $f(\text{Re}, k/D)$ stands for an unknown function of the Reynolds number and the dimensionless roughness.

More commonly, the head loss $h_L = \Delta p/\rho g$ is used rather than the pressure difference. Multiplying equation (5.8.2) by $V^2/2g$, we get the result

$$h_L = f(\text{Re}, k/D)(l/D)V^2/2g. \tag{5.8.3}$$

This is termed the "Darcy-Weisbach equation" and is used extensively in analyzing pipe-flow problems. The quantity f is called the pipe friction factor, and is a function of the Reynolds number and the relative roughness of the pipe. It must be determined experimentally.

Example 5.8.2. Wing lift and drag

The forces acting on a given airfoil (a wing or a turbine blade, for example) include lift, drag, and viscous and inertia forces. Find the important dimensionless parameters.

Sought: Dimensionless parameters given geometry and four forces.

Given: Geometrical factors include the shape, chord length, and angle of attack α. The shape will be represented by the wing thickness D, and the chord length will be represented by the length l. The forces are represented by the lift force F_L, the drag force F_D, the viscous force (μ, V, l), and the inertia force (ρ, V, l). These quantities have dimensions

$[F_D] = F,$ $[F_L] = F,$ $[\rho] = FT^2/L^4,$ $[V] = L/T,$ $[l] = L,$
$[\mu] = FT/L^2,$ $[\alpha] = 1.$

Assumptions: All quantities of importance to the problem have been included in the analysis.

Solution: With seven quantities and three dimensions there are thus at most four dimensionless parameters. From our knowledge of force ratios these parameters are

$C_D,$ $C_L,$ Re, $\alpha.$

Since we expect the lift and drag forces to be independent of one another, we can write

$$C_D = C_D(\text{Re}, \alpha), \tag{5.8.4}$$

$$C_L = C_L(\text{Re}, \alpha). \tag{5.8.5}$$

(We have simplified this problem in that we left out wing roughness, and assumed the Mach number is small enough that compressibility effects are not important. Roughness can be very important. It can affect the point where the flow changes from laminar to turbulent flow, which can produce major changes in the drag and lift forces. This was found on the first flight of the space shuttle, which returned from space cleaner than its model was during the wind tunnel tests, and consequently had a faster landing speed than had been anticipated.)

Many tables of C_D and C_L do not give the dependence on Re, and care must be taken to ascertain that the tables are used in the correct Reynolds number range. A C_L curve that is valid for a commercial aircraft wing is incorrect for a model airplane, for example. More details on this are given in Chapter 7.

Example 5.8.3. Ship drag

A ship traveling on the surface of a body of water experiences a drag due to viscosity and the energy that goes into making waves (gravity forces). Find the relevant dimensionless parameters.

Sought: Dimensionless parameters based on given geometry and two forces.

Given: We consider a family of hulls with the same shape distribution. The hull length is l. The drag force is F_D, the viscous force is represented by l, μ, ρ, the gravity force by $l, \rho,$ and g, and the inertia force by $l, \rho,$ and V. Thus the quantities of interest are $F_D, l, \mu, \rho, g,$ and V. Their dimensions are

$[F_D] = F,$ $[\rho] = FT^2/L^4,$ $[V] = L/T,$ $[l] = L,$ $[\mu] = FT/L^2,$ $[g] = L/T^2.$

Assumptions: All quantities of importance to the problem have been included in the analysis.

Solution: With six quantities and three dimensions there are at most three dimensionless parameters. From taking force ratios, these are

$C_D,$ Re,   Fr.

We expect then that

$$C_D = C_D(\mathrm{Re}, \mathrm{Fr}).$$ (5.8.6)

Example 5.8.4. Channel and river flows

The flow in a channel or river is due to gravity forces, with viscous and inertia forces also playing roles. If α is the slope of the river bed and l is a relevant length (for example, the average depth), find the dimensionless parameters.

Sought: Dimensionless parameters based on geometry and three forces.
Given: The important quantities representing gravity, viscous, and inertia forces are

$$\rho, \quad V, \quad l, \quad \mu, \quad g.$$

In addition, the slope of the river bed α is important. The dimensions of these quantities are

$$[\rho] = FT^2/L^4, \quad [V] = L/T, \quad [l] = L, \quad [\mu] = FT/L^2, \quad [g] = L/T^2,$$
$$[\alpha] = 1.$$

Assumptions: All quantities of importance to the problem have been included in the analysis.
Solution: With six parameters and three dimensions there are at most three independent dimensionless parameters. From knowledge of the force ratios these dimensionless parameters will then be the Reynolds and Froude numbers and the slope. We thus would have

$$\mathrm{Fr} = f_1(\mathrm{Re}, \alpha).$$ (5.8.7)

If we are interested in velocity or discharge, then we might want V to be a nonrepeating variable, rather than having it appear in both Fr and Re. In that case, we could write

$$\mathrm{Fr} = f(\mathrm{Re}/\mathrm{Fr}, \alpha),$$ (5.8.8)

where

$$\mathrm{Re}/\mathrm{Fr} = \sqrt{l^{3/2}g/\nu}.$$

---

9. Model Studies—Similitude

The use of dimensional analysis plays an important role in the interpretation of model studies as well as in developing the forms of formulas. It would be both too risky of human life and too costly to build an airplane from a paper or computer study and then to fly it without first testing a model in a wind tunnel. Trial and error would similarly be a nonproductive means of deciding how big a power plant a ship might need, hence the use of model studies in ship towing tanks. The principle behind the use of models is called similitude, and is nothing more than the application of dimensional analysis.

In conducting a model study, a geometrically scaled model is first constructed. Use of a geometrically scaled model gives us *geometric similitude*. By geometric scaling, we have made the ratio of corresponding lengths in the model and prototype to everywhere have the same value. *Dynamic similitude*, where the velocities and forces are scaled as well as the geometry, requires that the relevant force ratios be the same in model and prototype. This can be accomplished by making all the dimensionless parameters equal in model and prototype. When this is the case, results can be translated directly from model to prototype.

For example, consider the lift force on a wing, where we have seen that

$$C_L = C_L(\text{Re}, \alpha). \tag{5.9.1}$$

We would perform a model study by first building an airfoil of exactly the same geometric shape as the prototype, but with all dimensions reduced by a scale factor λ. We would then place this model in a wind tunnel, at an angle of attack α (one of our dimensionless parameters) that is exactly the same as we desire for the prototype. The wind tunnel would then be run at a speed that ensures that the Reynolds number for the model is the same as that for the prototype. By equating the Reynolds numbers, we find that

$$V_{\text{model}} = V_{\text{prototype}}\, l_{\text{prototype}}\, \nu_{\text{model}} / l_{\text{model}}\, \nu_{\text{prototype}}$$

$$= V_{\text{prototype}}\, \nu_{\text{model}} / \lambda \nu_{\text{prototype}}. \tag{5.9.2}$$

If we carry these experiments out successfully, and in the process we have not changed things so that factors that were not important for the prototype have not become important for the model, then even though we do not know the functional dependence of the lift coefficient on Reynolds number and angle of attack, we can say

$$C_{L\,\text{model}} = C_{L\,\text{prototype}} \tag{5.9.3}$$

for any given angle of attack. Thus the prototype forces are related to the model forces at that angle of attack by

$$F_{L\,\text{prototype}} = F_{L\,\text{model}}\, \rho_{\text{prototype}}\, v_{\text{prototype}}^2 / \rho_{\text{model}}\, v_{\text{model}}^2 \tag{5.9.4}$$

In principle, then, we should conduct all model tests using dynamic similitude, and our results will then be readily transferable from model to prototype. Unfortunately this principle is not always practical. Suppose, for instance, we wish to perform a model study on a 1,000-ft-long ship that travels at a speed of 25 ft/s. The model length will be dictated by the length of our towing tank. (We could tow our model in a river or lake, but then we would have little or no control on wave or wind conditions in our experiment.) If we are fortunate to have a tank several hundred feet long, we might decide to construct a 10-ft-long model, giving $\lambda = 0.01$.

Equation (5.8.6) tells us that if we make the Reynolds and Froude numbers equal in model and prototype, then the drag coefficients will be equal and our model results can easily be transferred to the prototype. Assuming the model tests will be constructed under the same gravitational conditions that the prototype operates in, i.e., on earth, so that the g's in the Froude numbers are the same value, equating the Froude number for model and prototype gives

$$V_{\text{model}} = \sqrt{\lambda}\, V_{\text{prototype}} = 0.1 V_{\text{prototype}}. \tag{5.9.5}$$

That is, since our prototype is moving at 25 ft/s, we want to conduct our model tests at 2.5 ft/s, a reasonable value.

Equality of the Reynolds number for model and prototype leads to the requirement that

$$\nu_{\text{model}} = \lambda \nu_{\text{prototype}}\, V_{\text{model}}/V_{\text{prototype}}. \tag{5.9.6}$$

For water, a typical value of kinematic viscosity might be 1.4×10^{-5} ft^2/s. Thus dynamic similitude requires that we conduct our tests in a fluid with a kinematic viscosity of 1.4×10^{-8} ft^2/s. If indeed such a fluid exists we would hope that it is not too toxic, too hazardous, and too expensive to use in our towing tank!

What to do in such a case? This is a place where engineering skill, experience, and judgment come into play. It is possible, using formulas discussed later in this book, to make good computations as to the viscous drag on the ship hull. If we make the **assumption** as did Froude that wave drag and viscous drag are independent so that

$$F_{D\text{ total}} = F_{D\text{ wave}} + F_{D\text{ viscous}}, \tag{5.9.7}$$

and that $F_{D\text{ wave}}$ depends only on the Froude number and $F_{D\text{ viscous}}$ only on the Reynolds number, then we can compute the viscous drag for model and prototype, obtain the wave drag for the model by taking the measured total drag, subtract the computed viscous drag from the total model drag, scale up the wave drag to the prototype using the drag coefficient, and add to this the computed viscous drag for the prototype, giving its total drag. This is perhaps not completely satisfactory from an esthetic point of view, but experience shows that it works quite well, and with a little engineering judgment (based, of course, on tests and experience) this is a reliable means of deciding the minimum thrust a propeller has to provide in order to propel a ship. The separation of wave and viscous forces as given by equation (5.9.7) is referred to as *Froude's hypothesis*.

Similar modeling difficulties arise in cases where the length scale becomes severe. Suppose, for instance, that we are interested in studying flood stages in the Mississippi River Valley. The river is of the order of 1,300 miles long, 1,300 ft wide, and 20 ft deep. If we were to construct a 1,000-ft model, the length scale becomes $\lambda = 1.45 \times 10^{-4}$, and our model depth would be of the order of 2.4×10^{-4} inch! Whatever effects surface tension has in the prototype, they would certainly be exaggerated in this model! Scaling the bottom contours would be impractical, as would be scaling the effects of the river sinuosity. Yet these factors play an important part in determining travel times for flood crests.

In this case, the only reasonable solution is to exaggerate the depth scale of the model, making the depth and width scaling ratios much greater than the length scale. By correspondingly exaggerating the time scale, it is possible to make reasonable predictions from such a model.

Modeling then is an art as well as a science. It requires a fundamental understanding of the physics of a problem, and many times has to rely on all of our theoretical and physical knowledge of fluid mechanics to obtain a desired answer.

10. Experimental Facilities

There are several types of facilities that are in common usage for conducting fluid experiments. The facilities can be roughly grouped by the main dimensionless parameter that they attempt to model.

a. Froude number facilities

Open channel flumes come in all sizes, from tabletop to channeled rivers. A typical flume looks somewhat like Figure 5.1. A head box (left side of figure) controls both the head of water available in the tank and the inlet depth to the channel through the action of two gates. An exit gate is also of use in controlling the flow and in positioning waves such as hydraulic jumps. The exiting water is either discharged or pumped back to the head box to be recirculated. Recirculating channels can have the recirculation path either on the side of or underneath the test section. Flumes are used to study open channel flows such as transients due to gate openings and closings, bores, hydraulic jumps, forces on structures, dispersion of pollutants, and deposition of sedimentation. They can be closed on both ends and a wavemaker added to convert them to a *wave tank*, enabling the study of wave forces on stationary objects such as offshore structures. *Towing basins* are usually very long tanks (typically hundreds of feet) containing stationary water. A carriage passes over the basin dragging a ship hull, for example, in order to measure the net force on the hull. *Maneuvering basins* are squarish tanks of stationary water using radio-controlled models to study the dynamic character of ship hulls, and to train ship pilots for harbor duty. The fluid used in flumes, basins, and wave tanks is usually water.

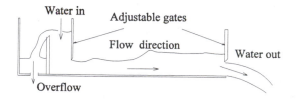

Figure 5.1. Cross section of a typical laboratory flume.

Figure 5.2. NASA–Ames Unitary Plan Wind Tunnel. *Legend*: A. Dry air storage spheres; B. aftercooler; C. 3-stage axial flow fan; D. drive motors; E. flow diversion valve; F. 8- by 7-ft supersonic test section; G. cooling tower; H. flow diversion valve; I. aftercooler; J. 11-stage axial flow compressor; K. 9- by 7-ft supersonic test section; L. 11- by 11-ft transonic flow section.

b. Mach number facilities

The air counterpart of the flume is the *wind tunnel*. Again, these come in various forms. For exclusively subsonic use they may be open circuit; air is taken from the atmosphere, passed through a test section, and then discharged back to the atmosphere. This has the disadvantage that the temperature, humidity, and pressure of the incoming air are not easily controlled. Closed circuit tunnels such as in Figure 5.2 have the advantage of complete control of the air used. At its Ames Research Center, the National Aeronautics and Space Administration (NASA) has subsonic tunnels large enough to test models in the 40-ft size. Their Unitary Plan Wind Tunnel is essentially three wind tunnels joined together to run off of one power drive. The supersonic (Mach number greater than 1) test sections, 8 by 7 and 9 by 7 ft in size, can be used to test models several feet in width. The transonic tunnel (Mach numbers ranging around 1), 11 by 11 ft in size, can test the difficulties involved in control of airplanes where the flow on the wing is supersonic in some regions and subsonic in others. Measurements in wind tunnels are typically made of three force components (drag, lift, and sidewise) and three moment components (pitch, roll, and yaw). Flow visualization is carried out through such means as tufts of yarn glued to the model, injection of smoke, and interferometers.

c. Cavitation number facilities

A *water tunnel* is essentially a wind tunnel with water as the working fluid. Water tunnels are used primarily for studying ship propellers and the effects of cavitation on

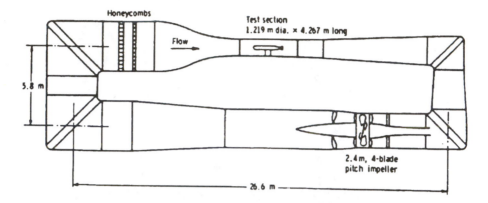

Figure 5.3. Garfield Thomas Cavitation Tunnel at the Applied Research Laboratory, The Pennsylvania State University. (Drawing courtesy the Applied Research Laboratory, The Pennsylvania State University.)

them. Figure 5.3 is a sketch of the original cavitation tunnel constructed at the Garfield Thomas Water Tunnel Facilities of The Pennsylvania State University. The tunnel can accommodate propellers up to 635 mm in diameter, and velocities of up to 18.29 m/s

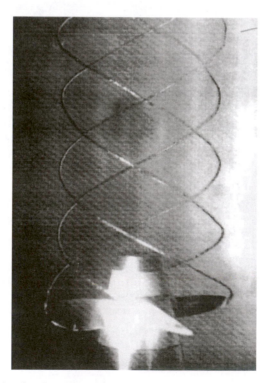

Figure 5.4. Cavitation farming at propeller blade tips.

can be achieved in the test section. The maximum motor power is 2,000 hp. The maximum and minimum pressures in the test section are 413.7 and 20.7 kPa. Cavitation numbers down to 0.1 are possible. Vanes at the corners turn the flow, and the honeycomb sections before the test section are used to reduce the scale of turbulence. Figure 5.4 shows a typical propeller test, with flow being from left to right. Cavitation occurs first at the propeller blade tips, where the velocity is fastest and the pressure is lowest. The cavitation bubbles are shed downstream from the propeller tips and form a spiral of pitch $V/R\Omega$, V being the flow speed, R the tip radius, and Ω the angular velocity of the propeller.

Suggestions for Further Reading

Baker, W. E., et al., *Similarity Methods in Engineering Dynamics*, Spartan, Rochelle Park, NJ, 1973.

Barenblatt, G. I., *Dimensional Analysis*, Gordon and Breach, New York, 1987.

Becker, H. A., *Dimensionless Parameters*, Halstead Press (Wiley), New York, 1976.

Boucher, D. F., and G. E. Alves, "Dimensionless numbers," *Chem. Eng. Prog.*, vol. **55**, pages 55–64, 1959.

Bridgman, P. W., *Dimensional Analysis*, Yale University Press, New Haven, 1931, 1963.

Buckingham, E., "On Physically Similar Systems: Illustrations of the Use of Dimensional Equations," *Phys. Rev.*, vol. **4**, no. 4, pp. 345–376, 1914.

Buckingham, E., "Model Experiments and the form of empirical equations," *Trans. ASME*, vol. **37**, pp 263–296, 1915.

Duncan, W. J., *Physical Similarity and Dimensional Analysis*, Arnold, London, 1953.

Focken, C. M., *Dimensional Methods and Their Applications*, Arnold, London, 1953.

Huntley, H. E., *Dimensional Analysis*, Rinehart, New York, 1951.

Inui, T., "Wavemaking Resistance of Ships," *Trans. Soc. Nav. Arch. Marine Eng.*, vol. **70**, pp. 283–326, 1962.

Ipsen, D. C., *Units, Dimensions, and Dimensionless Numbers*, McGraw-Hill, New York, 1960.

Jupp, E. E., *An Introduction to Dimensional Methods*, Cleaver-Hume, London, 1962.

Kline, S. J., *Similitude and Approximation Theory*, McGraw-Hill, New York, 1965.

Lanchester, F. M., *The Theory of Dimensions and its Applications for Engineers*, Crosby-Lockwood, London, 1940.

Langhaar, H. L., *Dimensional Analysis and Theory of Models*, Wiley, New York, 1951.

LeCorbeiller, P., *Dimensional Analysis*, Irvington, New York, 1966.

Massey, B. S., *Units, Dimensional Analysis, and Physical Similarity*, Van Nostrand Reinhold, New York, 1971.

Murphy, G., *Similitude in Engineering*, Ronald, New York, 1950.

Porter, A. W., *The Method of Dimensions*, Methuen, London, 1933.

Pankhurst, R., *Dimensional Analysis and Scale Factors*, Reinhold, New York, 1964.

Sedov, L. I., *Similarity and Dimensional Methods in Mechanics*, translated by M. Holt, Academic Press, New York, 1959.

Sharp, J. J., *Hydraulic Modeling*, Butterworth, London, 1981.

Skoglund, V. J., *Similitude—Theory and Applications*, International, Scranton, PA, 1967.

Taylor, E. S., *Dimensional Analysis for Engineers*, Clarendon Press, Oxford, England, 1974.

Yalin, M. S., *Theory of Hydraulic Models*, Macmillan, London, 1971.

Zierep, J., *Similarity Laws and Modeling*, Marcel Dekker, New York, 1971.

Problems for Chapter 5

Parameter formation

5.1. Arrange the following quantities in dimensionless parameters, using V, d as repeating variables: V (velocity), ρ (density), d (length), μ (viscosity), T (torque), g (gravity).

5.2. For the list of dimensional quantities V (velocity), ω (frequency), ρ (mass density), μ (viscosity), D (diameter), c (speed of sound), how many independent dimensionless parameters can be formed? Form an independent set of them using either algebraic or inspection methods.

5.3. Given a problem where velocity V, length b, viscosity μ, drag force F_D, and a frequency ω are known to be important, how many dimensionless parameters can be formed from these quantities? Give one possible set of dimensionless parameters.

5.4. For a particular problem the important parameters governing drag are force, dynamic viscosity, pressure, and length. How many dimensionless parameters could you form from these four quantities? What are the parameters?

5.5. Arrange the following quantities into dimensional parameters:

(a) τ (stress), ρ (mass density), V (velocity), p (pressure).

(b) h_L (head loss), V (velocity), ρ (mass density), g (gravitational acceleration), l (length), μ (viscosity).

(c) p (pressure), ρ (mass density), K (bulk modulus), g (gravitational acceleration), μ (viscosity).

5.6. Form the maximum number of independent dimensionless parameters possible in the following lists:

(a) D, V, μ, γ.

(b) Δp, D, V, μ.

(c) τ, ρ, D, μ.

5.7. Form as many dimensionless parameters as you can out of the following list: Q (discharge), D (diameter), p (pressure), ρ (mass density), μ (viscosity), g (gravity).

5.8. Arrange the following given quantities into dimensionless parameters: ρ (mass density), μ (viscosity), V (velocity), D (diameter), g (gravity), F (force).

5.9. Form as many independent dimensionless parameters as possible out of the quantities in each list:

(a) Δp, ρ, Q, D.

(b) Q, μ, g, D.

(c) ρ, V, g, D, h_L.

5.10. From the dimensional quantities ρ (mass density), ν (kinematic viscosity), V (velocity), l (length), g (gravity), p (pressure), F_D (drag force), how many dimensionless parameters can be formed? Form an appropriate set of dimensionless parameters.

5.11. Given the following quantities: Q (discharge), ρ (mass density), μ (viscosity), l (length), F_D (drag force). Form these quantities into dimensionless parameters. If a model test were to be run under which the above were the only important quantities, and the model was 1/10 scale and the fluid was the same for model and prototype, how would the force be scaled?

5.12. Arrange b (length), Q (volumetric discharge), μ (viscosity), ρ (mass density), and p (pressure) into the appropriate number of dimensionless parameters.

5.13. Arrange ρ (mass density), g (acceleration due to gravity), V (velocity), and T (torque) into a dimensionless parameter.

5.14. For a given problem, a pressure drop Δp depends on a length D, a kinematic viscosity ν, a discharge Q, and a specific weight γ. Using only these five quantities, find a set of dimensionless parameters that would be useful to the problem. If a model test were to be run using a fluid where $\gamma_m/\gamma_p = 1.3$ and $\nu_m/\nu_p = 20$ in a 1/10 scale model, how are the discharges scaled? For the previous part, how are the pressures scaled?

5.15. Given that in an experiment the drag force F_D depends on ρ (mass density), V (velocity), l (a length), μ (absolute viscosity), and g (gravity): What is the maximum number of dimensionless parameters that could be formed to describe this experiment? Form one possible set containing the number of parameters given in your answer. State in words how you would use your results to conduct an experiment on a model designed to determine F_D.

Interpretation

5.16. The Reynolds number, Froude number, drag coefficient, and pressure coefficient can be interpreted as ratios of certain forces. Write common dimensionless forms for each of these parameters, and show what they are equal to in terms of force ratios (i.e., $\Pi = X_{\text{force}}/Y_{\text{force}}$).

5.17. Of the problems listed below, which would be modeled using the Froude number? Which would be modeled using the Reynolds number?

(a) Wave drag on a ship.

(b) Drag on an underwater cable.

(c) Flow in a vertical water main.

(d) Flow through a partially opened valve.

(e) The rise of sap in a tree.

5.18. In which of the cases listed below would modeling be based on Reynolds number? In which of the above cases would modeling be based on Froude's number?

(a) Flow over a spillway.

(b) Flow in a horizontal water main.

(c) Waves breaking on a sea wall.

(d) Fish swimming in a lake.

(e) Slow flow of water in a capillary tube.

5.19. The drag force on a high-velocity projectile depends on the following forces: inertia, compressibility, viscous. Form the dimensionless parameters important to finding drag force. Give the general form of the expression for this drag.

Similarity

5.20. In a fluid rotating as a rigid body, it is desired to find by dimensional analysis a form for the pressure distribution. State what parameters are important, why you chose them (i.e., what forces, etc., do they represent), and obtain a relation for the pressure in terms of dimensionless terms.

5.21. A large venturi meter to be used to measure air flow is tested using a 1/5 scale model with water as the fluid. The flow rate is measured by means of the pressure drop from entrance to throat. What forces are important in this model study? Form dimensionless parameters scaling these forces. At a water flow rate of 3 ft^3/s the pressure drop is 14 psi in the model. What would be the corresponding air flow rate and pressure drop in the prototype? Take the temperature as 70°F.

5.22. An anchor cable 100 ft long and 0.1-ft diameter is to be used to hold a buoy in a channel where the average current is 3 ft/s. To determine the sidewise force on the cable due to the current, a 1/20 scale model is to be tested in the laboratory. Assume dynamical similarity, and that water is used in the test. What should be the velocity of the fluid in the test? What is the ratio of force in the prototype to force in the model? Comment on the feasibility of this model.

5.23. The steady ascent of a balloon in 30°C air is to be studied by a 1/50 scale model in 20°C water. List the important forces, tell why you included them, and give the relation between the velocities of the model and prototype.

5.24. A steel sphere (SG = 7) of 0.01-ft. diameter is dropped into still water at standard conditions. After steady conditions (terminal velocity) are reached, what is the drag force on the sphere? If a model study were to be in the same fluid made using a material with SG = 3, what would the diameter have to be to make the flows dynamically similar?

5.25. Two spheres are to be made so that one made of a material with specific gravity 2.5 is dropped in a fluid with specific gravity 0.9, and the other made of a material with specific gravity 4 is dropped in a fluid of specific gravity 1.2. They fall at the same velocity. What must be the ratios of their diameters and kinematic viscosities?

5.26. A spherical hailstone (SG = 0.8) of 6-mm diameter falls at a constant velocity through still air (temperature = 5°C). The falling hailstone is to be modeled in a tank of water at 20°C. If the model of the hailstone is 48 mm in diameter, what should be the specific gravity of the model?

5.27. A popular demonstration consists of supporting a sphere in a vertical jet of air discharging into the atmosphere. For this experiment, d = diameter of jet at nozzle = 2 in, D = diameter of sphere = 3 in, W = weight of sphere = 5.5×10^{-4} lb, V = velocity of air jet at nozzle = 30 ft/s, $v = 1.5 \times 10^{-4}$ ft^2/s, $\mu = 3 \times 10^{-7}$ lb-s/ft^2. What is the lift force on the sphere? Determine quantities for a dynamically similar experiment using water as the fluid, given $\mu = 2 \times 10^{-5}$ lb-s/ft^2, $v = 10^{-5}$ ft^2/s.

5.28. A sphere with SG = 0.7 and 6-mm diameter is dropped into still water at standard conditions. After steady conditions (terminal velocity) are reached, what is the drag force on the sphere? If a model study were to be made in the same fluid using a material with SG = 3, what would the sphere diameter have to be to make the flows dynamically similar?

5.29. The flow about a 150-mm-diameter projectile traveling in still air ($p = 1$ bar, $T = 30°C$) is to be modeled in a wind tunnel with a 25-mm model. If viscous effects are not important, and if $p = 0.7$ bar, $T = -20°C$ in the wind tunnel (air is the working fluid), what should be the wind tunnel velocity? If the drag on the model is 8 N, what is the drag on the prototype?

5.30. A high-speed airplane, designed to fly at 1,000 mph in air at a temperature of −30°F, is to be tested by means of a 1/50 scale model in a wind tunnel. Given that the wind tunnel temperature is 185°F and the pressure is 785 psfa, what should be the speed at which the wind tunnel is run? Has complete dimensional similitude been achieved? Explain.

5.31. It is desired to find the drag force on a light plane traveling at 150 mph by making a scale model (1/10) and testing it in a water tunnel. Given that $\rho_{water} = 1.9$ slugs/ft^3, ρ_{air} = 0.0019 slug/ft^3, $\mu_{water} = 3.8 \times 10^{-5}$ lb-s/ft^2, $\mu_{air} = 3.8 \times 10^{-7}$ lb-s/ft^2. What are the important dimensional parameters for this problem? How many independent dimensionless parameters can be formed from these? Form an appropriate set of dimensionless parameters. What should be the water velocity in the test be to correspond to 150 mph in the prototype? If the drag on the model is measured to be 50 lb, what would be the drag on the prototype?

5.32. On a calm autumn day, a leaf is observed to fall from a tree. Instead of falling straight down, it oscillates from side to side. It is desired to model this experiment by dropping disks of roughly the same shape in a container of water, with the intent of finding the period of oscillation. What forces do you feel are important? What quantities describe these forces? What other quantities would you need to do a dimensional analysis? How many pi parameters can you form? Form the dimensionless parameters.

5.33. A thin flat plate 3 ft long and 2 ft wide is towed fully submerged at 10 ft/s in a towing basin containing water with $\rho = 2$ slugs/ft^3, $\mu = 2 \times 10^{-5}$ lb-s/ft^2. The drag force is observed to be 3 lb. If the experiment were now to be run in a wind tunnel with an air stream velocity of 50 ft/s, $\rho = 0.0024$ slug/ft^3, $\mu = 3.8 \times 10^{-5}$ lb-s/ft^2, what should be the dimensions of the plate for dynamic similarity? What should be the drag force on this plate with dimensions as given? Is it possible to obtain complete dimensional similarity in this case? Comment on the feasibility of this model.

5.34. It is desired to study wind drag force on trucks by conducting model studies on 1/20 scale models in a water tunnel. If the truck is 50 ft long and is to travel at 70 ft/s, how long should the model be? At what speed should the model tests be conducted? What is the force on the truck? The temperature is 70°F.

5.35. A model study is to be performed on a 1/2 scale model of an automobile. The full-scale auto will run at 80 ft/s and is 15 ft long. Viscous and inertia forces are important factors in influencing the drag force. Air at the same conditions will be used for both model and prototype. At what speed should the model be run? A drag force of 250 lb is measured on the model. What drag force does this correspond to for the full-scale prototype?

5.36. A windmill is to be built for use as a large-scale power generator. At its proposed location, useful wind speeds vary from 20 to 60 ft/s. The diameter of the full-scale windmill fan is to be 150 ft. It is desired to determine the power generated by constructing a 1/50 scale model and testing it in water. What should be the water velocity range of the model tests? How should power output from the tests (torque times fan angular velocity) be converted to the full-scale prototype?

5.37. A model of a ship propeller is to be made at 1/2 geometrical scale. The same fluid is to be used for model and prototype. If the ship speed is 20 ft/s and the propeller rotates at 5 rpm, what are the corresponding velocities for the model? What is the ratio of prototype propeller torque to model torque?

5.38. A large fan for a wind generator (diameter = 150 ft) is to be tested in a wind tunnel using a 1/50 scale model. Assume that the fluid properties for the model and prototype are the same. If the prototype is to run at one revolution per second, what should be the speed of the model? What is the ratio of prototype torque to model torque? Comment on the feasibility of this model.

5.39. An airplane propeller is to be tested in a wind tunnel. In particular it is desired to study the thrust (a force) and the torque as a function of wind speed and the angular velocity of the propeller. Inertia and viscous forces are thought to be important. The dimensional quantities are then F (thrust), M (torque), ρ, μ, D, V, ω. How many independent dimensionless parameters can be formed? Form them. If a 1/3 scale model is to be built, and the prototype is to be operated at 200 ft/s and 350 rpm, what should be the linear and angular model velocities? (The same fluid properties are used for model and prototype.) What is the scaling ratio for the thrust forces? What is the scaling ratio for the torques?

5.40. A large 2-ft-diameter horizontal propeller is deeply submerged in a vat of 100°F kerosene and is used to keep it stirred. A model using a 6-in propeller in water at 50°F requires a torque of 1.5 lb-ft to turn the model propeller at 45 rpm. What prototype speed corresponds to the model speed of 45 rpm? What torque will the prototype exert? For kerosene at 100°F, SG = 0.82, $\mu = 2.8 \times 10^{-5}$ lb-s/ft^2 For water at 50°F, $\mu = 2.0 \times 10^{-5}$ lb-s/ft^2.

5.41. The resistance of a submarine is to be determined by measuring the force on a 1/20 scale model. If the velocity of the full-scale submarine is 15 ft/s, and the fluid is the same for model and prototype, at what velocity should the model be tested? What is the relation between the force on the model and the force on the prototype? Comment on the feasibility of this model.

5.42. A ship 400 ft long moves in freshwater ($\rho = 1.94$ slugs/ft^3, $\mu = 2.34 \times 10^{-5}$ lb-s/ft^2) at 20 mph. A 1/100 scale model of this ship is to be tested in a towing basin containing a liquid with SG = 0.92. What viscosity should this liquid have? At what velocity should the model be towed? What propulsive force on the ship corresponds to a force of 2 lb on the model?

5.43. A ship 360 ft long moves in freshwater at 60°F, at 30 mph. Find the kinematic viscosity of a fluid suitable for use with a model 12 ft long, if complete dynamical similarity is to be attained.

5.44. A pitot tube is modeled 1/5 scale. Knowing that inertia and viscous forces are important, if model and prototype are run in the same fluid, what is the scaling on the pressure differences measured by each pitot tube?

5.45. Waves occur on a large pond of depth h. The velocity of these waves is governed by gravity and surface tension. The fluid is relatively incompressible. What dimensionless parameters would be of importance in the study of these waves?

5.46. An outlet pipe leads from a large storage tank containing crude oil (SG = 0.855, $\nu = 7 \times 10^{-5}$ ft^2/s) to a pump, and then to an oil pipeline. When the depth of oil in the tank approaches the level of the outlet pipe, the pump malfunctions because air is drawn periodically into the pump due to sudden fluctuations in the oil surface. It is believed that gravitational and viscous forces are of equal importance. If a 1/4 scale model is to be constructed to study when pumping should stop, what should be the properties of the liquid to be used? What is the ratio between discharge in the model and discharge in the prototype?

5.47. The spillway of a dam is to be modeled by a 1/25 scale model. The prototype is 37.5 ft high, 50 ft wide, and the maximum head of water coming onto the spillway is 5 ft. Assuming water is to be used in the model, what should be the model height, and the water head? If the flow over the prototype is 2,000 ft^3/s, what should be the discharge over the model? If the model dissipates 0.15 hp at the base of the spillway, at what rate will energy be dissipated by the prototype?

5.48. It is proposed to deliver airmail to small cities by dropping packages by parachute from small planes. Naturally, the U.S. Postal Service wishes the packages to land undamaged. They propose to study how large and how heavy a package can be dropped with a given parachute, from a high height, by doing model tests in water. If 1/5 scale models are used, how are the prototype and model speeds related at terminal velocity? How should the densities of the model and prototype be related?

5.49. The pressure loss in a horizontal pipe of diameter D and length l is thought to be a function of inertia and viscous forces, as well as the roughness of the pipe (expressed as an average roughness height). What are the dimensionless parameters relevant to a study of pressure loss? Give a general expression for pressure loss as a function of these parameters.

5.50. The power input needed to drive a pump depends on the volumetric flow rate, the pump speed of rotation, the pump diameter, the fluid density, and the pressure against which it operates. Perform a dimensional analysis, giving pertinent dimensionless parameters. A 1-ft-diameter pump operating at 500 rpm is to be used to deliver 25 ft^3/s of water against a pressure of 100 psi. A model is to be built using a 0.1-hp motor turning a 0.5-ft-diameter pump at 100 rpm, and water as the fluid. What should be the flow rate and outlet pressure for the model? Determine the power needed for the 1-ft-diameter pump.

5.51. A flat plate 0.5 ft long and 0.1 ft wide is towed fully submerged at 10 ft/s in a towing basin containing water at 68°F ($\rho = 1.936$ slugs/ft^3, $\mu = 2.05 \times 10^{-5}$ lb-s/ft^2), and the drag force is observed to be 3 lb. Calculate the dimensions of a plate that will yield dynamically similar conditions in an air stream ($p = 14.7$ psi, 59°F, $\rho = 0.00238$ slug/ft^3, $\mu = 3.74 \times 10^{-5}$ lb-s/ft^2) having a velocity of 6 ft/s. What drag force will be exerted on the model plate? Has complete dynamic similarity been achieved? Explain your answer.

5.52. The moment exerted on the rudder of a submarine is studied by means of a 1/20 scale model in a water tunnel. The fluid conditions are the same for model and prototype. To what prototype speed does a water tunnel speed of 45 ft/s correspond? If the torque measured on the model is 15 lb-ft, what is the torque on the prototype?

5.53. To determine the wind load on a bridge under a 120-mph wind, a 1/100 scale model is tested in a wind tunnel, using air at the same ambient temperature and pressure as with the prototype. What should the wind tunnel velocity be to ensure dynamic similarity? What is the relation between the model wind force and the prototype?

5.54. The gravitational attraction of the moon causes tides in lakes and oceans. In channels and rivers it can cause oscillations called "seiches" that can be of quite large magnitude. Derive a dimensionless formula for the period of such oscillations in a long channel of constant depth h and width w.

chapter 6

Laminar Viscous Flow

Chapter Overview and Goals

In order to determine how velocity and pressure vary from point to point in laminar flows, it is necessary to have the momentum equation in the form of a differential equation. This enables us to integrate the momentum equation to find how the velocity and pressure vary with the coordinates. We use an infinitesimal control volume to develop this differential equation form throughout this chapter. We then apply this form to solve for the laminar flow between parallel plates. This parallel plate solution is then extended to the case where the plates are almost parallel, which leads us to the study of a slider bearing, and how fluid layers act as lubricants.

Rederiving the differential form of the momentum equation in cylindrical polar coordinates, an approach similar to that used in flow between parallel plates enables us to also find the equation for laminar flow in a circular conduit.

*At moderately high values of the Reynolds number, viscosity effects are confined to a very thin layer in the immediate vicinity of the boundary. This layer is called the **boundary layer**. Several useful definitions for boundary layer thickness are given, and an integral form of the momentum equation is developed that allows us to approximate the flow in the boundary layer, and to find the shear stress on the wall.*

The phenomenon of flow separation is discussed, and formulas for locating where flows separate are presented.

Up to this point in our study of fluid mechanics we have concentrated our attention on flows where momentum, pressure, and gravity forces dominate. There are many flows, however, in which viscous forces are just as important as these other forces. This can be true either throughout the entire flow regime, or only in isolated regions of the flow. In this chapter, we will consider several cases of flows where viscosity is important, and that are laminar. As stated in Chapter 3, laminar implies that the flow occurs in smooth overlapping sheets and is of an orderly nature. This generally means that we are considering flows in the lower Reynolds numbers range. The range of Reynolds number for which a flow is laminar varies, depending strongly on the flow geometry.

1. Flow between Parallel Plates

In introducing the concept of viscosity in Chapter 1, we used as a defining problem that of flow between parallel plates. We considered there the case where the plates were horizontal and the driving force for the flow was the motion of one of the plates. Pressure gradients, or gravity (if the plates are inclined), could also act as the driving forces. We shall now analyze this problem in more detail, using our basic principles as applied to an infinitesimal control volume of the fluid. This will allow us to obtain a detailed description of the flow.

We start with the situation shown in Figure 6.1. The two parallel plates are separated by a distance b, and are inclined at an angle θ with the horizontal. The x axis is taken along the bottom plate, and the y axis normal to the plate. A small rectangular element dx by dy by dz is chosen for analysis. This element is shown enlarged in Figure 6.2, along with all of the forces acting upon the element. The velocity is assumed to be parallel to the plate,[1] and the streamlines to be a family of parallel lines. Therefore the velocity will have a component only in the x direction. This component cannot vary with x, for if it did, continuity would be violated unless there were a y velocity component as well.

We consider the case where the plates are very wide in the z direction, so that there is no variation of any of the flow quantities with the z coordinate. The flow is also taken to be steady. Then to ensure continuity of mass flow the velocity can vary only with y. According to Newton's law of viscosity, presented in Chapter 1, the shear stress will be

$$\tau = \mu \, du/dy. \tag{6.1.1}$$

Since u does not vary with x, equation (6.1.1) tells us that the shear stress also will depend on only y, and not on x.

This shear stress will act on the four faces of our element in Figure 6.2 in the directions as shown. On face AB, the outward drawn normal to the face is in the plus y direction, and the sign convention for the positive direction for shear stress is in the

---

[1]If there were a velocity component perpendicular to the plate, fluid would be penetrating the plate.

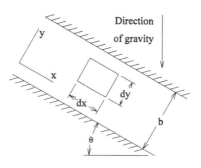

Figure 6.1. Flow between parallel plates geometry.

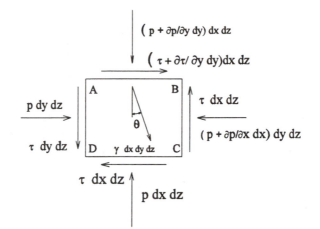

Figure 6.2. Control volume for flow between parallel plates.

plus x direction. Similarly on face BC, whose normal is in the plus x coordinate direction, the positive shear stress direction is in the plus y coordinate direction. On face CD, since the normal is in the negative y direction, the sign convention for shear stress is that it is positive if it acts in the negative x direction. On face DA with an outward normal in the negative x direction, the shear stress is positive if it acts in the negative y direction. This sign convention assures that the summation of moments is properly accounted for, for if we analyze two adjacent elements, when they are joined into a larger element, the shear stresses at the joining surfaces must cancel.

The pressure can vary with both x and y. Sufficiently far downstream from the entrance to the plates the flow will be fully established. This means that u is a function of only y, and the shear stress can vary with only the y coordinate. On faces CD and DA we will let the pressure be denoted by p and the shear stress by τ. As we consider other faces of our element, we will expand the various flow quantities in Taylor series about this face, and retain only the first two terms. For example, on face BC the pressure will be p plus its change in the x direction, or

$p + (\partial p / \partial x)\, dx.$

Since the shear stress τ does not vary with x, it will be the same on face BC as it is in AD. On face AB, we take values of the various quantities to be their values on face CD plus a small change due to variation in the x direction. This gives, for the pressure and shear stress, respectively,

$p + (\partial p / \partial y)\, dy \qquad \text{and} \qquad \tau + (\partial \tau / \partial y)\, dy.$

It may seem that we should be more detailed in our analysis, considering the variation of pressure and shear stress along a face as well as from face to face. It might also appear that choosing a different point about which to expand the Taylor series, for example the center of the element, might have an effect on the result. These forms of the analysis could also be carried out as above. They would lead to the conclusion that these alternate derivations lead to exactly the same result as we let our element become smaller and smaller. (For the student familiar with the Gauss and Green theorems of advanced calculus, an integral analysis similar to the momentum integral in Chapter 3 could be used. It avoids the questions raised by Taylor series expansions, and leads to our result in a quicker, cleaner manner.)

Starting with the control volume of Figure 6.2, we sum forces first in the x direction. The only contributors in this direction are the shear forces on faces AB and CD, the pressure forces on DA and BC, and the horizontal component of the body force. Writing their contributions in the order we have listed them, we have[2]

$$[\tau + (\partial \tau / \partial y)]\, dx\, dz - \tau\, dx\, dz + p\, dy\, dx - [p + (\partial p / \partial x)\, dx]\, dy\, dz$$

$$+ \gamma \sin \theta\, dx\, dy\, dx = 0.$$

Canceling common terms and dividing by the element volume, we are left with

$$d\tau / dy - \partial p / \partial x + \gamma \sin \theta = 0. \tag{6.1.2}$$

Summing forces in the y direction, the pressure forces on faces AB and CD, the shear forces on faces DA and BC, and the vertical component of the body force will be the only contributors. Since there is no momentum in the y direction, the result is

$$-[p + (\partial p / \partial y)\, dy]\, dx\, dz + \underline{p\, dx\, dz} - \underline{\tau\, dy\, dz} + \underline{\tau\, dy\, dz} - \gamma \cos \theta\, dx\, dy\, dz = 0.$$

The underlined terms cancel. Dividing by the volume we are left with

$$\partial p / \partial y + \gamma \cos \theta = 0, \tag{6.1.3}$$

which states that the pressure variation in the y direction is hydrostatic.

Of the two equations (6.1.2) and (6.1.3), it is easiest to solve equation (6.1.3) first. This gives

$$p = -y\gamma \cos \theta + f(x), \tag{6.1.4}$$

the "constant" of integration in this case being a function of x, since while taking partial derivatives with respect to one coordinate, we hold the other coordinate fixed. During integration of a partial derivative, the direction perpendicular to the path of integration is correspondingly held fixed.

Differentiating equation (6.1.4) with respect to x, we see that

$$\partial p / \partial x = df / dx. \tag{6.1.5}$$

That is, since the right-hand side of this equation depends at most on x and not on y, the pressure gradient in the x direction can vary at most with x, and not with y. Looking at equation (6.1.2), we see then that the first term can vary at most with y, the second term at most with x, and the third term stays constant. This equation must hold

---

[2]Because the velocity does not vary with x, it follows that there is no change in momentum in the x direction, and therefore the forces in the x direction must sum to zero.

throughout $0 \le y \le b$, and for many values of x. That can be true only if each term in the equation is a constant. That is, the shear stress must vary linearly with y and the pressure must vary linearly with x. This agrees with our physical understanding of the problem, since, from continuity considerations, we expect that if we are far enough away from any plate entrance and exit effects, our velocity profile will look the same no matter what value of x we choose. We would expect also that the driving force per unit length (pressure gradient in the x direction minus the gravity component) should be constant.

We can then integrate (6.1.2) with respect to y, giving for the shear stress

$$\tau(y) = (\partial p/\partial x - \gamma \sin \theta)y + A, \tag{6.1.6}$$

where A is a constant of integration. Using (6.1.1), we can integrate (6.1.6) again to find the velocity, giving

$$u(y) = (\partial p/\partial x - \gamma \sin \theta)y^2/2\mu + Ay/\mu + B, \tag{6.1.7}$$

where B is a second constant of integration. The quantity $(\partial p/\partial x - \gamma \sin \theta)$ represents the difference between the applied pressure gradient and the hydrostatic pressure gradient. It is the net pressure gradient that causes the flow to occur. The constants of integration A and B must be determined by conditions at the two plates that bound our flow.

a. Solid plates at both boundaries

For solid plates at both $y = 0$ and $y = b$, the appropriate boundary conditions are the no-slip ones. That is, the fluid velocity at a plate must equal the velocity of that plate. Let the lower plate velocity be U_L and the upper plate velocity be U_U. Then, according to (6.1.7), we have

$$U_L = u(y = 0) = B$$

and

$$U_U = u(y = b) = (\partial p/\partial x - \gamma \sin \theta)b^2/2\mu + Ab/\mu + B.$$

Solving for A and B and putting the results into (6.1.7), we have

$$u(y) = [(\partial p/\partial x - \gamma \sin \theta)(y^2 - yb)/2\mu] + [U_L + (U_U - U_L)y/b]. \tag{6.1.8}$$

The first square bracket term in (6.1.8) is referred to as "plane Poiseuille flow." It is the part of the flow that is driven by the pressure gradient and gravity. The velocity varies parabolically with the y coordinate, and the shear stress linearly with y. The second square bracket term in (6.1.8) is referred to as "plane Couette flow." This part of the flow is due solely to movement of the boundary, and it is the flow we used when we first introduced viscosity in Chapter 1. The velocity varies linearly with the y coordinate, and the shear stress is a constant.

The discharge can easily be found by integrating (6.1.8) between the two plates. If we use w as the width in the z direction, integration of the velocity over the area leads to

$$Q = w\int_0^b u\,dy = -[w(\partial p/\partial x - \gamma\sin\theta)b^3/12\mu] + [(U_L + U_U)bw/2]. \qquad (6.1.9)$$

This result shows that for Poiseuille flow to have a net flow in the positive x direction, the quantity $\partial p/\partial x - \gamma\sin\theta$ must be negative.

Example 6.1.1. Flow between parallel plates
Describe how the previous result can be used as an easy means of measuring the viscosities of liquids.

Sought: An experiment that affords a means of determining viscosity.
Given: Geometrical quantities of the experiment to be performed. Discharge and pressure gradient are to be measured.
Assumptions: Laminar steady viscous flow. Both the density and viscosity are considered to be constant. Entrance and exit effects are considered to be negligible, so the velocity depends on only the y coordinate.
Solution: For stationary horizontal plates, the relationship between discharge and pressure gradient is, from (6.1.9),

$$Q = -wb^3\partial p/\partial x/12\mu.$$

Setting up the plates as in Figure 6.3, and measuring the pressure at points A and B, we have for the pressure gradient

$$\partial p/\partial x = (p_B - p_A)/L,$$

and the viscosity is given by

$$\mu = wb^3(p_A - p_B)/12QL.$$

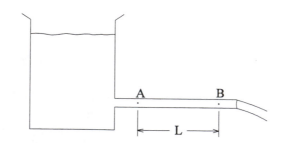

Figure 6.3. Method for measurement of viscosity.

By measuring the pressure difference with a manometer and the discharge with a stopwatch and a container of calibrated volume, we find the viscosity easily. For example, suppose that the pressure drop is 2 ft of water over a 5-ft length, and the flow rate is 4 gpm when $w = 0.2$ ft and $b = 0.05$ ft. Then we have

$$\partial p/\partial x = \Delta p/l = (p_B - p_A)/l = -2 \times 62.4/5 = -24.96 \text{ psf/ft},$$

$$Q = 4 \times 7.49/60 = 0.499 \text{ ft}^3/\text{s},$$

$$\mu = 0.2 \times (0.05)^3 \times (24.96/12 \times 0.499) = 1.042 \times 10^{-4} \text{ lb-s/ft}.$$

The method for measuring viscosity introduced in Chapter 1 involves the measurement of forces. Measuring viscosity as suggested in Chapter 1 thus necessitates more difficult measurements than needed in the present approach.

b. Solid plate plus a free surface

In the case of a liquid in the region $0 \le y \le b$ with the top plate absent, the liquid will still obey the no-slip condition with respect to the superposed gas. Also, the shear stress in the liquid must be the same as the shear stress in the gas; otherwise a concentrated force would result between the gas and the liquid. By this means the liquid will "drag" the gas along with it. Since the viscosity of a liquid is usually much greater than that of a gas, it is frequently a good approximation to simplify the condition of stress continuity by instead requiring that the shear stress at $y = b$ be zero. (This would be exactly true if there were a vacuum above the liquid.) The pressure gradient in the x direction must be zero, since the free surface will be a constant pressure surface (neglecting any hydrostatic pressure changes in the gas). Then from (6.1.7), we have

$$U_L = u(y = 0) = B$$

and

$$0 = \tau(y = b) = -\gamma b \sin \theta + A,$$

so that

$$u(y) = \gamma y(2b - y) \sin \theta/2\mu + U_L. \tag{6.1.10}$$

The discharge is

$$Q = w\int_0^b u \, dy = wb(gb^2 \sin\theta/3\nu + U_L). \tag{6.1.11}$$

Example 6.1.2. Parallel flow with a free surface

Moving belts are sometimes used to remove oil from the surface of settling ponds, as in Figure 6.4. Find the relationship between belt geometry, belt speed, and the rate at which the oil is removed.

Sought: A formula giving rate of oil removal as a function of the fluid properties and geometry.

Given: Belt angle, belt width, and belt speed. The pressure gradient parallel to the belt will be zero since the free surface of the oil will be at atmospheric pressure.

Assumptions: Laminar viscous flow. The viscosity of the surrounding air will be neglected, so that the shear stress on the oil at the free surface will be taken as zero. The region where the belt is in contact with the oil and water is taken to have negligible effect on the remainder of the flow.

Solution: For a vertical upward-moving belt, $\sin \theta = 1$ and $U_L < 0$. To have a net upward discharge ($Q < 0$), it is seen from (6.1.11) that we must have $U_L < -b^2 g / 3\nu$, the minus sign indicating that the belt must move in the $-y$ (upward) direction.

The thickness b of the oil film is controlled by the conditions at the end of the belt. By setting the derivative of Q with respect to b equal to zero, we see that the maximum discharge that can be attained at any given speed is when

$$-U_L = b^2 g / \nu.$$

This meets the requirement of upward flow. As an example, with oil having a kinematic viscosity of $2 \times 10^{-5} \text{m}^2/\text{s}$, we see that

$$-U_L = b^2 g / \nu = 9.8\ b^2 / 2 \times 10^{-5} = 4.9 \times 10^5\ b^2.$$

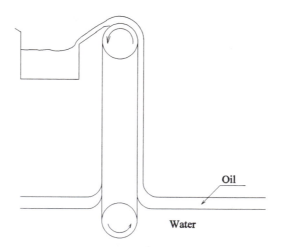

Figure 6.4. Recovery of oil from the surface of a pond.

Therefore to maintain an oil film thickness of 2 mm, the upward velocity must be greater than $4.9 \times 10^5 \times (2 \times 10^{-3})^2 = 1.96$ m/s. For these values, $Q/w = -2gb^3/3\nu = -2.61 \times 10^{-3}$ m^{2}s.

2. Lubrication

The purpose of lubrication is to replace dry friction by viscous friction, thereby reducing the friction force. In the process, the lubricant serves to keep the solid surfaces apart by building up a pressure between the surfaces.

Consider first a slider bearing as shown in Figure 6.5. The upper surface of the bearing is stationary, while the bottom slider moves with a velocity U. Each end of the bearing is exposed to the same environment, so the pressures at $x = 0$ and $x = L$ are equal. The two surfaces are taken to be slightly inclined with one another at an angle α that is small. The spacing b at any x is then given by

$$b(x) = b_0 - x \tan \alpha, \tag{6.2.1}$$

where

$$\tan \alpha = (b_0 - b_L)/L, \tag{6.2.2}$$

b_0 and b_L being the wall spacings at $x = 0$ and $x = L$, respectively.

Since the flow will generally be two-dimensional, the exact solution of this flow problem is difficult and complicated. However, we can have a very good approximation by assuming that entrance and exit effects are negligible, and that the dependency of the velocity distribution on y is the same as that for the flow between parallel plates. That is, from (6.1.8) we take

$$u = u(x, y) = (\partial p/\partial x)(y^2 - yb)/2\mu + U(1 - y/b), \tag{6.2.3}$$

where b is now a function of x as given by (6.2.1). Also, instead of taking the pressure gradient constant as before, we let it be a function of x yet to be determined.

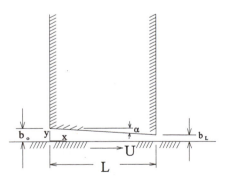

Figure 6.5. Defining figure for a slider bearing.

In finding the pressure gradient, we make use of global continuity. That is, we require that the discharge between the plates be the same at every value of x. From (6.1.9) the discharge per unit width is given by

$$Q/w = \int_0^b u \, dy = -(\partial p/\partial x)b^2/12\mu + Ub/2. \tag{6.2.4}$$

Solving (6.4.2) for the pressure gradient, we find that

$$\partial p/\partial x = -12\mu Q/wb^3 + 6U\mu/b^2. \tag{6.2.5}$$

This can be integrated to give the pressure at any x. This gives

$$p(x) = p_0 - \frac{6\mu Q}{w \tan \alpha}\left(\frac{1}{b^2} - \frac{1}{b_0^2}\right) + \frac{6\mu U}{\tan \alpha}\left(\frac{1}{b} - \frac{1}{b_0}\right), \tag{6.2.6}$$

where p_0 is the pressure at $x = 0$ and b_0 is the spacing there.

We can now find the discharge Q by applying to (6.2.6) the condition that the pressures at $x = 0$ and $x = L$ be equal to each other. This gives

$$\frac{Q}{w} = \frac{b_0 b_L U}{b_0 + b_L}, \tag{6.2.7}$$

where b_L is the bearing spacing at $x = L$. Substituting (6.2.7) into (6.2.5) and (6.2.6) gives

$$\frac{\partial p}{\partial x} = \frac{6U\mu}{b^3}\left(b - \frac{2b_0 b_L}{b_0 + b_L}\right) \tag{6.2.8}$$

and, after integration,

$$p(x) = p_0 + \frac{6U\mu}{\tan \alpha}\left[\frac{1}{b} - \frac{1}{b_0} - \frac{b_0 b_L}{b_0 + b_L}\right]. \tag{6.2.9}$$

We can find the point of maximum pressure by setting $\partial p/\partial x$ to zero, giving the result that the pressure is a maximum at the location where $b = 2b_0 b_L/(b_0 + b_L)$.

A question remaining is, how are b_0 and b_L determined? We have treated them as if they were known quantities, yet in a real situation they must be determined by outside factors. These factors are the load that the bearing carries, and how that load is positioned. By integrating the dynamic pressure over the length of the bearing we can find P, the load carried by the bearing. This is

$$P = w\int_0^L (p - p_0) \, dx = \frac{6wU\mu}{\tan^2\alpha}\left[\ln\left(\frac{b_0}{b_L}\right) + 2\left(\frac{b_L - b_0}{b_0 + b_L}\right)\right]. \tag{6.2.10}$$

In a similar fashion we compute the drag on the lower plate. This is

$$D = -w\int_0^L \mu\left(\frac{du}{dy}\right)_{y=0} dx = \frac{w\mu U}{\tan \alpha}\left[4\ln\left(\frac{b_0}{b_L}\right) + 6\left(\frac{b_L - b_0}{b_0 + b_L}\right)\right]. \tag{6.2.11}$$

The moment M of the pressure distribution about $x = 0$ is found by integrating the moment of the pressure according to

$$M = w \int_0^L x(p - p_0)dx = \frac{6wU\mu}{\tan^3 \alpha}\left[\frac{b_0(b_0 + 2b_L)\ln b_0/b_L + (b_L^2 + 4b_0 b_L - 5b_0^2)/2}{b_0 + b_L}\right].$$

(6.2.12)

Consequently, by knowing the load and where it acts, (6.2.11) and (6.2.12) allow the spacing and the angle α to be found.

To better see the dependency of the previous terms on the geometry of the bearing, it is helpful to write the previous equations in dimensionless form. Let $r = b_0/b_L$, and equation (6.2.10) can be written in dimensionless form as

$$C_P = Pb_L^2/wU\mu L^2 = 6 \left[\ln r - 2(r - 1)/(r + 1)\right]/(r - 1)^2.$$

(6.2.13)

Here C_P is a dimensionless quantity called the **load factor**, having a maximum value of 0.160 at $r = 2.2$ as seen from the plot of the load factor in Figure 6.6a. Similarly, for the **drag factor**, from (6.2.11) we have

$$C_D = Db_L^2/wU\mu L^2 = \tan \alpha \left[\ln r + 6(1 - r)/(r + 1)\right]/(r - 1)^2.$$

(6.2.14)

From (6.2.12) the **moment factor** is found to be

$$C_M = Mb_L^2/wL^3 U\mu = 6\left[\frac{r(r + 2)\ln r - 2.5(r - 1)^2 - 3(r - 1)}{(r - 1)^3(r + 1)}\right].$$

(6.2.15)

C_M has a maximum value of 0.0932 at $r = 2.4$, as seen in Figure 6.6a. The ratio of load factor to moment factor is shown in Figure 6.6b plotted as a function of the spacing ratio r.

To emphasize the advantage gained by using a fluid as a lubricant, we consider the ratio of drag to load forces. This ratio is solely a function of the spacing ratio r and the angle α the plates make with each other. After dividing (6.2.13) by (6.2.14), we find

$$C_D/C_P = D/P = \tan \alpha \left[4 \ln r + 6(1 - r)/(1 + r)\right]/6 \left[\ln r - 2(r - 1)/(r + 1)\right]. \quad (6.2.16)$$

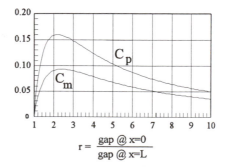

Figure 6.6a. Load and moment factors for slider bearings.

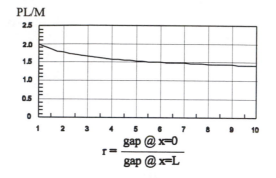

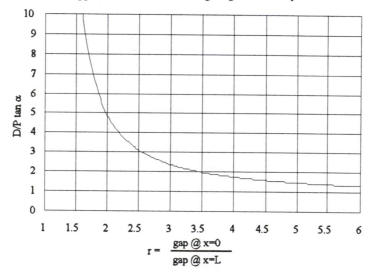

Figure 6.6b. Ratio of supported force times bearing length divided by the moment.

Figure 6.6c. Ratio of drag to support force × tan α as a function of *r*.

Since the angle α is quite small, the drag-to-load ratio will also be small, certainly much smaller than would be the case if we had solid-to-solid friction. The ratio $C_D/C_P \tan \alpha$ is plotted as a function of *r* in Figure 6.6c. Note that if the fluid were absent we would have solid-to-solid friction and the ratio D/P would equal the coefficient of friction.

Example 6.2.1. Slider bearing

Consider a slider bearing 30 cm long moving at a speed of 2 m/s. The oil lubricant has a viscosity of 2×10^{-5} N-s/m². The width of the bearing is 30 cm, $b_0 = 0.075$ mm, and $b_L = 0.04$ mm. What load can be carried, where must it be placed to maintain this spacing, and what is the drag-to-load ratio?

Sought: Load-carrying capacity of bearing, placement of load, and drag-to-load ratio.

Given: Bearing length (30 cm), bearing width (30 cm), bearing speed (2 m/s), lubricant viscosity (2×10^{-5} N-s/m^2), lubricant thickness at $x = 0$ (0.075 mm), lubricant thickness at $x = 30$ cm (0.04 mm).

Assumptions: Laminar viscous flow. The flow is assumed change gradually enough that it can be considered to be parallel flow.

Solution: Compute first the bearing angle from the given geometry as

$$\tan \alpha = (75 - 40) \times 10^{-6}/0.3 = 1.167 \times 10^{-4}.$$

Calculate next

$$6wU\mu/\tan^2 \alpha = 6 \times 0.3 \times 2 \times 2 \times 10^{-5}/(1.167 \times 10^{-4})^2 = 5{,}289.6.$$

From (6.2.10), the load carried is

$$P = 5{,}289.6 \times (\ln 1.875 - 2 \times 35/115) = 105.3 \text{ N,}$$

and from (6.2.11) the moment about the origin is

$$M = (5{,}289.6/1.167 \times 10^{-5})(75 \times 10^{-6} \times 155 \times 10^{-6} \ln 1.875$$

$$+ \ 10^{-12}(1600 + 4 \times 75 \times 40 - 5 \times 7^2)/2)/115 \times 10^{-6} = 17.77 \text{ N.}$$

Therefore the distance of the load from the front of the slider ($x = 0$) is

$$x_P = M/P = 16.9 \text{ cm.}$$

From (6.2.13), the drag-to-load ratio is

$$D/P = (2 \ln 1.875 - 3 \times 35/115) \times 1.167 \times 10^{-4}/3 \times (\ln 1.875 - 2 \times 35/115)$$

$$= 6.72 \times 10^{-4}.$$

The drag is then 0.0708 N.

The principle of the slider bearing is also used in some thrust bearings. These thrust bearings consist of wedge-shaped plates arranged in a circle. Each wedge is free to

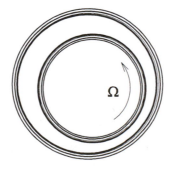

Figure 6.7. Journal bearing showing the shaft off-center in the hole.

rotate about a radial line. Each wedge thus acts as our slider bearing above, albeit with a more complicated geometry.

The journal bearing (Figure 6.7) is yet another version of the slider bearing. It consists of a rotating shaft supported by a fluid film inside a stationary circular support called the "bearing." The shaft center must always be off the center of the bearing for the bearing to support any load, so that the lubricant thickness varies over the circumference of the bearing. If the lubricant is a liquid, it is possible that lubricant will cavitate near the top of the shaft. This plus the more complicated geometry of the journal bearing makes it more difficult to analyze than the slider bearing, although the principle of operation is exactly the same.

The lubricant used in bearings depends very much on the application. We are accustomed to seeing oil used as a lubricant on most machine bearings, which provides the additional benefit of avoiding rust problems. For a ship's propeller shaft, water is used, and the bearing race may be made of wood. Modern low-drag bearings use a continuous flow of air or some other gas as a lubricant.

3.   Flow in a Circular Tube or Annulus

The analysis of Section 1 can also be applied to steady, fully developed, laminar flow in a circular tube or annulus. For this analysis cylindrical polar coordinates are appropriate. The control volume we choose is a ring (see Figures 6.8 and 6.9) of inner radius r, outer radius $r + dr$, and thickness dz. We expect that away from the entrance to the flow region, any effects of the entrance to the tube have died out, and therefore the streamlines will be straight lines parallel to the centerline of the tube. The pressure gradient will consist of a hydrostatic component plus an applied pressure gradient in the direction of the z axis of the tube or annulus. As in the case of flow between parallel plates, we expect also that for this fully developed flow (i.e., sufficiently far from the tube entrance) the velocity will be parallel to the wall. Thus there will be a velocity component only in the z direction. (In the vicinity of the entrance to the pipe, the flow will not be parallel to the pipe wall throughout the annulus, and there will be a small vertical velocity component as well.) In order that continuity be satisfied and the

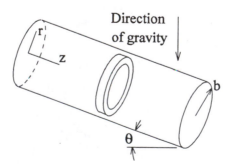

Figure 6.8. Defining figure of the control volume for flow in a circular tube.

discharge be independent of z, this velocity component, which we denote by w, must vary only with the radius. Since the shear stress is proportional to the velocity gradient, given by

$$\tau = \mu \, dw/dr, \tag{6.3.1}$$

it also must be a function only of radial position.

Figure 6.9 shows a cross section of the control volume on which we are going to sum forces. No shear forces are shown on the front and back faces. Because the shear stresses on these faces act in a radial direction, the net force that they exert on each face of our system is zero.

Summing forces in the z direction, we have

$$-\tau 2\pi r \, dz + [\tau + (d\tau/dr) \, dr]2\pi(r + dr) \, dz + p2\pi r \, dr - [p + (\partial p/\partial z) \, dz] \, 2\pi r \, dr$$

$$+ \gamma \sin \theta \, 2 \, \pi r \, dr \, dz = 0,$$

since the z component of momentum does not change. Canceling common terms, dividing by the volume, and neglecting terms of higher order, we are left with

$$d\tau/dr + \tau/r - \partial p/\partial z + \gamma \sin \theta = 0. \tag{6.3.2}$$

In appearance this equation is very close to (6.1.2), the extra term involving the shear stress being due to our annulus having an outer surface that has a larger area than does the inner surface.

We will not sum forces in other directions. Suffice it to say that, as you might expect from our study of the parallel plate case, the pressure varies hydrostatically across the cross section of the tube. The pressure gradient along the pipe axis in the direction of flow is constant, as it was in the plane case.

To integrate equation (6.3.2), it is convenient to first multiply it by r and then rearrange terms. Using the product rule of calculus it can be written in the form

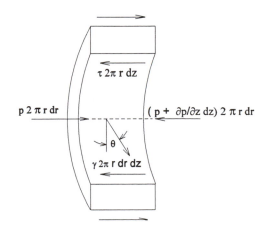

Figure 6.9. Control volume for flow in a circular tube.

$r\,d\tau/dr + \tau = d(r\tau)/dr = r(\partial p/\partial z - \gamma \sin \theta).$

Integration of this gives

$$\tau = 0.5r(\partial p/\partial z - \gamma \sin \theta) + A/r, \tag{6.3.3}$$

where A is a constant of integration. Introducing the velocity through equation (6.3.1), a second integration gives

$$w(r) = r^2(\partial p/\partial z - \gamma \sin \theta)/4\mu + (A/\mu)\ln r + B, \tag{6.3.4}$$

where B is the second constant of integration. Determination of A and B will depend on the conditions at the boundaries of our fluid.

a. Circular tube

We consider first flow inside a circular tube. Unlike the parallel flow case, moving walls are not of much interest here. When we consider flow between parallel plates where the fluid was contained between two walls, the no-slip condition was applied on each wall, resulting in two equations to be solved for A and B. For fluid filling the interior of a circular tube, we can still apply the no-slip condition at $r = b$, but in this case our flow domain has only one wall. The "missing" condition here is supplied by looking at our solution as given in (6.3.3) and (6.3.4). If A is anything other than zero, we see that both the shear stress and the velocity will be infinite along the axis of the pipe. This cannot be true, and is in fact, a peculiarity introduced by using cylindrical polar coordinates, which has a vanishing Jacobian on the axis, and also an ambiguity in the direction there. Our missing condition then becomes that of requiring both velocity and shear stress to be finite everywhere in $0 \le r \le b$, which will be true only if $A = 0$. Note that since the shear stress is proportional to the radial velocity gradient, having the shear stress zero at $r = 0$ (this follows from A being zero) means that the velocity is symmetric about the centerline of the cylinder. Applying the no-slip condition at $r = b$ then gives

$$0 = w(r = b) = b^2(\partial p/\partial z - \gamma \sin \theta)/4\mu + B,$$

resulting in the velocity distribution

$$w(r) = (-\partial p/\partial z + \gamma \sin \theta)(b^2 - r^2)/4\mu, \tag{6.3.5}$$

with

$$\tau(r) = -(-\partial p/\partial z + \gamma \sin \theta)r/2. \tag{6.3.6}$$

This velocity distribution is referred to simply as "Poiseuille flow."

The discharge associated with (6.3.5), found by integrating (6.3.5) across the pipe area, is

$$Q = 2\pi \int_0^b wr\,dr = \pi b^4(-\partial p/\partial z + \gamma \sin \theta)/8\mu. \tag{6.3.7}$$

Note that because Q depends on the fourth power of the radius, doubling the pipe radius with no change in pressure gradient increases the flow rate by a factor of 16.

An interesting result is the wall shear, which we can obtain from (6.3.6) by letting $r = b$. This gives

$$\tau|_{r=b} = -(\partial p/\partial z + \gamma \sin \theta)b/2. \tag{6.3.8}$$

We could also obtain this result (6.3.8) by considering a control volume consisting of a cylinder of fluid of radius b and length L. Summing forces in the axial direction as shown in Figure 6.10 yields

$$\tau|_{r=b} \pi bL + p\pi b^2 - [p + (\partial p/\partial z)L]\pi b^2 + \gamma \pi b^2 L \sin \theta = 0, \tag{6.3.9}$$

since the momentum out of the control volume is equal to the momentum into it. We cancel common terms and divide by the volume of the fluid, and (6.3.8) is again obtained. Notice, however, that in this derivation we have not said that the flow is laminar! In fact, (6.3.8) is true for both laminar and turbulent flows (providing the latter have mean quantities that do not vary with time), and states simply that the mean dynamic pressure gradient (the difference between the total pressure gradient and that due to gravity) balances the drag on the fluid by the walls.

A final result that can be obtained from (6.3.5), which will be of use in the next chapter, is the relationship of the head loss to the other parameters of the problem. Since the kinetic energy term does not change with distance along the pipe, head loss will be evidenced by pressure and potential energy changes. The head loss over a length of pipe is then

$$h_L = (-\partial p/\partial z + \gamma \sin \theta)l/\rho g, \tag{6.3.10}$$

and so (6.3.5) can be rewritten in the form

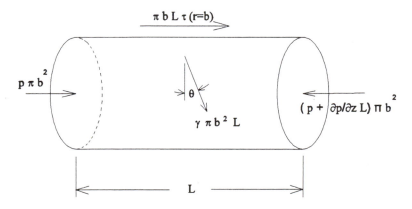

Figure 6.10. Control volume for flow in a circular pipe.

$$w(r) = \rho g h_L (b^2 - r^2)/4\mu l. \tag{6.3.11}$$

The average velocity is given by

$$w_{ave} = \int_0^b w(r)\, 2\pi r\, dr/\pi b^2 = \int_0^b \rho g h_L(b^2 - r^2) r\, dr/2\mu l b^2$$

$$= \rho g h_L(b^4/2 - b^4/4)/2\mu l b^2$$

$$= \rho g h_L D^2/32\mu l, \tag{6.3.12}$$

where $D = 2b$ is the tube diameter. Let $V = w_{ave}$ and define the Reynolds number by

$\mathrm{Re} = \rho V D/\mu.$

Then solving (6.3.12) for the head loss, we have

$$h_L = 32\mu l V/\rho g D^2 = 32 l V^2/\mathrm{Re}\, g D, \tag{6.3.13}$$

or in the form discussed in Chapter 4,

$$h_L = \frac{64}{\mathrm{Re}} \frac{l}{D} \frac{V^2}{2g}.$$

Therefore the friction factor f as defined in Chapter 5 is given for laminar pipe flow by

$$f = 64/\mathrm{Re}. \tag{6.3.14}$$

Example 6.3.1. Laminar flow in a tube

A fluid with viscosity 2.5×10^{-5} lb-s/ft^2 flows laminarly in a horizontal circular tube of radius 0.05 ft. The pressure drops 1 psi in 10 ft. Find the discharge and maximum velocity.

Sought: Volumetric discharge and maximum velocity for the given pressure gradient and geometry.

Given: Viscosity $(2.5 \times 10^{-5}$ lb-s/ft$^2)$, tube radius (0.05 ft), pressure drop (1 psi in 10 ft).

Assumptions: Laminar viscous flow. Entrance effects are neglected so that the parabolic velocity profile is assumed to be established throughout the fluid.

Solution: The pressure gradient is found by dividing the pressure drop over the distance in which it takes place. Then

$$\partial p/\partial z = -1 \times 144/10 = -14.4 \text{ lb/ft}^3.$$

The discharge from (6.3.7) is

$$Q = \pi \times 0.05^4 \times 14.4/8 \times 2.5 \times 10^{-5} = 1.414 \text{ ft}^3/\text{s}.$$

The maximum velocity in the pipe is, from (6.3.5),

$$w_{max} = w(r = 0) = 14.4 \times 0.05^2/4 \times 2.5 \times 10^{-5} = 360 \text{ ft/s}.$$

The average velocity is

$$w_{ave} = Q/A = 1.414/0.05^2\pi = 180 \text{ ft/s}.$$

---

b. Circular annulus

If the fluid is contained in the region $0 < a \le r \le b$, we do have our fluid between two solid boundaries, and we can now apply the no-slip condition at each wall. This gives

$$0 = w(r = a) = a^2(\partial p/\partial z - \gamma \sin \theta)/4\mu + A \ln a/\mu + B$$

and

$$0 = w(r = b) = b^2(\partial p/\partial z - \gamma \sin \theta)/4\mu + A \ln b/\mu + B.$$

Solving for A and B, we find the result as

$$w(r) = (-\partial p/\partial z + \gamma \sin \theta)[b^2 - r^2 - (b^2 - a^2)\ln r/b/\ln a/b]/4\mu \qquad (6.3.15)$$

and

$$\tau(r) = (-\partial p/\partial z + \gamma \sin \theta)[-2r - (b^2 - a^2)/r \ln a/b]/4. \qquad (6.3.16)$$

The discharge, found by integrating the velocity component given by (6.3.15) over the area, is thus

$$Q = 2\pi \int_a^b w(r)r \, dr = \pi(-\partial p/\partial z + \gamma \sin \theta)(b^2 - a^2)[b^2 + a^2 + (b^2 - a^2)/\ln a/b]/8\mu.$$

$$(6.3.17)$$

The logarithmic term in the velocity makes this velocity profile slightly more complicated-appearing than the simple parabolic and linear distributions we have previously seen. As the ratio $(b - a)/a$ gets small, however, the velocity distribution approaches the parabolic part of (6.1.8). This can be shown by making the change of variables $r = a + y$ and expanding the logarithm in a Taylor's series.

We introduce the variable $c = b - a$, which represents the spacing between the walls, and Taylor series expansions of the logarithmic terms give

$$\ln a/b = -\ln (b + a - a)/a = -\ln (1 + c/a) = -c/a + 0.5(c/a)^2 - (1/3)(c/a)^3 \pm \ldots,$$

$$\ln r/b = \ln ar/ba = \ln (1 + y/a) = \ln a/b + y/a - 0.5(y/a)^2 + (1/3)(y/a)^3 \pm \ldots.$$

Putting these into the velocity expression (6.3.10), we have then,

$$w(y) = (-\partial p/\partial z + \gamma \sin \theta)[(a + c)^2 - (a + y)^2 - (2ac + c^2)\ln r/b/\ln a/b]/4\mu$$

$$= \left(\frac{-\partial p/\partial z + \gamma \sin \theta}{4\mu}\right) \{2ac + c^2 - 2ay - (2ac + c^2)$$

$$\times\left[1+\frac{y/a-0.5(y/a)^2\pm\ldots}{-c/a+0.5(c/a)^2\pm\ldots}\right]\Big\}$$

$$=\left(\frac{-\partial p/\partial z+\gamma\sin\theta}{4\mu}\right)\left\{-2ay-y^2+(2a+c)\,y\left[\frac{1-0.5y/a+(1/3)(y/a)^2\pm\ldots}{1-0.5c/a+(1/3c/a)^2\pm\ldots}\right]\right\}$$

$$=\left(\frac{-\partial p/\partial z+\gamma\sin\theta}{4\mu}\right)[-y^2+y(c-y)+yc+\text{terms involving }a\text{ in the denominator}].$$

Therefore as a becomes large with c fixed, our velocity w approaches

$$w(r)\to\left(\frac{-\partial p/\partial z+\gamma\sin\theta}{2\mu}\right)(yc-y^2).\tag{6.3.18}$$

In words, in the limit as the spacing c becomes smaller and smaller compared to the inner radius a, (6.3.18) becomes (6.1.8). In applications where the spacing is small compared to the radius, then, plane Poiseuille flow is a very good approximation to circular Poiseuille flow.

Example 6.3.2. Annulus flow

A fluid with viscosity 2.5×10^{-5} lb-s/ft^2 flows laminarly in an annulus with inner radius 0.02 ft and outer radius 0.05 ft. The pressure drop is 1 psi in 20 ft. What is the discharge and maximum speed?

Sought: Volumetric discharge and maximum speed for the given pressure gradient and geometry.

Given: Viscosity (2.5×10^{-5} lb-s/ft^2), inner and outer radii (0.02 ft and 0.05 ft), pressure drop over 20 ft (1 psi).

Assumptions: Laminar viscous flow. Entrance effects are neglected so that the parabolic/logarithmic velocity profile is assumed to be established throughout the fluid.

Solution: The pressure gradient is the pressure drop over the length in which it occurs, or

$$\partial p/\partial z=-144/20=-7.2\text{ lb/ft}^3.$$

The maximum speed will occur where dw/dr (or τ) vanishes, which according to (6.3.11) is at

$$r=\sqrt{0.5(a^2-b^2)/\ln a/b}$$

$$=\sqrt{0.5(0.05^2-0.02^2)/\ln 2.5}=0.0339\text{ ft}.$$

The value for the maximum speed is then

$w(r = 0.0339) = 7.2[0.05^2 - 0.0339^2 - (0.05^2 - 0.02^2)\ln 0.678/\ln 0.4]/4 \times 2.5 \times 10^{-5}$
$\qquad\qquad = 33.13$ ft/s.

The discharge is, from (6.3.12),

$Q = \pi \times 7.2(0.05^2 - 0.02^2)[0.05^2 + 0.02^2 + (0.05^2 - 0.02^2) \ln 0.678/\ln 0.4]/8$
$\qquad \times 2.5 \times 10^{-5}$

$\qquad = 0.144$ ft^3/s.

The average velocity is then

$w_{\text{ave}} = Q/A = 0.144/\pi(0.05^2 - 0.02^2) = 21.8$ ft/s.

We have assumed in our analyses that the flow is fully developed, i.e., that entrance effects have taken place outside the region of our interest. We might ask how far from the entrance we must be in order for this to be true. While there is some effect of entrance shape, the entrance length L_{entrance} for the flow to be fully developed is primarily a function of the Reynolds number. The accepted correlation is

$$L_{\text{entrance}}/D = 0.06 \text{ Re}. \qquad\qquad (6.3.19)$$

The previous results hold only for circular tubes. In principle, similar analyses can be carried out for tubes of other shapes, but the mathematics becomes much more difficult, and in fact can be carried out analytically for only a few simple shapes like rectangles, ellipses, and triangles. Instead, it is usual to approximate the discharge results by introducing the concept of **hydraulic diameter** by

$$D_h = 4A/P, \qquad\qquad (6.3.20)$$

where, considering a cross section of the tube, A is the area of the fluid and P is the **wetted perimeter**. That is, P is the length of the line of contact between the fluid and the tube wall. P is different from the tube perimeter if the tube is not completely filled with fluid. Equation (6.3.7) could then be used to determine the relationship between discharge and pressure in a circular tube by replacing the radius b with $D_h/2$. The result is not exact, but in many cases it serves as a reasonable approximation.

Example 6.3.3. Hydraulic diameter of a circular cross section
Find the hydraulic diameter for a circular tube (radius b) running full and also half-full.

Sought: Hydraulic diameter of a circular tube running full and also half-full.
Given: Tube radius (b), and the fact that the tube is running either full or half-full.
Assumptions: None.
Solution: For the case where the tube is running full, the area is πb^2 and the wetted perimeter is $2\pi b$. Therefore

$D_h = 4(\pi b^2)/(2\pi b) = 2b.$

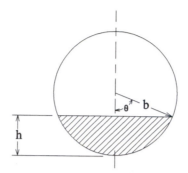

Figure 6.11. Partially filled circular tube.

If the same tube is half-full, the area is $0.5\,\pi b^2$ and the wetted perimeter is πb. Thus

$$D_h = 4(0.5\pi b^2)/(\pi b) = 2b.$$

On this basis, hydraulic diameter is the same for the tube running both full and half-full.

Example 6.3.4. Hydraulic diameter of a circular cross section
Find the hydraulic diameter for a circular tube (radius b) running with an arbitrary depth of fill h.

Sought: General formula for hydraulic diameter of a circular tube.
Given: Tube radius (b), and the fact that the tube is running with a depth of fill h.
Assumptions: None.
Solution: For the case where the depth of fill h is less than or equal to the radius b, the area is

$$A = b^2(\theta - \sin\theta\cos\theta)$$

and the wetted perimeter is

$$P = 2b\theta,$$

where

$$\theta = \tan^{-1}[\sqrt{b^2 - (b-h)^2}/(b-h)].$$

(See Figure 6.11 for the definition of the angle θ.) Then

$$b_h = 2A/P.$$

If the depth h is greater than b, the equivalent formulas are

$$A = b^2(\theta - \sin\theta\cos\theta),$$

$$P = 2b\theta,$$

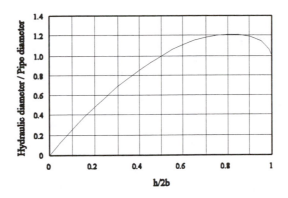

Figure 6.12. Hydraulic diameter versus depth of fill for a partially filled circular tube.

where $\theta = \tan^{-1}[\sqrt{b^2 - (h-b)^2}/(h-b)]$. A plot of hydraulic diameter as a function of h/b is given in Figure 6.12.

Example 6.3.5. Hydraulic diameter of a rectangular cross section
Find the hydraulic diameter for a rectangular tube with dimensions a by b running full.

Sought: Hydraulic diameter for a rectangular tube with dimensions a by b running full.
Given: Tube dimensions (a by b), and the fact that the tube is running full.
Assumptions: None.
Solution: A rectangular tube a by b running full has an area of ab and a wetted perimeter of $2(a + b)$. Its hydraulic diameter is then

$$D_h = 4(ab)/(2a + 2b) = 2ab/(a + b).$$

---

4. Stability of Tube Flow

In the previous section, we assumed that the flow was laminar, and that streamlines were all straight lines. In any real situation, however, there will be disturbances introduced into the flow by the surroundings. These disturbances could be caused by vehicles on neighboring streets, pedestrian traffic within a building, passing airplanes, loud nearby radios, and the like. When a small disturbance is transmitted to the flow, we say that the flow is *stable* if the flow is only momentarily disturbed and then settles back to its original laminar flow. If, on the other hand, the disturbance causes the flow to change character, either by becoming a more complicated laminar flow (a secondary flow) or by becoming a turbulent flow, the original flow is said to be *unstable*. The dividing case between these, where a disturbance neither grows nor decays, is said to be a *neutrally stable* flow.

The analytical determination of the stability of a flow is quite complicated, since some disturbance frequencies will grow and others will decay. The main region where energy transfer takes place is an internal boundary layer where the disturbance velocity matches the flow velocity, where the transfer of momentum is enhanced. This is a speciality topic in advanced fluid mechanics. The present status of this field is far from complete, since it generally requires a prior knowledge of what the post-stable state will look like. In many cases the multitude of parameters and the complexity of the mathematics renders a complete parameter study of a given problem impractical.

For the case of circular tube flow, Reynolds determined the stability of the flow by experimental means. (There is still no completely convincing analytic solution of this problem.) He attached a horizontal circular transparent tube to a large reservoir containing the fluid. The entrance to the tube was designed to provide a smooth introduction of the fluid. A valve was placed at the downstream end of the tube. A thin, hollow needle lying along the tube centerline allowed the introduction of dye. The flow was started a sufficiently long period of time after filling the reservoir, so that any disturbances introduced by filling had time to die out.

Reynolds found that for sufficiently slow flows (specifically, for a Reynolds number based on average velocity and diameter being less than 2,300) the flow was always stable and remained laminar. That is, the dye line would continue down the center of the pipe without breaking up. At somewhat faster flows, the flow appears to be in an unstable situation, where the dye stream changes abruptly and intermittently between being a coherent straight line and a rapidly dispersed cloud of dye. At faster flow rates, a coherent straight dye line never formed.

Reynolds' experiments have been repeated many times, both with his original equipment and with other versions of it. By allowing very long times for the filling disturbances to die out, and by isolating the tube from outside disturbances, laminar flows have been found for Reynolds numbers in the hundreds of thousands. For practical purposes, however, we will consider flow in a circular tube to be laminar if

$$\text{Re} = U_{ave}D/\nu \le \text{Re}_{cr} = 2,300. \tag{6.4.1}$$

Tube flow is an example of a flow that immediately becomes turbulent when it makes the transition from a stable to an unstable flow. Other flows can take on many intermediate laminar states before they become turbulent. An example is the stability of a thin horizontal film of fluid heated from below. When the temperature difference becomes sufficiently large, convection cells form that can look like hexagons when viewed from the top, but that can also take shapes such as long "roller" cells. These cells have been observed in the drying of metallic paints, and occasionally can be seen in a striking fashion in cloud layers in the atmosphere. This instability is called Rayleigh-Bénard instability, after the two people who first observed and reported it (Bénard 1900, 1901) and analyzed it (Rayleigh 1916). It is an example of a gravitational instability, in that the hotter, lighter fluid is buoyed up, only to be cooled, becoming heavier and subsequently falling.

An analogous phenomenon can be found in rotating flows. Taylor (1923) demonstrated that, for fluid contained between a stationary outer cylinder and a rotating inner one, at a sufficiently large rotational speed the flow breaks up into a stacked set of ring-shaped cells in which fluid particles move along spiral lines. Many other such secondary flows exist, their existence being made possible by the nonlinear nature of the convective acceleration terms in the governing equations. The consequences of such secondary flows can be important in engineering problems, since they can affect heat transfer characteristics, discharge-pressure relationships, and the like. It is important for the engineer to be aware that such complexities are possible and can cause unexpected difficulties.

It is useful to note that the laminar flow solutions we have considered do not depend on the Reynolds number; yet the transition to turbulence (or a secondary flow) does depend on either the Reynolds number, or on some dimensionless parameter acting in a role similar to the Reynolds number, such as what is referred to as the Taylor number in the case of rotating Couette flow. The explanation is that in the simple laminar flows we have considered there is no change in momentum, and hence no inertia forces. Thus an important ingredient of the Reynolds number is missing. When a small velocity is introduced into the flow by the surroundings, that velocity perturbation will have acceleration and, if it grows to be more than a small perturbation, will introduce appreciable inertia forces and transfer disturbance momentum between fluid layers. In this case, then, there is an inertia force with a Reynolds number to describe the flow. For lack of a better descriptor of the velocities connected with the perturbation, we still use the mean velocity as a representative of the inertia force in the Reynolds number.

5. Boundary Layer Theory

In the previous sections of this chapter we have considered problems where the momentum forces are absent, and viscous and pressure forces dominate the flow throughout the entire region of flow. In many problems involving flows past bodies, viscous effects are concentrated only in a thin region near the wall, where their magnitude is comparable to those of the momentum and pressure forces. Away from the wall, pressure and momentum forces dominate the flow, with viscous forces there being negligible.

For example, if we stand on the bank of a fast-flowing river, to us it may appear that the speed is roughly uniform across the river, and that the no-slip condition is violated at the river bank. So, too, if we consider the wind flowing past a wall. Yet these impressions are incorrect. If we were to use instrumentation capable of measuring velocity in a very thin region, we would find that the velocity rapidly goes to zero in a very thin region near the boundary so as to satisfy the no-slip condition. Such a thin region was first proposed by Prandtl in 1904. It was a revolutionary idea, not at first generally accepted, yet development of the concept made possible theoretical developments that have led to modern aviation and space flight. Other examples of boundary layer effects are the jet stream in the atmosphere and the Gulf Stream and Japanese Currents in the Atlantic and Pacific Oceans, although these cases not always tied to a physical boundary. The general concept of thin regions where rapid changes take place has found application in many fields other than fluid mechanics. In fluid mechanics it has provided the connection between the theoretical work of the nineteenth and earlier centuries, which largely neglected viscous effects, and problems of engineering concern such as drag.

Prandtl started with the concept that, near a wall, only the momentum equation parallel to the wall was going to be of importance. In this region, the velocity parallel to the wall will be much greater than the velocity normal to it. To formalize these ideas, consider a control volume consisting of a small region of height h and length dx near a wall, as shown in Figure 6.13. In this figure, $\delta(x)$ is the thickness of the boundary layer and h is a slightly larger (constant) number, used to keep our control volume rectangular. For now, we will adopt an imprecise definition of the boundary layer thickness, defining it as the thickness of the region at the wall in which viscous stresses are important and where the fluid velocity is less than the free stream velocity by a percentage point or so.

If U is the x velocity component at $y = h$, conservation of mass gives

$$\int_0^h u\, dy - \int_0^h [u + (\partial u/\partial x)\, dx]dy - v_h\, dx = 0$$

or, after canceling the integral of u and dividing by dx,

$$v_h = -\int_0^h \partial u/\partial x\, dy. \tag{6.5.1}$$

Since the net force on our control volume in the x direction is balanced by the net rate of transport of momentum in the x direction, we have

$$\int_0^h p\, dy - \int_0^h [p + (\partial p/\partial x)\, dx]\, dy - \tau_{\text{wall}}\, dx = \int_0^h \rho[u + (\partial u/\partial x)\, dx]^2\, dy - \int_0^h \rho u^2\, dy + \rho U v_h,$$

the last term being due to the momentum flow out of our control volume. Canceling terms, dividing by dx, using (6.5.1), and neglecting higher order terms, we are left with

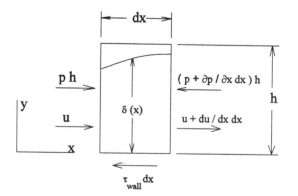

Figure 6.13. Control volume for boundary layer analysis.

$$-\int_0^h \partial p/\partial x \, dy - \tau_{\text{wall}} = \int_0^h \rho 2u(\partial u/\partial x) \, dy - \rho U v_h = \int_0^h \rho(\partial u/\partial x)(2u - U) \, dy, \quad (6.5.2)$$

the last line of (6.5.2) coming from the use of continuity in the form (6.5.1). We see, however, from differentiation and the chain rule of calculus that

$$d/dx \int_0^h \rho u(u - U) \, dy = \int_0^h \rho[\partial u/\partial x(2u - U) - u \, \partial u/\partial x] \, dx = \int_0^h \rho(\partial u/\partial x)(2u - U) \, dy$$

$$-\int_0^h \rho u \, dU/dx \, dy.$$

Then (6.5.2) can be rewritten as

$$-\int_0^h \partial p/\partial x \, dy - \tau_{\text{wall}} = d/dx \int_0^h \rho u(u - U) \, dy + \int_0^h \rho u \, dU/dx \, dy. \quad (6.5.3)$$

The upper end of our control volume ($y = h$) is just outside the thin boundary layer, and we started by assuming that viscous effects are unimportant here. Also, we said that the y component of velocity is small here. Consequently the Euler equations of Chapters 3 and 4 tell us that, at the outer edge of the boundary layer, our pressure gradient is balanced only by the acceleration, and that

$$-\frac{\partial p}{\partial x} = \rho U \frac{dU}{dx}. \quad (6.5.4)$$

Putting (6.5.4) into (6.5.3) and rearranging slightly, we have

$$\frac{d}{dx} \int_0^h \rho u(U - u) dy + \rho \frac{dU}{dx} \int_0^h (U - u) \, dy = \tau_{\text{wall}}. \quad (6.5.5)$$

Equation (6.5.5) is called the **momentum-integral form of the boundary layer equation**, and was first introduced by von Kármán (1921).

The differential form of the boundary layer equations can also be derived from the Navier-Stokes equations, derived in Chapter 4. For a case with a characteristic length L representing a finite body and a characteristic velocity U representing either the outer flow or the body speed, introduction of the dimensionless variables

$$x' = x/L, \quad y' = y \, \text{Re}/L, \quad u = v_x/U, \quad v = v_y \, \text{Re}/U, \quad p' = p/\rho U^2, \quad \text{Re} = UL/\nu$$

into the two-dimensional Navier-Stokes equations and then considering only the highest order terms gives

$$u \frac{\partial u}{\partial x'} + v \frac{\partial u}{\partial y'} = -\frac{\partial p'}{\partial x'} + \frac{\partial^2 u}{\partial y'^2} \qquad (x \text{ momentum}), \quad (6.5.6)$$

$$-\frac{\partial p'}{\partial y'} = 0 \qquad (y \text{ momentum}), \tag{6.5.7}$$

$$\frac{\partial u}{\partial x'} + \frac{\partial v}{\partial y'} = 0 \qquad (\text{continuity}). \tag{6.5.8}$$

Integration of these three equations over the boundary layer thickness would result in (6.5.5). Equations (6.5.6), (6.5.6), and (6.5.7) are, however, more general than (6.5.5).

We have not as yet given a precise definition as to what we mean by boundary layer thickness. The general concept we have used is that the outer edge of the boundary layer is where the velocity has the value it would have if the fluid were allowed to slip at the boundary. This, however, is a difficult definition to apply in practice, as it is very sensitive to how close to the non-slip value we consider is "close enough." For experimental data there is little likelihood that any two people would agree on the proper value of δ when analyzing the same data.

Instead, we use other definitions that are less subjective. Good choices are the two integrals appearing in (6.5.5). We define first the displacement thickness

$$\delta_D = \int_0^h (U - u)\, dy/U. \tag{6.5.9}$$

In Figure 6.14, we see that if we divide the area of the shaded region by U, the result is the integral in our definition of displacement thickness. By changing h slightly, or by having uncertainties or errors in our measurement of u and U, we would find relatively little change in the value of the integral. Thus the integral definition is relatively insensitive to errors in data or interpretation.

A physical interpretation of this definition is as follows. The actual discharge per unit width into the paper through an area of height h of a no-slip fluid is

$$\int_0^h u\, dy. \tag{6.5.10}$$

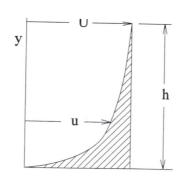

Figure 6.14. Horizontal velocity profile in the boundary layer.

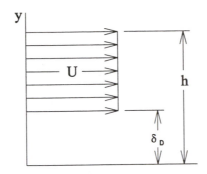

Figure 6.15. The effect of displacement thickness.

If, instead, the fluid were allowed to slip on the wall and the wall were displaced upward a distance δ_D (Figure 6.15), the discharge would be

$$U(h - \delta_D) = \int_0^h U\,dy - U\delta_D. \tag{6.5.11}$$

Equating the discharges given in (6.5.10) and (6.5.11) results in equation (6.5.9). It follows that the effect of viscosity and the no-slip condition on the flow rate is that the flow outside of the boundary layer is the same as if slip had occurred at the wall, but the body dimension was increased by an amount δ_D.

The second term in (6.5.5) allows us to define a momentum thickness

$$\delta_M = \int_0^h u(U - u)\,dy/U^2. \tag{6.5.12}$$

This has virtues regarding subjectivity that are similar to those of displacement thickness, and can be interpreted in a similar fashion. That is, we consider the momentum of a slipping flow with the same mass flow rate, but displaced a distance δ_M. We equate this to the momentum of our nonslipping flow and then solve for δ_M, obtaining (6.5.12) Obviously other thicknesses could be defined in a similar manner using energy and integrals involving still higher powers of u/U. In introducing them, we would want to be sure that the integral vanishes at and near the upper limit of integration, so that it is not critical what value we use for h.

Note: The usual notation for displacement and momentum thickness is δ^* for displacement thickness and θ for momentum thickness. The notation used here employs the symbol δ consistently as a reminder that it is a measure related to boundary layer thickness. The use of subscripts is also more explicit than the standard notation.

Equation (6.5.5) can now be conveniently rewritten in terms of the displacement and momentum thicknesses in the more convenient form

$$\frac{d}{dx}(\rho\delta_M U^2) + \rho\delta_D U\,\frac{dU}{dx} = \tau_{\text{wall}} \tag{6.5.13}$$

We will next see how (6.5.13) can be used for finding approximate values for wall shear stress. Before doing so, we note that the boundary layer equations in differential equation form have been solved for many cases and compared with the results found from the momentum integral formulation. The comparison generally is good up to the point where the shear stress becomes negative. Past this point (called the *flow separation point*, or simply, the *separation point*) the flow direction near the wall is reversed, the region near the wall where viscous effects are important is no longer thin, and boundary layer theory no longer applies.

One classic solution of the differential equation form of the laminar boundary layer equations is for the flat plate, where U = constant and $\partial p/\partial x = 0$. This flow was studied by Blasius (1908), a student of Prandtl. His results are shown in dimensionless form in Figure 6.16. For the wall shear, he found

$$\tau_{wall} = 0.332\rho U\sqrt{U\nu/x}. \tag{6.5.14}$$

The drag on a flat plate of length L and width w is then

$$w\int_0^L \tau_{wall}\ dx = 0.664\ \rho Uw\sqrt{U\nu L}, \tag{6.5.15}$$

or, in terms of a drag coefficient,

$$C_D = F/0.5\rho U^2 wL = 1.328/\sqrt{Re_L}. \tag{6.5.13}$$

These results provide a good reference point for comparison with approximate results we find from the integral form of the boundary layer equation.

Returning now to our integral momentum form (6.5.13) of the boundary layer equations, we first solve them for flow along a flat plate. We do this by assuming a "reasonable" form for the velocity distribution, and solve (6.5.13) for δ. By reasonable, we mean that the assumed velocity satisfies several necessary conditions at $y = 0$ and $y = \delta$. Because our integrals average our errors, our hope is that the results will not be sensitive to our guess.

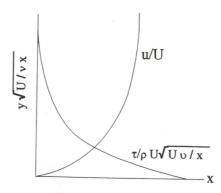

Figure 6.16. Variation of u and τ_{xy} in the boundary layer on a flat plate.

There are various velocity distributions we could try. For instance, we could let

$$u/U = \eta \quad \text{for } 0 \le \eta \le 1,$$

$$= 1 \quad \text{for } \eta > 1, \tag{6.5.17}$$

where $\eta = y/\delta$ is the dimensionless distance across the boundary layer. This satisfies the no-slip condition and $u = U$ at $y = \delta$. However, the shear stress computed from this is discontinuous at $y = \delta$, and we would hope to have a velocity distribution where the shear stress gradually dies out as we go away from the boundary. [Actually the numerical results obtained from using (6.5.17) for shear stress are not all that bad, being within about 13% of Blasius's results, demonstrating the power—and a degree of fortuity—of our method.]

An alternative assumed form giving somewhat better results is

$$u/U = 2\eta - \eta^2 \quad \text{for } 0 \le \eta \le 1,$$

$$= 1 \quad \text{for } \eta > 1. \tag{6.5.18}$$

This satisfies the same conditions as (6.5.16), and the shear stress now goes smoothly to zero at the outer edge of the boundary layer as well. Still another alternative form for the velocity profile having similar virtues is

$$u/U = \sin(\pi\eta/2) \quad \text{for } 0 \le \eta \le 1,$$

$$= 1 \quad \text{for } \eta > 1. \tag{6.5.19}$$

The results for (6.5.18) and (6.5.19) differ little, and either one can be used with about the same degree of success. All three of (6.5.17), (6.5.18), and (6.5.19) are best suited when pressure gradient effects are minor.

To demonstrate the solution procedure, we will use (6.5.18). Computing first δ_D and δ_M, and letting $D = h/\delta$, we get

$$\delta_D = \delta \int_0^1 (1 - 2\eta + \eta^2) d\eta + \delta \int_1^D 0 \, d\eta = \delta/3 \tag{6.5.20}$$

and

$$\delta_M = \delta \int_0^1 (2\eta - \eta^2)(1 - 2\eta + \eta^2) \, d\eta + \delta \int_1^D 0 \, d\eta = 2\delta/15. \tag{6.5.21}$$

Since from (6.5.17) the wall shear is

$$\tau_{\text{wall}} = \mu(\partial u/\partial y)_0 = 2\mu U/\delta, \tag{6.5.22}$$

putting (6.5.19) and (6.5.20) into (6.5.21), and using the facts that the x pressure gradient is zero and U is constant, we have

$$\rho \frac{d}{dx}\left(\frac{2\delta U^2}{15}\right) = \frac{2\mu U}{\delta}.$$

Upon simplifying and separating variables, this becomes

$$\delta\, d\delta = 15\nu\, dx/U. \tag{6.5.23}$$

Integration of (6.5.23) together with the condition that $\delta = 0$ at $x = 0$, the leading edge of the plate, gives

$$\delta = \sqrt{30\nu x/U} = 5.4777\sqrt{\nu x/U}. \tag{6.5.24}$$

For the wall shear we have, from (6.5.22),

$$\tau_{\text{wall}} = (2/\sqrt{30})\rho U\sqrt{U\nu/x} = 0.365\,\rho U\sqrt{U\nu/x}, \tag{6.5.25}$$

which differs numerically by a factor of 10% from Blasius's solution, but is correct in terms of the U, ν, and x dependency.

The general results of the previous calculation reveal some interesting and useful facts about the boundary layer for this flow. We see from (6.5.24) that the boundary layer thickness grows as $\sqrt{x}$ and that the shear stress is proportional to $1/\sqrt{x}$. The infinite value for the shear stress at the leading edge of the plate is unreasonable—in fact boundary layer theory is not a valid approximation near the leading edge. As the boundary layer gets thicker, the wall shear stress drops, since the velocity has a greater distance to change from zero to U. As a result, the drag force increases only as $\sqrt{L}$. (Even though the wall shear stress was unrealistic near the leading edge, its integral does give a realistic—and finite—force. The error contribution to the total force is small.) Thus, doubling the length of a plate increases the drag force by only 41%.

In summary, then, our crude approximation to the boundary layer velocity profile gives an adequate result when compared with Blasius's exact solution, and in fact, predicts exactly the same interdependency between the important parameters of the problem.

Pohlhausen (1921) attempted to give a more accurate approximation of u in the boundary layer that would also include flows with pressure gradients. He chose a fourth-order polynomial for u, and required that

$$u(x, 0) = 0, \qquad\qquad u(x, \delta) = U,$$

$$\mu\frac{\partial^2 u}{\partial y^2}(x, 0) = \frac{\partial p}{\partial x}, \qquad \frac{\partial u}{\partial y}(x, \delta) = 0,$$

$$\frac{\partial^2 u}{\partial y^2}(x, \delta) = 0.$$

Besides the obvious requirement on u at the wall and the outer edge of the boundary layer, these conditions specify that the balance of forces be fully satisfied right at the wall (where the acceleration vanishes), and that both the shear stress and its gradient vanish at the outer edge of the boundary layer. (Pohlhausen tried other boundary conditions and orders of polynomials as well. When compared with the exact solutions known at the time, the above seemed to give the best results.)

The fourth-order polynomial that satisfies the above Pohlhausen conditions is given by

$$u/U = 2\eta - 2\eta^3 + \eta^4 + 2\Lambda(\eta - 3\eta^2 + 3\eta^3 - \eta^4) \qquad \text{for } 0 \le \eta \le 1,$$

$$= 1 \qquad\qquad\qquad \text{for } \eta > 1, \qquad\qquad (6.5.26)$$

with

$$\eta = y/\delta, \qquad \Lambda = \frac{\delta^2}{12\nu}\frac{dU}{dx}.$$

Then

$$\delta_D = \delta(3 - \Lambda)/10, \qquad\qquad\qquad\qquad\qquad (6.5.27)$$

$$\delta_M = \delta(37 - 4\Lambda - 5\Lambda^2)/315, \qquad\qquad\qquad (6.5.28)$$

and

$$\tau_{wall} = 2\mu U(1 + \Lambda)/\delta. \qquad\qquad\qquad\qquad (6.5.29)$$

The approximation (6.5.26) for u breaks down if $\Lambda > 1$, since (6.5.26) predicts u/U is greater than 1 for values of η near 1. It also breaks down if Λ is less than -1, when it predicts u/U is negative for values of η near zero. Therefore it should be used only where $-1 \le \Lambda \le +1$.

A comparison of the calculations for a flat plate using the various approximate profiles is given in Table 6.1. The error column compares the dimensionless wall shear with the exact analytical result obtained by Blasius. His results are given in Table 6.2.

Table 6.1. Results using various approximate flat-plate velocity profiles in the momentum-integral equation as compared to the Blasius solution

Velocity profile	δ_D/δ	δ_M/δ	$\delta/\sqrt{\nu x/U}$	$\tau_{wall}/\rho U\sqrt{U\nu/x}$	Error in τ_{wall}
Linear (6.5.17)	$1/2$ $= 0.5$	$1/6$ $= 0.167$	$\sqrt{12}$ $= 3.46$	$1/\sqrt{12}$ $= 0.288$	13%
Parabolic (6.5.18)	$1/3$ $= 0.333$	$2/15$ $= 0.133$	$\sqrt{30}$ $= 5.48,$	$\sqrt{2/15}$ $= 0.365$	9.9%
Sinusoidal (6.5.19)	$1 - 2/\pi$ $= 0.363$	$2/\pi - 1/2$ $= 0.137$	$\pi\sqrt{2}/(4 - \pi)$ $= 4.80$	$\sqrt{(2 - \pi/2)/2}$ $= 0.328$	1.5%
Pohlhausen (6.5.26)	0.3	$37/315$ $= 0.117$	$\sqrt{1260/37}$ $= 5.836$	$\sqrt{37/315}$ $= 0.343$	3.3%
Exact (Blasius)	0.344	0.1328	5.0	0.332	0%

Table 6.2. Blasius's results for flow in the boundary layer of a flat plate with no pressure gradient

$y(U/2x\nu)^{1/2}$	u/U	$\tau(2x/\rho\mu U^3)^{1/2}$	$y(U/2x\nu)^{1/2}$	u/U	$\tau(2x/\rho\mu U^3)^{1/2}$
0.0	0.00000	0.46960	1.8	0.76106	0.30045
0.1	0.04696	0.46956	2.0	0.81669	0.25567
0.2	0.09391	0.46931	2.2	0.86330	0.21058
0.3	0.14081	0.46861	2.4	0.90107	0.16756
0.4	0.18761	0.46725	2.6	0.93060	0.12861
0.5	0.23423	0.46503	2.8	0.95288	0.09511
0.6	0.28058	0.46173	3.0	0.96905	0.06771
0.7	0.32653	0.45718	3.4	0.98797	0.03054
0.8	0.37196	0.45119	3.8	0.99594	0.01176
0.9	0.41672	0.44363	4.2	0.99882	0.00386
1.0	0.46063	0.43438	4.6	0.99970	0.00108
1.2	0.54525	0.41057	5.0	0.99994	0.00026
1.4	0.62439	0.37969	5.4	0.99999	0.00005
1.6	0.69670	0.34249	6.0	1.00000	0.00000

Boundary layers for flat plates are laminar if the Reynolds number based on local distance from the leading edge is less than 300,000–600,000, the exact value depending on the degree of turbulence in the stream outside of the boundary layer.

Example 6.5.1. Flat-plate boundary layer

A flat plate 2 m long and 0.5 m wide is in a uniform stream of velocity 3 m/s in a fluid with kinematic viscosity 2×10^{-5} m²/s and density 900 kg/m³. Using (6.5.17) as the assumed velocity distribution, find the thickness of the boundary layer as a function of distance from the leading edge, as well as the total drag force on the plate.

Sought: Formulas for boundary layer thickness and drag on a flat plate.

Given: The plate dimensions (2 m long and 0.5 m wide), constant uniform stream velocity (3 m/s), fluid kinematic viscosity (2×10^{-5} m²/s), fluid density (900 kg/m³). For the incoming flow parallel to the flat plate the pressure gradient is zero.

Assumptions: Laminar viscous flow. The linear velocity distribution given by (6.5.17) holds in the boundary layer.

Solution: According to Table 6.1 the displacement thickness is $\delta/2$ and the momentum thickness is $\delta/6$. From (6.5.13) and (6.5.17) we have

$$\rho U^2 \, d(\delta/6)/dx = \tau_{\text{wall}} = \mu U/\delta,$$

which upon rearranging and dividing by $\rho U^2/6$ becomes

$$\delta \frac{d\delta}{dx} = \frac{\partial \delta^2}{2 \, dx} = \frac{6\nu}{U}.$$

Integrating and saying that $\delta = 0$ at $x = 0$ gives

$$\delta = \sqrt{12\nu x/U} = \sqrt{12 \times 2 \times 10^{-5}x/3} = 0.00894\sqrt{x}.$$

The boundary layer thickness at the downstream edge of the plate is then

$$\delta(x = 2) = 0.00894\sqrt{2} = 12.6 \text{ mm}.$$

The wall shear stress is given from the above by

$$\tau_{\text{wall}} = \mu U/\delta = 900 \times 2 \times 10^{-5} \times 3/0.00894\sqrt{x} = 6.04/\sqrt{x}.$$

From this we see that at the downstream edge of the plate the shear stress is 4.27 Pa. (The boundary layer approximation is not valid at the leading edge, so we cannot evaluate the shear there from our results.) The drag force on one side of our plate is

$$F = w \int_0^L \tau_{\text{wall}} \, dx = 0.5 \int_0^2 6.04/\sqrt{x} \, dx = 8.54 \text{ N}.$$

Since the Reynolds number at the downstream edge of the plate is

$$\text{Re} = 3 \times 2/2 \times 10^{-5} = 3 \times 10^5,$$

the flow will be laminar all along the plate.

Example 6.5.2. Entrance flow

Flow enters a channel with the velocity a uniform value U_0 across the channel. Evaluate the entrance length (the distance which the flow takes to change from the uniform velocity to the plane Poiseuille parabolic profile) if the channel is of width $2b$. A boundary layer will form at each boundary and grow until the two boundary layers meet and the velocity profile becomes parabolic. (See Figure 6.17.) Note that in order to satisfy continuity requirements, the flow outside of the boundary layers must increase with x until the boundary layers meet.

Sought: A formula for the developing boundary layer at the entrance to a wide channel.
Given: The plate spacing ($2b$) and the incoming uniform velocity U_0. The viscosity will be taken as known as well.

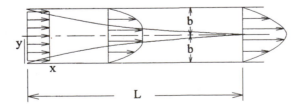

Figure 6.17. Developing flow-velocity profiles at an entrance to a channel.

Assumptions: Laminar viscous flow. It is assumed that the entering pressure gradient is zero, since no pressure gradient is needed for a uniform flow. Also, it is assumed that sufficiently far downstream from the entrance the velocity profile will be parabolic, with the appropriate constant pressure gradient. Between these two regions, the pressure gradient will decrease monotonically.

Solution: Consider the half of the flow in $0 \leq y \leq b$. Take the velocity distribution to be parabolic in y in the boundary layer, and uniform (but increasing as a function of distance from the entrance). This form for the velocity is suggested by the velocity profile that occurs when the top and bottom boundary layers meet. With $\eta = y/\delta$, we have

$$u/U(x) = 2\eta - \eta^2 \quad \text{for } 0 \leq \eta \leq 1,$$

$$= 1 \quad \text{for } \eta > 1.$$

In order to have flow continuity across the channel, the discharge at any x must be the same. The rate of flow entering the channel is $U_0 b$. The flow rate at any point inside the channel is the integral of u with respect to y, with y going from 0 to b. Since the two flow rates must be equal, use of the above assumed velocity profile gives

$$U_0 b = U(b - \delta) + U\delta \int_0^1 (2\eta - \eta^2) \, d\eta,$$

which upon solving for U gives

$$U = U_0/(1 - \delta/3b).$$

From (6.5.13),

$$\tau_{\text{wall}} = 2\mu U/\delta = d/dx(2\delta\rho U^2/15) + \rho U \, (dU/dx) \, \delta/3.$$

Putting in the expression for U obtained from continuity and rearranging,

$$30\nu \, dx/U_0 \, b = (6b + 7\delta)\delta \, d\delta/3(b - \delta/3)^2.$$

To find the entrance length, we integrate this equation with respect to x, letting x range from 0 to L, with δ going from 0 to b. The result is

$$L = 0.1038U_0 \, b^2/\nu.$$

At $x = L$, we have the familiar parabolic velocity profile of Poiseuille flow. The centerline velocity increases from U_0 to $1.5U_0$, indicating that the flow is being accelerated.

Equation (6.5.13) can be rewritten in various forms, each of which has advantages for particular problems. For example, in (6.5.13) we have already introduced the displacement and momentum thickness into (6.5.5) to obtain

$$\frac{d}{dx} (\rho U^2 \delta_M) + \rho U \delta_D \frac{dU}{dx} = \tau_{\text{wall}}. \tag{6.5.13}$$

Expanding the derivative of the first term gives

$$\frac{d\delta_M}{dx} + (2\delta_M + \delta_D)\frac{dU}{U\,dx} = \frac{\tau_{wall}}{\rho U^2}.$$ (6.5.30)

Frequently a "shape factor" $H = \delta_D/\delta_M$ is introduced, together with the wall shear coefficient $c_f = 2\tau_{wall}/\rho U^2$. Then (6.5.30) can be written in the form

$$\frac{d\delta_M}{dx} + \delta_M(2 + H)\frac{dU}{U\,dx} = c_f/2.$$ (6.5.31)

Comparison of these three forms of (6.5.5) reveals that all give a first-order ordinary differential equation that is nonlinear.

Many forms of approximate velocity profiles other than Pohlhausen's polynomial have been introduced. A summary of some of these is given in Rosenhead (1963). These other forms have various advantages over Pohlhausen's quartic polynomial, but lead to the same point—the need to solve a first-order ordinary differential equation. The profiles differ in the ease by which the solutions to these differential equations can be obtained by analytic means or by simple numerical means. With the present accessibility of personal computers, the ordinary differential equation resulting from the use of any of these profiles can easily by programmed using a Runge-Kutta or similar scheme. In fact, the local two-dimensional form of the boundary layer equations, whose solution requires no assumed velocity profile, is also easily programmed. Today, the use of the momentum integral method is perhaps best suited to obtaining a rough analytic approximation with a minimum of calculation. It does reveal the functional dependence of the shear stress and boundary layer thicknesses on the other parameters of the problem, but it may not be accurate as to the details of the flow.

6. Flow Separation

For the flat plate considered in the previous section, there is no pressure gradient in the flow, and the boundary layer flow is driven only by the uniform stream. The simplicity of the flat plate is a very special case, involving flow without pressure gradients. The flow near the wall is retarded (decelerated) by the viscous stress at the wall, and the boundary layer therefore continues to grow in thickness to the end of the plate.

In contrast, for flows past curved bodies, or even for a flat plate when the uniform stream is not parallel to the plate, pressure gradients will exist due to accelerations and decelerations driven by body shape. On upstream portions of the body the pressure gradients are often negative (the pressure decreasing along the wall), or even zero. In these regions we say that we have *favorable pressure gradients*, since the pressure gradient is acting to accelerate the flow along the wall, against the wall shear stress.

In other regions along the body, the pressure gradient can be negative. (The pressure is then increasing along the body wall.) These regions we call regions of *adverse pressure gradients*, since here the pressure gradient and viscous forces act together to oppose the flow and decrease the momentum, thus causing the flow to decelerate. The boundary layer thickness in this region increases rapidly because of this deceleration. Since viscous forces are generally greatest near the wall, an adverse pressure gradient can cause a region of reverse flow to develop next to the wall. When this happens, we say that the flow has *separated*. The point at which the reversed flow region starts is called the *separation point*. For an unseparated flow, the velocity increases with distance from the wall; hence $\partial u/\partial y > 0$. For the reversed flow in the separated region, the fluid velocity becomes negative away from the wall, and so there is a region where $\partial u/\partial y$ is negative. The separation point is then located where $\tau_{wall} = 0$, or equivalently, where $(\partial u/\partial y)_{wall} = 0$.

The separation point is always located downstream from where the pressure gradient changes from *favorable* ($\partial p/\partial x < 0$) to *unfavorable* ($\partial p/\partial x > 0$), that is, somewhere past the point where $\partial p/\partial x = 0$. Boundary layer theory can predict the flow up to, but not past, the separation point, **provided** U is known up to the separation point. The differential equation form of the boundary layer equations are partial differential equations that are said to be of parabolic form. This allows us to find the solution of the boundary layer equations by means of marching-type methods. Therefore the upstream conditions determine the downstream conditions. For separated flows, "upstream" changes direction once the separation point is passed, and we do not have any information in this wake region to start our solution. Since the flow past the separation point influences U, boundary layer theory alone is not sufficient. We must either resort to the full differential equation form of the momentum equation, including viscous effects, find U by experimental means in the region before separation, or use another approximation akin to the boundary layer approximation, to be able to continue our solution.

The situation is in fact somewhat worse than described above. The boundary layer equations can have singularities near the separation point, and in the unsteady flow case, numerical investigations have found singularities in the interior flow. These singularities seem to be peculiarities of the boundary layer approximation, and are not physical effects that occur in the flow. Presumably the full momentum equations do not have these singularities.

We next look at several approximate schemes for predicting the location of the separation point. We restrict our attention to two-dimensional laminar boundary layers, since the results become much more involved for turbulent flows. Three-dimensional laminar separation is usually predicted using the two-dimensional criteria, since they seem to work fairly successfully in three dimensions as well, providing that the two-dimensional reference frame is aligned in the flow direction. We expect that these schemes will work reasonably well for bodies with gradually changing geometries and with flows that are mostly two dimensional in nature. There do not appear to be any generally accepted simple formulas for complicated flows.

These schemes require a knowledge of $U(x)$, the flow at the outer edge of the boundary layer. Since separated flows usually have wakes, we may not be able to find U by analytic means in these cases, since inviscid flow theory has difficulties in dealing with wakes. When that happens, the most reliable results for U are obtained from experiment.

Thwaites (1949) correlated a number of solutions of the momentum integral form of the boundary layer equations and found that displacement thickness could be predicted fairly closely by the approximation

$$\delta_M^2(x) = \delta_{M_0}^2 + [(0.45\nu/U^6(x)] \int_0^x U^5(x) \, dx, \tag{6.6.1}$$

where δ_{M_0} is the displacement thickness at $x = 0$. He found that separation occurred at the point where

$$\delta_M^2 = -0.09\nu/dU/dx. \tag{6.6.2}$$

To use Thwaites' method, the displacement thickness is computed from (6.6.1) and then compared with the right-hand side of (6.6.2) until (6.6.2) is satisfied, locating the separation point. In many cases the coordinate x can be chosen so that the initial value of the momentum thickness δ_{M_0} is either zero, or else so that δ_{M_0} will be negligible compared to δ_M near the separation point. In either case the criterion simplifies to

$$-0.2U^6/(dU/dx) = \int_0^x U^5 \, dx. \tag{6.6.3}$$

Example 6.6.1. Flow past a circular cylinder—Thwaites' criterion

Use as the potential solution for flow past a cylinder $U = 2U_0 \sin (x/a)$. This is a fairly good representation of the velocity for a region approximately $\pm 30°$ from the upstream stagnation point, but increasingly inaccurate for points beyond. Nevertheless, it is useful as an example of the method. Find where separation occurs using Thwaites' criterion.

Sought: Location of flow separation of the laminar boundary layer on a long circular cylinder.

Given: Velocity profile at outer edge of boundary layer [$U = 2U_0 \sin (x/a)$], cylinder radius (a). Fluid viscosity is assumed known.

Assumptions: Laminar viscous flow. Assumed formula for flow speed at outer edge of boundary layer. Assumed that Thwaites' criterion is valid.

Solution: Taking $x = 0$ at the upstream stagnation point, where we will neglect the displacement thickness, Thwaites' criterion in the form (6.6.3) gives

$$-0.2U^6/(dU/dx) = \int_0^x U^5 \, dx,$$

or with $s = x/a$, upon carrying out the required operations, we have

$$-0.2a(2U_0)^5 \sin^6 s/\cos s = a(2U_0)^5[-\sin^4 s \cos s - 4 \cos s + 4/3 \cos^3 s + 8/3]/5.$$

After canceling the $a(2U_0)^5$ on both sides of the equation and multiplying both sides by 5, we have

$$-\sin^6 s = [-\sin^4 s \cos s - 4 \cos s + 4/3 \cos^3 s + 8/3] \cos s.$$

Replacing $\sin^2 s$ by the familiar equivalent $1 - \cos^2 s$ leads to

$$-(1 - \cos^2 s)^3 = [-\cos s + 2 \cos^3 s - \cos^5 s - 4 \cos s + 4/3 \cos^3 s + 8/3] \cos s,$$

or

$$\cos^6 s - 3 \cos^4 s + 3 \cos^2 s - 1 = -\cos^6 s + 10/3 \cos^4 s - 5 \cos^2 s + 8/3 \cos s.$$

Putting all terms on the left-hand side, we have

$$2 \cos^6 s - 19/3 \cos^4 s + 8 \cos^2 s - 8/3 \cos s - 1 = 0.$$

The roots of this sixth-order equation in $\cos s$ must be found numerically, using a scheme such as evaluation of the right-hand side for a sequence of s, starting at, say, $s = \pi/2$. The result is

$$s = 1.7996,$$

which corresponds to separation at $103.11°$ downstream from the upstream stagnation point.

A second criterion used for predicting separation is due to Stratford (1954), who suggested that separation will occur in a laminar boundary layer where

$$(x - x_0)^2 c_p (dc_p/dx)^2 = 0.0104. \tag{6.6.4}$$

Here x_0 is the point where $dU/dx = 0$, and $c_p = 1 - U^2/U_{max}^2$. This criterion is often simpler to use than Thwaites' criterion, since no integration is necessary, and the two give approximately the same results.

Example 6.6.2. Flow past a circular cylinder—Stratford's criterion
Repeat the previous example using Stratford's criterion.

Sought: Location of flow separation of the laminar boundary layer on a long circular cylinder.

Given: Velocity profile at outer edge of boundary layer [$U = 2U_0 \sin(x/a)$], cylinder radius (a).

Assumptions: Laminar viscous flow. Assumed formula for flow speed at outer edge of boundary layer. Assumed that Stratford's criterion is valid.

Solution: Since from the velocity profile at the outer edge of the boundary layer

$$U_{max} = 2U_0,$$

$$c_p = 1 - U^2/U_{max2} = 1 - \sin^2 s,$$

then $dc_p/dx = -(2/a) \sin s \cos s$, where s is x/a. The Stratford criterion predicts separation where

$$(x - x_0)^2 c_p (dc_p/dx)^2 = 0.0104.$$

Since $x_0 = \pi/2$, after dividing the above by a, we have

$(s - 0.5\pi)^2(1 - \sin^2 s)(- 2 \sin s \cos s)^2 = 0.0104.$

The value of s that satisfies this equation and is near 0.5π is

$s = 1.9588,$

predicting that separation occurs at $112.19°$ past the upstream separation point.

Example 6.6.3. Flow past a circular cylinder
The previous two examples both predict that separation occurs much further downstream than is found in experiments. This is because the value of U that we used does not give a very good description for U after $30°$ or so. A better velocity profile was found from experiment by Schmidt and Wenner (1941) to be

$U/U_0 = 2s - 0.451\, s^3 - 0.00578s^5,$

where s is x/a (a is the cylinder radius) as in the previous example. Find the separation point using the Stratford criterion.

Sought: Location of flow separation of the laminar boundary layer on a long circular cylinder.
Given: The velocity at the outer edge of the boundary layer,

$U/U_0 = 2s - 0.451s^3 - 0.00578s^5,$

and the cylinder radius (a).
Assumptions: Laminar viscous flow. Schmidt and Wenner's formula for flow speed at outer edge of boundary layer. Thwaites' criterion is assumed to hold.
Solution: The maximum velocity occurs where

$$\frac{d(U/U_0)}{ds} = 2 - 1.353s^2 - 0.0289s^4 = 0.$$

This is a quadratic equation in s^2; hence its roots can be found by the usual quadratic formula. The positive root is $s_{max} = 1.1976$, giving $U_{max}/U_0 = 1.6063$. Write Stratford's criterion in the dimensionless form

$$(s - s_{max})^2[1 - (U/U_{max})^2]\left[\frac{-2U}{U_{max}^2}\frac{dU}{dx}\right]^2 - 0.0104 = 0.$$

Numerical evaluation of this equation by substituting successive values of $s > s_{max}$ gives the root $s = 1.4463$, which corresponds to a separation point at $82.67°$.

Avoidance or delay of separation is frequently an important engineering problem, since it can minimize wake thickness and therefore drag. On airplane wings, delay of separation enhances lift. Separation delay can be accomplished in many ways. In automobiles, the location of the separated flow regions can be delayed to positions

toward the rear by making window glass flush with the body and avoiding protuber-
ances such as side-view mirrors. Many aircraft wings have rows of short (about 1 in
or more long) vanes placed vertically on their top surface. These vanes are called
"vortex generators." These vortex generators induce an artificial turbulence in the
boundary layer. This turbulent boundary layer has much less of a tendency to separate
than a laminar one would, allowing the plane to climb at a steeper angle than it could
without the vortex generators, while still maintaining lift.

Suggestions for Further Reading

Bénard, H., "Les Tourbillons cellulaires dans une nappe liquide," *Revue générale des Sciences pures et appliquées*, vol. **11**, pp. 1261–1271 and 1309–1328, 1900.

Bénard, H., "Les Tourbillons cellulaires dans une nappe liquide transportant de la chaleur par convection en régime permanent," *Ann. Chim. Phys.*, vol. **23**, pp. 62–144, 1901.

Blasius, H., "Grenzschichten in Flussigkeiten mit kleiner Reibung," *Z. angew. Math. Phys.*, vol. **56**, pp. 1–37, 1908. (An English translation can be found in NACA TM 1256.)

Kármán, T. von, "Uber laminare und turbulent Reibung," *Z. angew. Math. Mech.,* vol. **6**, pp. 233–252, 1921.

Prandtl. L., "Uber Flüssigkeitsbewegung bei sehr kleiner Reibung," *Proc. Third Int. Math. Congress*, Heidelberg, 1904.

Rayleigh, Lord, "On convective currents in a horizontal layer of fluid when the higher temperature is on the under side," *Phil. Mag.*, vol. **32**, pp. 529–546, 1916.

Rosenhead, L., editor, *Laminar Boundary Layers*, Clarendon Press, Oxford, 1963.

Schlichting, H., *Boundary Layer Theory*, 7th edition, McGraw-Hill, New York, 1979.

Schmidt, E., and Wenner, K., "Warmeabgabe über den Umfangeinesangeblasenen geheizten Zylinders," *Forsch. Gebiete Ingenieurwesens*, VDI, vol. **12**, pp. 65–73, 1941.

Stratford, B. S., "Flow in the laminar boundary layer near separation," Aeronautical Research Council London, RM-3002, 1954.

Taylor, G. I., "Stability of a viscous liquid contained between two rotating cylinders," *Phil. Trans. Roy. Soc. London, Ser. A*, vol. **223**, pp. 289–343, 1923.

Thwaites, B., "Approximate calculation of the laminar boundary layer," *Aero. Quart.*, vol. **1**, pp. 245–280, 1949.

Thwaites, B., editor, *Incompressible Aerodynamics*, Clarendon Press, Oxford, 1960.

White, F., *Viscous Fluid Flow*, McGraw-Hill, New York, 1974.

Problems for Chapter 6

Flow between parallel plates

6.1. For laminar flow between parallel plates with the geometry shown in the figure, find
where the velocity is a maximum. What is the discharge? Indicate whether it is up or
down the plate. $\mu = 3 \times 10^{-5}$ lb-s/ft^2.

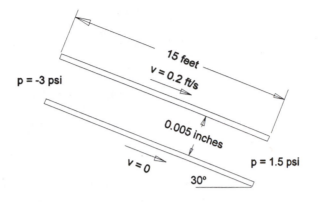

Problem 6.1

6.2. For flow down a vertical plane at $y = 0$ with a free surface at $y = b$, find the velocity as a function of y. What is the discharge?

6.3. For flow between parallel plates with the configuration shown in the figure, what must U be so that at 0.0025 ft above the bottom plate the shear stress is zero? For this value of U, what is the discharge? Is the net flow to the right or left? $\mu = 3 \times 10^{-5}$ lb-s/ft^2.

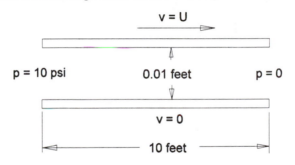

Problem 6.3

6.4. Given that for flow between two parallel plates $u = u_{max}(1 - y^2/b^2)$, where the plates are at $\pm b$, find the volume rate of flow, the rate at which momentum is transferred between the plates, and the rate of energy transfer.

6.5. Find the absolute viscosity of a fluid given the information in the figure.

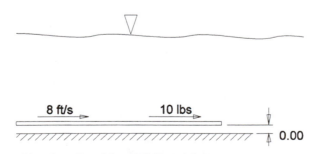

Plate length = 3 feet, width = 4 feet

Problem 6.5

6.6. For flow between horizontal parallel plates, with the bottom plate stationary and the top plate moving, what must be the relation between U and the pressure gradient to have the net discharge be zero?

6.7. For flow between parallel plates 0.006 ft apart and making an angle of 30° with the horizontal, if $p_A = 6$ psi and $p_B = 8$ psi, where point B is 4 ft farther down the plate than point A, $U_{top} = 30$ ft/s, $U_{bottom} = 0$, $\gamma = 50$ lb/ft^3, $\mu = 10^{-5}$ lb-s/ft^2, what is the discharge per unit width?

6.8. Water runs down a 45° plane with a depth of 0.1 ft ($\nu = 1.2 \times 10^{-5}$ ft^2/s). The flow is laminar and $\gamma = 62.4$ lb/ft^3. The upper surface is a free surface. What is the pressure gradient? What is the pressure on the plane? What is the shear stress on the plane? What is the discharge per unit width?

6.9. Flow takes place between two horizontal, stationary parallel plates placed 0.1 ft apart. The kinematic viscosity is 10^{-3} ft/s^2 and the density is 2 slugs/ft^3. The maximum velocity is 3 ft/s. What is the pressure gradient? What is the flow rate? What is the momentum flux?

6.10. Rainwater running down a plane inclined 30° with the horizontal is observed to have a maximum velocity of 1 in/s. What is the thickness of the water layer? What is the pressure on the plane? $\mu = 2.5 \times 10^{-5}$ lb-s/ft^2.

6.11. A fluid of SG = 0.9 and a viscosity of 1 poise (P) is contained between a vertical moving belt and a wall as shown in the figure. The spacing is 0.02 ft between the belt and the wall. The fluid is exposed to the atmosphere at each end. What is the minimum belt speed needed to bring fluid upward? (That is, Q should be positive upward.)

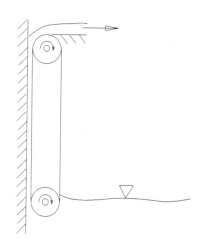

Problem 6.11

6.12. A vertical belt is used to remove oil from the surface of a settling pond. It is observed to pick up a layer of fluid of thickness b. The belt moves at an upward velocity V. What is the net volume rate of flow per unit width? What is the minimum belt velocity needed to have a net upward rate of flow, and hence, to keep fluid on the belt? For a fixed belt velocity, what thickness b maximizes the discharge for a given oil layer thickness?

6.13. A solid cylinder of weight W, length a, and diameter D is falling in a fluid-filled cylinder of diameter $D + 2c$ and length b. See the figure. The fluid has specific weight γ and viscosity μ. The specific weight of the cylinder is $s\gamma$. The ratios c/D and b/a are very small, and the flow between the walls may be considered to be the flow between parallel plates. Assume that the cylinder has reached terminal (constant) velocity V. Find the speed of fall V in terms of properties and geometric parameters, realizing that the flow is caused by the cylinder displacing fluid.

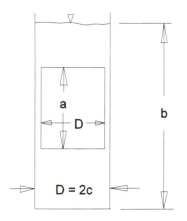

Problem 6.13

6.14. The 2-in-diameter piston shown in the figure is pushed to the left with a constant speed of 0.002 ft/s. The pressure in the chamber on the left stays at constant pressure, and the fluid leaves this chamber at a rate equal to the rate of volume displacement by the piston. The clearance between the piston and cylinder is sufficiently small that the flow can be regarded as being between two parallel planes. What is the pressure gradient $\partial p/\partial z$ developed by the flow? If $\mu = 10^{-6}$ lb-s/ft^2, what force F is necessary to make the piston move a this rate?

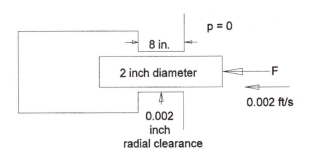

Problem 6.14

6.15. The 2-in-diameter piston shown in the figure for Problem 6.14 is pushed to the left with a constant velocity V and is resisted by a constant pressure of 5 psi. The fluid leaves this chamber at a rate equal to the rate of volume displacement by the piston. The clearance between the piston and cylinder is sufficiently small that the flow can be regarded as being between two parallel planes. What is the pressure gradient $\partial p/\partial z$ developed by the flow? If $\mu = 10^{-6}$ lb-s/ft^2, what force F is necessary to make the piston move at this rate?

6.16.   Two fluids flow between vertical plates at $y = \pm b$. The upper fluid ($0 \le y \le b$) has density ρ_{up} and viscosity μ_{up}, while the lower fluid ($-b \le y \le 0$) has density ρ_{low} and viscosity μ_{low}. Find the velocity profile, noting that the velocities and shear stresses must be continuous at $y = 0$.

Rotating boundaries

6.17.   The rotating plate-oil bath combination shown in the figure is sometimes used as a speed control. If the disk has a diameter of 6 in, $SG_{oil} = 0.75$, and $\nu_{oil} = 0.02$ ft^2/s, calculate the resisting torque as a function of the angular velocity Ω assuming laminar flow. The spacing is sufficiently small that the velocity can be assumed to vary as the product of r and z.

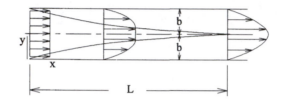

Problem 6.17

Couette flows

6.18.   For flow between two concentric vertical cylinders as shown in the figure, the streamlines are concentric circles and the velocity component v_θ is in the form $Ar + B/r$, where A and B are constants of integration. Find A and B given that the outer

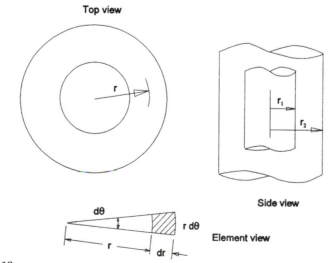

Problem 6.18

cylinder of radius r_o rotates with angular velocity Ω_0, and the inner cylinder of radius r_i rotates with angular velocity Ω_i.

6.19. A device consisting of a vertical cylinder mounted on a torque-measuring device and immersed in a rotating cup containing fluid is frequently used to measure the viscosity of a fluid. (See the figure.) The apparatus is usually designed so that the gap $c = b - a$ is small compared to the radius a, and the flow can be thought of as taking place between parallel planes. What is the pressure gradient in the circumferential direction? What is the velocity as a function of radius? What is the radial pressure gradient? What is the torque exerted on the side of the inner cylinder?

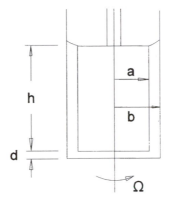

Problem 6.19

6.20. In the flow of the figure for Problem 6.19, the inner cylinder is held fixed and the outer cylinder is rotated at a speed of 10 rev/s. What is the moment exerted on the inner cylinder, assuming that the flow is laminar and that the ratio of the gap thickness $b - a$ to the radius of the inner cylinder a is sufficiently small that the flow can be taken as between parallel flat planes?

Lubrication

6.21. A slider bearing 3 cm long and 15 cm wide moves at a speed of 0.2 m/s. The lubricant has a viscosity of 2×10^{-3} N-s/m². What is the maximum load the bearing can carry if the spacing b_L at $x = L$ is 0.01 mm?

6.22. A slider bearing 3 cm long and 15 cm wide moves at a speed of 0.2 m/s. The lubricant has a viscosity of 2×10^{-3} N-s/m². What is the maximum load the bearing can carry if the drag force is to be kept to 3% of the load if the spacing b_L at $x = L$ is 10^{-5} m? What is the spacings b_0 if the spacing $b_L = 10^{-5}$ m?

6.23. A slider bearing is made in the form of a step where all walls are parallel to one another, as shown in the figure. Derive an expression for the pressure in each part of the bearing. Let the pressure be zero at entrance and exit, and assume that both the pressure and velocity are continuous and well-behaved at the step. Compute the total discharge and the total load P per unit width on the bottom plate.

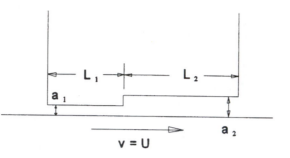

Problem 6.23

Circular pipe flows

6.24. Water flows in a horizontal circular pipe of 1-in-diameter cross section. What must be the pressure drop in 100 ft in order to have a Reynolds number based on the diameter of 2,000? Use water properties at 70°F.

6.25. Water ($\mu = 3 \times 10^{-5}$ lb-s/ft^2, $\nu = 10^{-5}$ ft/s) flows through a 0.01-ft-diameter horizontal tube that is 10 ft long. The Reynolds number based on diameter is 1,800. What is the pressure change along the length of the tube? What is the average velocity?

6.26. For the flow shown in the figure, assuming that the flow is laminar and that the Reynolds number is 1,500, what is the height h? Given: $\rho = 1.94$ slugs/ft^3, $\nu = 10^{-5}$ ft^2/s.

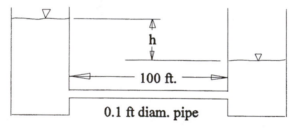

Problem 6.26

6.27. For water at 20°C flowing in a circular pipe of diameter 15 mm, what is the maximum pressure gradient allowable that guarantees that the flow is laminar? How does this change if the pipe is vertical?

6.28. Assuming the flow to be laminar, estimate the discharge for the flow shown in the figure.

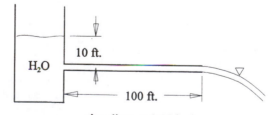

pipe diam. = 0.25 inch

Fluid properties: $\nu = 0.00001$, S.G. = 1

Problem 6.28

6.29. A vertical tube 2 m long and 10 mm in diameter carries a liquid having a viscosity of 0.02 Pa-s and a density 800 kg/m^3. The tube is supplied by a very shallow tank, and entrance losses are negligible. What is the discharge and Reynolds number?

6.30. In laminar flow in a horizontal 12-in-diameter pipe, two instruments for reading stagnation pressure are installed. One is on the centerline and the other is placed 3 in from the centerline but in the same horizontal plane. If the specific weight of the liquid flowing is 50 lb/ft^3 and the flow rate is 10 ft^3/s, calculate the reading of a differential manometer connected to the two tubes if the manometer contains mercury. SG$_{mercury}$ = 13.6. ν = 0.007 ft^2/s. How does viscosity affect your result?

6.31. Glycerin at 100°F flows laminarly through a 3/8-in-diameter horizontal pipe with a pressure drop of 5 psi/ft. Find the discharge and Reynolds number.

6.32. The piston shown in the figure of 100-mm diameter moves slowly to the right, pushing the water (at temperature 20°C) out through the capillary tube (0.001 m in diameter, 0.5 m in length). How fast can the piston move and still guarantee laminar flow in the capillary tube? What is the pressure on the face of the piston assuming laminar flow? What is the total force exerted on the walls of the capillary tube? No leakage occurs past the piston.

Problem 6.32

Flow in an annulus

6.33. A piston is forced to move with a constant velocity V into a closed chamber containing a fluid of viscosity μ and density ρ. The radius of the piston is a. The radius of the cylinder wall is b. Find the velocity profile in the annulus. Notice that the motion of the fluid is due to both the motion of the wall and a pressure gradient that will develop. Solve for the discharge Q in terms of the pressure gradient, the piston velocity V, and the fluid properties and geometric parameters.

Boundary layer flow

6.34. Using the parabolic approximation for the flow in the boundary layer on a flat plate, find how large x must be to have $\delta = 0.001$? What is the wall shear at the point? How much bigger does x have to be to have δ double? ρ = 1,000 kg/m^3, $\nu = 10^{-6}$ m^2/s, U = 1.5 m/s.

6.35. Using the Blasius solution for the flow in the boundary layer on a flat plate, find the wall shear at x = 0.05 m. ρ = 1,000 kg/m^3, $\nu = 10^{-6}$ m^2/s, U = 1.5 m/s.

Flow in an annulus

6.36. Annular flow tubes are often used in heat exchangers, where an important considera- tion is the head loss per unit length. Plot $f_{annulus}/f_{circular\ pipe}$ versus the radius ratio a/b = r_{inner}/r_{outer}, where f is the friction factor.

6.37. A horizontal annulus pipe has an outer diameter of 250 mm and an inner diameter of 150 mm. The average velocity is 0.5 m/s. What is the pressure gradient if the fluid is SAE 10W motor oil at 20°C?

6.38. A hydraulic lift in an automobile service station supports 8,000 lb. The cylinder has a diameter of 8 in, and the radial clearance between piston and cylinder is 0.002 in. The piston length is 8 ft. If the oil used is SAE 20W at 85°F, what is the rate of oil leakage past the piston?

Hydraulic diameter

6.39. Air at 20°C flows through a 300-mm by 450-mm rectangular duct. The discharge is 0.04 m^3/min. What is the Reynolds number?

6.40. What is the hydraulic diameter of a circular annulus with outer diameter 250 mm and inner diameter 150 mm?

Boundary layer flow

6.41. Prepare a table giving values of the velocity u on a flat plate as a function of x at $y = 3$ mm. $U = 2$ m/s, $\nu = 10^{-4}$ m^2/s. First use the Blasius results from Table 6.2; then compare his results with that obtained from the approximate parabolic velocity profile.

6.42. A velocity field outside the boundary layer that has a favorable pressure gradient is $U = U_0(1 + x/L)$. Using the parabolic approximation for the velocity within the boundary layer and equation (6.5.13), determine the boundary layer thickness as a function of x. Show that near $x = 0$ your solution approaches the flat plate solution. How would your solution change if you used a different approximation for the flow within the boundary layer? *Note*: The differential equation you obtain is linear in δ^2. The homogeneous solution is a power of U. You should be able to reduce the problem to a simple integration.

6.43. A velocity field outside the boundary layer that has an adverse pressure gradient is $U = U_0(1 - x/L)$. This was proposed by Howarth in 1938. Using the parabolic approximation for the velocity within the boundary layer and equation (6.5.10), determine the boundary layer thickness as a function of x. The note at the end of the previous problem will be helpful.

6.44. A velocity field outside the boundary layer that has an adverse pressure gradient is $U = U_0/(1 + x/L)$. This is one of a family of profiles proposed by H. Görtler in 1957. Using the parabolic approximation for the velocity within the boundary layer and equation (6.5.10), determine the boundary layer thickness as a function of x. The note at the end of Problem 6.42 will be helpful.

6.45. The wing of a model airplane has a chord length of 250 mm and a span of 1,500 mm. They can be considered to be flat plates for the purposes of drag calculations. The plane flies at 0.5 m/s. For an air temperature of 30°C, find the boundary layer thickness at the trailing edge of the wing, and the total viscous drag on the wing. Use the parabolic approximation for the velocity profile in the boundary layer.

Separation

6.46. A velocity field outside the boundary layer that has an adverse pressure gradient is $U = U_0(1 - x/L)$. This was proposed by Howarth in 1938. Separation has been found to occur for this velocity at $x = 0.1199 L$. Use Thwaites' method with $\delta_{M0} = 0$ to compute the separation point. Compare your result with that of Howarth.

6.47. Repeat the previous problem, this time using the Stratford separation criterion. Compare your result with that of Howarth.

6.48. A velocity field outside the boundary layer that has an adverse pressure gradient is $U = U_0/(1 + x/L)$. This is one of a family of profiles proposed by H. Görtler in 1957. He found separation to occur at $x = 0.159L$. Use Thwaites' method with $\delta_{M0} = 0$ to compute the separation point. Compare your result with that of Görtler.

6.49. Repeat the previous problem, this time using the Stratford separation criterion. Compare your result with that of Görtler.

6.50. A velocity field outside the boundary layer which has an adverse pressure gradient is $U = U_0 [(x/L) - (x/L)^3]$. Use the Stratford separation criterion to compute the separation point.

chapter 7

Turbulent Viscous Flow

Chapter Overview and Goals

Turbulent flows have velocity components and pressures that exhibit random and chaotic behavior. These flows are generally described in terms of their statistics rather than the instantaneous flow itself. To accomplish this, the velocities and pressures are split into mean and fluctuating components. The statistical quantities that are of primary interest are the root-mean-square values and correlations of the velocity components. Particular correlations of importance are the Reynolds stresses, which represent the momentum exchange introduced by the turbulent fluctuations, and are the turbulent counterpart of the molecular stresses.

In order to both correlate and predict the Reynolds stresses, the concepts of eddy viscosity and mixing length are introduced, along with the concept of turbulent length scales.

Special cases treated in some detail are those of turbulent pipe flow, turbulent open channel flow, and boundary layers on walls. The introduction of the concepts of admissible roughness and minor losses allows for the solution of a wide class of practical problems.

Dimensionless parameters such as drag and lift coefficients are applied to the calculation of problems such as vehicle drag and airplane wings.

In our study of laminar flows in Chapter 6, we were able to find analytic solutions for the simple flow cases considered there. For turbulent flows there are no simple solutions for any flows. We are instead forced to deal with a combination of empirical results combined with general guidelines from theory. Many cases of engineering interest have been sufficiently studied so that we are able to make reliable predictions for those situations. There are still, however, large gaps in our fundamental knowledge, which make turbulent flows a mysterious region waiting to be further explained and explored. Numerical studies of turbulent flows for the atmosphere, oceans, and general fluid mechanics research have been in fact been major uses of supercomputers.

1. Reynolds Stresses

In dealing with the measurement of turbulent flows, there is such a wealth of information regarding the variations with space and time that it is impossible to make any deductions as to detailed flow behavior. It is customary instead to treat quantities in a statistical fashion. An example of a randomly fluctuating quantity is shown in Figure 7.1. Considering a quantity such as pressure, for example, we would write

$$p = \bar{p} + p'. \tag{7.1.1}$$

A barred quantity such as $\bar{p}$ represents the time mean value of that quantity, i.e.,

$$\bar{p} = (1/T) \int_0^T p(t)\, dt. \tag{7.1.2}$$

Here T is the sample time. In a mathematical sense we might want to make the sample time very large, perhaps even infinite. In practice, the frequencies of turbulent fluctuations are sufficiently large that a sample time of the order of seconds or less is a sufficient averaging time. We will use the terms "mean" and "average" interchangeably here.

The primed quantity is what is left over after we have performed the averaging process. These are called the ***turbulent fluctuations***, or sometimes, ***disturbance quantities***. By definition the time average of any turbulent fluctuation quantity is zero. To measure the magnitude of a turbulent fluctuation, a useful concept is that of the root-mean-square (rms) value. For pressure, the rms value would be

$$\text{rms } p' = \sqrt{\overline{p'^2}} = \sqrt{(1/T) \int_0^T p'(t)^2\, dt} \tag{7.1.3}$$

Note that the operations of squaring and averaging are not interchangeable.

Using rectangular Cartesian coordinates, we denote the x, y, z velocity components by u, v, w, respectively. The fluctuations in the velocity components result in particle

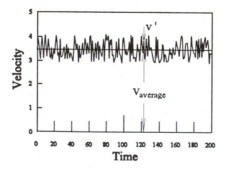

Figure 7.1. Turbulent velocity versus time. The average velocity is 3.5, the rms velocity is 0.34.

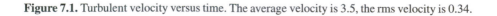

mixing and an interchange of momentum between the mean and fluctuating components of the flow. We write then,

$$u = \overline{u} + u', \qquad v = \overline{v} + v', \qquad w = \overline{w} + w'. \tag{7.1.4}$$

The rms value of a quantity is called the ***intensity*** of that quantity. The intensity of a quantity gives us an idea of how much that quantity departs from its mean value. As an example, the kinetic energy per unit mass associated with the mean flow is

$$ke_{\text{mean flow}} = (\overline{u}^2 + \overline{v}^2 + \overline{w}^2)/2. \tag{7.1.5}$$

The kinetic energy per unit mass associated with the turbulent flow is

$$ke_{\text{turbulent flow}} = (\overline{u'^2} + \overline{v'^2} + \overline{w'^2})/2. \tag{7.1.6}$$

There is a seemingly infinite amount of information contained in a turbulent quantity such as u'. In going to a statistical description of u', we hope that by giving only a few of its descriptors such as the mean and rms values, we have presented the important information in the data. Obviously two quantities are not enough to describe the complicated behavior of the quantity with time. (Try having someone identify a friend of yours after you have specified only the friend's height and age!) Averages and rms values are examples of statistical ***moments*** of the quantity. They are the first and second moments, respectively. We can form a moment of any order, so there is a never-ending string of them. The third and fourth moments are measures of ***skewness*** (departure from symmetry) and ***kurtosis*** (peakedness). To try to reconstruct a quantity from its moments, in principle we need to know all the moments to do a reliable job. In fluid mechanics, one almost never measures higher than the fourth moment. It is unusual in fact to measure beyond the second moment.

Besides knowing the moments of a particular quantity, we often wish to know how changes in one quantity relate to changes in another quantity. Such a relationship is called a ***correlation***. A typical correlation is that between the x and y velocity components,

$$\overline{u'v'} = (1/T) \int_0^T u'(t)v'(t)\, dt. \tag{7.1.7}$$

If the value of a correlation is large, then it is expected that the same cause is affecting both. (A large value in this case means large compared to $[\overline{u'^2}\,\overline{v'^2}]^{1/2}$, the maximum value that this correlation can take.) We see that an intensity is the correlation between a turbulent velocity component and itself. Intensities are also called ***autocorrelations***.

Multiplying the equation for acceleration developed in Chapter 3 by the mass density, we have

$$\rho D\mathbf{v}/Dt = \rho[\partial \mathbf{v}/\partial t + (\mathbf{v} \cdot \nabla \mathbf{v})] = \partial(\rho \mathbf{v})/\partial t + \nabla \cdot (\rho \mathbf{v}\mathbf{v}) - \mathbf{v}[\partial \rho/\partial t + \nabla \cdot (\rho \mathbf{v})]. \tag{7.1.8}$$

When the decomposition form (7.1.4) for the velocities is inserted into the term involving the product $\rho \mathbf{v}\mathbf{v}$ and then averaged, terms like

$$\overline{\rho u'^2}, \qquad \overline{\rho u'v'}, \qquad \overline{\rho u'w'}, \qquad \overline{\rho v'^2}, \qquad \overline{\rho w'^2}, \qquad \overline{\rho v'w'} \qquad (7.1.9)$$

arise. The negative of these terms are called the **Reynolds stresses**. They have dimensions of stress, and represent the **turbulent momentum flux**. When they are moved to the force side of the momentum equation, their gradients, like the gradients of the mean viscous stresses, can change the momentum of the mean flow. These Reynolds stress components represent the momentum interchange that takes place between the mean and turbulent flow components.

2. Eddy Viscosity and Mixing Length Concepts

It is natural in a complicated subject such as turbulent flow to hope that somehow a simplifying assumption can be found that will reduce a new complex problem to an older one that is better understood. So too with turbulence. One of the first attempts at simplification was by Boussinesq (1877, 1896), who said that, since the molecular stresses were the molecular viscosity times the mean rates of deformation, perhaps the Reynolds stresses could be written as a turbulent, or eddy, viscosity times the mean rates of deformation. This eddy viscosity would be a property of the turbulent flow, and not just a property of the fluid as is the case for the molecular viscosity. Except near walls where the Reynolds stresses must vanish because of the no-slip condition, the eddy viscosity so defined is many orders of magnitude larger than the molecular viscosity. The concept of such an eddy viscosity is useful for many flows, and is used extensively in meteorology and oceanography as well as in engineering. The difficulties with it, however, include the following:

1. We do not know how to find it from theory. Presumably experimental data can give us results for a particular flow situation.
2. By saying that the Reynolds stresses are the eddy viscosity times the mean rates of deformation of the flow, we are implying that Reynolds stresses vanish where the rates of deformation vanish. Experiments do not bear this out.

Prandtl (1925) tried to make this concept more concrete by introducing the concept of **mixing length**. Influenced by the concept of mean free path in atomic physics, he postulated that the average distance a fluid particle had to travel before its velocity changed to that of the new region was related to the average velocity gradient, and so in a shear flow

$$u' \cong l \, d\bar{u}/dy, \qquad v' \cong l \, d\bar{u}/dy.$$

Then a typical Reynolds stress component would be

$$-\overline{\rho u'v'} = \rho l^2 (d\bar{u}/dy)^2, \qquad (7.2.1)$$

where we have let the mixing length l absorb the proportionality constant. Therefore we have

$$\mu_{eddy} = \rho l^2 \, d\bar{u}/dy. \tag{7.2.2}$$

Using a different model for turbulence, von Kármán later suggested the model

$$l = \kappa \, d\bar{u}/dy/d^2\bar{u}/dy^2, \tag{7.2.3}$$

where κ was a universal constant with a value in the range 0.4–0.41. Subsequent work has shown that these ideas work well in some simple shear flows, but are not necessarily accurate in other more complicated flows. The concepts are nevertheless useful, and appear today in more sophisticated form in computer modeling of turbulent flows.

3. Turbulent Pipe Flow

One of the most often encountered engineering situations dealing with turbulent flow is flow in pipelines. These flows have been studied extensively in many series of experiments, and of all the turbulent flows that can exist, our state of knowledge of these turbulent flows is the most complete. Extensive measurements of turbulent flow quantities were carried out by Laufer (1954) for NACA; a summary of some of these results for mean velocity distribution and turbulent fluctuations is shown in Figures 7.2 to 7.5.

In Chapter 5 we showed [equation (5.6.2)] that by assuming that the head loss was a function of the pipe diameter, average velocity, pipe roughness, and the kinematic viscosity. Just from dimensional considerations, then, the head loss could be written in the form

$$h_L = f(L/D)V^2/2g = f(L/D)8Q^2/\pi^2 D^4 g, \tag{7.3.1}$$

where V is the average velocity, Q is the discharge, and A is the cross-sectional pipe area. The friction factor f depends on the local Reynolds number and the relative roughness k/D, k being the average roughness height. Equation (7.3.1) is called the

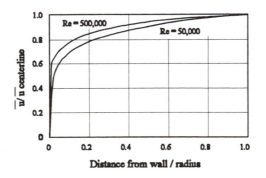

Figure 7.2a. Mean velocity distribution for two Reynolds numbers. Adapted from J. Laufer, 1954.

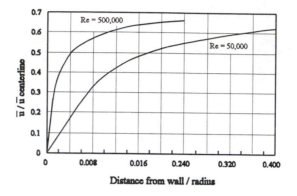

Figure 7.2b. Mean velocity distribution near the wall for two Reynolds numbers. Adapted from J. Laufer, 1954.

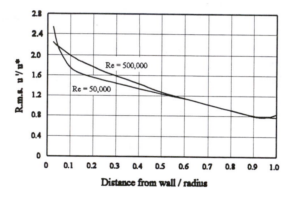

Figure 7.3a. Root-mean-square u' distribution for two Reynolds numbers. Adapted from J. Laufer, 1954.

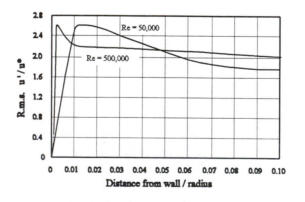

Figure 7.3b. Root-mean-square u' distribution near the wall for two Reynolds numbers. Adapted from J. Laufer, 1954.

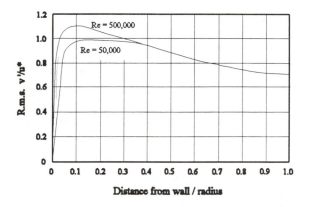

Figure 7.4a. Root-mean-square v' distribution for two Reynolds numbers. Adapted from J. Laufer, 1954.

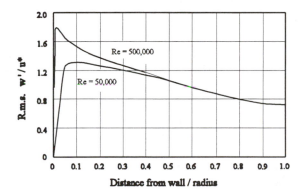

Figure 7.4b. Root-mean-square w' distribution for two Reynolds numbers. Adapted from J. Laufer, 1954.

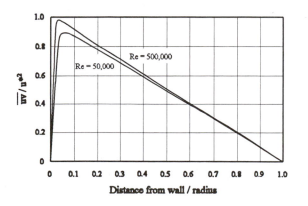

Figure 7.5. Reynolds stresses for two Reynolds numbers. Adapted from J. Laufer, 1954.

Darcy-Weisbach equation. The friction factor f was measured by Nikuradse (1933), who took sand grains of known height and carefully glued them on the inside of pipes. Knowing the roughness height, the pipe diameter and length, and the fluid kinematic viscosity, he could determine f from (7.3.1) by measuring discharge as a function of pressure drop. By repeating the experiments with commercial pipe he was able to determine an equivalent sand-grain roughness. Later, Colebrook (1938–1939) found that Nikuradse's experiments could be empirically described by the formula

$$1/\sqrt{f} = -0.8 \ln (k/3.7D + 2.51/\mathrm{Re}\sqrt{f}). \tag{7.3.2}$$

Moody (1944) summarized this equation in graphical form (Figure 7.6). This graph is referred to as the ***Moody diagram***.

Looking at the Moody diagram, it is seen that there are four well-defined regimes. Starting at the low Reynolds number end of the diagram, we see that all data falls on a straight line for laminar flow. In this ***laminar zone*** region, head loss will vary linearly with the velocity. Pipe roughness is not a significant parameter in this region.

To the right of the value Re = 2,300, we have a region called the ***critical zone***. This is a region of transition from laminar to turbulent flow, and experimental results tend to vary widely in this region, depending on small details of the physical setup of the experiment and the disturbances in the environment, as well as with time. At best, we can say that the flow characteristics in this zone are going to be very sensitive to small changes in settings of the system.

The third region, the ***transition zone***, starts around a Reynolds number of 4,000. It is bounded on the left by the smooth pipe curve. In this region results for commercial pipes do not agree well with Nikuradse's experiments, probably because the roughness in the commercial pipes occurs in a much more random pattern than in his carefully placed sand-roughened pipes. In regions 2 and 3, the head loss will vary with a power of the velocity lying between 1 and 2.

In the last, and highest, Reynolds number region of the Moody diagram, the curves are very flat, and the friction factor is almost completely independent of Reynolds number. This is called the ***fully developed turbulent region***. Here the head loss is proportional to the square of the velocity, and pipe roughness governs the friction factor, not viscosity.

When a pipe is new, its roughness depends primarily on the material of which the pipe is manufactured. Representative values of roughness are given in Table 7.1. As a pipe ages in use, the roughness and diameter will generally change. Dirt, scale, and foreign particles will build up on the walls, and suspended chemicals in the fluid will be deposited out. When a real system is designed, the system must be overdesigned to make allowances for this aging.

Before explaining the use of the Moody diagram, let us examine the results for laminar flow. From (6.3.7), with V as the average velocity, the head loss is due to a combination of static and dynamic pressure loss, and can be expressed by

$$Q = \pi b^2 V = \pi b^4 (-\partial p/\partial z + \gamma \sin \theta)/8 = \pi b^4 g h_L/8\mu L. \tag{7.3.3}$$

Table 7.1. Pipe roughness

Material	k (ft)	k (mm)
Asphalted cast iron	4×10^{-4}	0.12
Cast iron	85×10^{-5}	0.25
Concrete	10^{-3}–10^{-2}	0.3–3.0
Drawn tubing	5×10^{-6}	0.0015
Galvanized iron	5×10^{-4}	0.15
Glass, plastic	0.0	0.0
Riveted steel	3×10^{-3}–3×10^{-2}	0.9–9.0
Wood stave	6×10^{-4}–3×10^{-3}	0.18 to 0.9
Wrought iron, steel	15×10^{-5}	0.046

Solving for the head loss and putting the result in the general form of (7.3.1), we have after rearranging,

$$h_L = (8\mu/b)(L/b)(V/g) = (64\mu/DV)(L/D)(V^2/2g). \tag{7.3.4}$$

Comparing this with (7.3.1), the friction factor for laminar flow in a pipe is seen to be

$$f = 64/\text{Re}, \tag{7.3.5}$$

a result that is independent of pipe roughness.[1] This equation plots as a straight line on the log–log coordinates of the Moody diagram with a slope of minus one, and describes the straight-line behavior of the low Reynolds number region.

In pipe flows where pipe friction is the major reason for energy loss, there are three general types of design problems that are frequently encountered:

Type 1. Given Q, L, D, k, ν, find h_L. This can be thought of as a situation where we have an existing pipeline and we want to provide a predetermined flow rate of a given fluid. The design question is, what size of pump or other driving force is needed?

Type 2. Given h_L, L, D, k, ν, find Q. Here we have both the pipeline and the pump in place, and we wish to determine what the flow rate will be.

Type 3. Given h_L, Q, L, ν, k, find D. Here we have the pump, and wish to deliver a given flow rate. We now must determine a pipe size (diameter) to accomplish this.

The Moody diagram is ideally set up for the problems of type 1. We can easily form Re and k/D from the given quantities, and simply enter the Moody diagram to find f. Equation (7.3.1) then gives us the head loss. If we want the rate of energy (power) loss, we multiply head loss by γQ. Note that if we were to compute f directly from equation (7.3.2) an iterative solution would be necessary.

The Moody diagram is less well-suited for problems of types 2 and 3. We have in fact chosen the wrong repeating variables for this type of problem. What we must do

---

[1]This was also derived in Chapter 6, equation (6.3.14).

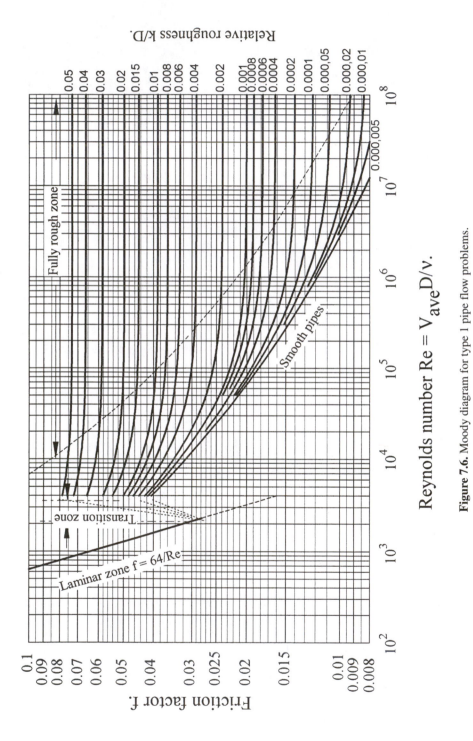

Figure 7.6. Moody diagram for type 1 pipe flow problems.

356

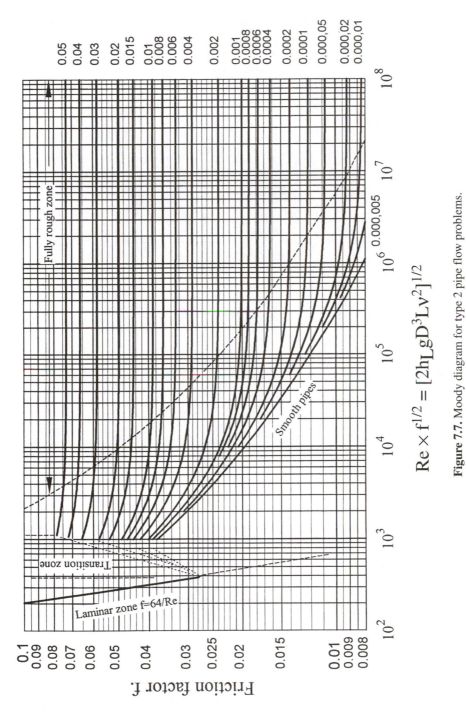

Figure 7.7. Moody diagram for type 2 pipe flow problems.

$$Re \times f^{1/2} = [2h_{LG}gD^3Lv^2]^{1/2}$$

357

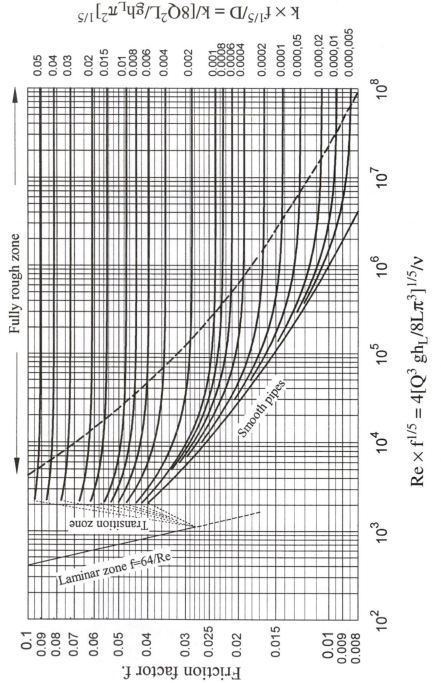

$$\text{Re} \times f^{1/5} = 4[Q^3 \, gh_L/8L\pi^3]^{1/5}/\nu$$

Figure 7.8. Moody diagram for Type 3 pipe flow problems.

is rewrite (7.3.2) in terms of new variables. For problems of type 2 we do this by arranging (7.3.1) so as to solve for the unknown Q in the form

$$Q = \sqrt{2h_L A^2 gD/fL} \tag{7.3.6}$$

and use this to eliminate Q in the Reynolds number. Writing the Reynolds number in terms of Q gives

$$Re = 4Q/\pi\nu D. \tag{7.3.7}$$

Using (7.3.6) to eliminate Q in this expression and rearranging, we have

$$Re\sqrt{f} = \sqrt{2h_L gD^3/\nu^2 L} = \text{REFHALF} \quad \text{(say)}, \tag{7.3.8}$$

where REFHALF contains only known quantities for problems of type 2. Thus $Re\sqrt{f} = \text{REFHALF}$ would be a better dimensionless quantity to use for the horizontal axis of the Moody diagram for these problems. Rewriting (7.3.2) using this, we have

$$1/\sqrt{f} = -0.86 \ln (k/3.7D + 2.51/\text{REFHALF}). \tag{7.3.2a}$$

Note that now the right-hand side of (7.3.2a) contains only known quantities, and a direct, noniterative solution for f from equation (7.3.2a) is possible.

A similar procedure works for problems of type 3. Solving (7.3.6) for D, we have

$$D = (8Q^2 fL/gh_L\pi^2)^{1/5}. \tag{7.3.9}$$

This time we must eliminate D in both the Reynolds number and in k/D. For the Reynolds number we eliminate D from (7.3.7) using (7.3.9) to obtain

$$Re = 4/\nu(8fL\pi^3/Q^3 gh_L)^{1/5}.$$

Upon rearranging we find that we are able to compute the quantity

$$Re\, f^{1/5} = 4(Q^3 gh_L/8L\pi^3)^{1/5}/\nu = \text{REFFIFTH} \quad \text{(say)}. \tag{7.3.10}$$

In a similar fashion, eliminating D from k/D gives

$$k/D = k/(8Q^2 fL/gh_L\pi^2)^{1/5}.$$

Rearranging this we find that we are also able to compute the quantity

$$kf^{1/5}/D = k/(8Q^2 L/gh_L\pi^2)^{1/5} = \text{KFFIFTH} \quad \text{(say)}. \tag{7.3.11}$$

Again the right-hand side consists of known quantities for this type of problem. Thus $Re\, f^{1/5}$ and $f^{1/5}k/D$ are quantities better suited to our calculations than the original set of dimensionless parameters. In terms of these variables, (7.3.2) can be rewritten as

$$1/\sqrt{f} = -0.86 \ln (\text{KFFIFTH}/3.7 f^{1/5} + 2.51/\text{REFFIFTH}\, f^{3/10}). \tag{7.3.2b}$$

However, an iterative solution of (7.3.2b) is necessary to find f.

To simplify the solution of type 2 and type 3 problems, equations (7.3.2a) and (7.3.2b) have been plotted in terms of their appropriate dimensionless parameters in Figures 7.7 and 7.8.

Good design practice in many cases recommends that the fluid velocities in conduits not be too high. For instance, for water in general service conditions, reasonable velocities are 4 to 10 ft/s, while in a city water supply system 7 ft/s is a reasonable velocity. When plastic pipe is used, 5 ft/s is generally the maximum speed to be used.

Before looking at specific examples of how to deal with each of these three cases, it is useful to first put the procedures involved in stepwise forms.

Problem type 1.   Given Q, L, D, k, ν, find h_L. This type of problem can be solved directly from the Moody diagram, Figure 7.6.

1. Compute the area $A = \pi D^2/4$ and the velocity $V = Q/A$.
2. Look up k in Table 7.1 and compute the relative roughness k/D and the Reynolds number $\text{Re} = VD/\nu$.
3. Either look up the friction factor f from the Moody diagram or use equation (7.3.2) to compute it. For the latter procedure, an iterative procedure such as Newton's method, described in Appendix C, must be used to solve the equation.
4. Compute the head loss from (7.3.1) as $h_L = f(L/D)V/2g = f(L/D)Q^2/2gA^2$.
5. Compute the power loss from $P = \rho g Q h_L$.

Problem type 2.   Given h_L, L, D, k, ν, find Q.

1. Compute the area by $A = \pi D^2/4$. Look up k in Table 7.1 and form the dimensionless roughness parameter k/D.
2. Form the dimensionless parameter

$$\text{Re}\sqrt{f} = \sqrt{2h_L g D^3/\nu^2 L} = \text{REFHALF.}$$

3. Use either Figure 7.7 or equation (7.3.2a) to find f.
4. Find the Reynolds number from $\text{Re} = \text{REFHALF}/\sqrt{f}$. Since D and ν are known you can now find V.
5. Compute the head loss from (7.3.1) as $h_L = f(L/D)V^2/2g = f(L/D)Q^2/2gA^2$.
6. Compute the power loss from $P = \rho g Q h_L$.

Problem type 3.   Given h_L, L, Q, k, ν, find D.

1. Look up k in Table 7.1.
2. Compute the dimensionless parameters $\text{Re}\, f^{1/5} = 4(Q^3 g h_L/8L\pi^3)^{1/5}/\nu = \text{REFFIFTH}$ and $kf^{1/5}/D = k/(8Q^2 L/g h_L \pi^2)^{1/5} = \text{KFFIFTH.}$
3. Solve for f using either Figure 7.8 or equation (7.3.2b). Note that if you use (7.3.2b) an iterative procedure is necessary.
4. Find the Reynolds number from $\text{Re} = \text{REFFIFTH}/f^{1/5}$. Since you know Q and ν you can now find D and V.
5. Compute the head loss from (7.3.1) as $h_L = f(L/D)V^2/2g = f(L/D)Q^2/2gA^2$.
6. Compute the power loss from $P = \rho g Q h_L$.

The transition zone in the Reynolds number range $2,300 \leq Re \leq 4,000$ is not well-defined. For computational purposes we draw a straight line on Figure 7.6, intercepting the laminar flow line at $Re = 2,300$ and the turbulent flow curves at $Re = 4,000$ and the appropriate value of k/D. This line is given by the formula

$$f = (64/2,300) \times (Re/2,300)^{\alpha}, \tag{7.3.12}$$

where

$$\alpha = \log (2,300 \times f_{4,000}/64)/\log (4,000/2,300). \tag{7.3.13}$$

Here $f_{4,000}$ is the friction factor at a Reynolds number of 4,000 for the appropriate k/D. This has the virtue of giving a ballpark approximation, as long as we appreciate that in this range of Reynolds numbers, local conditions can have large effects.

The following example problems were solved using Figures 7.6, 7.7, and 7.8.

Example 7.3.1. Type 1 pipe problem.
Given a horizontal galvanized iron pipe 100 ft long, 0.1 ft in diameter, and carrying 68°F water at an average velocity of 5 ft/s, what is the head loss and the power loss?

Sought: Head and power losses for the given conditions.

Given: From Appendix B the kinematic viscosity is 1.1×10^{-5} ft^2/s. The pipe diameter D is 0.1 ft, the pipe length is 100 ft, the average velocity is 15 ft/s, and the pipe roughness is, from Table 7.1, 0.0005 ft. The discharge is given by $Q = V \times$ area $= 0.1178$.

Assumptions: Uniform and established turbulent viscous flow.

Solution: The appropriate dimensionless variables are found from the given conditions to be

$$Re = 5 \times 0.1/1.1 \times 10^{-5} = 2.72 \times 10^5$$

and

$$k/D = 0.0005/0.1 = 0.005.$$

Interpolating from the Moody diagram, we find the friction factor to be

$$f = 0.0311.$$

Note also that the flow is in the fully turbulent range. From (7.3.1), the head loss is found to be

$$h_L = 0.0311(100/0.1)(15^2/2 \times 32.2) = 108.6 \text{ ft}.$$

The power loss is then

$$P = h_L \rho g Q = 108.6 \times 62.4(\pi \times 0.01 \times 15/4) = 798.3 \text{ lb-ft/s}$$

$$= 798.3/550 = 1.451 \text{ hp}.$$

Note that since the average velocity is over 10 ft/s, these conditions are not within usual design practice.

Example 7.3.2. Type 2 pipe problem

A cast iron pipe of length 200 m, 0.05 m in diameter carries a fluid of viscosity 10^{-6} m^2/s with a head loss of 40 m. Find the discharge.

Sought: The discharge Q.

Given: The pipe diameter is 0.05 m, the pipe length is 200 m, the fluid kinematic viscosity is 10^{-6} m^2/s, and the head loss is 40 m. For a cast iron pipe, $k = 0.25$ mm.

Assumptions: Uniform and established turbulent viscous flow.

Solution: The dimensionless parameters we can easily form are

$$k/D = 0.25 \times 10^{-3}/0.05 = 5 \times 10^{-3}$$

and

$$\mathrm{Re}\sqrt{f} = \sqrt{2} \times 40 \times 9.8 \times 0.05^3/10^{-12} \times 200 = 2.21 \times 10^4.$$

Use of Figure 7.7 gives $f = 0.0311$. We can thus complete calculation of the Reynolds number, giving

$$\mathrm{Re} = 2.21 \times 10^4/\sqrt{0.0311} = 1.253 \times 10^5.$$

From this,

$$V = \nu\,\mathrm{Re}/D = 10^{-6} \times 1.253 \times 10^5/0.05 = 2.51 \text{ m/s.}$$

The discharge is then

$$Q = V \times \text{area} = 0.00493 \text{ ft}^3/\text{s.}$$

Example 7.3.3. Type 3 pipe problem

Given a 200-ft pipe length with a head loss of 10 ft, what should be the diameter of a concrete pipe of roughness 0.001 ft in order to have a discharge of 10 ft^3/s for a fluid with kinematic viscosity 2×10^{-5} ft^2/s.

Sought: Pipe diameter for given flow conditions.

Given: Pipe length 100 ft, pipe roughness 0.001 ft, fluid kinematic viscosity 2×10^{-5} ft^2/s, discharge 10 ft^3/s.

Assumptions: Uniform and established turbulent viscous flow.

Solution: First form the two dimensionless parameters

$$\mathrm{Re}\, f^{1/5} = 4(Q^3 g h_L/8L\pi^3)^{1/5}/\nu = 4(10^3 \times 32.2 \times 10/8 \times 200 \times \pi^3)^{1/5}/2 \times 10^{-5}$$

$$= 2.907 \times 10^5$$

and

$$k f^{1/5}/D = k/(8Q^2 L/gh_L\pi^2)^{1/5} = 10^{-3}/(8 \times 10^2 \times 200/32.2 \times 10 \times \pi^2)^{1/5}$$

$= 0.4567 \times 10^{-3}$.

From Figure 7.8, $f = 0.0211$. Then

$$Re = 2.907 \times 10^5/0.0211^{1/5} = 6.29 \times 10^5$$

and

$$k/D = 0.4567 \times 10^{-3}/0.0211^{1/5} = 0.001.$$

Solving for D gives

$$D = 0.001/0.001 = 1 \text{ ft},$$

$$V = Re \times \nu/D = 6.29 \times 10^5 \times 2 \times 10^{-5}/1 = 12.6 \text{ ft/s}.$$

---

a. Minor losses

In the previous discussion we dealt solely with the friction loss that occurs in a pipe due to contact of the fluid with a pipe wall. In any real system, the pipe sections will be joined by connectors, elbows, unions, tees, valves, contractions, expansions, etc., all of which will contribute additional losses. These losses are termed ***minor losses*** and are usually written in the form

$$h_L = KV^2/2g. \tag{7.3.14}$$

The name minor loss does not necessarily convey the correct information about the numerical size of these losses, since they well might be larger in magnitude than the pipe friction losses.

Values of the resistance coefficient K for some popular fittings are shown in Tables 7.2, 7.3, and 7.4. Since fittings and valves usually do not satisfy geometric similarity from size to size, the value of K generally depends on the size (diameter) of the pipe. Also, there can be substantial differences in fittings from differing manufacturers.

It is sometimes useful in a given system to express K as an equivalent length of pipe that would incur the same losses as does the fitting alone. This length, introduced according to

$$K = fL_{\text{equivalent}}/D, \tag{7.3.15}$$

depends on the friction factor for our flow, and thereby depends on the flow itself.

Note that since K in Table 7.2 is assumed independent of velocity, and since therefore the form of equation (7.3.14) states that head loss depends on the square of the velocity, there is an implicit assumption that the flow is fully turbulent.

The values of K given for valves are for the valves in their fully open position. Fully closed valves have an infinite K.

Valves serve a number of different purposes and come in many configurations. Gate valves use a gate or a wedge that is moved at right angles to the flow direction. They

Table 7.2. K factors for pipe fittings

Fitting threaded (T) or flanged (F)	Nominal pipe diameter (in)									
	1/2	3/4	1	2	4	5	6	8–10	12–16	18–24
45° standard elbow (T)	0.43	0.40	0.37	0.30	0.27	0.26	0.24	0.22	0.21	0.19
90° standard elbow (T)	0.81	0.75	0.66	0.57	0.51	0.48	0.45	0.42	0.39	0.36
180° return bend (T)	1.35	1.25	1.10	0.95	0.85	0.80	0.75	0.70	0.65	0.60
Standard T—straight through flow (T)	0.54	0.50	0.44	0.38	0.34	0.32	0.30	0.28	0.26	0.24
Standard T—flow through branch (T)	1.62	1.50	1.32	1.14	1.02	0.96	0.90	0.84	0.78	0.72
Entrance—flush	0.021	0.020	0.018	0.015	0.013	0.013	0.012	0.011	0.010	9.4E-3
Globe valve—fully opened (T, F)	9.18	8.50	7.82	6.46	5.78	5.44	5.10	4.76	4.42	4.08
Gate valve—fully opened (T, F)	0.22	0.20	0.18	0.15	0.14	0.13	0.12	0.11	0.10	0.096
Angle valve—fully opened (T)	1.49	1.38	1.27	10.5	0.94	0.88	0.83	0.77	0.72	0.66
Swing check valve (T)	2.70	2.50	2.30	1.90	1.70	1.60	1.50	1.40	1.30	1.20
Swing check valve (F)	1.35	1.25	1.15	0.95	0.85	0.80	0.75	0.70	0.65	0.60
Lift check valve (T)	16.2	15.0	13.8	11.4	10.2	9.60	9.00	8.40	7.80	7.20

Data from Crane Corporation Technical Report 410.

364

Table 7.3. *K* factor for flanged 90° elbows. *R* is the bend radius measured to fitting center-line; *D* is the actual pipe inner diameter

	K									
	Nominal pipe diameter (in)									
R/D	1/2	3/4	1	2	4	5	6	8–10	12–16	18–24
1	0.54	0.50	0.46	0.38	0.34	0.32	0.30	0.28	0.26	0.24
2	0.32	0.30	0.28	0.22	0.20	0.19	0.18	0.17	0.16	0.14
3	0.32	0.30	0.28	0.22	0.20	0.19	0.18	0.17	0.16	0.14
4	0.38	0.35	0.32	0.27	0.24	0.22	0.21	0.20	0.18	0.17
6	0.46	0.43	0.39	0.32	0.29	0.27	0.26	0.224	0.22	0.20
8	0.65	0.60	0.55	0.46	0.41	0.38	0.36	0.034	0.31	0.29
10	0.81	0.75	0.69	0.57	0.51	0.48	0.45	0.42	0.39	0.36
12	0.92	0.85	0.78	0.65	0.58	0.54	0.51	0.48	0.44	0.41
14	1.03	0.95	0.87	0.72	0.65	0.61	0.57	0.53	0.49	0.46
16	1.13	1.05	0.97	0.80	0.71	0.67	0.63	0.59	0.55	0.50
18	1.24	1.15	1.06	0.87	0.78	0.74	0.69	0.64	0.60	0.55
20	1.35	1.25	1.15	0.95	0.85	0.80	0.75	0.70	0.65	0.60

Data from Crane Corporation Technical Report 410.

are usually used in either a fully open or fully closed position since they are not recommended as a throttling valve. Ball valves are also used for closure purposes.

Globe valves get their name from their spherical body shape. They have a baffle separating the interior of the valve into two parts. A handwheel moves a plug perpendicular to the flow direction to open and close the valve. If the plug is moved in a nonperpendicular direction it is called an "angle valve." Both globe and angle valves can be used to throttle the flow. Other throttling valves are needle valves and butterfly valves.

Check valves are used to ensure that flow occurs in only one direction. The flow causes a hinged gate (swing check valve) or a spring-mounted gate or ball (lift check valve) to open when the pressure difference across the valve reaches a given level. Some lift check valves rely on gravity rather than on a spring, and must be positioned so that the gate moves vertically.

Table 7.4. *K* for flush entrance

Entrance radius/*D*	0.0	0.02	0.04	0.06	0.1	≥0.15
K	0.50	0.28	0.24	0.15	0.09	0.04

Example 7.3.4. Minor losses in pipe flow

A galvanized iron pipeline consists of 200 ft of 1-in-diameter pipe with three 90° elbows, seven couplings, and one fully opened globe valve. The mean velocity is 9 ft/s, the kinematic viscosity is 3×10^{-5} ft²/s, and the specific gravity is 0.9. What is the power loss?

Sought: Power loss due to fluid friction and minor losses for the given conditions.

Given: Pipe diameter (1 in), pipe length (200 ft), pipe roughness (0.0005 ft), mean velocity (9 ft/s), specific density (0.9 × 62.4 lb/ft³), 3 90° elbows $(K = 30 \times f_T)$, 7 couplings $(K = 0.08 \times f_T)$, 1 fully opened globe valve $(K = 340 \times f_T)$.

Assumptions: Uniform and established turbulent viscous flow.

Solution: The Reynolds number is

$$\text{Re} = (1/12) \times 9/3 \times 10^{-5} = 25{,}000$$

and the dimensionless roughness is

$$k/D = 0.0005/(1/12) = 0.006.$$

From the Moody diagram of Figure 7.6, f_T, the friction factor at complete turbulence, is 0.032. From Table 7.2 and the Darcy-Weisbach equation,

$$h_L = \left(f\frac{L}{D} + \Sigma K \right)\frac{V^2}{2g} = [f \times 200 \times 12 + 3 \times (30 \times 0.032) + 7 \times (0.08 \times 0.032)$$

$$+ 340 \times 0.032] \times 9^2/(2 \times 32.2) = 3{,}019f + 17.33.$$

From the Moody diagram of Figure 7.6 for Re = 25,000, k/D = 0.006, we find that f = 0.0355. Thus the head loss is 124.5 ft, with corresponding power loss

$$P = \gamma Q h_L = 0.9 \times 62.4 \times (9 \times \pi/4 \times 144) \times 124.5 = 343.2 \text{ ft-lb/s} = 0.62 \text{ hp.}$$

---

b. Multiple pipe circuits

In a piping system of any size, there will usually be a complex circuit of piping involving branching, parallel lines, and the like. The effect is analogous to an electric circuit, but more difficult to analyze since the problem is inherently nonlinear in both velocity and friction factor. At a node (connection point) where branching of the piping takes place, pressure (corresponding to voltage in our analogy) will be the same for all the pipes connected at that node. The net incoming and departing discharges (corresponding to current in our analogy) at each node must sum to zero in order to satisfy continuity. A flow chart for the procedure is much as for the previous three systems, but it is complicated further by the fact of multiple discharges. An adaptation of the flow diagram of type 2 problems might be as follows:

Type 2 problems for piping circuits.   Given h_L, L, D, k, ν in each pipe and a total Q, find Q in each pipe. An iterative solution is needed. The following procedure is to be complied with at a node:

1. Compute the area for each pipe connecting to this node by $A_i = \pi D_i^2/4$ and form the dimensionless roughness parameters k_i/D_i.
2. Make initial guesses for the f_i. Your guesses should lie somewhere between the value of f for fully developed turbulent flow (corresponding to the dimensionless roughness parameter found in step 1) and 0.1.
3. Rearrange (7.3.1) as

$$Q = A\sqrt{2gh_L/[f(L/D) + \Sigma K]}$$

and compute an estimated Q_i from this expression for each pipe using the estimated (guessed) f_i.
4. Correct the Q_i found in the previous step so that global continuity is satisfied at this node. This gives

$$Q_{i \text{ corrected}} = Q_{\text{total}} \, Q_i / \sum_{j=1}^{n} Q_j$$

5. Using the Q_i's found in step 4, compute the estimated velocities V_i from $V_i = Q_i/A_i$ and the estimated Reynolds numbers from $\text{Re}_i = V_i D_i/\nu$.
6. Look up the corrected friction factors f_i from the Moody diagram using the appropriate relative roughness parameters and the estimated Reynolds numbers from step 6.
7. If the friction factors found in step 6 are sufficiently close to your previous estimated f_i's (close enough so that the discharges found in step 3 do not differ substantially from the discharges in step 4), then the process has reached a satisfactory conclusion. Otherwise go to step 3, using the friction factor from step 6.

We will not delve into the details of the analysis of such a system since the complexity and number of iterations increases rapidly with the number of pipes and nodes, but the principles involved should be clear to anyone who has analyzed simple electrical resistance circuits. Engineering offices that perform such calculations now frequently rely on commercial computer codes for the design process.

Example 7.3.5. Piping circuit
A 6-in-diameter galvanized iron pipe has a total flow of 90 gpm of water at 75°F. It in turn feeds three 3-in-diameter galvanized iron pipes of lengths 210, 250, and 300 ft. The minor losses in the three pipes correspond to K's of 10, 23, and 30, and the head loss over the length of the three pipes is 1.3 ft of water. Each of the three flows discharges into the atmosphere with a negligible kinetic energy. Find the flows in each pipe.

Sought: Flow rates in each of the three legs of a piping system.
Given: $Q_{\text{total}} = 90$ gpm $= (90/7.48)/60 = 0.2005$ ft^3/s, $k = 0.0005$ ft, $k/D = 0.0005/(3/12) = 0.002$, $\nu = 10^{-5}$ ft^2/s.

Assumptions: No elevation change, no kinetic energy at the exit, no energy lost in the transition from a 6-in pipe to a 3-in pipe. Uniform and established turbulent viscous flow.

Solution: The steps in the solution corresponding to the above are:

1. $A = \pi D^2/4 = \pi(0.25)^2/4 = 0.049087 \text{ ft}^2$.

2. Guess all $f_i = 0.023$. From the Moody diagram this is seen to be the fully developed turbulent value for pipes of this relative roughness.

3. Since

$$h_{Li} = [f_i(L_i/D_i) + K_i]Q_i^2/2gA_i^2,$$

$$Q_i = A_i\sqrt{2gh_{Li}/[f_i(L_i/D_i) + K_i]}.$$

Then

$$Q_1 = A_1\sqrt{2gh_{L1}/[f_1(L_1/D_1) + K_1]} = 0.049087\sqrt{2 \times 32.2h_{L1}/[0.023(210/0.25) + 10]}$$

$$= 0.0829 \text{ ft}^3/\text{s},$$

$$Q_2 = A_2\sqrt{2gh_{L2}/[f_2(L_2/D_2) + K_2]}$$

$$= 0.049087\sqrt{2 \times 32.2h_{L2}/[0.023(250/0.5) + 23]} = 0.0662 \text{ ft}^3/\text{s},$$

$$Q_3 = A_3\sqrt{2gh_{L3}/[f_3(L_3/D_3) + K_3]}.$$

$$= 0.049087\sqrt{2 \times 32.2h_{L3}/[0.023(300/0.5) + 30]} = 0.0592 \text{ ft}^3/\text{s}.$$

The sum of Q_1, Q_2, and Q_3 is 0.2083 ft^3/s, which is slightly larger than the 0.2005 ft^3/s supplied.

4. Correct the discharges according to $Q_i' = Q_{\text{total}}Q_i/\Sigma Q_i$, obtaining

$$Q_1' = 0.798 \text{ ft}^3/\text{s}, \qquad Q_2' = 0.0637 \text{ ft}^3/\text{s}, \qquad Q_3' = 0.0570 \text{ ft}^3/\text{s}.$$

5. Compute the velocities from $V_i = Q_i/A_i$. Then

$$V_1 = 1.626 \text{ ft/s}, \qquad V_2 = 1.298 \text{ ft/s}, \qquad V_3 = 1.16 \text{ ft/s}.$$

Compute the Reynolds numbers from $\text{Re}_i = V_iD_i/\nu$. Thus

$$\text{Re}_1 = 40,660, \qquad \text{Re}_2 = 32,462, \qquad \text{Re}_3 = 29,010.$$

6. From the Moody diagram the corrected friction factors are

$$f_1 = 0.027, \qquad f_2 = 0.028, \qquad f_3 = 0.028.$$

7. The error in the total Q as computed in step 3 is 3.85%.

Further iteration is needed to achieve acceptable accuracy. Go back to step 3.

3. Computation with the new values of f gives

$$Q_1 = 0.0785 \text{ ft}^3/\text{s}, \qquad Q_2 = 0.0629 \text{ ft}^3/\text{s}, \qquad Q_3 = 0.0563 \text{ ft}^3/\text{s}.$$

The total discharge corresponding to these is 0.1977 ft^3/s.

4. The corrected Q's are

$Q_1 = 0.0797$ ft³/s, $Q_2 = 0.0638$ ft³/s, $Q_3 = 0.0571$ ft³/s.

5. The corrected velocities are

$V_1 = 1.623$ ft/s, $V_2 = 1.299$ ft/s, $V_3 = 1.163$ ft/s,

and the corrected Reynolds numbers are

$Re_1 = 40{,}572$, $Re_2 = 32{,}477$, $Re_3 = 29{,}083$.

6. The friction factors corresponding to the new Reynolds numbers are essentially unchanged.
7. The error in the discharge is 1.42%. The discharges then are reasonably close to convergence, and we will end the iteration here.

c. Hydraulic and energy grade lines

In analyzing a flow system it is often helpful to have a graphical depiction of what is happening. Two such graphical tools are the ***hydraulic grade line*** and the ***energy grade line***. The hydraulic grade line is a plot of $p/\rho g + h$ along a pipeline or channel, and represents the height to which the liquid would rise in a vertical tube (piezometer) inserted in the pipeline. The energy grade line is a plot of the total energy $p/\rho g + h + v^2/2g$ along the pipeline. Its slope represents the rate at which energy head is being lost.

4. Turbulent Boundary Layer Flows

a. Fully established turbulent flow in a smooth pipe

In Chapter 6 we discussed laminar flow boundary layers and derived the momentum-integral expression to find the details of the flow. Turbulent boundary layers are usually much thicker than laminar boundary layers, and their behavior is also much more complicated. Analytic solutions for the flow are not possible. Instead, we use a combination of experimental results and theory to arrive at our results.

In laminar boundary layers we used U, the velocity at the outer edge of the boundary layer, and y, the distance normal to the wall, as appropriate variables. For turbulent boundary layers, these are suitable descriptors at the outer edge of the boundary layer, but not near the wall. Instead, near the wall we define an appropriate velocity u^* called the shear velocity, where

$$u^* \equiv \sqrt{\tau_{\text{wall}}/\rho}. \tag{7.4.1}$$

Experience and experiments have shown that an appropriate dimensionless coordinate near the wall is

$$y^+ = u^* y/\nu, \tag{7.4.2}$$

where y is measured from the wall. For a pipe of radius R and with r being the distance of a position from the pipe centerline, we have

$$y = R - r. \tag{7.4.3}$$

Since we have taken the flow to be fully established, the boundary layer fills the pipe, hence the boundary layer thickness δ equals the pipe radius R.

Measurements show that the turbulent boundary layer is made up of three distinct zones:

Viscous sublayer	$0 \leq y^+ \leq 5$, where the shear stress due to the average velocity dominates,
Buffer zone	$5 < y^+ < 70$, where the shear stress due to the average velocity and Reynolds stresses are of equal importance,
Turbulent logarithmic zone	$70 > y^+ > u^*R/\nu$, where the shear stress due to the Reynolds stresses dominate the shear stress due to the average velocity.

In the **viscous sublayer** the Reynolds stress components are negligible and the mean velocity gradient is constant. It follows that the time-averaged velocity component parallel to the pipe centerline varies with position according to the relation

$$\bar{u} = T_{\text{wall}} y/\mu = u^* y^+. \tag{7.4.4}$$

The laminar sublayer is generally extremely thin, and obtaining reliable measurements of quantities in the laminar sublayer is a formidable challenge to experimenters.

The **buffer zone** is a region dividing the laminar sublayer from the logarithmic zone. It has no particular distinguishing characteristics, other than the fact that Reynolds stresses have now grown to the point where they are of the order of the stresses due to the mean velocity gradient. A purely empirical formula that fits experimental data, that agrees with (7.4.4) at $y^+ = 5$, and that agrees with (7.4.8) at $y^+ = 30$, is

$$\bar{u} = u^*[(2/\kappa) \ln y^+ - 3.05] = u^*[(4.6/\kappa) \log y^+ - 3.05]. \tag{7.4.5}$$

Here κ is **von Kármán's universal constant**, and has a value based on experimental measurements in the range 0.4–0.41.[2]

In the **logarithmic region**, also called the near-wall region, the mixing length has been found to vary linearly with distance from the wall. In this region the Reynolds shear stress dominates the shear stress, and the shear stress is approximately equal to the wall shear stress. That is, we have

$$l = \kappa y \tag{7.4.6}$$

---

[2]For the purposes of mathematical manipulation, the natural logarithm (ln) is most convenient. However, for plotting experimental results, logarithms to the base 10 (log) are more convenient. Recall that $\log F = 0.434 \ln F$.

and

$$\bar{\tau} + \mu \, d\bar{u}/dy - \overline{\rho u'v'} \approx \rho \kappa^2 y^2 \, |du/dy| \, \overline{du/dy} \approx \tau_{wall} \tag{7.4.7}$$

in this region. Rewriting this by solving for $d\bar{u}/dy$, we find that

$$d\bar{u}/dy = \sqrt{\tau_{wall}/\rho} \, \kappa y = u^*/\kappa y.$$

Integration of this gives

$$\bar{u} = (u^*/\kappa) \ln y + C_1.$$

L. Prandtl found that an empirical fit with experimental data for smooth pipes gave a value for C_1 of approximately

$$C_1 = 5.5u^* + (u^*/\kappa) \ln u^*/\nu.$$

Putting this into the expression for $\bar{u}$, we have

$$\bar{u}/u^* = (1/\kappa) \ln y^+ + 5.5 = (2.3/\kappa) \log y^+ + 5.5. \tag{7.4.8}$$

Evaluating this at the centerline $y = R$ we find that the velocity on the pipe centerline is given by

$$U_{centerline}/u^* = (1/\kappa) \ln Ru^*/\nu + 5.5 = (2.3/\kappa) \log Ru^*/\nu + 5.5. \tag{7.4.9}$$

The velocity profile given by (7.4.4), (7.4.5), and (7.4.8) is shown in Figure 7.9.

The above results involve quantities such as τ_{wall} and $U_{centerline}$ that are inconvenient to measure directly. We next connect them with more easily measured and familiar variables.

In Chapter 6 we found that for a horizontal pipe the wall shear and pressure gradient are related by

$$\tau_{wall} = -(\partial p/\partial x)R/2 = \Delta p \, D/4RL, \tag{6.3.8}$$

where L is the length of pipe over which Δp is measured. As derived earlier in this chapter, the Darcy-Weisbach equation for a horizontal pipe is

$$h_L = -(\partial p/\partial x)L/\rho g = \Delta p/\rho g L = f(L/D)U_{ave}^2/2g, \tag{7.3.1}$$

where U_{ave} is the velocity averaged over the pipe diameter. In terms of the wall shear, this gives

$$-(\partial p/\partial z)L = 2\tau_{wall}L/\rho g R = 4u^{*2}L/gD = f(L/D)U_{ave}^2/2g. \tag{7.4.10}$$

Solving this equation for the friction factor f in terms of u^* gives

$$f = 8(u^*/U_{ave})^2. \tag{7.4.11}$$

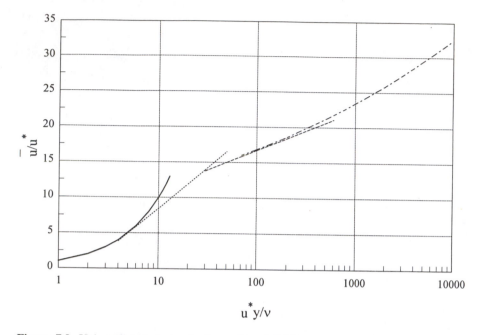

Figure 7.9. Universal turbulent velocity profile. Solid line—equation 7.4.4; short dashed line—equation 7.4.5; long dashed line—7.4.8; dot-dashed line—equation 7.4.24.

Equation (6.3.8) relates the wall shear to the easily measured pressure difference over a length of pipe L, and (7.3.1) relates this in turn to the velocity averaged over cross-sectional area.

To relate the centerline velocity to the mean velocity, we use still another experimental result from Prandtl. He found that in the central region of the pipe away from the wall a good fit to experimental data is

$$\bar{u} = U_{\text{centerline}} + (u^*/\kappa) \ln y/R = U_{\text{centerline}} + (2.3u^*/\kappa) \log y/R. \tag{7.4.12}$$

If we make the assumption that the sublayer and buffer zones are so thin that the above velocity profile can be used across the entire pipe senction, computation of the average velocity gives

$$U_{\text{ave}} = \int_0^R \bar{u}\, 2\pi r\, dr/\pi R^2 = 2\int_0^R [U_{\text{centerline}} + (u^*/\kappa) \ln y/R]\, r\, dr/R^2$$

$$= U_{\text{centerline}} - 1.5u^*/\kappa. \tag{7.4.13}$$

From (7.4.9) we have

$$U_{\text{centerline}}/u^* = (1/\kappa) \ln Ru^*/\nu + 5.5.$$

Elimination of $U_{\text{centerline}}$ between these two results gives

$$U_{ave} = u^*[(1/\kappa) \ln Ru^*/\nu + 1.75],$$ (7.4.14)

where we have used $\kappa = 0.4$. Combining this with (7.4.14) and using $\kappa = 0.4$ gives

$$1/f^{1/2} = U_{ave}/8^{1/2} u^* = (1/8^{1/2})[(1/\kappa) \ln Ru^*/\nu + 1.75]$$

$$= 0.884 \ln (Ru^*/\nu) + 0.619 = 0.884 \ln (DU_{ave}/2 \times 8^{1/2}\nu) + 0.619$$

$$= 0.884 \ln (Re_D f^{1/2}/2 \times 8^{1/2}) + 0.619 = 0.884 \ln (Re_D f^{1/2}) - 0.91$$

$$= 2.036 \log (Re_D f^{1/2}) - 0.91,$$ (7.4.15)

where Re_D is the Reynolds number based on the diameter and average velocity. Comparing this with (7.3.2), Colebrook's fit to Moody's data, shows that in the limit as the roughness goes to zero, (7.3.2) gives

$$1/f^{1/2} = 1.98 \log (Re_D f^{1/2}) - 0.79.$$ (7.4.16)

Therefore Colebrook's strictly empirical results derived solely from head loss/pressure gradient data and the present analysis based on velocity measurements and some theory give reasonable agreement.

The forms we have used for the mean velocities in the various regions come from a combination of theory, experiment, and fitting correlations to experimental data. Different experimenters have arrived at slightly different values for the various constants.

b. Momentum integral formulation

The velocity distribution given in piecewise manner by (7.4.4), (7.4.5), (7.4.8), and (7.4.12), while giving a fairly detailed and accurate picture of the flow, are not particularly convenient for computation such as might be required by the momentum-integral method. For such purposes J. Nikuradse proposed a polynomial approximation in the form

$$\bar{u}/U_{centerline} = (y/\delta)^{1/n},$$ (7.4.17)

where n varies with the Reynolds number. For pipe flow, $\delta = R$. He assumed that this representation of the time-averaged velocity held throughout the boundary layer. The average velocity U_{ave} is then given by

$$U_{ave} = \int_0^R \bar{u}\, 2\pi r\, dr/\pi R^2 = 2\, U_{centerline} \int_R^0 (y/\delta)^{1/n}(R - y)\,(-dy)$$

$$= 2n^2 U_{centerline}/(1 + n)(1 + 2n).$$ (7.4.18)

Fitting his experimental results with the velocity profiles given by (7.4.17), Nikuradse found the correlation between Re_D and n as summarized in the first two rows of Table

7.5. The third row was obtained from (7.4.18), and the fourth and fifth rows were obtained using (7.4.9).

Note that the power law velocity profile (7.4.17) cannot be used to determine wall shear, since its derivative is infinite at the wall. To overcome this, Blasius in 1913 surveyed all the existing experimental data for turbulent flow in smooth pipes and found that they could be represented by the empirical correlation

$$f = 0.2661(\nu/RU_{ave})^{1/4}. \tag{7.4.19}$$

If (7.4.19) is combined with (7.4.11), we have

$$f = 8\,(u^*/U_{ave})^2 = 0.2661(\nu/RU_{ave})^{1/4}.$$

Solving this for U_{ave}/u^*, we have

$$U_{ave}/u^* = (8/0.2661)^{4/7}\,(Ru^*/\nu)^{1/7} = 6.99(Ru^*/\nu)^{1/7}. \tag{7.4.20}$$

To extend this further, Prandtl in 1921 noticed that with a few changes this last result could be used to extend pipe flow results to flat plates. He noted that for a power law of 1/7, the ratio of average velocity to centerline velocity is approximately 0.8 (0.817 in Table 7.5). Rewriting (7.4.20) in terms of the centerline velocity gives

$$U_{centerline}/u^* = 8.74(Ru^*/\nu)^{1/7}. \tag{7.4.21}$$

From this, the friction velocity u^* can be found as

$$u^* = U_{centerline}^{7/8}(\nu/R)^{1/8}/8.74^{7/8}$$

$$= 0.150U_{centerline}(\nu/U_{centerline}R)^{1/8}. \tag{7.4.22}$$

This gives for the wall shear stress

$$\tau_{wall} = \rho u^{*2} = \rho(0.150)^2 U_{centerline}^2(\nu/U_{centerline}R)^{1/4}$$

$$= 0.0225\rho U_{centerline}^2(\nu/U_{centerline}R)^{1/4}. \tag{7.4.23}$$

Prandtl then proposed that equations (7.4.21) and (7.4.23) were valid with the centerline velocity replaced by the velocity at the outer edge of the boundary layer,

Table 7.5. Power law correlation of velocity profile with smooth pipe data

Re_D	4×10^3	2.3×10^4	1.1×10^5	1.1×10^6	2×10^6	3.24×10^6
n	6	6.6	7	8.8	10	10
$U_{ave}/U_{centerline}$	0.791	0.807	0.817	0.850	0.865	0.865
$u^*/U_{centerline}$	0.0559	0.0461	0.0396	0.0329	0.0315	0.0304
$\tau_{wall}/\rho U_{ave}^2$	0.00500	0.00325	0.00236	0.00150	0.00132	0.00123

and that it held at any point within the boundary layer. With $U_{\text{centerline}}$ replaced by the local velocity u, and R by the distance y from the wall, (7.4.21) becomes

$$u/u^* = 8.74\,(yu^*/\nu)^{1/7}. \qquad (7.4.24)$$

Similarly (7.4.23) becomes

$$\tau_{\text{wall}} = 0.0225\rho U^2(\nu/U\delta)^{1/4}. \qquad (7.4.25)$$

This result does in fact give very good results for Reynolds numbers less than 100,000. For higher Reynolds numbers, smaller values of n give better results.

c. Turbulent flow past smooth flat plates with zero pressure gradient

The momentum integral form of the boundary layer equations developed in the previous chapter [equation (6.5.3)] can also be used in turbulent flow. The velocity distribution given in piecewise manner by (7.4.4), (7.4.5), and (7.4.8) together with the wall shear as given by (7.4.21) could be used as good approximations for flat plates as well as pipe flow. The only change needed would be the replacement of the pipe radius with the boundary layer thickness δ, which is now an unknown.

We will instead use (7.4.17) as an approximation of the turbulent velocity profile that is valid over the logarithmic portions of the boundary layer on a flat plate (which includes most of the boundary layer except very near the wall) with zero pressure gradient. Then

$$\delta_D = \int_0^\delta (U - \bar{u})\,dy/(n+1) \qquad (7.4.26)$$

and

$$\delta_M = \int_0^\delta \bar{u}(U - \bar{u})\,dy/U^2 = n\delta/(n+1)(n+2). \qquad (7.4.27)$$

We wish to use this with (6.5.10), namely,

$$\frac{d(\rho\delta_M U^2)}{dx} + \rho\delta_D\,U\,\frac{dU}{dx} = \tau_{\text{wall}}. \qquad (7.4.28)$$

Since U is a constant by virtue of the zero pressure gradient, (7.4.28) along with (7.4.21) and (7.4.23) gives

$$\tau_{\text{wall}} = \rho U^2\,\frac{d(\delta_M)}{dx} = \frac{n\rho U^2}{(n+1)(n+2)}\,\frac{d\delta}{dx}. \qquad (7.4.29)$$

From (7.4.25) we have

$$\tau_{\text{wall}} = 0.0225\rho U^2(\nu/U\delta)^{1/4}.$$

Combining this and using $n = 7$ [recall that equation (7.4.25) was developed for $n = 7$] gives

$$\tau_{\text{wall}} = \frac{7\rho U^2}{72}\frac{d\delta}{dx} = 0.0225\rho U^2 (\nu/U\delta)^{1/4}.$$

Canceling the density and the velocity U and rearranging so as to separate variables gives

$$\delta^{1/4}d\delta = \frac{72(0.0225)}{7}(\nu/U)^{1/4}\,dx = 0.2314\,(\nu/U)^{1/4}\,dx.$$

Integration gives

$$0.8\delta^{5/4} = 0.2314(\nu/U)^{1/4}\,x,$$

or finally,

$$\delta = 0.371x\,(\nu/Ux)^{1/5}. \tag{7.4.30}$$

Putting this in (7.4.25) gives

$$\tau_{\text{wall}} = 0.0225\rho U^2 (\nu/U0.371\,x\,(\nu/Ux)^{1/5})^{1/4}$$

$$= 0.0288\,\rho U^2\,(\nu/Ux)^{1/5}. \tag{7.4.31}$$

Assuming that this relation holds from the leading edge of the plate, the drag force on a plate of length L and width W is

$$F_{\text{drag}} = W\int_0^L \tau_{\text{wall}}dx = W\,0.0288\,\rho U^2\,1.25(\nu/U)^{1/5}\,L^{4/5}$$

$$= WLC_{\text{drag}}\rho U^2/2, \tag{7.4.32}$$

where the drag coefficient C_{drag} is given by

$$C_{\text{drag}} = 0.072\,(\nu/UL)^{1/5}. \tag{7.4.33}$$

If the 0.072 is replaced by 0.074, this is in good agreement with experimental results in the range $5 \times 10^5 < (\nu/UL)^{1/5} < 10^7$ for plates whose boundary layers start as turbulent from the leading edge.

For the drag coefficient for a laminar boundary layer on a flat plate, we found in Chapter 6, (6.5.13), that the drag coefficient grows over a length L of the plate measured from the leading edge according to $\text{Re}_L^{-1/2}$, where $\text{Re}_L = UL/\nu$, and the boundary layer thickness grows as $x^{1/2}$. For a turbulent flow, (7.4.33) states that the turbulent drag coefficient grows as $\text{Re}_L^{-1/5}$, and (7.4.30) shows that the boundary layer grows as $x^{4/5}$.

The above analysis completely neglects the laminar portion of the boundary layer near the leading edge of the plate. An empirical formula that includes both the effects of a laminar boundary layer near the leading edge of the plate plus those of the following turbulent boundary layer after transition is

$$C_{\text{drag}} = 0.427/(\log \text{Re}_L - 0.407)^{2.64} - A/\text{Re}_L, \qquad (7.4.34)$$

where the constant A depends on what value we choose for the critical value of the Reynolds number at which the turbulent boundary layer starts. A reasonable approximation due to Schultz-Grunow and Prandtl is

$$A = 1{,}060 + 340(10^{-5}\,\text{Re}_{\text{critical}} - 3), \qquad (7.4.35)$$

with the critical Reynolds number lying in the range $3 \times 10^5 \geq \text{Re}_{\text{critical}} \geq 10^6$.

Example 7.4.1. Flat plate boundary layer

A flat plate 2 ft wide is in a wind tunnel with an air speed of 500 ft/s. The air kinematic viscosity is 2×10^{-4} ft^2/s and its density is 10^{-3} slug/ft^3. At a distance 4 ft from the leading edge of the plate, compute the thicknesses of the laminar sublayer, the logarithmic zones, and the boundary layer. Find the velocity at the edges of these regions and the drag on the plate.

Sought: Velocity at several points in the turbulent boundary layer on a flat plate.

Given: Air kinematic viscosity (2×10^{-4} ft^2/s), air density (10^{-3} slug/ft^3), air velocity at outer edge of boundary layer (500 ft/s).

Assumptions: Assume the critical Reynolds number to be 5×10^5, since this is somewhere in the middle of the range where transition is known to occur. Then transition to turbulence occurs at a distance of

$$x = \nu\,\text{Re}/U = 2 \times 10^{-4} \times 5 \times 10^5/500 = 0.2 \text{ ft}$$

from the leading edge of the plate.

Solution: From (7.4.31), since at 4 ft $\text{Re}_x = Ux/\nu = 500 \times 4/2 \times 10^{-4} = 10^7$,

$$\tau_{\text{wall}}/\rho U^2 = 0.0288\,(\nu/Ux)^{1/5} = 0.0288\,(2 \times 10^{-4}/500 \times 4)^{1/5} = 1.147 \times 10^{-3}.$$

From (7.4.1),

$$u^* = \sqrt{\tau_{\text{wall}}/\rho} = 1.147 \times 10^{-3}\,U = 16.93 \text{ ft/s}.$$

The thickness of the viscous sublayer is $y^+ = 4$, or in dimensional terms,

$$y_{\text{viscous sublayer}} = 4\nu/u^* = 4 \times 2 \times 10^{-4}/16.93 = 4.73 \times 10^{-5} \text{ ft}.$$

The mean velocity at the outer edge of the viscous sublayer is

$$u = 4u^* = 67.72 \text{ ft/s}.$$

The logarithmic zone starts at y^+ between 30 and 70, which is equivalent to y between 3.54×10^{-4} and 8.27×10^{-4} ft. From (7.4.30), the boundary layer thickness at $x = 4$ is given by

$$\delta = 0.371\, x\, (\nu/Ux)^{1/5} = 0.371 \times 4 \times (2 \times 10^{-4}/16.93 \times 4)^{1/5} = 0.116 \text{ ft}.$$

At 15% of the boundary layer thickness $(1.74 \times 10^{-2}$ ft$)$ the velocity is, from (7.4.8),

$$\bar{u} = u^*[5.5 + (2.3/\kappa) \log (u^*y/\nu)]$$

$$= 16.93[5.5 + (2.3/0.4) \log (16.93 \times 1.74 \times 10^{-2}/2 \times 10^{-4})] = 401.5 \text{ ft/s}.$$

For a critical Reynolds number of 5×10^5, A is found from (7.4.35) to be

$$A = 1{,}060 + 340 \times (10^{-5}\, \text{Re}_{\text{critical}} - 3)$$

$$= 1{,}060 + 340 \times (10^{-5} \times 5 \times 10^5 - 3) = 1{,}060 + 340 \times (5 - 3) = 1{,}740.$$

From (7.4.34) the drag coefficient for this 4-ft length is

$$C_{\text{drag}} = 0.427/(\log \text{Re}_L - 0.407)^{2.64} - A/\text{Re}_L$$

$$= 0.427/(\log 10^7 - 0.407)^{2.64} - 1{,}740/10^7 = 2.764 \times 10^{-3}.$$

The force on this plate with width 2 ft is then

$$F_D = C_D \rho U^2 Lw/2 = 2.764 \times 10^{-3} \times 10^{-3} \times 500^2 \times 4 \times 2/2 = 2.764 \text{ lb}.$$

If we had assumed that the flow was turbulent starting at the leading edge, the drag would be from (7.4.33)

$$C_{\text{drag}} = 0.072(\nu/UL)^{1/5} = 0.072(2 \times 10^{-4}/16.93 \times 4)^{1/5} = 5.64 \times 10^{-3},$$

and thus

$$F_D = C_D \rho U^2 Lw/2 = 5.64 \times 10^{-3} \times 10^{-3} \times 500^2 \times 4 \times 2/2 = 5.64 \text{ lb}.$$

An interesting point of this problem is that at the outer edge of the laminar sublayer, which makes up only 0.04% of the boundary layer thickness, the mean velocity has achieved 13.5% of its final value. At 15% of the boundary layer thickness it is already at 80% of its final value.

Example 7.4.2. Flat plate boundary layer

Using the data of Example 7.4.1 together with the momentum integral method, find the boundary layer thickness, displacement thickness, momentum thickness, and u^*.

Sought: Boundary layer properties on a flat plate using the momentum integral method.
Given: Air kinematic viscosity $(2 \times 10^{-4}$ ft^2/s$)$, air density $(10^{-3}$ slug/ft$^3)$, air velocity at outer edge of boundary layer (500 ft/s).
Assumptions: Turbulent viscous flow.

Solution: Using $n = 7$ we have from (7.4.30), (7.4.26), and (7.4.27),

$$\delta = 0.371x/(\nu/Ux)^{1/5} = 0.371x/(2 \times 10^4/500x)^{1/5} = 0.0195x^{4/5},$$

$$\delta_D = \delta/8 = 0.00244x^{4/5},$$

$$\delta_M = 7\delta/72 = 0.00189x^{4/5}.$$

From (7.4.25),

$$\tau_{\text{wall}}/\rho U^2 = 0.0288 \, (\nu/U\delta)^{1/4} = 0.0288 \times (2 \times 10^{-4}/500 \times 0.0195x^{4/5})^{1/4}$$

$$= 1.51 \times 10^{-3}x^{-1/5},$$

$$\tau_{\text{wall}} = 0.378x^{-1/5} \, \rho U^2 = 1.51 \times 10^{-3} \times x^{-1/5} \times 10^{-3} \times 500^2 = 0.378x^{-1/5},$$

$$u^* = \sqrt{\tau_{\text{wall}}/\rho} = \sqrt{0.378x^{-1/5}/10^{-3}} = 19.45x^{-1/10}.$$

When we carry out numerical computations of turbulent flows, the mean velocity in the boundary layer could be found by direct numerical integration of the momentum equations if the Reynolds stresses were known. Therefore a means for modeling the Reynolds stresses is necessary before this integration can be carried out. A simple and frequently used device is to use the Prandtl mixing length concept of (7.2.1) together with

$$l = \kappa y[1 - \exp (-y^+/A^+)] \qquad \text{for } 0 \leq y \leq y_c, \tag{7.4.36a}$$

$$= 0.0.68U\delta_D \qquad \text{for } y_c \leq y \leq \delta. \tag{7.4.36b}$$

Here y_c is called the cross-over distance, and is found by matching the two expressions for l during the numerical integration. A^+ is an empirical constant, with investigators using values in the range $25 \leq A^+ \leq 36$. The constant κ is von Kármán's constant, with a value of about 0.41. The exponential form for the mixing length of (7.4.36a) was first proposed by van Driest (1955) as an empirical fit to experimental data. It provides a good model for the Reynolds stress from the buffer region to the outer edge of the boundary layer. The form (7.4.36a) is also well suited to numerical computation, although to find the mean velocity it is necessary to carry out the integration

$$\bar{u} = 2\int dy^+/[1 + \sqrt{1 + l^2}],$$

which must be done numerically.

d. Rough walls

The previous results hold for smooth walls. If the height of the wall roughness is comparable to the laminar sublayer thickness, the roughness can have an appreciable effect on the flow. Reportedly the first flight of the space shuttle Enterprise had a faster landing speed than expected because the exposure to outer space and the heat of reentry cleaned its surface to a point not previously achieved in tests. Also, windmills used for electric power generation have been found to lose from 30 to 50% of their energy productive capacity from increased roughness due to insects and debris impacting the blades.

An *admissible roughness* is defined as a roughness equal to the thickness of the viscous sublayer. Since the sublayer thickness increases with increasing distance from the leading edge of the plate, a suitable location choice for a definition is the thickness of the laminar sublayer where transition from laminar to turbulent flow takes place. A formula frequently used for admissible roughness height is

$$k_{adm} = 100\nu/U. \tag{7.4.37}$$

This gives a good estimate of the degree of machining needed to produce a surface sufficiently smooth for given conditions.

Letting k denote the roughness height, when k exceeds the admissible roughness value, the formulas presented for a smooth plate no longer hold. Experiments similar to those for sand-roughened pipes have been carried out. It was found that appropriate values for boundary layer thickness and the shear stress coefficient are given by either

$$c_f = \tau_{wall}/0.5\rho U^2 = (2.87 + 1.58 \log x/k)^{-2.5} \tag{7.4.38}$$

or alternatively,

$$c_f = (3.96 \log \delta/k + 5.8)^{-2}. \tag{7.4.39}$$

Eliminating c_f between these two results, δ can be easily found to be given by

$$\delta = 0.0343 \times k \times 10^{(2.87 + 1.58 \log x/k)^{1.25}}. \tag{7.4.40}$$

Example 7.4.3. Rough flat plate boundary layer

For a sand-roughened flat plate in a wind tunnel with an air speed of 500 ft/s, the air kinematic viscosity is 2×10^{-4} ft^2/s and the air density is 10^{-3} slug/ft^3. Given that the roughness is twice the admissible roughness, find the boundary layer thickness and shear stress coefficient at a distance of 8 ft from the leading edge.

Sought: Boundary layer thickness and shear stress for a turbulent boundary layer on a flat plate.

Given: Air speed at outer edge of boundary layer (500 ft/s), air kinematic viscosity (2×10^{-4} ft^2/s), air density (10^{-3} slug/ft^3), the roughness is 2 times the admissible roughness.

Assumptions: Fully turbulent viscous flow.

Solution: According to (7.4.37) the admissible roughness is

$$k_{adm} = 100 \times 2 \times 10^{-4}/500 = 4 \times 10^{-5} \text{ ft};$$

hence the roughness for this case is

$k = 8 \times 10^{-5}$ ft.

From (7.4.38),

$c_f = (2.87 + 1.58 \log 10^5)^{-2.5} = 2.63 \times 10^{-3}$.

From (7.4.39),

$\log \delta/k = (-5.8 + 1/\sqrt{2.63 \times 10^{-3}})/3.96 = 3.46$.

Thus $\delta = 8 \times 10^{-5} \times 2{,}884 = 0.23$ ft. Comparing this with the results of Example 7.4.1, we see that the boundary layer thickness for this roughened plate is approximately 51% higher than for the corresponding smooth plate.

Note: In this analysis we did assume that the length of the laminar part of the boundary layer was negligible compared to the 8-ft length of the plate.

---

5. Drag and Lift Forces

In the previous sections of this chapter, we have investigated the special cases of flow in pipes and past flat plates. For flows past other shapes, the flow patterns generally are even more complicated, and we have to content ourselves with being able to forecast the total force and moment on a body by consideration of the drag, lift, and moment coefficients. These coefficients will depend on Reynolds number as well as on other dimensionless parameters.

The total drag force on a body comes directly from the action of pressure and viscous stresses on the body. The viscous stress, particularly that part due to viscous shear, is referred to as the **skin friction** component of the drag force. The portion due to pressure is variously referred to as the **form drag**, **shape drag**, or just **pressure drag**. Particularly when flow separation occurs, leading to the formation of a wake, the pressure difference across a body can mean that form drag is the principle portion of the drag force.

The pressure drag is also subdivided into **wave drag** when there is a free surface present, and the motion past the body results in changes in potential energy, and **unsteady drag** when the flow past the body is not steady. While most drag forces are proportional to the velocity to the second power for turbulent flows, unsteady drag is proportional to the acceleration of the body.

The drag coefficient for steady flow with no free surface effects was defined in Chapter 5 as

$$C_D = F_D/0.5\rho AV^2, \qquad (7.5.1)$$

where F_D is the drag force. The area A used in (7.5.1) is usually the projected area in the direction of V, although for a thin body such as a flat plate or an airfoil aligned with the flow the area of the plate itself is used.

Similarly the lift coefficient was defined as

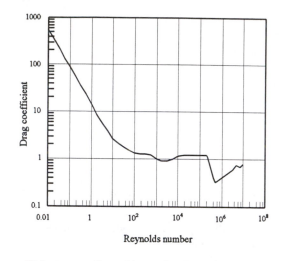

Figure 7.10. Drag coefficient versus Reynolds number for a circular cylinder.

$$C_L = F_L/0.5\rho AV^2, \tag{7.5.2}$$

the lift force being perpendicular to the drag force. The area used in (7.5.2) is usually a projected area perpendicular to V.

Two standard drag coefficient curves (C_D versus Reynolds number) are those for circular cylinders and spheres, shown in Figures 7.10 and 7.11. For very low Reynolds numbers (Re < 1), the drag coefficient for a sphere is

$$C_D = 24/\text{Re}, \qquad \text{Re} < 1, \tag{7.5.3}$$

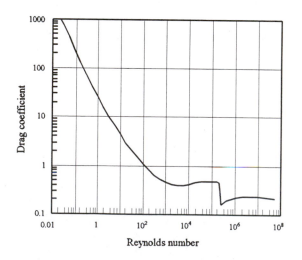

Figure 7.11. Drag coefficient versus Reynolds number for a sphere.

the Reynolds number being based on the sphere diameter. This result was found from an analytic analysis by Stokes (1851), and it has been used for finding the viscosity of fluids to use in Millikan's famous oil-drop experiment for finding the charge on an electron.

The corresponding result for drag coefficient of a circular cylinder is

$$C_D = 8\pi(1 - 0.87/S^2)/S \, \text{Re}, \qquad \text{Re} < 1, \tag{7.5.4}$$

where

$$S = -0.07721 + \ln(8/\text{Re}).$$

Here the Reynolds number is based on the cylinder diameter. This result was found by S. Kaplun (1957).

For values of the Reynolds number higher than about one, (7.5.3) and (7.5.4) are not valid. Figures 7.10 and 7.11 show that in this case the actual drag coefficient is higher than predicted by these equations. For both the cylinder and sphere (and indeed for most rounded body shapes) drag coefficient curves are qualitatively the same. That is, the drag coefficient decreases less and less as the Reynolds number is increased, until a region is reached where the drag coefficient remains relatively constant. As the Reynolds number increases from 10^5 to 10^6, the drag coefficient is seen to drop suddenly and then rise slightly, when the curve again appears to flatten out. This region of sudden dip in the drag coefficient is called the ***drag crisis***. In this limited range of Reynolds numbers, if we were to plot F_D versus V we would see that for the same force three different velocities are possible!

This drag crisis is in fact a sudden change from laminar to turbulent flow in the boundary layer flow near the body surface. For the laminar boundary layer on a circular cylinder, separation occurs approximately 78° from the upstream stagnation point. Past the separation point the pressure is relatively constant. For the turbulent boundary layer, separation occurs near 108°. Therefore, in a turbulent boundary layer some of the pressure on the downstream side of the cylinder counterbalances the pressure on the upstream side. For at least a small range of Reynolds number turbulence in the boundary layer gives us a reduction in the drag force over that found for laminar flow.

Tables 7.6 and 7.7 give representative drag coefficients for a number of two- and three-dimensional shapes in a region of Reynolds number where the drag coefficient is relatively constant.

Example 7.5.1. Drag force on a chimney

A 10-m-high chimney is of circular cross section with a diameter of 2 m. The nominal wind speed is said to be $U_{\text{nom}} = 5$ m/s, but careful measurement shows that the wind velocity profile is given by $u = 1.16 U_{\text{nom}} \log y/1.3 \times 10^{-3}$ for $y > 1$ m. Find the force and moment on the chimney, using $\eta = 1.26$ kg/m^3 and $\nu = 2 \times 10^{-5}$ m^2/s.

Sought: Forces and moments on a chimney due to a turbulent wind.
Given: Chimney height (10 m) and diameter (2 m), wind velocity profile

Table 7.6. Drag coefficients for two-dimensional shapes (10^4 < Re < 10^6). The drag coefficients in all cases are based on the projected area

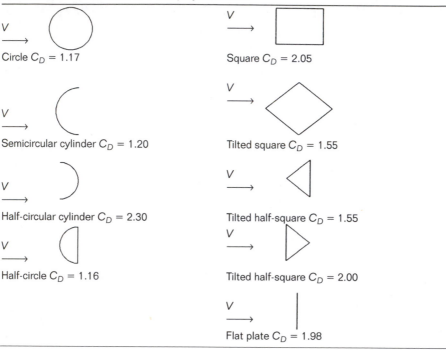

Circle C_D = 1.17	Square C_D = 2.05
Semicircular cylinder C_D = 1.20	Tilted square C_D = 1.55
Half-circular cylinder C_D = 2.30	Tilted half-square C_D = 1.55
Half-circle C_D = 1.16	Tilted half-square C_D = 2.00
	Flat plate C_D = 1.98

$u = 1.16 \times 5 \log y/1.3 \times 10^{-3}$ for $y > 1$ m.

Assumptions: The effects of the wind on the first 1-m height of the chimney have negligible effect on the force and moment. Uniform and established turbulent viscous flow.

Solution: At 1 m,

$u = 5.8 \log (1/1.3 \times 10^{-3}) = 16.74$ m/s,

Re = $16.74 \times 2/2 \times 10^{-5} = 1.67 \times 10^6$.

At 10 m,

$u = 5.8 \log (10/1.3 \times 10^{-3}) = 22.54$ m/s,

Re = $22.54 \times 2/2 \times 10^{-5} = 2.25 \times 10^6$.

In this range of Reynolds numbers, $C_D = 0.34$. Thus

$dF_D = 0.5 \times C_D \rho U^2 \times 5 \, dy = 0.5 \times 0.34 \times 1.26 \times (5.8 \times \log y/1.3 \times 10^{-3})^2 \times 5 \times dy$

$= 6.786 \, (\ln y/1.3 \times 10^{-3})^2 \, dy.$

Table 7.7. Drag coefficients for two-dimensional shapes (10^4 < Re < 10^6). The drag coefficients in all cases are based on the projected area

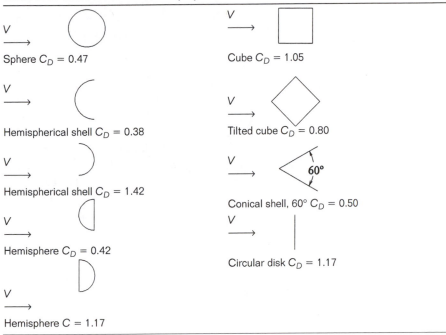

Sphere C_D = 0.47	Cube C_D = 1.05
Hemispherical shell C_D = 0.38	Tilted cube C_D = 0.80
Hemispherical shell C_D = 1.42	Conical shell, 60° C_D = 0.50
Hemisphere C_D = 0.42	Circular disk C_D = 1.17
Hemisphere C = 1.17	

Then

$$F_D = \int_1^{10} 6.786(\ln y/1.3 \times 10^{-3})^2 \, dy = 4{,}131 \text{ N}.$$

The moment about the base is

$$M_D = \int_1^{10} y \times 6.786(\ln y/1.3 \times 10^{-3})^2 \times 5 \times dy = 24{,}171 \text{ N-m}.$$

a. Vehicle drag

Vehicle drag reduction has gained prominence during the last few decades as a way of improving fuel economy, and automobile manufacturers now regularly do wind tunnel testing of their cars. A representative formula for vehicle drag is

$$F_D = KW + C_D \rho A V^2/2, \tag{7.5.5}$$

where W is the weight of the vehicle and K is a static drag coefficient due to tire drag. K depends somewhat on tire inflation pressure. We see from (7.5.5) that at speeds of

$$V_{crossover} = \sqrt{2KW/\rho C_D A} \qquad (7.5.6)$$

the aerodynamic drag is equal to the tire drag. A typical value of K is of the order of 0.01. Table 7.8 contrasts some values of the crossover velocity using this value of K.

Automobile drag coefficients over the decades have been reduced from values of 0.6 down to around 0.31 in the 1980s, and experimental cars with drag coefficients of the order of 0.14 (for extreme designs, values as low as 0.06 have been reached) have been achieved. Trucks, however, typically still have drag coefficient values of the order of 0.8–0.95. Spoilers on cab roofs can reduce these drag coefficients as much as 15% if properly positioned. They are an attempt to eliminate a large eddy on the top of the truck cab, the eddy being the cause of a large region of separated flow on the top of the trailer and also causing a large form drag. Experimental trucks have been designed that smoothly blend together the tractor and trailer, and also eliminate much of the open space below the trailer. Such designs lower the drag coefficient appreciably, particularly when there is no cross wind. With strong cross winds, however, the dependence of the drag coefficient on the wind direction is also needed to describe the forces on the vehicle. Much more detail is necessary to understand what happens in what is a very complicated problem.

The drag coefficients quoted above are for the case where there is no cross wind. Cross-wind forces can be quite large, and cross winds can even change the frontal drag coefficient by altering the way the flow separates. (Increases in drag forces of 30% or more have been found in such situations for trucks and buses.) Cross-wind forces can play a major role in the handling capabilities of a vehicle. As an example, the famous Volkswagon beetle was particularly susceptible to cross winds. When a 40-mph cross wind was suddenly encountered when traveling at 60 mph, the vehicle tended to move in the wind direction with a speed of 1 ft/s unless corrective measures were taken.

Example 7.5.2. Automobile drag

A passenger car weighing 2,500 lb has a drag coefficient of 0.31 and a cross-sectional area of 30 ft^2. The static drag coefficient is 0.025. Find the crossover speed, and write out the expression for both drag force and power required to overcome drag when ρ = 0.00244 slug/ft^3.

Table 7.8. Crossover velocities for typical vehicles

Vehicle type	W	CD	A (ft^2)	V (ft/s)	V (mph)
Passenger car	3,000	0.4	32	45.0	30.6
Pickup truck	8,000	0.8	90	46.4	31.7
Line haul truck	70,000	0.8	107	84.0	57.2

Sought: Crossover speed and expressions for drag force and power for drag on a passenger car.

Given: Car weight (2,500 lb), car drag coefficient (0.31), car cross-sectional area (30 ft^2), static drag coefficient (0.025), air density (0.00244 slug/ft^3).

Assumptions: Uniform and established turbulent viscous flow.

Solution: From (7.5.6),

$$V_{\text{crossover}} = \sqrt{2 \times 0.025 \times 2{,}500/0.00244 \times 0.31 \times 30} = 74.2 \text{ ft/s} = 50.6 \text{ mph.}$$

The drag force is given by the expression

$$F_D = 0.025 \times 2{,}500 + 0.31 \times 0.00244 \times 30V^2/2 = 62.5 + 0.0133V^2 \text{ lb.}$$

Power $= V \times F_D = 62.5V + 0.0133V^3$ lb-ft/s.

Note: In the above formulas, V must be in feet per second.

b. Lift forces—the Magnus effect

Why do lift forces develop? We all know the shape of airfoils and propeller blades, but it may not be obvious how this shape contributes to the generation of lift forces.

An elementary answer was given by G. Magnus as early as 1853. Imagine a uniform stream of velocity U flowing perpendicular to the axis of a long horizontal circular cylinder. We would expect from symmetry that there would be no net lift on the cylinder on the average.

Now suppose that the cylinder is rotated by some external motor. Suppose that the rotation is such that the velocity it imposes acts to oppose the uniform stream at the bottom of the cylinder and supplement it at the top. From our energy equation this would say that since the cylinder's rotation increases the fluid speed on the top of the cylinder, the pressure at the top of the cylinder is lower than the corresponding pressure at the bottom of the cylinder, where the rotation lowers the speed. This difference in pressure gives rise to a net upward force that will be proportional to the mean stream velocity U times the angular velocity.

The curving trajectory of a spinning baseball has frequently been described as due to the Magnus effect, although the three-dimensionality of the baseball and the stitches and seams raise many additional questions and this cannot be accepted as a complete explanation. More directly, several ships were built by Flettner, who mounted rotating cylinders on tall vertical masts. A small Flettner ship is shown in Figure 7.12. (The flat disks seen on the top and bottom of the cylinder help the flow to be more like its two-dimensional idealization and thereby improve the performance.) Flettner also built a large ocean-going ship with three rotating cylinders, each 27 ft in diameter and 98 ft tall, but the need for a wind to be present (to provide the U), the need for a power plant for driving the cylinders, and the economics of such a ship soon led to the abandonment of this mode of propulsion. More recently, Captain Jacques Costeau has adapted the Magnus effect (Figure 7.13) as a concept friendly to the environment.

Figure 7.12. Small Flettner rotor craft. The largest ship built using rotating cylinders, the *Barbara*, had 3 cylinders 23 ft in diameter and 98 ft tall, rotating at 150 rpm. The *Barbara* made two Atlantic Ocean crossings in 1927.

Figure 7.13. Captain Jacques Cousteau's ship *Alcone*. The two "turbosails" use the Magnus effect, but achieve the circulation by sucking air through perforated lateral vents rather than by rotation of cylinders. With winds over 25 knots (12.5 m/s) the diesel engines can be turned off entirely. The suction-producing fan still must be powered. (Courtesy the Cousteau Society.)

Of course it is not necessary to rotate something to achieve a lift force. The purpose of the rotation in Magnus's work was to increase the fluid speed on the top surface and decrease it on the bottom. This can be accomplished by other means as well. For instance, Thwaites in England found that by putting a small flap on the lower right-hand quadrant of a nonrotating cylinder a similar effect could be induced. Observation of airplane wings, propeller blades, and other ***lifting surfaces*** reveal that what they have in common is a rounded leading edge and a sharp trailing edge. These accomplish the desired lift effect by positioning the rear stagnation point at the trailing edge.

c. Airfoils

An airfoil is a lifting surface with curved upper and lower surfaces characterized by a sharp trailing edge. (You have no doubt seen venetian blind slats "lift" and oscillate in a cross wind when the slat separation was small. The slat is an airfoil shape with a sharp leading edge as well as a sharp trailing edge.) Representative results for a symmetrically shaped airfoil (suited for a boat rudder, for example) are shown in Figures 7.14 and 7.15 for an NACA 0006 wing section. The shape of

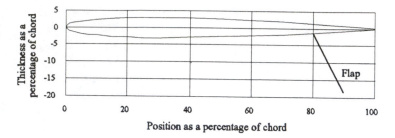

Figure 7.14. NACA 0006 symmetric airfoil with flap.

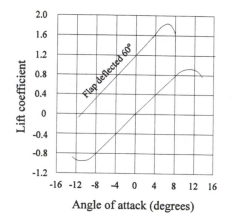

Figure 7.15. Lift coefficient for an NACA 0006 airfoil.

this airfoil is shown in Figure 7.16. The lift coefficient (the area used is the cord length times the span) is seen to increase almost linearly with the angle of attack until an attack angle of 13.5° is reached. Increasing the angle further results in a sharp drop in the lift coefficient. This is due to *stall* developing. At stall, the boundary layer on the top surface of the wing separates completely from the upper surface of the wing, and the ability to support lift is greatly decreased. Blades in turbines, torque converters, propellers, and the like are airfoils also, and the angle of attack must be kept sufficiently small so that stall does not occur. Windmills used for power generation are usually designed for wind speeds around 15 mph. At faster speeds, the blades are designed to turn so that stall is induced, thereby protecting the blades from unduly large forces.

Many devices have been developed to increase the angle of attack at which stall commences. These devices act in various ways to control the boundary layer on the upper surface of the wing. We have already mentioned the use of vortex generators to induce an artificial turbulence into the boundary layer, thereby delaying separation. Leading edge slats and trailing edge slotted flaps also accomplish this (usually with an increase in the drag coefficient as well), and are commonly used on large aircraft during takeoff and landing. Some experimental military

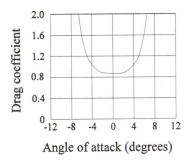

Figure 7.16. Drag coefficient for an NACA 0006 airfoil.

planes have installed slots along the wing whereby they can suck in air from the boundary layer, reducing the boundary layer thickness as well as its tendency toward separated flow regions.

The type of lifting surface used is very much a function of Reynolds numbers. Model airplane wings typically have a Reynolds number that is 100 to 1,000 times smaller than the prototype. If the model wing were built to geometric scale, the lifting power of the model wing would not be nearly as good as for the prototype wing. Generally, the model wings are made thicker, and to a different shape, to compensate for this.

An illustration of the effect of Reynolds number on flight has been given by Lissaman (1983) as shown in Figure 7.17. The figure would look somewhat magnified but roughly the same if we plotted just chord length on the horizontal scale. Thus we can think of this figure as relating speed to size. Propulsion at Reynolds numbers below 100 is strongly influenced by viscous effects. For example, microorganisms propel themselves by waving cilia and flagella in a whiplike manner. If the same organisms were placed in a fluid of smaller viscosity, they either would have more trouble propelling themselves or would be completely immobilized. Further up the Reynolds number scale, fish wave their tails in a flapping manner, and insects[3] and birds flap their wings. Model airplanes overlap both insects and small birds up to a Reynolds number of 10^5, and can be made to fly by either wing flapping or by more conventional means. Above these Reynolds numbers, where we anticipate at least some of the boundary layer will be turbulent, lies the range of high-performance flight, occupied by large soaring birds, human-powered aircraft, sail planes, and commercial aircraft. Human-designed flight was possible only after a thorough understanding of the fluid dynamics fundamentals was at hand.

---

[3]"It is told that such are the aerodynamics and wing loading of the bumblebee that, in principle, it cannot fly—If all this be true—life among bumblebees must bear a remarkable resemblance to life in the United States." *American Capitalism: The Concept of Countervailing Power*, John Kenneth Galbraith. Dr. Galbraith's conclusion may be correct on economics and power, but his credentials as an aerodynamicist are suspect. Bees—and humming birds—have very low Reynolds numbers, so if they tried to fly in the manner of an airplane they would experience an extremely high degree of frustration. Fortunately they have adapted to their Reynolds number and fly very well by flapping their wings vertically at very high speeds. Larger birds—at higher Reynolds numbers—use different flapping motions to achieve flight. And so on up the Reynolds number range.

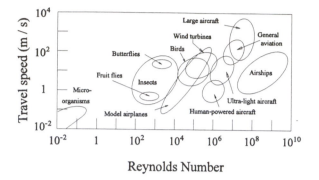

Figure 7.17. Speeds (swimming and flight) versus Reynolds numbers for a number of different propulsion means. (After P.B.S. Lissaman, *Ann. Rev. Fluid Mech.*, vol. **15**, 1983.)

References and Suggestions for Further Reading

Alciatore, D., and W. S. Janna. "Modified Pipe Friction Diagrams That Eliminate Trial and Error Solutions," *Proc. 1st Natl. Fluid Dynamics Congress*, pt. 2, pp. 911–916, AIAA, Washington, 1988.

Bearman, P. W., et al., "The Effect of a Moving Floor on Wind-Tunnel Simulation of Road Vehicles," Paper No. 880245, SAE Transactions, *J. Passenger Cars*, vol. **97**, sec. 4, 1988, pp. 4.200–4.214.

Blasius, H. "Das Ähnlichkeitsgesetz bei Reibungsvorgängen in Flüssigkeited," *Forsch. Arb. Ingwes.*, Number 134, Berlin (1913).

Colebrook, C. F. "Turbulent flow in pipes, with particular reference to the transition region between the smooth and rough pipe laws," *J. Inst. Civ. Eng. Lond.*, vol. **11**, pages 133–156, 1938–1939.

Crane Company, The, *Flow of Fluids Through Valves, Fittings, and Pipes*, Technical Paper No. 410, Chicago, 1982.

Haaland, S. E., "Simple and Explicit Formulas for the Friction Factor in Turbulent Pipe Flow," *Fluids Eng.*, pp. 89–90, March 1983.

Hoerner, S. F., *Fluid-Dynamic Drag*, published by the author, 148 Busteed Drive, Midland Park, NJ 07432, 1965.

Hucho, W. H., *Aerodynamics of Road Vehicles*, Butterworth, Boston, 1986.

Hydraulic Institute, *Pipe Friction Manual*, 3d edition, The Hydraulic Institute, New York, 1961.

Idelchik, I. E., *Handbook of Hydraulic Resistance*, 2d edition, Hemisphere, New York, 1986.

Inui, T., "Wavemaking Resistance of Ships," *Trans. Soc. Nav. Arch. Marine Eng.*, vol. **70**, pp. 283–326, 1962.

Jepson, R. W., *Analysis of Flow in Pipe Networks*. Ann Arbor Pub., Ann Arbor, MI, 1976.

Johnson, Jr., R. C., G. E. Ramey, and D. S. O'Hagen, "Wind Induced Forces on Trees," *J. Fluids Eng.*, vol. **104**, pp. 25–30, March 1982.

Kaplun, S., "Low Reynolds Number Flow Past a Circular Cylinder," *J. Math. Mech.*, vol. **6**, pages 595–603, 1957.

Larsson, L., et al., "A Method for Resistance and Flow Prediction in Ship Design," *Trans. Soc. Nav. Arch. Marine Eng.*, vol. **98**, 1990, pp. 495–535.

Laufer, J., "The Structure of Turbulence in Fully Developed Pipe Flow," NACA TR 1174, 1954.

Lissaman, P. B. S., "Low-Reynolds-Number Airfoils," *Ann. Rev. Fluid Mechs.*, vol. **15**, pages 223–239, 1983.

Lyons, J. L., *Lyons' Valve Designers Handbook*, Van Nostrand Reinhold, New York, 1982.

Magnus, G., "Über die Abweichung der Geschosse u. auffallende Erscheinungen bei rotirenden Köpern," *Annalen der Physik*, vol. **88**, p. 1, 1853.

Moody, L. F., "Friction Factors for Pipe Flow." *Trans. ASME*, vol. **66**, pp. 671–684, Nov. 1944.

Nikuradse, J., "Stromungsgestze in rauhen Rohren," *Ver. Deut. Ing. Forschungsh.*, vol. **361**, 1933. (English translation in NACA TM 1292.)

Patton, K. T., "Tables of Hydrodynamic Mass Factors for Translational Motion," ASME Winter Annual Meeting, Paper 65-WA/UNT-2. 1965.

Prandtl, L., "Über den Reibungswiderstand strömender," *Luft. Ergebn.* AVA Göttingen, 1st series, 136 (1921); see also *Collected Works II*, pp 620–626.

Reynolds, O., "An Experimental Investigation of the Circumstances which Determine Whether the Motion of Water Shall Be Direct or Sinuous and of the Law of Resistance in Parallel Channels," *Phil. Trans. Roy. Soc.*, vol. **174**, pp. 935–982, 1883.

Runstadler, Jr., P. W., et al., "Diffuser Data Book," *Creare Inc. Tech. Note 186*, Hanover, NH, 1975.

Schultz-Grunow, F., "Neues Widerstandsgesetz fur glatte Platten," *Luftfahrtforschung*, vol. **17**, page 239, 1940. (English translation in NACA TM 986, 1941.)

Sovran, G., T. Morel, and W. T. Mason, Jr., editors, *Aerodynamic Drag Mechanisms of Blunt Bodies and Road Vehicles*, Plenum, New York, 1978.

Uram, E. M., and H. E. Weber, editors, "Laminar and Turbulent Boundary Layers," *ASME Symp. Proc.*, vol. **100167**, February 1984.

Driest, E. R., "On Turbulent Flow Near a Wall," *J. Aero. Sci.*, vol. **23**, pages 1007–1011, 1956.

Problems for Chapter 7

Pipe flow

7.1. In a 24-in-diameter horizontal pipeline, the pressure drops 10 psi in 100 ft. What is the average shear stress at the wall of the pipe? The fluid is water at 68°F.

7.2. For pipe flow, the head loss due to friction is 34 ft when 2 ft^3/s of a liquid flows in 500 ft of 6-in-diameter pipe. The mass density of the liquid is 2.46 slugs/ft^3 and the viscosity is 0.00025 lb-s/ft^2. What is the pipe roughness?

Pipe—type 1

7.3. SAE 10W oil at a temperature of 60°F is flowing in a horizontal galvanized iron pipe of 24-in-diameter and 350-ft length. The flow rate is 21 ft^3/s. The pressure at the downstream end of the pipe is atmospheric. What must the pressure be at the upstream end of the pipe?

7.4.   For a horizontal galvanized 0.1-ft-diameter pipe 100 ft long with water (50°F) flowing at a velocity of 9 ft/s, what is the head loss due to viscosity? What would be the head loss if 2 elbows were inserted into the system? ($K_{elbow} = 0.9$.)

7.5.   For the figure, find H for a flow rate of 0.5 ft^3/s, $\mu = 0.0002$ lb-s/ft^2, $\gamma = 60$ lb/ft^3. The head loss coefficient for the valve is 5. Neglect entrance, exit, and elbow losses. The pipe is made of steel. What is the power loss?

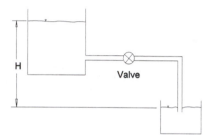

Pipe diameter 3 inch, pipe length 210 feet

Problem 7.5

7.6.   A 6-in horizontal pipeline (roughness = 0.0004 ft) is used to transport various petroleum products, one following the other in the pipeline. Gasoline (SG = 0.68, $\nu = 6 \times 10^{-5}$ ft^2/s) is being sent through the pipeline following kerosene (SG = 0.81, $\nu = 3 \times 10^{-5}$ ft^2/s) with a fairly sharp interface between the two liquids. If the flow rate is 1.6 ft^3/s and at a given instant of time the pressure difference between two points located 3,600 ft apart is 50 psi, locate the position of the interface from the high-pressure end of the pipe. What is the pressure gradient in the gasoline? What is the pressure gradient in the kerosene?

7.7.   In the figure, what should be the value of H to have 6 ft^3/s of water flowing through the 1-ft-diameter steel pipe? Neglect any inlet and exit losses. What should be the value of H to double this discharge? $\nu = 1.2 \times 10^{-5}$ ft^2/s.

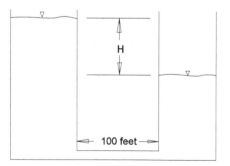

100 feet

Problem 7.7

7.8. Water at 60°F flows through a 3.25-in-diameter pipe of length 7 ft and roughness factor 0.013 ft with a velocity of 10 ft/s. Find the head loss.

7.9. Oil (SG = 0.9, μ = 0.002 lb-s/ft^2) is pumped at a rate of 0.1 ft/s through the system in the figure, consisting of 300 ft of 2-in-diameter galvanized iron pipe. Considering only friction losses, find the rate at which energy must be supplied by the pump.

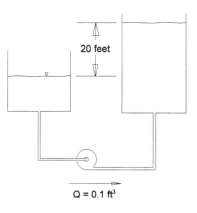

20 feet

Q = 0.1 ft^3

Problem 7.9

7.10. Consider a wind tunnel to be a round constant-diameter pipe, and suppose that the design is such that losses due to turns are negligible. What horsepower motor for a fan must be purchased to circulate air at 60°F in a wind tunnel at 300 ft/s? The tunnel is a closed loop 200 ft long, 6 ft in diameter, and is made of galvanized iron. Assume 100% efficiency. ρ_{air} = 0.0022 slug/ft^3.

7.11. Calculate the pump horsepower required to maintain a flow rate of 0.33 ft^3/s in the flow described by the figure.

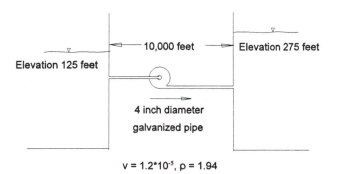

10,000 feet   Elevation 275 feet

Elevation 125 feet

4 inch diameter

galvanized pipe

v = 1.2*10^{-5}, ρ = 1.94

Problem 7.11

7.12. An oil with $\nu = 0.0001$ ft²/s, SG = 0.85, flows at a rate of 2,000 gpm through 1,000 ft of an 8-in-diameter cast iron pipe. What are the Reynolds number and friction factor? What is the head loss? Express the energy loss in horsepower.

7.13. A pump is used to drain a swimming pool. If the properties of the water are $\rho = 1.94$ slugs/ft³ and $\nu = 1.2 \times 10^{-4}$ ft²/s, and the flow rate through 100 ft of 2-in-diameter galvanized pipe is 0.2 ft³/s when the elevation difference between the surface of the pool and the pipe outlet is 10 ft, what useful horsepower is the pump providing?

7.14. If water at 60°F is flowing through a 1-ft-diameter commercial steel pipe 100 ft long at a Reynolds number of 10^6, what is the head loss?

7.15. Two reservoirs 200 ft apart are connected by a 2-in-diameter smooth pipe. It is desired to transfer 0.25 ft³/s water at 60°F ($\nu = 1.4 \times 10^{-5}$ ft²/s, $\rho = 1.94$ slugs/ft³) by means of a pump halfway between the two levels. Give the pressure distribution along the pipe as a function of x. What should be the horsepower of the pump if it is 80% efficient? Take both reservoirs to be at the same level and pressure.

7.16. A welded steel pipeline of 24-in diameter is designed to carry a flow of 12,600 gpd of oil across the country, pumping stations being located every 50 miles. If the overall pump efficiency is 85%, the viscosity of the oil is 5×10^{-4} lb-s/ft², the specific weight of the oil is 55 lb/ft³, and an average roughness is 0.002 ft, determine the minimum horsepower input required at each station when the system is new.

7.17. A 0.5-ft-diameter, 50-ft-long galvanized pipe is used to provide cooling water. For $\nu = 10^{-5}$ ft²/s, what power must the pump provide to maintain a flow velocity of 10 ft/s?

7.18. For the flow depicted by the figure, given that the discharge is 0.01 ft³/s, find the pressure at point A and the total energy head loss from A to C. The fluid is water.

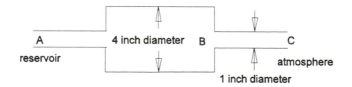

Problem 7.18

7.19. If all the conditions of Problem 7.9 are kept as is, what is the maximum discharge you can have and guarantee that the flow will be laminar? For this discharge, what pump power is required?

7.20. A 3-in steel pipe 400 ft long conveys 200 gpm of water ($\rho = 1.938$ slugs/ft³, $\nu = 10^{-5}$ ft²/s) from a water main ($p = 90$ psi), to the top of a building 60 ft above the main. What pressure can be maintained at the top of the building?

7.21. Water with $\nu = 10^{-5}$ ft²/s is flowing through a cast iron pipe 300 ft long and 0.1 ft in diameter. The pipe connects two reservoirs, the downstream one being lower by 15 ft. If entrance, exit, and minor losses can be neglected, what is the head loss? What is the discharge through the pipe?

Pipe—type 2

7.22. A vertical pipe of 0.11-ft diameter and 100-ft length is used to empty an open storage tank at the top of a building. If it is a galvanized iron pipe and the storage tank is filled with water at 60°F to a depth of 10 ft, what is the average flow velocity in the pipe? What is the rate of energy loss? The pipe discharges into the atmosphere.

7.23. A piping system carries 70°F water from a reservoir and discharges it as a free jet. How much flow is to be expected through an 8-in-diameter commercial steel pipe with the fittings shown in the figure? Use $K = 0.9$ for a 90° elbow.

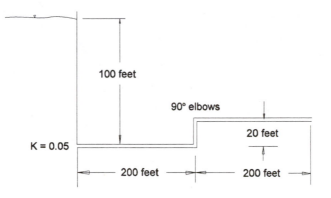

Problem 7.23

7.24. A large tank discharges water at 60°F through a 1-ft-diameter horizontal concrete pipe (roughness = 0.01 ft) that is 10 ft long. The other end of the pipe is open to the atmosphere. If the water depth in the tank is initially 6 ft, what is the velocity of the fluid in the pipe?

7.25. Find the velocity in the pipe shown in the figure. Given: roughness = 0.00012 ft, $D = 0.4$ ft, $L = 10$ ft, $\nu = 2 \times 10^{-5}$ ft^2/s. Neglect any minor losses.

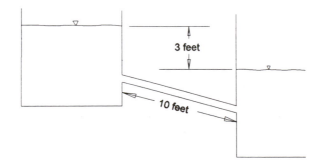

Problem 7.25

7.26. Calculate the rate of flow through the pipeline of the figure, and the pressures at A and B. Neglect minor losses. $k = 10^{-4}$ ft, $\gamma = 62.4$ lb/ft^3, $\nu = 10^{-5}$ ft^{2}s.

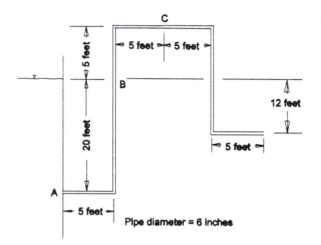

Problem 7.26

7.27. Find the discharge through a pipe where the head loss is 2.9 ft in a 100-ft length. Roughness $= 0.0005$ ft, $D = 6$ in. The fluid is water at 60°F. Neglect entrance losses.

7.28. Find the discharge per unit width for the flow situation shown in the figure. $\nu = 10^{-5}$ ft^2/s, $k = 0.0001$ ft.

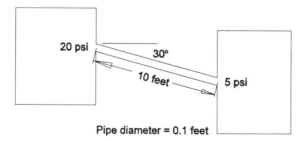

Problem 7.28

7.29. For a horizontal new galvanized iron pipe of diameter 25 mm and length 10 m discharging into the atmosphere, find the discharge. The pipe inlet is at 100 kPa. Use $\nu = 10^{-6}$ m^2/s, $\rho = 1,000$ kg/m^3. Neglect entrance losses.

7.30. What is the discharge for the system shown in the figure? The pipe is smooth. SG $= 0.8$, $\mu = 0.1$ P.

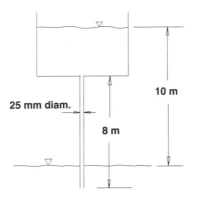

Problem 7.30

7.31. A gas is pumped through a smooth horizontal 10-in-diameter pipe with a pressure head drop of 1/2 in of water per 1,000 ft. $\mu = 2.1 \times 10^{-7}$ lb-s/ft^2, $\gamma = 0.06$ lb/ft^3. Find the discharge Q.

7.32. Flow in a horizontal, commercial steel pipe occurs with the fluid kinematic viscosity of 4×10^{-5} ft^2/s and a pipe diameter of 600 mm. The pressure drop is 1.2 kPa in 60 m. What is the discharge through the pipe?

7.33. Water (20°C) is drained from a large reservoir on the roof of a building by a vertical copper pipe (drawn tubing) 40 m in length and 30 mm diameter. Find the discharge.

7.34. Compute the discharge for water at 60°F for 1-in-diameter galvanized iron pipe. The pressure at A and the pipe exit in the figure may be considered to be atmospheric, and losses at A, B, and the exit can be neglected.

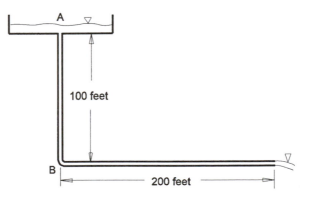

Problem 7.34

7.35. Find the rate of discharge through a 12-in-diameter galvanized pipe where the pressure drop in 100 ft is 0.17 psi. $\nu = 5 \times 10^{-5}$ ft^2/s, SG = 1.

7.36. Find the discharge through a 6-in horizontal galvanized iron pipe neglecting minor losses. $\nu = 10^{-5}$ ft^2/s, $L = 100$ ft, $h_L = 0.54$ ft. Recompute these results including the following minor losses:

2 standard 90° elbows

1 fully open gate valve

3 standard tees (flow-through branch).

For minor losses included, what is the rate of total energy loss, expressed in horse-power?

7.37. In a smooth horizontal pipe with an inside diameter of 0.1 ft, pressure-measuring devices 400 ft apart read a pressure difference of 1.4 psi. Determine the average velocity in the pipe when water is flowing. $\nu = 1.4 \times 10^{-5}$ ft^2/s.

7.38. For water ($\nu = 10^{-6}$ m^2/s) flowing in a new 10-mm-diameter galvanized iron pipe with a head loss of 2 kPa per 100 m of pipe length, what is the discharge?

7.39. A vertical galvanized iron pipe 10 ft long and 0.025 ft in diameter is connected to a shallow reservoir containing water at the top end, and discharges into a large reservoir at the bottom end. What is the discharge rate? $\nu = 10^{-5}$ ft^2/s.

7.40. Water at 120°F ($\nu = 6 \times 10^{-6}$ ft^2/s) is carried from a reservoir on top of a building to a point in the basement 120 ft below the reservoir surface through 115 ft of a 1.0-in-diameter smooth pipe. At what velocity will the water leave the pipe when it is open at the bottom end?

7.41. A 3-in-diameter cast iron, horizontal pipe 400 ft long has a pressure drop of 30 psi. For $\nu = 1.2 \times 10^{-5}$ ft^2/s, what is the discharge?

7.42. Water at 20°C is pumped from a ground-level reservoir to a second reservoir on top of a building through a 20-m length of vertical smooth pipe. The reservoir pressures can be considered to be atmospheric. If a 500-hp pump (efficiency = 70%) is used, what should the pipe diameter be to ensure a flow rate of 1 m^3/s?

Pipe—type 3

7.43. Crude oil (SG = 0.86) at 50°C flows in a concrete pipe with roughness 0.6 mm and length 200 m. The desired flow rate is 0.15 m^3/s, and a pump that delivers 6 hp is available. What should the diameter of the pipe be?

7.44. If the flow in the previous problem occurred in a smooth pipe, what pipe diameter would be needed?

7.45. SAE 10 oil at 20°C ($\rho = 869$ kg/m^3, $\mu = 0.0814$ N-s/m^2) is to be delivered in a smooth pipe 100 m long at a flow rate of 0.016 m^3/s. An 85% efficient 40-hp pump is available. What should the diameter of the pipe be?

7.46. Galvanized pipe is to be used to deliver hot water at 80°C in a pipe 30 m long. A flow rate of 2.5 m^3/s is desired. If an 80% efficient 4-hp pump is to be used, what pipe diameter should be selected? The conditions at either end of the pipe can be considered to be reservoirs. $\rho = 500$ kg/m^3.

7.47. A 210-ft piping system is made up of 10 lengths of 21-ft sections of galvanized pipe. There are six 90° standard elbows, 2 standard tees (straight through), and 2 fully opened globe valves. A 125-hp 80% efficient pump is to be used. What should be the pipe diameter to have a flow of 0.5 ft^3/s of water at 60°F? $K_{connection} = 0.1$.

Boundary layer flows

7.48. An airplane wing has a cord length of 20 ft. For the plane flying at a speed of 500 mph at 30,000 ft elevation, estimate the following quantities: length of laminar boundary layer, thickness of turbulent boundary layer at the trailing edge, thickness of laminar sublayer at the trailing edge.

7.49. Water (80°F) flows over a smooth plate 12 ft long and 3 ft wide. The flow speed is 7 ft/s. If transition to turbulence occurs at a Reynolds number of 5×10^5, what is the drag on the portion of the plate where the boundary layer is laminar? What is the drag on the portion of the plate where the boundary layer is turbulent?

7.50. If a boundary layer "trip" was placed near the leading edge of the plate in the previous problem so that the boundary layer was everywhere turbulent, what would be the drag on the plate?

Drag forces

7.51. A solid sphere of 1-ft diameter is dropped in air at 70°F and falls so that the Reynolds number based on diameter is 10^4. At what velocity does the sphere fall? What is the weight of the sphere?

7.52. If a 1-in-diameter steel sphere (SG = 7.8) attains a velocity of 2.78 ft/s when falling through oil (SG = 0.9 and $\mu = 0.0075$ lb-s/ft^2), what are the Reynolds number and drag coefficient?

7.53. A baseball weighing 0.3 lb and having a diameter of 2.9 in travels at 80 mph on a sunny summer day when the temperature is 90°F. What is the air drag on the ball?

7.54. A 1-mm-diameter solid sphere falls in 30°C glycerine. Its terminal velocity is 1.3 mm/s. What is the mass density of the sphere?

7.55. R. A. Millikan was awarded a Nobel prize for the first accurate measurement of the charge on an electron. He sprayed a few very tiny droplets of an oil between two electrically charged plates. He first determined the diameter of the small drops by measuring their terminal velocity V_{wo} and using Stokes' formula. He then applied a voltage E across the plates, resulting in an upward force Eq/d on the drop and a terminal velocity V_w. E is the electric potential in volts (N-m/C), d is the plate spacing in meters, and q is the charge in coulombs (C). Making repeated measurements, he found that the q he determined were always integer multiples of a specific number q_1, the charge due to a single electron. Determine first a formula for the droplet radius when no electric field is present. Then determine the terminal velocity when the electric field is present. Solve for q in terms of the mass of the droplet, the terminal velocities with and without an electric field, and the ratio field E/d.

7.56. In a repeat of Millikan's experiment mentioned in the previous problem, the voltage was 115 V, the plate spacing was 4.53 mm, the oil density was 920 kg/m^3, and the air viscosity was 1.82×10^{-4} P. Two horizontal lines were drawn 1.73 mm apart, and the time of traverse of these two lines were measured with a stopwatch. By changing the direction of the electric field it was possible to use the same droplet for each measurement, although each set of measurements did have different numbers of electrons attached to the droplet. Times of transit were as follows:

72.2 s averaged over 10 measurements, no electric field, direction down,

41.1 s averaged over 8 measurements, electric field present, direction up,

23.5 s averaged over 2 measurements, electric field present, direction up,

16.5 s averaged over 5 measurements, electric field present, direction up,

12.5 s averaged over 3 measurements, electric field present, direction up,

8.06 s averaged over 7 measurements, electric field present, direction up.

Find the terminal velocities for the six sets of data, and estimate the charge due to a single electron. (*Note*: Millikan also found that, due to the small size of these droplets,

Brownian motion gives errors in measuring velocity, resulting in charge estimates that are about 27% too high.)

7.57. A solid circular cylinder 7 ft long and with a radius of 1 ft is pulled horizontally in a direction perpendicular to its longitudinal axis through a viscous fluid of infinite extent at a constant velocity of 1.5 ft/s. If $\mu = 3 \times 10^{-4}$ lb-s/ft^2, SG = 0.9, determine the force needed to sustain this motion.

7.58. A sphere 300 mm in diameter is towed through water at 20°C at 3.6 km/h. What size sphere has the same drag coefficient in an air stream having a velocity of 36 km/h? Calculate the drag forces on these spheres. Consider the air to be at standard atmospheric pressure and 15°C.

7.59. When a body accelerates in a fluid, an additional force proportional to the acceleration is added to the drag. This drag force is written as $F_a = M_{added}\, dV/dt$, where M_{added} is called the added mass and depends on the shape of the body. For a sphere the added mass is equal to the mass of the fluid displaced by the sphere. If a sphere of radius 15 mm and SG 1.7 is released from rest in water (SG = 1), what is its initial acceleration?

Automobile drag

7.60. A passenger car weighing 3,800 lb has a projected area of 30 ft^2, a drag coefficient of 0.35, and a static drag coefficient of 0.025. What is the crossover speed, and the power needed to overcome drag at 30 mph and at 70 mph?

7.61. If the drag coefficient in the previous problem were lowered to 0.3, what would be the percentage of power saved at the two speeds?

chapter 8

Open Channel Flows

Chapter Overview and Goals

The presence of a free surface on a channel simplifies some aspects of flow analysis, since the pressure variation is hydrostatic in most flow regimes of interest. Flows in open channels are strongly influenced by gravity and momentum forces; hence the Froude number is the important descriptor for these flows.

This chapter utilizes the control volume analysis techniques developed in Chapter 3 along with the dimensionless parameters considered in Chapter 5. Since the depth of the flow in a channel can vary along the length of a channel, the Froude number changes along the channel length, and both standing and traveling waves are possible. Small- and large-amplitude waves are considered, the relation between upstream and downstream depths across the wave are computed, and formulas for the propagation speed of the wave are developed.

1. Introduction

Canals, rivers, and lakes have played vital roles in the development of the history of our world. They have played vital roles in transportation in many countries, and are still actively used both in commerce and pleasure boating in countries such as the U.S., France, England, Germany, and The Netherlands. The Welland Canal that bypasses Niagara Falls allows ocean-going vessels to travel to the Great Lakes region, providing economical transportation without the need for changing mode or vehicle of transportation. In California and Arizona, canals are important to agriculture and for transportation of water for human consumption, opening the desert to farming and living, and providing the U.S. with fresh food supplies year-round.

The flow of liquids in open channels, rivers, and partially filled conduits is distinguished from other fluid flows in that it is driven entirely by gravity, since it is not possible to maintain a pressure gradient along the free surface of the flow. This means that the primary dimensionless parameter describing such flows is the Froude number, representing the ratio of inertia to gravity forces. The Reynolds number plays a secondary role, describing losses that can take place in these flows.

The presence of a free surface in these flows allows the formation and propagation of waves on this surface. Waves can have both beneficial and detrimental effects. They are pleasant to watch at the beach in the summer, the bow wave of a ship is esthetically pleasing, shore waves make possible sports such as surfboarding, tidal waves have been harnessed for energy generation, and small waves have been used in manufacturing processes such as wave soldering of printed circuit boards. However, ship bow waves are major sources of ship drag, storm-caused waves erode beach fronts, waves caused by opening sluice gates can damage hydroelectric installations, and tsunamis, ocean waves caused by earthquakes and landslides on the ocean floor, have caused the loss of many lives over the years. Knowledge of the behavior of these waves is thus an important engineering subject.

2. Forces on Spillways

Water flowing over the spillway of a dam creates a force on the dam that can be analyzed by our basic equations. If the surface of the spillway is designed so that changes in spillway shape occur smoothly, good assumptions are that: (1) the pressure distribution is hydrostatic far upstream and downstream of the spillway, and (2) energy is being conserved as the water flows over the spillway. From the control volume of Figure 8.1, we have the following results for a spillway of constant width w:

Continuity: $Q = y_1 w v_1 = y_2 w v_2,$ (8.2.1)

Energy: $gy_1 + v_1^2/2 = gy_2 + v_2^2/2,$ (8.2.2)

Momentum in horizontal direction:

$(0.5\rho gy_1)\, wy_1 - (0.5\rho gy_2)\, wy_2 - F_s = \rho Q(v_2 - v_1),$ (8.2.3)

with F_s being the force that the spillway exerts on the water.

If y_1 and v_1 are known, we can solve for y_2 and v_2 from the continuity and energy equations. We then have

$v_2 = v_1 y_1/y_2,$ (8.2.4)

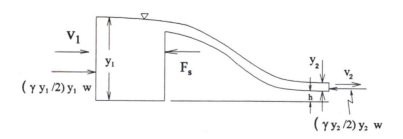

Figure 8.1. Control volume for flow over a spillway showing velocities and horizontal force components. Vertical force components are omitted for clarity.

where y_2 is found from the energy equation (8.2.2) with v_2 eliminated by (8.2.4); namely,

$$[gy_1 + v_1^2/2]\, y_2^2 = gy_2^3 + (y_1 v_1)^2/2. \tag{8.2.5}$$

This is seen to be a cubic equation in y_2 provided y_1 and v_1 are known.

Once the depths and velocities are determined, F_s is found from (8.2.2) as

$$F_s = 0.5\rho gw(y_1^2 - y_2^2) - \rho Q(v_2 - v_1). \tag{8.2.6}$$

In computing the depths we must make a choice between the three roots of the cubic equation (8.2.5). Generally the geometry of the spillway will be of help in ruling out some of the roots.

Note that the actual shape of the spillway does not enter directly into our calculations. The shape, however, does play an indirect role in our equations, in that it must be a smooth enough shape that energy losses are negligible.

Example 8.2.1. Force on a spillway

For the spillway shown in Figure 8.1 with an upstream depth of 8 m, a spillway width of 7 m, and an upstream velocity of 1.5 m/s, find the force on the spillway. The exit floor is at the same elevation as the floor ahead of the spillway. Energy losses can be neglected.

Sought: Horizontal force on a spillway.

Given: $v_1 = 1.5$ m/s, $y_1 = 8$ m, $w = 7$ m, $\rho = 1{,}000$ kg/m^3.

Assumptions: Hydrostatic pressure variation both upstream and downstream of the spillway. Uniform flow before and after the spillway, constant density. Head losses are zero.

Solution: The continuity, energy, and momentum equations corresponding to equations (8.2.1), (8.2.2), and (8.2.3) are

Continuity: $Q = y_1 wv_1 = 8 \times 7 \times 1.5 = y_2 wv_2 = 84$ m^3/s,

Energy: $gy_1 + v_1^2/2 = 9.8 \times 8 + 1.5^2/2 = gy_2 + v_2^2/2 = 79.525,$

Momentum: $(0.5\rho gy_1)\, wy_1 - (0.5\rho gy_2)\, wy_2 - F_s = (0.5 \times 62.4 \times 8) \times 7 \times 8$

$$- (0.5 \times 62.4 \times y_2) \times 7 \times y_2 - F_s = \rho Q\,(v_2 - v_1) = 1.94 \times 84\,(v_2 - 1.5).$$

Then from continuity

$$v_2 = 12/y_2,$$

and from energy

$$9.8y_2 + 12^2/2y_2^2 = 79.525.$$

Rewriting the last equation by multiplying by y_2^2 and rearranging, we have

$$9.8y_2^3 - 79.525y_2^2 + 72 = 0.$$

According to Descartes' rule of signs (Appendix C) there are two positive roots to this equation. The smallest positive real root of this equation is found by Newton's method (Appendix C) to be $y_2 = 1.017$ m. The larger root is near 8. Then

$v_2 = 12/1.017 = 11.80$ m/s.

From (8.2.6) we have for F_s

$$F_s = 0.5(\rho g y_1)(w y_1) - 0.5(\rho g y_2)(w y_2) - \rho Q(V_2 - V_1)$$

$$= (0.5 \times 1,000 \times 9.8 \times 8) \times 7 \times 8 - (0.5 \times 1,000 \times 9.8 \times 1.017) \times 7 \times 1.017$$

$$- 1,000 \times 84(11.80 - 1.5)$$

$$= 2,159,723 - 865,000 = 1,295,026 \text{ N}.$$

3. Forces on Gates

As we saw in the previous example and discussion, the simplifying feature in open channel flows is that in regions where the flow is nearly parallel and uniform, the pressure distribution is known to be very nearly hydrostatic. If we place a gate across an open channel (the gate does not reach to the bottom of the channel, so that flow occurs under the gate), we create a change in the stream elevation across the gate. At the top of the gate, the pressure is of course atmospheric. This must also be true at the bottom of the gate, since the free surface of the stream downstream of the gate is at atmospheric pressure also. If we were to measure the pressure distribution on the gate, we would find that the pressure starts out varying in a nearly hydrostatic manner at the top of the gate, since there is little or no flow there. However, near the bottom of the gate the pressure must depart from the hydrostatic condition and decrease to atmospheric pressure again. Without detailed knowledge of the way in which the pressure variation occurs, it is impossible to obtain the force on the gate by integration of the pressure distribution.

If we consider a control volume (see Figure 8.2) that includes the gate and extends upstream and downstream of the gate far enough so that the pressure variations on the

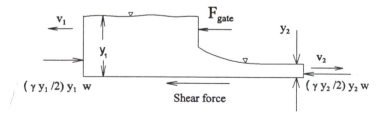

Figure 8.2. Control volume for flow under a gate showing velocities and horizontal force components. Vertical force components have been omitted for clarity.

ends of the control volume are hydrostatic, the continuity equation for a gate of constant width is

$$Q = wy_1v_1 = wy_2v_2. \tag{8.3.1}$$

Therefore, if the velocity is known upstream from the gate and the depth is known both upstream and downstream of the gate, the velocity downstream can be determined by continuity.

We see from the control volume that the only horizontal forces acting on the control volume are the two forces due to the pressure on the ends of the control volume, the force F_g exerted on the control volume by the gate, and the viscous force F_v on the channel bottom. The momentum equation thus tells us that

$$(0.5\rho gy_1)wy_1 - (0.5\rho gy_2)wy_2 - F_v - F_g = \rho Q(v_2 - v_1). \tag{8.3.2}$$

Considering the energy equation, it is reasonable to expect that energy will be lost as the flow passes under the gate. Thus the energy equation including this energy loss is

$$(y_1 + v_1^2/2g) - (y_2 + v_2^2/2g) = h_L. \tag{8.3.3}$$

This equation is used to solve for the head loss once all depths and velocities are known.

Since F_v is the viscous shear force on the channel bottom and h_L represents the energy lost through viscous dissipation in the interior of our control volume, both represent viscous effects. The most reasonable assumption that we can make to simplify the problem further is that F_v is negligible compared to the pressure forces on the right-hand side of (8.3.2), and can be assumed zero. This is born out by experiments. In that case, using the continuity condition (8.3.1) to eliminate v_1 in (8.3.2), we are left with

$$F_g = (0.5\rho gy_1)wy_1 - (0.5\rho gy_2)wy_2 - \rho Qv_1(y_1/y_2 - 1). \tag{8.3.4}$$

Example 8.3.1. Force on a gate
For the situation shown in Figure 8.2 with water flowing, the upstream depth is 6 m, the downstream depth is 2 m, the spillway width is 5 m, and the upstream velocity is 2 m/s. Find the force on the spillway.

Sought: The horizontal force that the water exerts on the spillway.
Given: $v_1 = 2$ m/s, $y_1 = 6$ m, $y_2 = 2$ m, $w = 5$ m, $\rho = 1{,}000$ kg/m³.
Assumptions: Hydrostatic pressure variation both upstream and downstream of the gate. The viscous shear force on the channel bottom is negligible compared to the pressure forces. Constant density. Uniform flow before and after the gate.
Solution: The continuity equation corresponding to equation (8.3.1) is

$$Q = wy_1v_1 = 5 \times 6 \times 2 = wy_2v_2 = 60 \text{ m}^3/\text{s}.$$

Solving for v_2 gives

$v_2 = 60/5 \times 2 = 6$ m/s.

From the momentum equation (8.3.2) we find

$$F_g = 0.5\rho gw(y_1^2 - y_2^2) - \rho Qv_1(y_1/y_2 - 1) = 0.5 \times 1,000 \times 9.8 \times 5 \times (6^2 - 2^2)$$

$$- 1,000 \times 60 \times 2(6/2 - 1) = 784,000 - 240,000 = 544,000 \text{ N}.$$

From (8.3.3) we see that the head loss is

$$h_L = (y_1 + v_1^2/2g) - (y_2 + v_2^2/2g) = (6 + 2^2/2 \times 9.8) - (2 + 6^2/2 \times 9.8)$$

$$= 6.204 - 3.837 = 2.367 \text{ m of water}.$$

The rate of energy loss due to the flow under the gate is

$$\rho gQh_L = 1,000 \times 9.8 \times 60 \times 2.367 = 1,392 \text{ kW}.$$

4. Flow Over a Small Bump in a Channel

Small bumps or objects on the floor of a channel can have differing effects, depending on the speed of the flow. We consider a rectangular channel of constant width w. The channel bottom is horizontal except in a small region where the change in bottom elevation is $y = h(x)$. (See Figure 8.3.) If h and its change with x are small enough, energy losses will be negligible and we can write the continuity and energy equations for the control volume in simple form. The average pressure on each end is the average hydrostatic pressure (0.5 × specific weight × depth), and the average potential energy per unit mass is one-half the depth times g. With y denoting the water depth at any point, we see that

$$p/\rho + yg = 0.5 \times y \times g + 0.5 \times y \times g = y \times g.$$

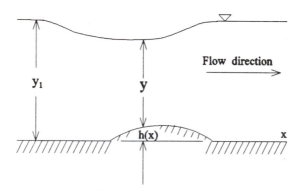

Figure 8.3. Flow over a small bump in the floor of a channel.

We therefore have

Continuity: $Q = wv_1y_1 = wvy,$ (8.4.1)

Energy: $v_1^2/2 + y_1g = v^2/2 + (y + h)g.$ (8.4.2)

Differentiating (8.4.1) with respect to x gives after a little rearrangement

$0 = y \, dv/dx + v \, dy/dx.$

Upon solving this for dv/dx we have

$dv/dx = - (v/y) \, dy/dx.$ (8.4.3)

Differentiating (8.4.2) with respect to x gives us

$0 = v \, dv/dx + (dy/dx + dh/dx)g.$ (8.4.4)

Substituting dv/dx from (8.4.3) and (8.4.4), dividing by g, and solving for dy/dx, we have

$$\frac{dy}{dx} = \frac{dh/dx}{v^2/yg - 1}.$$ (8.4.5)

The dimensionless quantity $v/\sqrt{yg}$ is familiar to us from Chapter 5 as the Froude number. We will use the symbol Fr to refer to it. We note in Table 8.1 the following consequences from (8.4.5) depending on whether the Froude number is greater or less than unity:

Table 8.1. The effect of Froude number and bottom slope on free surface slope

	Fr < 1	Fr > 1
$dh/dx > 0$	$dy/dx < 0$: the depth *decreases* in this region.	$dy/dx > 0$: the depth *increases* in this region
$dh/dx < 0$	$dy/dx > 0$: the depth *increases* in this region	$dy/dx < 0$: the depth *decreases* in this region

Note also from (8.4.5) that where $dh/dx = 0$ the only possibility that the slope dy/dx of the free surface will be finite is if Fr = 1. In that case (8.4.5) is zero divided by zero, and further analysis is necessary to determine the result.

If the Froude number of a flow is less than unity, the flow is said to be a ***subcritical flow***. If the Froude number of the flow is equal to unity, the flow is said to be a ***critical flow***, and if the Froude number of the flow is greater than unity, the flow is said to be a ***supercritical flow***. By looking at the free surface over a bump in a channel to see how the depth is changing, we can determine whether the Froude number is greater or less than unity, and consequently whether the flow is subcritical or supercritical. The next section will further develop the consequences of flow criticality.

Example 8.4.1. Flow over a bump in a channel
Water flowing in a flat rectangular channel of width 4 m has a depth of 2 m and a speed of 8 m/s. A bump occurs in the channel in the region $-0.5 < x < 0.5$. The shape of the bump is given by $h(x) = 0.1 \cos(\pi x)$. Find the shape of the free surface.

Sought: An equation giving the shape of the free surface.
Given: $v_1 = 8$ m/s, $y_1 = 2$ m, $w = 4$ m, $Q = 5 \times 2 \times 4 = 40$ m³/s, $h(x) = 0.1 \cos(\pi x)$, the incoming Froude number is $v_1/\sqrt{y_1 g} = 8/\sqrt{2 \times 9.8} = 1.81$; therefore the incoming flow is supercritical. Thus the depth will increase in the region $-0.5 < x < 0$ where $dh/dx > 0$ and decrease in the region $0 < x < 0.5$ where $dh/dx < 0$.
Assumptions: Parallel flow upstream and downstream of the bump, no energy loss, constant density.
Solution: From the continuity equation we have

$$v = Q/yw = 64/4y = 16/y.$$

Then equation (8.4.5) becomes

$$\frac{dy}{dx} = \frac{dh/dx}{v^2/yg - 1} = -\frac{0.1 \times \pi \times \sin(\pi x)}{256/9.8y^3 - 1}.$$

To find the shape of the free surface, we separate variables and integrate over the channel depth. This results in

$$\int_2^y (256/9.8y^3 - 1)\, dy = \int_{-0.5}^x -0.1 \times \pi \times \sin(\pi x)\, dx,$$

giving

$$(-256/2 \times 9.8y^2 - 1)\,|_2^y = 0.1 \cos(\pi x)\,|_{-0.5}^x,$$

or

$$(-128/9.8)(y^{-2} - 1/4) - y + 2 = 0.1 \cos(\pi x).$$

Rearranging to solve for x, we find x as a function of y in the form

$$x = \cos^{-1}\{10[(-128/9.8)(y^{-2} - 1/4) - y + 2]\}/\pi.$$

To proceed further, numerical evaluation of this function is necessary. Results of such computations are shown in Figure 8.4.

Example 8.4.2. Flow over a bump in a channel
Water flowing in a flat rectangular channel of width 4 m has a depth of 2 m and a speed of 2 m/s. A bump occurs in the channel in the region $-0.5 < x < 0.5$. The shape of the bump is given by $h(x) = 0.1 \cos(\pi x)$. Find the shape of the free surface.

Sought: Equation of the free surface.

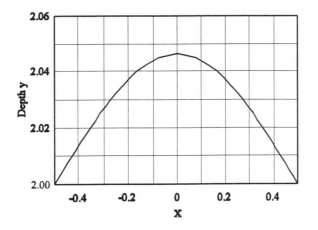

Figure 8.4. Depth results for Example 8.4.1.

Given: $v_1 = 2$ m/s, $y_1 = 2$ m, $w = 4$ m, $Q = 2 \times 2 \times 4 = 16$ m³/s, $h(x) = 0.1 \cos{(\pi x)}$, the incoming Froude number is $v_1/\sqrt{y_1 g} = 2/\sqrt{2 \times 9.8} = 0.452$. Since the Froude number is less than one the incoming flow is subcritical. Therefore the depth will decrease in the region $-0.5 < x < 0$ where $dh/dx > 0$, and increase in the region $0 < x < 0.5$, where $dh/dx < 0$.

Assumptions: Parallel flow upstream and downstream of the bump, no energy loss.

Solution: From the continuity equation,

$v = Q/yw = 16/4y = 4/y.$

Then equation (8.4.5) becomes

$$\frac{dy}{dx} = \frac{dh/dx}{v^2/yg - 1} = -\frac{0.1 \times \pi \times \sin{(\pi x)}}{16/9.8y^3 - 1}.$$

To find the shape of the free surface, we separate variables and integrate as in the previous example, giving

$$\int_2^y (16/9.8y^3 - 1)\, dy = \int_{-0.5}^x -0.1 \times \pi \times \sin{(\pi x)}\, dx,$$

or

$$(-16/2 \times 9.88y^2 - 1)\,|_2^y = 0.1 \cos{(\pi x)}\,|_{-0.5}^x ,$$

or finally

$$(-8/9.8)(y^{-2} - 1/4) - y + 2 = 0.1 \cos{(\pi x)}.$$

Solving for x as a function of y, we find

$$x = \cos^{-1}\{10[(-8/9.8)(y^{-2} - 1/4) - y + 2]\}/\pi.$$

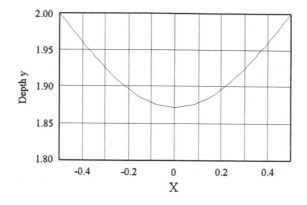

Figure 8.5. Depth results for Example 8.4.2.

To proceed further, numerical evaluation of this function is necessary. Results of such computations are shown in Figure 8.5.

5. Small-Amplitude Standing Gravity Waves

Frequently, small waves are seen on the surface when water flows in a channel. When these waves are perpendicular to the direction of flow they can be analyzed by the methods we have discussed in Chapter 3. We will first consider the special case where the waves are very small in height and are stationary, i.e., when they do not move relative to a stationary observer.

Referring to Figure 8.6, we take the height and velocity downstream of the wave to differ from the upstream values by only infinitesimal amounts. Upstream and downstream from the wave, the flow will be assumed to be parallel and uniform across the cross section of the channel. Then pressure will vary only in the direction of gravity, being the hydrostatic pressure variation described in Chapter 2. Applying the continu-

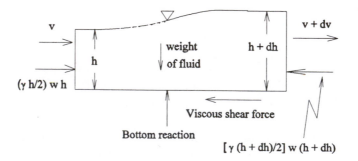

Figure 8.6. Control volume for an infinitesimal stationary wave.

ity and momentum equations and neglecting any shear force on the channel bottom, for a channel of unit width w we have from continuity

$$vwh = w(v + dv)(h + dh). \tag{8.5.1}$$

For the momentum equation, the only relevant forces acting across the jump are the hydrostatic forces p_{ave} times the area, since the viscous forces at the channel bottom are negligibly small in comparison. Then the sum of hydrostatic forces across the wave equals the change in momentum across the wave, or

$$w[\rho gh^2/2 - \rho g(h + dh)^2/2] = \rho vwh(v + dv - v). \tag{8.5.2}$$

Eliminating common terms and neglecting terms of higher order than the first, these reduce to

$$h\,dv + v\,dh = 0 \tag{8.5.3}$$

and

$$-g\,dh = v\,dv. \tag{8.5.4}$$

Eliminating the infinitesimals, we see that the condition for a standing wave is

$$v = \sqrt{gh}. \tag{8.5.5}$$

[*Note*: Equation (8.5.5) could also have been obtained from using the continuity and energy equation, with the assumption of no loss of energy across the wave.]

Equation (8.5.5) states that to have the wave remain in place the channel flow must move with the velocity $\sqrt{gh}$. It is also the speed at which the wave would travel on the surface of a quiescent channel. We will discuss this further in the next section of this chapter.

While we have assumed that the amplitude of the wave is infinitesimal, a more detailed two-dimensional analysis would show that (8.5.5) is valid as long as the channel depth is larger than the wavelength of the waves. When this is not the case the wave speed depends on the wavelength and on surface tension. For cases where the wave speed depends on wavelength, the wave is said to be ***dispersive***, since a wave consisting of more than one wavelength will change shape as it advances. The long waves we have considered here will advance in a constant depth channel with little change in shape.

We note also that (8.5.5) tells us that in very deep water the waves can travel at extremely high speeds. Earthquakes on or below the ocean floor cause waves called ***tsunamis***, which have been clocked at speeds of 600 mph where the ocean is deep. In these regions the wave height is very small, perhaps several inches at the most. As these waves encounter shallow water, the kinetic energy is converted to potential energy with little loss, and the waves can become 100 ft or more in height. Prediction of where tsunamis will come ashore is important all around the Pacific rim, where many deaths have occurred from these waves.

Example 8.5.1. Small-amplitude gravity waves
What is the wave speed in a 10-ft-deep rectangular channel?

Sought: Wave speed in a rectangular channel.
Given: Depth = 10 ft, rectangular channel.
Assumption: Constant density parallel flow, small-amplitude wave, uniform flow conditions upstream and downstream from the wave.
Solution: According to equation (8.5.5), a small-amplitude wave will travel at a speed of

$$v = \sqrt{32.2 \times 10} = 17.9 \text{ ft/s.}$$

---

6. Hydraulic Jumps

Hydraulic jumps are waves of finite amplitude that occur on the free surface of a liquid. They can be seen in many places other than channels, e.g., drinking fountains, kitchen sinks, street gutters during a rain, and tidal waves on rivers and bays adjoining oceans. They are frequently used at the bottom of a dam spillway to dissipate energy and slow the flow so that it will not damage river banks or bottoms. To ensure that the jump occurs on the concrete apron of the spillway, large boulders or concrete blocks are placed so as to trigger the occurrence of a jump.

The analysis for the finite-size wave is in principle much the same as for the infinitesimal wave considered in a previous section. Referring to Figure 8.7 and taking the channel to be rectangular in cross section with a width of w, we get from the continuity and momentum equations

$$v_1 w h_1 = v_2 w h_2 \qquad \text{(continuity)} \tag{8.6.1}$$

and

$$\rho g w h_1^2 / 2 - \rho g w h_2^2 / 2 = \rho v_1 w h_1 (v_2 - v_1) \qquad \text{(momentum).} \tag{8.6.2}$$

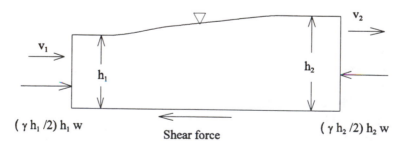

Figure 8.7. Control volume for a hydraulic jump showing velocities and horizontal force components. Vertical force components are omitted for clarity.

Solving (8.4.1) for the downstream velocity gives

$$v_2 = v_1 h_1 / h_2. \tag{8.6.3}$$

Using this to eliminate v_2 in (8.6.2) gives

$$g(h_1 - h_2)(h_1 + h_2) = v_1^2 h_1 (h_1 - h_2)/h_2. \tag{8.6.4}$$

This is seen to be a cubic equation for the unknown depth h_2, with one root being

$$h_2 = h_1; \tag{8.6.5}$$

that is, it is entirely possible that no jump occurs. Dividing out this root gives the quadratic equation

$$g(h_1 + h_2) = v_1^2 h_1 / h_2.$$

The other roots are given by the quadratic formula to be

$$h_2 = h_1(-1 \pm \sqrt{1 + 8v_1^2/h_1 g})/2. \tag{8.6.6}$$

The root with the negative sign in front of the square root sign is seen to always give h_2 as a negative value, which has no physical meaning. The second root will have h_2 greater or less than h_1, depending on whether $v_1^2/h_1 g$ is greater than or less than unity.

To assist in sorting out what can physically happen, it is helpful to consider the energy equation. Because the wave is of a finite height a significant energy loss can be expected to take place across the jump. This is due to turbulence and viscous effects in and near the wave.[1]

As in the case of a bump in a channel that we just considered, the pressure and potential energy terms in the energy equation sum together to give the depth at that point. The energy equation then is, after dividing by the mass rate of flow,

$$(gh_2 + v_2^2/2) - (gh_1 + v_1^2/2) = -gh_L. \tag{8.6.7}$$

Eliminating v_2 leads to

$$h_L = (h_2 - h_1)[-1 + v_1^2(h_1 + h_2)/2gh_2^2]. \tag{8.6.8}$$

To put this entirely in terms of the depths, we can solve (8.4.6) for the velocity, giving

$$v_1^2/gh_1 = (h_2/h_1 + 1)(h_2/2h_1). \tag{8.6.9}$$

Substitution of this into (8.4.8) yields, after a bit of algebraic rearrangement,

$$h_L = (h_2 - h_1)^3/4h_1 h_2. \tag{8.6.10}$$

The power loss P_L is given by

---

[1] We neglect viscous effects in the momentum equation, but we now are saying that we must include them in the energy equation. This is not inconsistent. In the momentum equation, the only external force due to viscosity is at the channel bottom, well away from the wave, and this force is indeed small compared to the pressure forces. In the energy equation, we are including viscous losses throughout the entire control volume, which are not small compared to the other energy terms.

$$P_L = gh_L \dot{m} = g\dot{m}(h_2 - h_1)^3/4h_1h_2. \tag{8.6.11}$$

We are now finally in a position to state the possible outcomes of our analysis. First, if the incoming flow is subcritical flow, equation (8.6.6) predicts that the downstream depth is less than the upstream depth, and according to equation (8.6.6) the head loss is negative. In other words, an energy gain is predicted. This does not agree with intuition, and indeed is physically not possible according to the entropy inequality. Thus if the incoming flow is subcritical the only possible solution is

$$h_2 = h_1. \tag{8.6.12}$$

When, however, the incoming flow is supercritical flow, either

$$h_2 = h_1 \tag{8.6.12}$$

or

$$h_2 = h_1(-1 + \sqrt{1 + 8v_1^2/gh_1})/2 \tag{8.6.13}$$

is a possible solution. According to (8.6.10), equation (8.6.12) has no energy loss and equation (8.6.13) has a positive energy loss. Both then are possible. Which one will occur?

The answer to that question is—it depends on the circumstances! If we have a long channel or river, the flow may continue without a jump for quite a distance, and so (8.6.12) is the proper solution. If, however, there is an upstream obstruction (a bend, rocks, blockage, or whatever), or if the channel is long enough that bottom friction changes the flow sufficiently, a jump will occur and (8.6.13) is the proper solution. The dividing case, v_1^2/gh_1 equal to unity (critical flow), is seen to be the infinitesimal wave of the previous section with zero head loss.

Before leaving this topic, it is instructive to do one more bit of manipulation of our equations. We wish to have a relation between the Froude numbers v_1^2/gh_1 and v_2^2/gh_2 before and after the jump. Starting with equation (8.6.9), and noting that the continuity equation can be rewritten in the form

$$(v_1^2/gh_1)/(v_2^2/gh_2) = (h_2/h_1)^3,$$

after some straightforward algebra to eliminate the appearance of h_2 in (8.6.9) other than in the Froude number combination, it is found that

$$v_2^2/gh_2 = [(1 + \sqrt{1 + 8v_1^2/gh_1})/4(v_1^2/gh_1)^{2/3}]^3$$

or in terms of the Froude numbers,

$$\mathrm{Fr}_2^2 = [(1 + \sqrt{1 + 8\,\mathrm{Fr}_1^2})/4]^3/\mathrm{Fr}_1^2. \tag{8.6.14}$$

Investigating this equation by inserting values for Fr_1, we find that whenever Fr_1 is greater than unity, Fr_2 will be less than unity, and vice versa. This is seen in Figure 8.8. Therefore, across a hydraulic jump the flow always goes from supercritical flow to subcritical flow.

The appearance of the Froude number, encountered previously in our analysis of flow over a bump, is not surprising, since the Froude number is an important parameter in most problems involving free surface flows. We have purposely put our equations

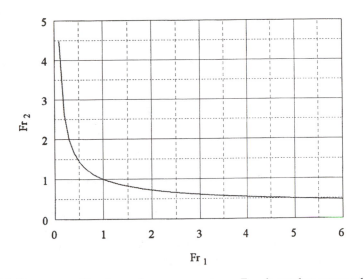

Figure 8.8. Downstream Froude number versus upstream Froude number across a hydraulic jump.

in dimensionless form in this example, since it makes interpretation of the terms easier and frees us from a particular system of units when plotting results.

The preceding analysis considered a normal hydraulic jump, that is, one that is perpendicular to the flow direction. Oblique jumps can also occur. The above analysis holds, providing that the velocities used in the equations are the velocity components normal to the jump itself. Velocity components parallel to the jump go through the jump unchanged. If the upstream Froude number is large enough, it is possible that the downstream Froude number based on the total speed can be greater than unity. Thus, successive jumps are possible, enabling a supercritical flow to turn a corner. An analogy with shock waves will be seen in Chapter 10.

Example 8.6.1. Hydraulic jump
A standing wave is seen to occur in a rectangular channel, with the elevation changing abruptly from 3 to 4 m. What are the velocities before and after the jump?

Sought: Velocities before and after a hydraulic jump.
Given: Rectangular channel, $h_1 = 3$ m, $h_2 = 4$ m.
Assumptions: Incompressible constant density flow that is parallel and uniform before and after the jump. Hydrostatic pressure variation before and after the jump.
Solution: Rearranging equation (8.6.13) by solving for $8v_1^2/h_1 g$, we have

$$8v_1^2/h_1 g = -1 + (2h_2/h_1 + 1)^2 = -1 + (2 \times 4/3 + 1)^2 = 12.44.$$

Therefore

$v_1 = \sqrt{12.44 \times h_1 \times g}/8 = \sqrt{12.44 \times 3 \times 9.8}/8 = 6.76$ m/s.

From the continuity equation $v_2 = v_1 h_1/h_2 = 5.07$ m/s.

7. Unsteady Flows—Moving Waves—Bores

In analyzing infinitesimal waves and hydraulic jumps, we considered them to be stationary. Often they move with respect to the flow, as in tidal waves entering a bay or river, or in waves due to the sudden opening or closing of a gate. Our previous analysis can be easily extended by using the moving control volume concept of Chapter 3. Figure 8.9a shows a wave moving speed c with respect to a fixed observer. An observer moving with the wave sees the velocities shown in Figure 8.9b, which will be the control volume we use for our analysis. Applying continuity and momentum to this control volume for a channel constant width w, we have

$$wh_1(v_1 - c) = wh_2(v_2 - c), \tag{8.7.1}$$

and for the momentum equation [with the help of (8.7.1)],

$$\rho gwh_1^2/2 - \rho gwh_2^2/2 = 0.5\rho gw(h_1^2 - h_2^2)$$

$$= \rho wh_2(v_2 - c)^2 - \rho wh_1(v_1 - c)^2$$

$$= \rho wh_1(v_1 - c)[(v_2 - c) - (v_1 - c)]. \tag{8.7.2}$$

Solving for c from (8.7.1) we have

$$c = (h_2 v_2 - h_1 v_1)/(h_2 - h_1). \tag{8.7.3}$$

We use this expression to eliminate c and (8.7.2) becomes

$$0.5g(h_1^2 - h_2^2) = -h_1 h_2(v_2 - v_1)^2/(h_2 - h_1). \tag{8.7.4}$$

Equation (8.7.4) is seen to be cubic in the quantity h_2. We simplify the solution of this equation by introducing the dimensionless parameters

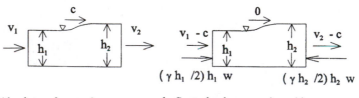

a. Absolute reference frame   b. Control volume moving with wave

Figure 8.9. Absolute reference frame picture and control volume for a moving wave of finite amplitude.

$$s = h_2/h_1, \qquad F = (v_1^2/gh_1) \times (v_2/v_1 - 1)^2.$$

Then after some rearrangement equation (8.7.4) becomes

$$s^3 - s^2 - (1 + 2 \times F)s + 1 = 0. \tag{8.7.5}$$

Introducing the parameter

$$G^2 = 1 + 1.5 \times F,$$

making the change of variables

$$s = 1/3 - (4/3) \times G \times \cos \delta,$$

and recalling the trigonometric relation

$$4 \cos^3 \delta = 3 \cos \delta + \cos 3 \delta,$$

we find that equation (8.7.2) reduces to

$$\cos 3\delta = (7 - 3G^2)/4G^3,$$

which has the three roots

$$\delta_k = \{\cos^{-1}[(7 - 3G^2)/4G^3] + 2\pi k\}/3, \qquad k = 1, 2, \text{ or } 3.$$

The three roots for finding the depth ratio s are then given by

$$s = 1/3 - (4/3) \times G \times \cos \delta_k, \qquad k = 1, 2, \text{ or } 3,$$

or

$$h_2/h_1 = 1/3 - (4/3)\sqrt{1 + 1.5 \times F} \cos \delta_k, \tag{8.7.6}$$

with

$$\delta_k = \{\cos^{-1}[(1 - 1.125F)/(1 + 1.5F)^{3/2}] + 2\pi k\}/3. \qquad \text{with } k = 1, 2, \text{ or } 3.$$

Equation (8.7.6) gives three different roots. Which of these actually corresponds to the depth of the fluid? As in the case of the hydraulic jump the answer is the one that gives a positive real root and a loss, not gain, in energy.

For the positive roots, we consider the head loss from 1 to 2. From the energy equation, this is

$$h_L = h_1 + (v_1 - c)^2/2g - h_2 - (v_2 - c)^2/2g.$$

Introducing the dimensionless parameters and the continuity equation, we have

$$h_L = [h_1/(s - 1)][-(s - 1)^2 + 0.5F(s + 1)]. \tag{8.7.7}$$

We require that for this flow to be physically realizable, the head loss be a positive quantity.

Example 8.7.1. Moving bore

A water channel initially at rest has flowing water introduced at a speed of 5 ft/s and with a depth of 2 ft. What is the depth of the water downstream of the wave and the wave speed?

Sought: Depth h_2 and the wave speed.

Given: $v_1 = 5$ ft/s, $h_1 = 2$, $v_2 = 0$ ft/s, $\rho = 1.94$ slugs/ft^3.

Assumptions: Constant density. Parallel and uniform flow upstream and far downstream from the bore.

Solution: From continuity,

$$c = (v_1 h_1 - v_2 h_2)/(h_1 - h_2) = 5 \times 2/(2 - h_2).$$

Since the bore is moving into still water,

$$F = v_1^2/gh_1 = 5^2/2 \times 32.17 = 0.3886,$$

Putting this into the momentum equation, and letting $s = h_1/2$, the result is equation (8.9.11), with now $G = (1 + 1.5 \cdot F)^{1/2} = 1.2581$. Since the bore is moving into still water $v_2 = 0$, and so

$$\cos 3\delta = (7 - 3G^2)/G^3 = 0.2826$$

and

$$\delta = 0.4281, 2.522, 4.617 \text{ radians},$$

$$s = 1/3 - (4 \times 1.258/3) \times \cos (0.4281 + 2\pi k/3) = -1.1928, +0.4933, +1.6995,$$

$$h_2 = 2s = -2.3856, 0.9866, 3.399.$$

The energy equation gives for the head losses

$$h_L = h_1 - h_2 + [(v_1 - c)^2 - (v_2 - c)^2]/2g = 4.4197, -0.1318, 0.1006.$$

The negative root $h_2 = -2.3856$ obviously has no physical meaning. The first of the two positive roots $h_2 = 0.9866$ gives a negative h_L, that is, an energy gain, which is not possible. The proper root is then $h_2 = 3.399$ ft. The wave speed corresponding to this depth is

$$c = 5 \times 2/(2 - 3.399) = -7.148 \text{ ft/s},$$

the minus sign telling us that the wave moves upstream (to the left).

Table 8.2. Commonly used open channel cross sections

	Shape	Area A	Wetted perimeter P	Hydraulic radius $2A/P$
Rectangular		yb	$b + 2y$	$\dfrac{2yb}{b + 2y}$
Wide (infinite) rectangle		yb	$\approx b$	$\approx 2y$
Triangular		$y^2 \cot \alpha$	$\dfrac{2y}{\sin \alpha}$	$y \cos \alpha$
Trapezoidal		$y(b + y \cot \alpha)$	$b + \dfrac{2y}{\sin \alpha}$	$\dfrac{2y(b + y \cot \alpha)}{b + 2y/\sin \alpha}$
Circular		$\dfrac{D^2}{4}(\alpha - \sin \alpha)$ $\alpha = \sin^{-1}(h/D)$ $0 \le \alpha \le \pi$	αD	$\dfrac{D}{2}\left(1 - \dfrac{\sin \alpha}{\alpha}\right)$

8. Channel Cross Section Shape Effects in Open Channel Flow

Historically, open channel flow was studied earlier than pipe flow, and many of the results still in use go back well over 200 years. It is customary in open channel flow to use $S_b = \sin \alpha$ as a measure of the bottom slope, S for head loss per unit length of channel, and R_h for the hydraulic radius.[2] Several commonly used open channel cross sections are shown in Table 8.2 along with their areas, wetted perimeters, and hydraulic radius.

In open channel flow we still expect that the head loss is given by the Darcy-Weisbach equation in the form

$$h_L = f \frac{L}{2R_h} \frac{V^2}{2g} \tag{8.8.1}$$

that we developed in Chapter 4. Here we have substituted twice the hydraulic radius for the pipe diameter. Solving this equation for velocity, we have

[2]In open channel flow we generally deal with small slopes, and frequently the tangent of the angle is used instead of the sine. Also, no distinction is usually made between the depths measured perpendicular to the channel bottom and in the direction of gravity. For angles less than about 6° the error in making such approximations is very small.

$$V = \sqrt{8g/f}\ \sqrt{R_h h_L/2L} = \sqrt{8g/f}\ \sqrt{R_h S/2}, \tag{8.8.1a}$$

where $S = h_L/L$ is the head loss per unit length along the channel. In 1769, Chezy proposed the formula

$$V = C\sqrt{R_h S/2}, \tag{8.8.2}$$

where C is called the flow resistant factor, R_h is the hydraulic radius, and S is the head loss per unit length of channel. C is not dimensionless in this formulation (Chezy preceded the awareness of the importance of dimensional analysis), and has the dimensions $(\text{length})^{1/2}/\text{s}$. Comparison of the Chezy equation (8.8.1) and its Darcy form (8.8.2) gives

$$C = \sqrt{8g/f}. \tag{8.8.3}$$

Therefore, since f depends on flow properties, so does C.

In 1889, Manning presented a further formulation of the problem as

$$V = (1/n)(R_h/2)^{2/3}\sqrt{S}, \tag{8.8.4}$$

where n is called "Manning's roughness coefficient." Solving this for S (head loss per unit length) gives

$$S = (nV)^2(2/R_h)^{4/3}. \tag{8.8.5}$$

Comparing (8.8.2) and (8.8.4), we see that

$$C = (1/n)(R_h/2)^{1/6}, \tag{8.8.6}$$

where n has the dimensions $(\text{length})^{1/6}/\text{s}$. Typical values for n are given in Table 8.3 in SI units. If it is desired to use British units, the table results can be converted using

$$n_{\text{British units}} = 1.486 \times n_{\text{SI units}}. \tag{8.8.7}$$

Equations (8.8.1) and (8.8.4) are very much like the Darcy-Weisbach formula used in pipe flow in the fully developed range.

For further values see the book by V. T. Chow referenced at the end of the chapter.

a. Uniform flow

Fully developed flow in a channel at constant depth (and, by continuity, constant velocity as well), is referred to as flow at *normal depth*, or *uniform flow*. In uniform flow the head loss per unit length is exactly equal to the change in elevation of the channel bottom, and $S = S_b$.

Example 8.8.1. Discharge in a trapezoidal channel

Find the discharge in a poured concrete channel with trapezoidal cross section. The bottom slope $S_b = 0.001$. The bottom of the trapezoid has a width of 5 m, the depth is 3 m, and the slope of the sides is 45°.

Sought: Volumetric discharge Q.

Table 8.3. Manning's factor n

Surface	Manning's n (SI)		Manning's n (British)	
	Range	Nominal value	Range	Nominal value
Lucite, glass, or plastic film		0.010		0.0149
Planed wood	0.010–0.014	0.012	0.0149–0.0208	0.0178
Unplaned wood	0.011–0.015	0.013	0.0163–0.0223	0.0193
Finished concrete	0.011–0.013	0.012	0.0163–0.0193	0.0178
Unfinished concrete	0.013–0.016	0.015	0.0193–0.0238	0.0223
Vitrified sewer pipe	0.010–0.017	0.013	0.0149–0.0253	0.0193
Brick in mortar	0.012–0.017	0.015	0.0178–0.0253	0.0223
Corrugated metal flumes		0.025		0.0371
Concrete pipe	0.012–0.016	0.015	0.0178–0.0238	0.0223
Cast iron and wrought iron	0.012–0.017	0.015	0.0178–0.0253	0.0223
Riveted and spiral steel pipe	0.013–0.020	0.017	0.0193–0.0297	0.0253
Earth—smooth	0.017–0.025	0.022	0.0253–0.0371	0.0327
Earth—with weeds and stone	0.020–0.040	0.035	0.0297–0.0594	0.0520
Dry rubble	0.025–0.035	0.030	0.0371–0.0520	0.0446
Rock cuts	0.025–0.035	0.033	0.0371–0.0520	0.0490
Gravel		0.029		0.0431

Given: $S_b = 0.001$, $n = 0.015$, $b = 5$ m, $\alpha = 45°$, $y = 3$ m.

Assumptions: Uniform flow.

Solution: First calculate the area, wetted perimeter, and hydraulic radius. From Table 8.2,

$$A = (5 + 3 \cot 45°)(3) = 24 \text{ m}^2,$$

$$P = (5 + 2)(3/\sin 45°) = 13.485 \text{ m},$$

$$R_h = 2A/P = (2)(24/13.485) = 3.56 \text{ m}.$$

From equation (8.8.4),

$$Q = VA = (A/n)(R_h/2)^{2/3}S_b^{1/2} = (24/0.015) \times (3.56/2)^{2/3} \times (0.001)^{1/2} = 74.3 \text{ m}^3/\text{s}.$$

9. Channels with Optimum Shape

A practical question to ask is what channel configuration has a minimum hydraulic radius for a given flow rate, since the cost of construction and maintenance of a channel will depend on its size. Assume uniform flow and use (8.8.4) to write the velocity in the form

$$V = Q/A = (1/n)(R_h/2)^{2/3}\sqrt{S_b} = (1/n)(A/P)^{2/3}\sqrt{S_b},$$

with Q the discharge.

We can solve this expression for A to obtain

$$A = (\sqrt{S_b}/Qn)^{3/5}P^{2/5} = cP^{2/5} \quad \text{(say),} \qquad (8.9.1)$$

where $c = (\sqrt{S_b}/Qn)^{3/5}$ is a constant for fixed Q, S_b, and n. For constant c, then, the area will be minimum when the wetted perimeter is a minimum.

The problem of finding which shape is the optimum cross section for given flow conditions is a difficult problem in the branch of mathematics called "the calculus of variations." We content ourselves here with looking at the standard cross sections and determining the optimum cross sections for each. (The cross section that does in fact have the optimum shape is the circle.) The idea is to minimize either area or wetted perimeter, but to do so while still satisfying (8.9.1). Since A and P will depend on the geometric parameters of the shape, there are basically two methods for minimizing the hydraulic radius:

1. Solve A (or P) for one of the geometric parameters and substitute this into P (or A). Then put this expression into (6.4.7), obtaining an equation in P (or A). Minimize this expression by differentiating with respect to the remaining geometric parameters and setting the derivative of P (or A) with respect to these variables to zero.
2. A technique called "Lagrange multipliers" is useful for minimizing a quantity such as A (or P) with a restraint condition like (8.9.1). Write either $F = A + \lambda(A - cP^{2/5})$ or $G = P + \lambda(A - cP^{2/5})$. Differentiate either F or G with respect to the geometric parameters and set these derivatives to zero.

Examples 8.9.1 to 8.9.4 illustrate these methods for finding the optimum channel, while Table 8.4 summarizes the results.

Table 8.4. Optimum conditions for open channel cross sections

	Shape	Optimum Conditions	Area A	Wetted perimeter P	Hydraulic radius $2A/P$
Rectangular		$b = 2y$	$2y^2$	$4y$	y
Wide (infinite) rectangle		No optimum conditions	yb	$\approx b$	$\approx 2y$
Triangular		$\alpha = 45°$	y^2	$2y\sqrt{2}$	$y/\sqrt{2}$
Trapezoidal		$\alpha = 60°$ $b = 2y/\sqrt{3}$	$y^2\sqrt{3}$	$2y\sqrt{3}$	y
Circular		$\alpha = 180°$	$\dfrac{\pi D^2}{4}$	D	$\dfrac{D}{2}$

Example 8.9.1. Optimum shape for a rectangular channel
Find the optimum shape for a rectangular channel with given Q, S, and n.

Sought: Optimum shape for a channel of rectangular cross section.
Given: From Table 8.2 for a rectangular channel, $A = by$, $P = b + 2y$.
Assumptions: Uniform flow.
Solution: From (8.9.1) we know that $A = cP^{2/5}$, where c is a constant. First solve for b from the expression for wetted perimeter, obtaining

$$b = P - 2y.$$

Putting this in A we get

$$A = y(P - 2y).$$

Since $A = cP^{2/5}$, this means that

$$A = y(P - 2y) = cP^{2/5}.$$

Differentiating this with respect to y, we have

$$(P - 2y) + y(dP/dy - 2) = c(2/5)P^{-3/5}\,dP/dy.$$

Setting $dP/dy = 0$ for the minimum condition, we are left with

$$(P - 2y) + y(-2) = 0,$$

or upon solving for P,

$$P = 4y.$$

This gives

$$P = 4y, \qquad b = 2y, \qquad \text{and} \qquad A = 2y^2.$$

Example 8.9.2. Optimum shape for a triangular channel
Find the optimum shape for a triangular channel with given Q, S, and n.

Sought: Optimum shape for a channel of triangular cross section.
Given: From Table 8.2 for a triangular channel, $A = y^2 \cot \alpha$, $P = 2y/\sin \alpha$.
Assumptions: Uniform flow.
Solution: From (8.9.1) we know that the expression $A - cP^{2/5}$, where c is a constant, must be zero. We will use the method of Lagrange multipliers for this problem to illustrate its use. Write

$$F = A + \lambda(A - cP^{2/5}),$$

where λ, a constant, is the Lagrange multiplier. Since

$\partial A/\partial y = 2y \cot \alpha, \qquad \partial A/\partial \alpha = -y^2/\sin^2\alpha, \qquad \partial P/\partial y = 2/\sin \alpha,$

$\partial P/\partial \alpha = -2y \cos \alpha/\sin^2\alpha,$

to minimize F with respect to D and α we differentiate F with respect to y and have

$\partial F/\partial y + (2y \cot \alpha)(1 + \lambda) - (2c\lambda/5)2/P^{3/5} \sin \alpha = 0$

and

$\partial F/\partial \alpha = (-y^2 \sin^2 \alpha)(1 + \lambda) + (2c\lambda/5)2y \cos \alpha/P^{3/5} \sin^2 \alpha = 0.$

Solving these two equations for terms involving P, we obtain

$$\frac{5P^{3/5}(1 + \lambda)}{2c\lambda} = \frac{1}{y \sin \alpha \cot \alpha} = \frac{2 \cos \alpha}{y}.$$

From the expressions on either side of the second equals sign we see that y cancels and we are left with

$\cos^2 \alpha = 1/2, \qquad$ which implies $\alpha = 45°$.

This value of α gives

$P = 2y\sqrt{2} \qquad$ and $\qquad A = y^2.$

Example 8.9.3. Optimum shape for a trapezoidal channel
Find the optimum shape for a trapezoidal channel with given Q, S, and n.

Sought: Optimum shape for a channel of trapezoidal cross section.
Given: From Table 8.2 for a trapezoidal channel, $A = y(b + y \cot \alpha)$, $P = b + 2y/\sin \alpha$.
Assumptions: Uniform flow.
Solution: From (6.4.7) we know that $A = cP^{2/5}$, where c is a constant. First solve for b from the expression for wetted perimeter, obtaining

$b = P - 2y/\sin \alpha.$

Putting this in A we get

$A = y(P - 2y/\sin \alpha + y \cot \alpha).$

Since $A = cP^{2/5}$, this means that

$A = y(P - 2y/\sin \alpha + y \cot \alpha) = cP^{2/5}.$

Differentiating this with respect to y, we have

$(P - 2y/\sin \alpha + y \cot \alpha) + y(dP/dy - 2/\sin \alpha + \cot \alpha) = c(2/5)P^{-3/5}dP/dy.$

Setting $dP/dy = 0$ for the minimum condition, we are left with

$(P - 2y/\sin \alpha + y \cot \alpha) + y(-2/\sin \alpha + \cot \alpha) = 0,$

or upon solving for P,

$$P = 2y(2 \sin \alpha - \cot \alpha) = 2y(2 - \cos \alpha)/\sin \alpha.$$

If in this expression we hold y constant and differentiate with respect to α, we find that

$$dP/d\alpha = 2y[1 - (2 - \cos \alpha) \cos \alpha/\sin^2 \alpha] = 2y[\sin^2 \alpha - 2 \cos \alpha + \cos^2 \alpha]/\sin^2 \alpha$$

$$= 2y[1 - 2 \cos \alpha]/\sin^2 \alpha.$$

We note then that $dP/d\alpha = 0$ when $\alpha = \cos^{-1} 0.5 = 60°$, which means the trapezoid is half of a hexagon. This value of the angle gives

$$P = 2y\sqrt{3}, \qquad b = P - 4y/\sqrt{3} = 2y\sqrt{3}/3, \qquad \text{and} \qquad A = y(b + y/\sqrt{3}) = y^2\sqrt{3}.$$

Example 8.9.4. Optimum shape for a circular channel

Find the optimum shape for a circular channel with given Q, S, and n.

Sought: Optimum shape for a channel of circular cross section.
Given: From Table 8.2 for a circular channel, $A = (\alpha - \sin \alpha)D^2/8$, $P = \alpha D/2$.
Assumptions: Uniform flow.
Solution: From (6.4.7) we know that $A = cP^{2/5}$, where c is a constant. We will again use the method of Lagrange multipliers for this problem. Write

$$F = A + \lambda (A - cP^{2/5}).$$

Since

$$\partial A/\partial D = (\alpha - \sin \alpha)D/4, \qquad \partial A/\partial \alpha(1 - \cos \alpha) \, \partial P/\partial D = \alpha/2, \qquad \partial P/\partial \alpha = D/2,$$

to minimize F with respect to D and α, we have

$$\partial F/\partial D = (\alpha - \sin \alpha)(D/4)(1 + \lambda) - (2c\lambda/5)\alpha/2P^{3/5} = 0$$

and

$$\partial F/\partial \alpha = (1 - \cos \alpha)(D^2/8)(1 + \lambda) - (2c\lambda/5)D/2P^{3/5} = 0.$$

Solving this, we obtain

$$\frac{5P^{3/5}(1 + \lambda)}{2c\lambda} = \frac{(\alpha - \sin \alpha)D/4}{\alpha/2} = \frac{(1 - \cos \alpha)D^2/8}{D/2} .$$

From the expression on either side of the second equals sign, we see that D drops out and we are left with

$$\alpha - 2 \sin \alpha + \alpha \cos \alpha = 0.$$

The only root of this expression for $0 \leq \alpha \leq 2\pi$ is $\alpha = \pi$ radians or, equivalently, $180°$. This value of the angle gives

$$P = \pi D \quad \text{and} \quad A = \pi D^2/8.$$

10. Channels with Gradual Slope

For an open channel with a small bottom slope S_b, the energy equation written for a control volume of length L going from x_1 where the bottom elevation, depth, and velocity are z_1, y_1, and V_1, to x_2 where the corresponding quantities are z_2, y_2, and V_2 (Figure 8.10) is

$$[V_1^2/2g + y_1 + z_1] - [V_2^2/2g + y_2 + z_2] = h_L \equiv SL,$$

where S is the head loss per unit channel length. Since $z_1 - z_2 = S_b L$,

$$[V_1^2/2g + y_1] - [V_2^2/2g + y_2] + S_b L = SL. \tag{8.10.1}$$

If we consider the two points x_1 and x_2 to be an infinitesimal distance dx apart, then

$$L = dx = x_2 - x_1, \quad dy = y_2 - y_1, \quad dV = V_2 - V_1,$$

and the energy equation can be rewritten as

$$[V_1^2/2g + y_1] - [(V_1 + dV)^2/2g + y_1 + dy] + S_b dx = S\,dx. \tag{8.10.2}$$

Canceling common terms and neglecting differentials of second order, we are left with

$$-V\,dV/g - dy + S_b\,dx = S\,dx. \tag{8.10.3}$$

After dividing by dx and some rearrangement, (8.10.3) becomes

$$S - S_b = -dy/dx - V(dV/dx)/g. \tag{8.10.4}$$

By continuity, at any section we have

$$VA = V_1 A_1 = Q, \tag{8.10.5}$$

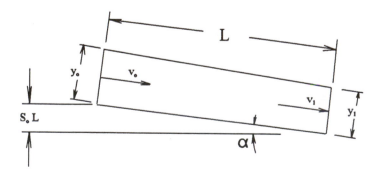

Figure 8.10. Control volume for open channel flow.

which upon differentiating with respect to x becomes

$A\ dV/dx + V\ dA/dx = 0.$ (8.10.6)

Since A varies directly with y that in turn varies with x, using the product rule of calculus we can rewrite this as

$A\ dV/dx + V(dA/dy)(dy/dx) = 0.$

Solving this for dV/dx, we have

$dV/dx = -[V^2\ dA/dy](dy/dx)/gA.$ (8.10.7)

Using this to eliminate dV/dx in (8.10.4), we have the form

$S - S_b = [-1 + (V^2/gA)(dA/dy)]\ dy/dx.$

This can be solved for dy/dx, giving

$$\frac{dy}{dx} = \frac{(S_b - S)}{1 - (V^2/gA)(dA/dy)} = \frac{S_b - S}{1 - \mathrm{Fr}^2},$$ (8.10.8)

where Fr is the Froude number in the form

$$\mathrm{Fr}^2 = \frac{V^2}{gA}\frac{dA}{dy}.$$

We see from (8.10.8) that the question of whether the depth increases or decreases depends on both the Froude number and the ratio S/S_b. Recall that flow with Fr < 1 is called subcritical flow, flow with Fr = 1 is called critical flow, and flow with Fr > 1 is called supercritical flow. Additional descriptive terminology commonly used is the following:

The slope is ***mild*** if $S < S_b$.

The slope is ***critical*** if $S = S_b$.

The slope is ***steep*** if $S > S_b$.

The bed is ***horizontal*** if $S_b = 0$.

The bed is said to be ***adverse*** if $S_b < 0$.

Consequently, these deductions follow from (8.10.8):

y increases if:

1. Fr < 1 (subcritical flow) and $S < S_b$ (mild slope).

2. Fr > 1 (supercritical flow) and $S > S_b$ (steep slope).

3. Fr > 1 (supercritical flow) and the bed is horizontal.

4. Fr > 1 (supercritical flow) and the slope is adverse.

y decreases if:

1. Fr > 1 (supercritical flow) and $S < S_b$ (mild slope).

2. Fr < 1 (subcritical flow) and $S > S_b$ (steep slope).

3. Fr < 1 (subcritical flow) and the bed is horizontal.

4. Fr < 1 (subcritical flow) and the slope is adverse.

In order to be able to solve (8.10.8), it is necessary to substitute an explicit expression for S. Using the Chezy form (8.8.5) for S, we find that the result is

$$\frac{dy}{dx} = \frac{S_b - S}{1 - (V^2/g)(dA/Ady)} = \frac{S_b - (Vn)^2(2/R_h)^{4/3}}{1 - (V^2/g)(dA/Ady)}$$

$$= \frac{S_b - (Qn/A)^2(2/R_h)^{4/3}}{1 - (Q^2/A^2g)(dA/Ady)}, \tag{8.10.9}$$

where $Q = VA$.

When the area of the channel is specified, it is possible to see the trend for the depth y from (8.10.9). For instance, for a rectangular channel of width b, $A = yb$ and $dA/dy = b$. Then (8.10.9) becomes

$$\frac{dy}{dx} = \frac{S_b - (Qn)^2 P^{4/3}/A^{10/3}}{1 - (Q^2/y^2b^2g)(1/y)} = \frac{S_b - (Qn)^2(b + 2y)^{4/3}/(by)^{10/3}}{1 - (Q^2/b^2g)(1/y^3)}$$

$$\approx \frac{S_b[1 - (Qn)^2/S_bb^2y^{10/3}]}{1 - (Q^2/b^2g)(1/y^3)} \tag{8.10.10}$$

if $b \gg y$ (a very wide channel). To make the discussion easier, let

$$y_n = [(Qn)^2/b^2S_b]^{3/10} \quad \text{and} \quad y_c = (Q^2/b^2g)^{1/3},$$

where y_n is called the **normal depth**. Then (6.4.15) can be rewritten as

$$\frac{dy}{dx} \approx \frac{S_b[1 - (y_n/y)^{10/3}]}{[1 - (y_c/y)^3]}. \tag{8.10.11}$$

It is seen from (8.10.11) that when y is at the normal depth, there will be no further change in the depth.

Note that the above definition of normal depth is restricted to very wide channels. More generally, it would vary with y, and be defined as the value of y for which

$$S = S_b. \tag{8.10.12}$$

This reduces to the previous definition when $R_h = y$ and $A = yb$.

The solution to (8.10.11) will depend on the relative magnitudes of y_c and y_n. The following summarizes the trends in y for a wide rectangular channel:

$y_n > y_c$, $S_b > 0$ (mild slope):

$0 < y < y_c$: Supercritical flow. $dy/dx > 0$, $y \to y_c$.
$y_c < y < y_n$: Subcritical flow. $dy/dx < 0$, $y \to y_c$.
$y_n < y$: Subcritical flow. $dy/dx > 0$, y increases.

$y_n = y_c$, $S_b > 0$ (critical slope):

$0 < y < y_c$: Supercritical flow. $dy/dx > 0$, $y \to y_c$.
$y_c < y$: Subcritical flow. $dy/dx > 0$, y increases.

$y_n < y_c$, $S_b > 0$ (steep slope):

$0 < y < y_n$: Supercritical flow. $dy/dx > 0$, $y \to y_n$.
$y_n < y < y_c$: Supercritical flow. $dy/dx < 0$, $y \to y_n$.
$y_c < y$: Subcritical flow. $dy/dx > 0$, y increases.

$S_b = 0$ (horizontal slope):

$0 < y < y_c$: Supercritical flow. $dy/dx > 0$, $y \to y_c$.
$y_c < y$: Subcritical flow. $dy/dx < 0$, $y \to y_c$.

$S_b < 0$ (adverse slope):

$0 < y < y_c$: Supercritical flow. $dy/dx > 0$, $y \to y_c$.
$y_c < y$: Subcritical flow. $dy/dx < 0$, $y \to y_c$.

To see how the depth changes with x, it is simplest to rearrange (8.10.9) to solve for dx. Doing so we have

$$dx = \frac{1 - Q^2(dA/dy)/A^3 g}{S_b - (2/R_h)^{4/3}(Qn/A)^2} \, dy,$$

which upon integration over a channel length L with depths y_1 at the start and y_2 at the end gives

$$L = \int_{y_1}^{y_2} \frac{1 - Q^2(dA/dy)/A^3 g}{S_b - (2/R_h)^{4/3}(Qn/A)^2} \, dy. \tag{8.10.13}$$

The integrand is a known function of y for a given problem, and hence we have found how the distance along the channel varies as a function of y. We therefore have our solution (y as a function of distance x) in inverse form.

Table 8.5. Results of numerical integration of equation (8.10.14)

y	x	R_h	Froude number	Integrand
3.00	0	2.367	0.538	Infinite
3.05	684.3	2.416	0.525	5,788.3
3.10	1,193.7	2.413	0.512	4,545.4
3.15	1,607.6	2.446	0.500	3,801.5
3.20	1,961.0	2.461	0.489	3,306.5
3.25	2,272.7	2.476	0.477	2,953.5
3.30	2,554.0	2.491	0.466	2,689.1
3.35	2,812.0	2.504	0.456	2,483.8
3.40	3,051.7	2.518	0.446	2,319.7

To be able to carry out the integration in (8.10.13) the shape of the channel cross section must be specified so that dA/dy is known. Generally it is necessary to carry out the integration numerically, using a numerical scheme such as the trapezoidal rule. (See Appendix C.) For example, in a rectangular channel where $A = by$, $P = b + 2y$, and $R_h = 2by/(b + 2y)$, (8.10.13) becomes

$$L = \int_{y_1}^{y_2} \frac{1 - Q^2/b^2y^3g}{S_b - (Qn)^2(b + 2y)^{4/3}/(by)^{10/3}} \, dy. \tag{8.10.14}$$

Numerical integration is necessary to proceed further.

Example 8.10.1. Slowly varying open channel flow
A planed wood channel having a rectangular cross section and width 4 m has a slope of 0.06°. The discharge is 35 m³/s, and at the initial location the depth is 3 m. Find the depth as a function of distance down the channel.

Sought: Variation of depth in an open channel with a rectangular cross section.
Given: Channel is rectangular with width 4 m, channel slope is 0.06°, channel is made of planed wood, discharge is 35 m³/s, initial depth is 3 m. From Table 8.3 Manning's $n = 0.012$.
Assumptions: Slowly varying flow.
Solution: Here

$$A = 4y, \qquad S_b = \sin 0.06 = 0.001047,$$

$$R_h = 2A/P = 8y/(2y + 8) = 4y/(y + 4),$$

$$S = (Qn/by)^2 \times [2/R_h]^{4/3} = (35 \times 0.012/4y)^2 \times [(y + 4)/2y]^{4/3},$$

$$Fr^2 = Q^2/gb^2y^3 = (35^2/9.8 \times 4^2y^3,$$

$$y_c = \left(\frac{Q^2}{b^2 g}\right)^{1/3} = \left(\frac{35^2}{4^2 \times 9.8}\right)^{1/3} = 1.984 \text{ m},$$

and y_n is the root that makes $S_b = [(Qn/\text{area})^2 \times (2/R_h)^{4/3}]$. A numerical solution of this latter equation gives $y_n = 2.8997$ m. Thus $y > y_n > y_c$, and the flow is subcritical with a mild slope. We expect then that as x increases, y and the hydraulic radius will increase and the Froude number will decrease.

We put the above results into (6.4.18) and the result is

$$x = \int_3^y \frac{1 - \text{Fr}^2}{S_b - S} \, dy,$$

with the various quantities given above. This integral must be integrated numerically. Since the integrand varies smoothly with y, the integration can be carried out using the trapezoidal rule. A summary of the results of the numerical integration is shown in Table 8.5. The integration was carried out using a step size of $\Delta y = 0.001$.

---

11. Dams

Dams are the principal users of large hydraulic turbines, and by themselves are interesting structures in the manner in which they use many of the topics we have heretofore discussed. They also demonstrate the benefits and uses of this branch of fluid mechanics. One of the early major dams, Hoover Dam, completed in 1935, serves as a good illustration of the many features of such a large structure.

The Colorado River is one of the principal rivers of the southwestern United States. Its basin drains seven states (Wyoming, Colorado, Utah, New Mexico, Arizona, Nevada, and California) before flowing through Mexico into the Gulf of California. Its waters are vital to United States food production and also to the viability of the population of the southwestern states. But a river, besides providing benefits, also is a source of problems caused by occasional flooding, which can block canals and irrigation headworks with sediment, change the course of the river, and cause animal and human deaths. During low-flow years, special diversion works are necessary for agriculture to be successful, and low flows limit the amount of land that can be irrigated and populated. The chief driving force that initiated the construction of Hoover Dam was a particularly devastating series of floods starting in 1905. They caused a change in course of the Colorado River severe enough to flood the Imperial Valley in southern California for over a year, creating in the process the Salton Sea.

The federal government's response to this problem was the authorization of the construction of Hoover Dam, whose construction took place between 1931 and 1935. The short time period involved in its construction, together with the remoteness and ruggedness of this desert region, the relative primitiveness of the equipment then available, the severe temperatures encountered, and the fact that the dam was completed 2 years before the scheduled completion date, make it even more of an engineering miracle than its size alone would indicate. (The American Society of Civil

Engineers has designated Hoover Dam as one of the seven modern civil engineering wonders of the United States.) At the time of construction it was the world's largest dam. For over 10 years it was the largest hydraulic power provider in the world. As in the case of many of the world's largest dams, the primary motivation of the dam was control of a river flow, and the hydraulic power produced by the water flowing through it was a secondary, but still important, benefit. Lake Mead, the reservoir created by the construction of the dam, extends about 110 miles upstream of Hoover Dam to the exit of the Grand Canyon of the Colorado. Besides serving as a water supply to nearby cities and water storage for flood control, it is an important water recreational resource in a desert region in which natural lakes are absent. Power provided by Hoover Dam during World War II made possible the creation of the aviation industry in southern California.

Figure 8.11 shows a picture of the dam and Figure 8.12 shows a typical turbine installation in the turbine rooms. The dam itself is of concrete-arch gravity construction, and is 726 ft high from the foundation rock to the two-lane highway on its crest. Four 50-ft-diameter diversion tunnels, two on each side of the river, were tunneled out of the rock at an early stage in the construction of the dam. They allowed the river to bypass the site until the dam was completed. Originally these tunnels had entrances far upstream of the dam. Today, the upstream entrances of the outer pair of tunnels have been plugged, and their downstream portions are fed by large spillways whose levels are controlled by adjustable gates. These spillways, together with exit gates, are used during times of severe flooding to cause a large portion of the river flow to bypass the dam.

The upstream entrances to the second pair of tunnels were also plugged after construction was completed. The inner pair of tunnels now contain 30-ft-diameter steel pipes, and connect the intake towers to the penstock tubes. In emergency situations needle valves in the plugs and exit gates can be opened to divert water past the dam.

Four intake towers on the upstream side of the dam feed water through the inner tunnels to 30-ft-diameter penstock tubes. From these, a series of 13-ft-diameter penstock tubes, one for each turbine, direct the water to the turbine entrance chambers. The water then proceeds through the vanes and out the draft tubes back into the river. The turbines are housed in a horseshoe-shaped building at the foot of the dam. The maximum usable head is 590 ft, and the minimum is 420 ft. During the period of 1937 through 1982, the average annual net power generated was about 3.5 billion kilowatt-hours, with a maximum during this period of almost 6.5 billion kilowatt-hours in 1953. (An oil-fired generating plant would need about 6 million barrels of oil to provide the same average amount of energy.) Additional turbines have been added since that time. The energy produced is rigidly allocated to the various states surrounding Lake Mead, as shown in Table 8.6.

Since the construction of Hoover Dam, two other dams have been added to the Colorado River downstream of Hoover Dam, and one upstream. Davis Dam, completed in 1949, is 67 river-miles downstream of Hoover Dam and is an earth and rockfill embankment type. It rises 128 ft above the river surface; the waters

Figure 8.11. Hoover Dam as seen from the downstream side. Lake Mead, formed by the dam, is in the background, and the Colorado River is in the foreground. The buildings at the foot of the dam house the turbines, control room, and shop. Dams further upstream and downstream the Colorado River are also controlled from this location. (Photo courtesy the Bureau of Reclamation, Department of the Interior, U.S. Government.)

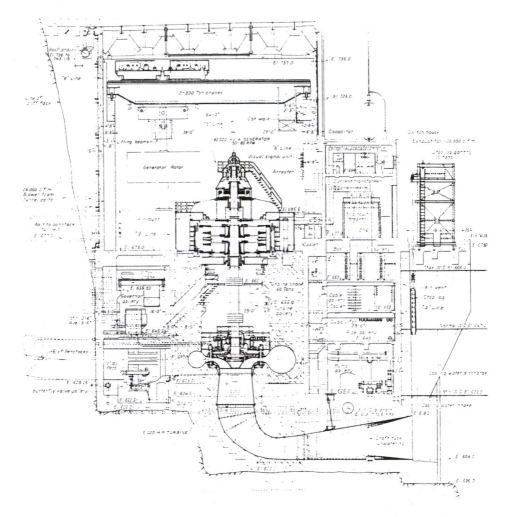

Figure 8.12. A cross section of a turbine installation in one of the turbine rooms, along with connecting water lines, for Hoover Dam. (Drawing courtesy the Bureau of Relcamation, Department of the Interior, U.S. Government.)

impounded by it created Lake Mohave. Further downstream is Parker Dam, which is a concrete-arch structure extending 85 ft above the riverbed. It created Lake Havasu. Both dams and the lakes they created provide electric power, serve important roles for flood control, provide water supplies for both drinking and irrigation water, and provide important recreational sites and wildlife habitats. Upstream of Hoover Dam and the Grand Canyon is the Glen Canyon Dam, which created Lake Powell.

Table 8.6. Power allocation for Hoover Dam

Total power production (%)	Power receiver
35.3	Metropolitan Water District of Southern California
17.6	State of Nevada
17.6	State of Arizona
17.6	City of Los Angeles, CA
7.9	Southern California Edison Company
1.8	City of Glendale, CA
1.6	City of Pasadena, CA
0.6	City of Burbank, CA

Suggestions for Further Reading

Ackers, P., et al., *Weirs and Flumes for Flow Measurement*, Wiley, New York, 1978.

ASCE, "Friction Factors in Open Channels," Report of the Committee on Hydromechanics, *ASCE J. Hydraul. Div.*, March 1963, pp. 97–143.

Bakhmeteff, B. A., *Hydraulics of Open Channels*, McGraw-Hill, New York, 1932.

Bos, M. G., I. A. Replogle, and A. I. Clemmens, *Flow Measuring Flumes for Open Channel Systems*, Wiley, New York, 1984.

Bos, M. G., *Long-Throated Flumes and Broad-Crested Weirs*, Martinus Nijhoff (Kluwer), Dordrecht, The Netherlands, 1985.

Boussinesq, J., "Essai sur la theorie des eaux courants," Memoires presentes par divers savants a l'Academie des Sciences, vol. **23**, 1877.

Boussinesq, J., "Theorie de l'ecoulement tourbillonant et tumultueux des liquides dans les lits rectilignes a grande section (tuyaux de conduite et cannaux decouverts), quand cet ecoulement c'est regularise en un regime uniforme, c'est-a-dire, moyennement pareil a travers toutes les sections normales du lit," *Comptes Rendus de l'Academie des Sciences*, vol. **CXXII**, pages 1290–1295, 1896.

Brater, E. F., *Handbook of Hydraulics*, 6th edition, McGraw-Hill, New York, 1976.

Chezy, A., 1769. Discussed in "Antoine Chezy, histoire d'une formule d'hydraulique," by G. Mouret, *Annales des Ponts et Chaussees*, vol. **II**, 1921.

Chow, V. T., *Open Channel Hydraulics*, McGraw-Hill, New York, 1959.

French, R. H., *Open-Channel Hydraulics*, McGraw-Hill, New York, 1985.

Henderson, F. M., *Open Channel Flow*, Macmillan, New York, 1966.

Hoover Dam, Fifty Years, U.S. Department of the Interior, Bureau of Reclamation, S/N 024-003-00159-7, 1985.

Manning, R., "On the Flow of Water in Open Channels and Pipes," *Trans. I.C.E. Ireland*, vol. **20**, pp. 161–207, 1891.

Powell, R. W., "Resistance to Flow in Rough Channels," *Trans. Am. Geophys. Union*, vol. **31**, no. 4, pp. 575–582, August 1950.

Robertson, J. M., and H. Rouse. "The Four Regimes of Open Channel Flow," *Civ. Eng.*, vol. **11**, no. 3, pp. 169–171, March 1941.

Sellin, R. H. J., *Flow in Channels*, Gordon & Breach, London, 1970.

U.S. Bureau of Reclamation, "Research Studies on Stilling Basins, Energy Dissipators, and Associated Appurtenances," Hydraulic Lab. Rep. Hyd-399, June 1, 1955.

Warring, R. H., editor, *Hydraulic Handbook*, 8th edition, Gulf Pub., Houston, 1983.

Problems for Chapter 8

Gates

8.1. Determine the net force on the gate in the figure and the energy loss. The fluid is water with specific weight = 62.4 lb/ft^3.

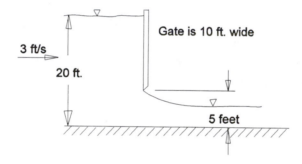

Problem 8.1

8.2. For a discharge of 40 ft^3/s per foot width in the figure, determine the downstream velocity, the rate at which energy is lost, and the force on the gate per unit channel width. The fluid is water with specific weight = 62.4 lb/ft^3.

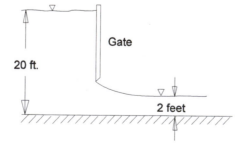

Problem 8.2

8.3. Given that $v = 0.8$ ft/s in the figure, what is the force/width on the gate? Is it possible for a hydraulic jump to occur after the gate as shown? The fluid is water with specific weight = 62.4 lb/ft^3.

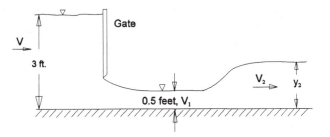

Problem 8.3

8.4. A gate in a 5-ft-wide river allows water to flow under it at a flow rate of 35 ft^3/s. The gate controls the upstream depth. See the figure. What is the force on the gate? At what rate is energy lost? The fluid is water with specific weight = 62.4 lb/ft^3.

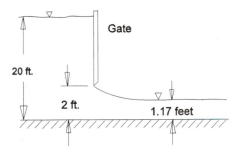

Problem 8.4

8.5. Calculate the net force per unit width exerted on the gate in the figure. The fluid is water with specific weight = 62.4 lb/ft^3. $Q = 50$ ft^3/s.

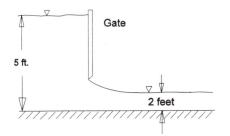

Problem 8.5

8.6. If $y = 0.5$ m in the figure, find the discharge per unit width over the dam and the depth y, assuming no losses before the hydraulic jump occurs. The fluid is water.

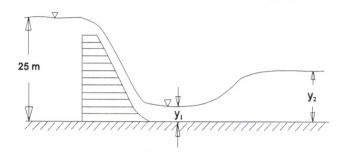

Problem 8.6

8.7. Determine the force exerted by the water per unit length on the gate in the figure. The discharge is 12 ft³/s per foot width, and the pressure far upstream and downstream from the gate can be considered hydrostatic.

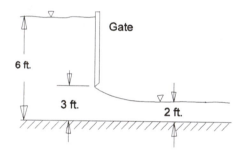

Problem 8.7

8.8. Determine the net force on the gate in the figure per unit width if $Q = 40$ ft³/s per foot width of gate. At what rate is energy lost?

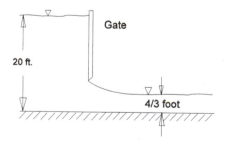

Problem 8.8

8.9. Find the force acting on the gate shown in the figure for water flowing in a 5-ft-wide rectangular channel at a volume flow rate of 75 ft³/s. Compute the power loss due to the gate.

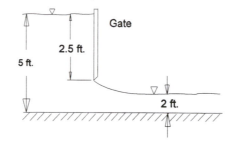

Problem 8.9

8.10. Calculate the total force on the gate in the figure. Assume no energy loss. The fluid is water.

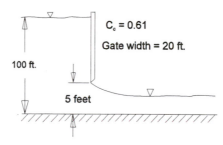

Problem 8.10

8.11. If the flow rate through the gate in the figure is 50 ft³/s per foot width of the gate, calculate the horizontal force component per unit width on the gate. The fluid is water.

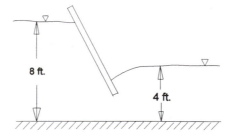

Problem 8.11

Spillways

8.12. For a flow rate of 10 ft³/s per width of channel with water flowing as shown in the figure, find the rate of energy loss per unit width when $y = 5$ ft. Express the answer both as a head loss (in feet of water) and in lb-ft/s. Find the net horizontal force exerted on the dam, neglecting bottom friction. (When streamlines are horizontal, there is no vertical acceleration and thus the vertical pressure change is hydrostatic.)

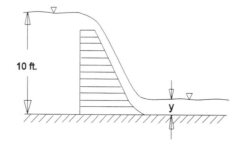

Problem 8.12

8.13. If in the previous problem y is not given, but the dam is designed so that energy loss is negligible, show that there is more than one possible value for y satisfying all conditions.

8.14. Water flows over a dam as shown in the figure at a flow rate of $Q = 128$ ft³/s. The width of the dam is 6 ft. What is the horizontal force on the dam? Is it possible for a hydraulic jump to form downstream of the dam? If so, what is its conjugate height?

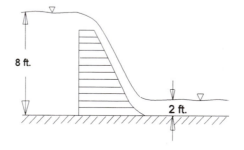

Problem 8.14

Bumps

8.15. A river with a depth of 4 ft and an average velocity of 6 ft/s passes over a log 1 ft in diameter and 10 ft long. If the depth is 3 ft after the log, and the streamlines can be assumed to be parallel both upstream (4-ft depth) and downstream (3-ft depth) of the log, what is the total horizontal force on the log? What is the total rate of energy loss due to the log?

8.16. Water flows in a wide unfinished concrete channel with a slope of 0.04°. The water depth over a 6-mm bump in the channel is found to be 60 mm. Find the depth and velocity upstream of the bump, knowing that the flow is critical over the bump.

8.17. Water flows in a 10-ft-wide smooth channel with a depth of 1 ft and a velocity of 1.55 ft/s. A 4-in-high bump occurs in the channel. Find the depth and velocity over the bump assuming no losses.

8.18. Water flows in a 10-ft-wide smooth channel with a depth of 1 ft and a velocity of 1.62 ft/s. A 0.2-ft-high bump occurs in the channel. Find the depth and velocity over the bump assuming no losses.

8.19. A channel with rectangular cross section has water flowing subcritically at the speed of 1.5 m/s, 1 m deep. The flow over a 2-m-long bump in the channel occurs smoothly, and returns to a floor at the original elevation. The downstream flow depth is less than 1 m. Find the depth and velocity of the supercritical flow after the bump.

8.20. In the previous problem, the flow over the bump is critical. Find the height of the bump.

8.21. A venturi flume is often used to measure the flow rate in a rectangular channel. This flume consists of a gradually varying width change in the channel, going from a width w_1 to a narrower width w_2, and then back again to the original width w_1. Assuming no energy losses, derive a formula for the discharge Q in terms of the two widths and their accompanying depths.

Small-amplitude waves

8.22. A small pebble thrown into a pond produces a wave that travels outward at a speed of 3 ft/s. What is the depth of the pond?

8.23. A small pebble thrown into a stream produces a wave that travels upstream at a speed of 4 ft/s and downstream at a speed of 16 ft/s. What is the speed and depth of the river?

8.24. Dropping a continuous series of pebbles into a river produces no upstream waves, but does produce a downstream wave pattern that is included within a triangular region making angles of ± 60° with the direction of the current. Determine the river speed and the Froude number. The depth is 6 m. *Hint*: Each pebble will produce a circular wave whose center moves downstream with the current, and whose radius increases at a rate equal to the wave speed. The wave pattern seen is bounded between lines tangent to these circles and passing through the point where the pebbles are inserted.

Hydraulic jumps

8.25. Flow occurs in a 5-ft-wide channel at the rate of 0.4 ft³/s and a velocity of 0.22 ft/s. The channel bottom smoothly drops in elevation 6 in with no appreciable loss in energy. What are the possible depths and velocities downstream of the elevation change?

8.26. Flow occurs in a 5-ft-wide channel at the rate of 0.4 ft³/s and a velocity of 0.22 ft/s. The channel bottom smoothly rises in elevation 3 in with no appreciable loss in energy. What are the possible depths and velocities downstream of the elevation change?

8.27. A hydraulic jump occurs in a rectangular open channel of width 15 ft. The water depths before and after the jump are 2 ft and 5 ft, respectively. Calculate the critical depth and the rate of energy loss.

8.28. A hydraulic jump occurs downstream from a spillway in a rectangular channel 20 m wide. The channel depth is 0.3 m and the current is 2.5 m/s before the jump. What is the depth and velocity after the jump? What is the rate of power loss in the jump?

8.29. Derive the formula relating the depths before and after a hydraulic jump when the channel cross section is a triangle, symmetric about the vertical axis and with apex downward.

Moving waves

8.30. A vertical flat plate is moved upstream in a river at a speed of 3 ft/s. The current in the river is 7 ft/s at a depth of 2 ft upstream from the plate. Downstream from the plate, the depth has been reduced by the plate to 0.5 ft. Neglecting viscous effects at the river bottom, find the force required to move the plate (per unit width of plate).

8.31. In a rectangular channel with velocity 6 ft/s flowing at a depth of 6 ft, a surge wave 1 ft high travels upstream. What is the speed of the wave?

8.32. A rectangular channel 5 ft deep and 10 ft wide has the flow completely stopped downstream by the closure of a gate. Compute the height and speed of the resulting wave for the following discharges: (a) 1,000 ft³/s; (b) 100 ft³/s.

8.33. Water flows in a rectangular channel at a depth of 5 ft. A gate is suddenly closed, causing the water to deepen to 8 ft at the gate. What was the original velocity of the flow?

8.34. A river 30 ft wide is flowing at a depth of 2 ft. What is the minimum velocity for a hydraulic jump to occur? What are the pressure force and the rate of momentum flow on the upstream side of the jump?

8.35. For a water flow rate of 10 ft³/s per unit width in the figure, what is the energy loss expressed as a head loss? What is the power loss expressed as ft-lb/s/ft width? What is the net horizontal force on the dam, neglecting viscous friction on the dam walls?

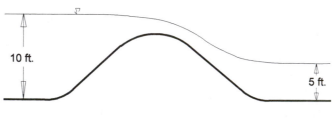

Problem 8.35

8.36. Water flows in a rectangular open channel at a depth of 4 ft. A gate is suddenly closed, causing the water to deepen to 8 ft at the gate. Find the original velocity of flow in the channel, and the speed at which the wave moves.

8.37. A stream of water flows in a rectangular channel at a depth of 2 ft with an average velocity of 10 ft/s. A gate is suddenly closed, bringing the water at the gate to rest and causing a wave to move upstream. Find the speed at which the wave travels.

8.38. In a rectangular channel with water flowing at an average velocity of 6 ft/s and depth of 3 ft, a surge wave 1-ft high moves upstream. What is the speed of the wave and the charge behind the wave?

8.39. A stream flowing at a velocity of 10 ft/s and a depth of 3.1 ft is suddenly blocked by a cave-in. A wave is seen to move upstream. Find the wave speed and the rate at which energy is lost.

8.40. A surge wave is started in a channel with triangular cross section (shown in the figure) by the sudden closure of a gate. Find the speed of the wave in terms of the original depth and velocity. Leave your answer in the form of an algebraic equation containing h, V, c.

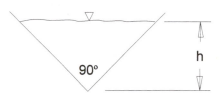

90°   h

Problem 8.40

8.41. A river with a rectangular cross section is flowing at a depth $h = 5$ ft and a velocity $V = 3$ ft/s, when a landslide far downstream completely blocks the river. This causes the flow to suddenly deepen to a depth H. Compute the new depth in terms of the given quantities.

8.42. Flow occurs in a 3-ft-wide channel at a depth of 2 ft and a velocity of 20 ft/s. A gate is suddenly closed, stopping the flow and causing a wave to move upstream. Find the velocity at which the wave moves.

Optimal-shape channel cross sections

8.43. Find the hydraulic radius for a channel with the cross section of an equilateral triangle. The width of the triangle is b, and the height is H. The apex of the triangle is at the top of the channel. At what depth ratio h/H is the hydraulic radius a maximum?

8.44. For the channel of problem 8.43, what depth ratio h/H maximizes the discharge?

8.45. Flow occurs in a finished concrete channel of trapezoidal shape with a bottom slope of 5°. The bottom width is 10 m, the sides are inclined at angles of 10° with the vertical, and the flow depth is 3 m at the channel center. Find the velocity using the Chezy formula.

Gradually sloping channels

8.46. For an unfinished concrete channel of rectangular cross section with a width of 10 m, a depth of 0.2 m, and a slope S_b of +0.04, compute the critical and normal depths for each of the three cases given below. Indicate whether the flow is critical or subcritical, whether the depth will increase or decrease, and to what value the depth approaches.
(a) $Q = 2$ m³/s,   (b) $Q = 4$ m³/s,   (c) $Q = 10$ m³/s.

8.47. Find the normal and critical depths in a smooth earth channel having a cross section in the shape of an equilateral triangle. The discharge is 4 m³/s and the bottom slope is 0.001.

8.48. By what percent do the normal and critical depths found in the previous problem change if the channel is made of finished concrete?

8.49. A channel with a triangular cross section, the sides making angles of 45° with the horizontal, is made of brick in mortar. It is desired to have a normal depth of 0.9 m. What should be the slope of the channel? The discharge is 3 m³/s.

8.50. What is the critical depth of a channel with a cross section in the shape of an equilateral triangle, apex pointed downward, having a discharge of 1 m³/s?

8.51. If water is flowing in the triangular channel shown in the figure to a depth of 6 in, and the channel slopes 1 ft in 100 ft of length, what depth y gives maximum discharge for given Manning n?

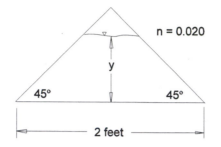

Problem 8.51

8.52. Determine the normal depth for uniform flow in a smooth earth rectangular open channel of width 10 m carrying a flow of 300 m³/s.

8.53. For a triangular open channel with the vertex pointing down, find the expressions for the normal and critical depths. Let α be the angle the channel side makes with the horizontal.

8.54. A trapezoidal channel with a bottom width of 5 ft and with side slopes of 2H:1V has a slope of 1 ft in 300 ft. Derive the formula for average shear stress at the walls of the channel. If the depth of flow is 8 ft, what is the average shear stress at the walls?

8.55. Channels used to carry off flood waters often consist of a deep narrow main channel with a wider flood plain channel above, as shown in the figure. The main channel is often constructed to have a smaller roughness coefficient than the flood-plain channel. For a channel with slope 0.02°, find the discharge if $y_1 = 5$ m, $y_2 = 3$ m, $b_1 = 6$ m, $b_2 = 20$ m, $n_1 = 0.015$, $n_2 = 0.03$. Solve for the discharge in each channel separately, then add the two discharges to get the total discharge.

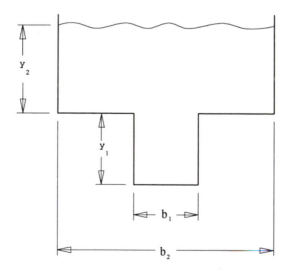

Problem 8.55

8.56. For channel flows whose depths vary slowly, equation (8.10.11) can often be integrated using large step sizes in x. For a rectangular channel with $b = 10$ m, $n = 0.05$, $Q = 40$ m³/s, channel slope = 0.05°, and an initial depth of 1 m, replace dy/dx by $\Delta y/\Delta x$ in (8.10.11) and find the location where the critical depth is reached.

chapter 9

Compressible Flows

Chapter Overview and Goals

Compressibility of the fluid in the high-speed flow of gases and liquids introduces a new class of flow effects differing from those we so far have seen. We first introduce the speed of sound and the Mach number. The Mach number is the dimensionless parameter that appears repeatedly throughout this chapter. Compressibility effects in liquids are next studied, including the effect of dissolved gases on the sonic speed and compressibility effects in pipe flow. The thermodynamics of ideal gases is briefly reviewed, and the isentropic flows of these gases is then considered. The possibility of nonisentropic flow regions such as shock waves is introduced, and compressible flow in a nozzle is studied. Nozzle flow offers the possibilities of subsonic, sonic, and supersonic flow existing in the same device, along with the possibility of isolated shock waves.

Compressible gas flow in a pipe, where wall friction and/or external heating can be important, is also presented as an illustration of how mechanisms other than area changes can affect the direction in which the Mach number changes.

The previous chapters have dealt almost exclusively with constant density flows. Compressibility effects in high-speed flows can cause a dramatically different range of phenomena to appear. These phenomena are important in many physical processes, both in gases and in liquids. In the case of gases we will restrict our attention largely to the flow of ideal gases, for besides being of engineering importance, these flows illustrate with a minimum of mathematics most of the important physics involved with compressible flows.

Another feature of compressible flow prompts a note of caution to the student. In incompressible flows we usually deal with pressure differences, and gage pressures can be used. In this chapter, it will *always* be necessary to deal with absolute pressures and absolute temperatures, since properties and state equations are functions of absolute temperatures and pressures. Keep this foremost in your mind when working problems.

In dealing with compressible flows, we will find that there are more flow variables to be considered than there were for the case of incompressible flows. Besides the usual flow variables of velocity and pressure, we also have to include the variables of temperature and density as primary variables. It will be convenient to define secondary variables such as the speed of sound and the Mach number. Because of the greater number of variables, the algebra needed to arrive at a final form of an equation will necessarily be greater than for a comparable incompressible flow. Our three conservation laws for mass, momentum, and energy, will be supplemented by state equations, process equations, and the definitions of sonic speed and Mach number. The algebraic details of the derivations will necessarily be more involved than we have seen in previous chapters of the book.

1. The Speed of Sound

If we cause a very small pressure disturbance, for example, the pressure disturbance we introduce in the air when we speak, this pressure disturbance will travel at a velocity called the *sonic velocity*, or speed of sound. We can determine what this velocity is in a manner very similar to that used in Chapter 5 for finding the speed at which a small gravity wave travels. Consider the situation in Figure 9.1a, where we have a fluid moving in a conduit with velocity v. A sound wave travels down the conduit with velocity c. This is an unsteady flow, so we use the concept of moving with the wave, as introduced in Chapter 3, Section 8, to render the flow stationary to the observer, as in Figure 9.1b. Applying the continuity equation to this control volume, we have

$$(v - c)\rho A = (v + dv - c)(\rho + d\rho)A$$

or, after cancellations and dropping terms of second and higher order in the various differentials,

$$(v - c)\, d\rho + \rho\, dv = 0. \tag{9.1.1}$$

Similarly, applying the momentum equation gives

$$(p + dp)A - pA = \rho A(v - c)[v - c - (v + dv - c)]$$

or, again after cancellations and dropping terms of second and higher order in the various differentials,

$$- dp = \rho(v - c)\, dv. \tag{9.1.2}$$

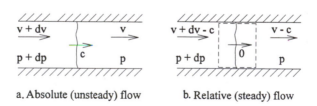

a. Absolute (unsteady) flow   b. Relative (steady) flow

Figures 9.1a. and 9.1b. Velocities and pressures for a sonic wave progressing in a duct. Dashed line indicates control volume used in the text.

Eliminating dv between (9.1.1) and (9.1.2), we have

$$dp = (v - c)^2 d\rho,$$

or

$$|v - c| = \sqrt{dp/d\rho}. \tag{9.1.3}$$

Thus in a case where the fluid in the conduit is originally at rest ($v = 0$), the sound wave travels with a velocity

$$c = \sqrt{dp/d\rho} = \sqrt{K/\rho}, \tag{9.1.4}$$

where K is the bulk modulus of the fluid. The above result applies to both liquids and gases.

Example 9.1.1. Sonic speed in water
Find the speed of sound in water, given that its bulk modulus is 311,000 psi, and its density is 1.94 slugs/ft^3.

Sought: The speed of sound in water at the given conditions.
Given: The bulk modulus for water is 311,000 psi, and the density is 1.94 slugs/ft^3.
Assumptions: The water contains no dissolved gases and has uniform properties.
Solution: The speed of sound in water is, from (9.1.4),

$$c = \sqrt{K/\rho} = \sqrt{311,000 \times 144/1.95} = 4,805 \text{ ft/s},$$

or in SI units,

$$c = \sqrt{2.14 \times 10^9/10^3} = 1,463 \text{ m/s}.$$

Note that the definition of the speed of sound in terms of the bulk modulus is valid both in liquids and in gases. The subsequent formulas that we developed using the ideal gas laws are of course restricted to use with gases only.

Example 9.1.2. Sonic speed in air
Find the speed of sound in air, given that the bulk modulus for air is 20.6 psi with a density of 0.00238 slug/ft^3.

Sought: The speed of sound in air at the given conditions using the bulk modulus.
Given: The bulk modulus for air is 20.6 psi, with a density of 0.00238 slug/ft^3.
Assumptions: Air can be considered to be an ideal gas.
Solution: The speed of sound is, from (9.1.4),

$$c = \sqrt{K/\rho} = \sqrt{20.6 \times 144/0.00238} = 1,116 \text{ ft/s}.$$

In SI units,

$$c = \sqrt{0.142 \times 10^6/1.226} = 340 \text{ m/s}.$$

If temperature or pressure had been given, we could have instead used either $c = \sqrt{kRT}$ or $c = \sqrt{kp/\rho}$, since air can usually be considered to be an ideal gas.

The pressure changes due to a sound wave passing through a fluid are so small (infinitesimal) that heat transfer is negligible, and the process is reversible. Thus a sound wave represents an isentropic process. For ideal gases equation (9.6.2) can be simplified by use of the isentropic relation p/ρ^k and the ideal gas equation $p = \rho RT$ to

$$c = \sqrt{kp/\rho} = \sqrt{kRT}. \qquad (9.1.5)$$

The sonic velocity of an ideal gas thus depends only on temperature.

Example 9.1.3. Sonic speed in an ideal gas
Find the sonic speed in air at 70°F.

Sought: The speed of sound in air at the given conditions using the temperature.
Given: Temperature is 70°F, with $k = 1.4$ and
$R = 1{,}716$ ft-lb/slug -°R $= 287$ kJ/kg-K,
Assumptions: Air is an ideal gas under these conditions.
Solution: From (9.1.5),

$$c = \sqrt{kRT} = \sqrt{1.4 \times 1716 \times (460 + 273)} = 1{,}128 \text{ ft/s},$$

or in SI units,

$$c = \sqrt{1.4 \times 287 \times (273 + 20)} = 343 \text{ m/s}.$$

Of the various parameters we have introduced so far, only k, the ratio of specific heats, is dimensionless. This ratio is a fluid property, not a flow property. The important dimensionless flow property that appears in compressible flows is the *Mach number*, named after Ernst Mach. It is defined as the ratio of the local flow speed to the local sonic velocity, or

$$M = v/c. \qquad (9.1.6)$$

A physical interpretation of the Mach number is that it represents the ratio of momentum forces to the compressible forces in the flow.

Our definition of the speed of sound is a very technical one. It is the velocity at which a very small (infinitesimally so) pressure wave travels in a fluid at rest. Some "sounds" travel at speeds that are greater than the local sonic speed if the pressure difference across their wave front no longer is infinitesimal, but is sufficiently big that it must be considered to be finite. Such pressure disturbances are called *shock waves*. Typical examples of shock waves are the bursting of an inflated paper bag or a tire, the sound heard when a whip is cracked, and the disturbances caused by very fast objects such as high-velocity bullets, jet aircraft, and rockets.

To see the difference in subsonic and supersonic behavior, consider a sound source traveling in a straight line with a velocity v. Suppose that, starting at time zero and at every fixed time interval Δt after, this source emits a spherical sound wave that travels outward with a speed c. There are three important cases to be considered:

$v < c$ (M < 1, subsonic),

$v = c$ (M = 1, sonic),

$v > c$ (M > 1, supersonic).

Consider first the subsonic case. At time 5 Δt, for example, the situation is as shown in Figure 9.2a. The spherical wave fronts are nested, with no intersection. An observer in front of the source would notice a Doppler effect: that is, because the wave fronts are no longer all spaced a constant distance apart, the apparent frequency would be higher as the source approached the observer, and lower once the source passed the observer.

In the sonic case (Figure 9.2b), the wave fronts all touch at the source. An observer directly ahead of the source would suddenly hear all the source as an abrupt noise when the source arrived at the observer, and then, because of the Doppler effect, as a lowered frequency once it passed the observer.

This phenomenon develops further as the velocity of the sender becomes supersonic (Figure 9.2c). The single point of contact of all the spherical waves seen in the sonic case now develops instead into a cone of angle $\sin^{-1}(1/M)$ that contains the intersecting spherical waves. Our concentrated point of sound as seen by an observer has been replaced by this concentrated cone of sound, whose angle is called the **Mach**

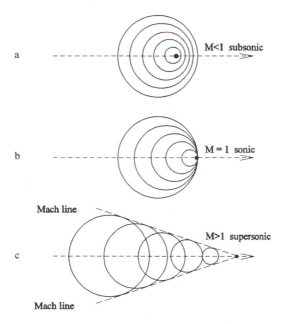

Figures 9.2a, 9.2b, and 9.2c. Sound waves from a moving source. The filled-in circle indicates the present position of the source.

angle. No sound reaches the observer until the Mach cone passes. The sonic boom is an example of this.

2. Effects of Gas Entrainment on Bulk Modulus and Sonic Speed

Propagation of sound waves in a liquid containing gas bubbles dispersed throughout the liquid can be profoundly effected by the presence of the gas. Suppose that we have a liquid volume V_{liquid} and a gas volume V_{gas}. The total volume of the combination is then

$$V_{\text{total}} = V_{\text{liquid}} + V_{\text{gas}}. \tag{9.2.1}$$

The bulk modulus, defined in Chapter 1 as

$$K = \rho \, dp/d\rho = -V \, dp/dV,$$

is

$$K_{\text{liquid}} = -V_{\text{liquid}} \, dp/dV_{\text{liquid}} \tag{9.2.2}$$

for the liquid by itself and

$$K_{\text{gas}} = -V_{\text{gas}} \, dp/dV_{\text{gas}} \tag{9.2.3}$$

for the gas by itself. For the combination the bulk modulus is

$$K_{\text{total}} = -V_{\text{total}} \, dp/dV_{\text{total}} = -(V_{\text{liquid}} + V_{\text{gas}}) \, dp/d(V_{\text{liquid}} + V_{\text{gas}})$$

$$= \frac{-(V_{\text{liquid}} + V_{\text{gas}})}{V_{\text{liquid}}} \frac{V_{\text{liquid}} \, dp/dV_{\text{liquid}}}{\dfrac{d(V_{\text{liquid}} + V_{\text{gas}})}{dV_{\text{liquid}}}}$$

$$= \frac{-(V_{\text{liquid}} + V_{\text{gas}})}{V_{\text{liquid}}} \frac{V_{\text{liquid}} \, dp/dV_{\text{liquid}}}{1 + \dfrac{dV_{\text{gas}}}{dV_{\text{liquid}}}}$$

$$= \frac{-(V_{\text{liquid}} + V_{\text{gas}})}{V_{\text{liquid}}} \frac{V_{\text{liquid}} \, dp/dV_{\text{liquid}}}{1 + \dfrac{V_{\text{liquid}} \dfrac{dp}{dV_{\text{liquid}}}}{V_{\text{gas}} \dfrac{dp}{dV_{\text{gas}}}} \dfrac{V_{\text{gas}}}{V_{\text{liquid}}}}$$

$$= \frac{V_{\text{liquid}} + V_{\text{gas}}}{V_{\text{liquid}}} \frac{K_{\text{liquid}}}{1 + \dfrac{K_{\text{liquid}} V_{\text{gas}}}{K_{\text{gas}} V_{\text{liquid}}}}$$

$$= \frac{1}{\dfrac{V_{liquid}}{V_{total}}} \frac{K_{liquid}}{1 + \dfrac{K_{liquid}}{K_{gas}} \dfrac{V_{gas}}{V_{liquid}}} = \frac{1}{\dfrac{V_{total} - V_{gas}}{V_{total}}} \frac{K_{liquid}}{1 + \dfrac{K_{liquid}}{K_{gas}} \dfrac{V_{gas}}{(V_{total} - V_{gas})}}$$

$$= \frac{1}{\dfrac{V_{liquid}}{V_{total}}} \frac{K_{liquid}}{1 + \dfrac{K_{liquid}}{K_{gas}} \dfrac{V_{gas}}{V_{liquid}}} = \frac{K_{liquid}}{\dfrac{V_{total} - V_{gas}}{V_{total}} + \dfrac{K_{liquid}}{K_{gas}} \dfrac{V_{gas}}{V_{total}}}$$

$$= \frac{K_{liquid}}{1 + \dfrac{V_{gas}}{V_{total}} \left(\dfrac{K_{liquid}}{K_{gas}} - 1 \right)}. \tag{9.2.4}$$

The mass density of the combination is given by

$$\rho_{total} = (\rho_{gas} V_{gas} + \rho_{liquid} V_{liquid})/V_{total}. \tag{9.2.5}$$

The sonic speed in the combination is then

$$c^2 = K_{total}/\rho_{total} = \frac{V_{total} K_{liquid}}{1 + \dfrac{V_{gas}}{V_{total}} \left(\dfrac{K_{liquid}}{K_{gas}} - 1 \right)} \frac{1}{\rho_{gas} V_{gas} + \rho_{liquid} V_{liquid}}.$$

Letting $G = V_{gas}/V_{total}$ and $r = \rho_{gas}/\rho_{total}$, this becomes

$$c^2 = \frac{c_{liquid}^2}{[1 + G(rc_{liquid}^2/c_{gas}^2 - 1)][G(1/r - 1) + 1]}. \tag{9.2.6}$$

Figure 9.3 shows a plot of c as a function of G for air dissolved in water. Note that for $G = 0$ the sonic speed is that of the liquid, and for $G = 1$ it is the sonic speed of the gas. For the range of G values between these the speed of sound drops rapidly to a value much below either sonic speed of the pure substances.

Example 9.2.1. Sonic speed in a mixture of oil and air
SAE 10 motor oil has a mass density of 917 kg/m^3 and a bulk modulus of 1.31×10^9 N/m^2 at 20°C. Air at a volume concentration of 1% is mixed in with the oil. What is the sonic speed?

Sought: Sonic speed of an oil/air mixture.
Given: Bulk modulus and density of the oil. The sonic speed of air is 326 m/s.
Assumptions: The air is uniformly distributed throughout the oil.
Solution: The acoustic speed in the oil and air are given by

$$c_{oil} = \sqrt{K/\rho} = \sqrt{1.31 \times 10^9/917} = 1,195 \text{ m/s},$$

$$c_{air} = 326 \text{ m/s}.$$

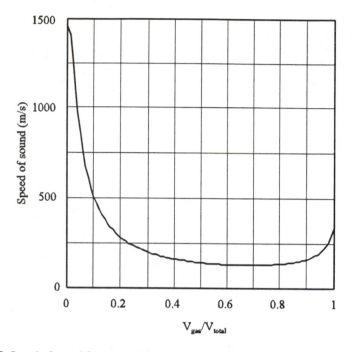

Figure 9.3. Speed of sound for a water/air mixture.

The volume ratio $G = 0.01$ and the mass density ratio $r = 1.2/917 = 0.00131$. Then

$$c^2 = \frac{c_{\text{liquid}}^2}{[1 + G(rc_{\text{liquid}}^2/c_{\text{gas}}^2 - 1)][G(1/r - 1) + 1]}$$

$$= \frac{1{,}195^2}{[1 + 0.01(0.0013 \times 1{,}195^2/326^2 - 1)][0.00131 \times (100 - 1) + 1]}$$

$$= \frac{1{,}195^2}{0.99 \times 1.13},$$

$$c = 1{,}130 \text{ m/s}.$$

3. Water Hammer

The propagation of sound waves in liquid-containing conduits has been called *water hammer*, although it can occur in any liquid-bearing pipe line. It is familiar in the home, where sudden closure of water valves often causes a "banging," or hammering, noise due to the sound waves hitting bends and being reflected. It can occur in

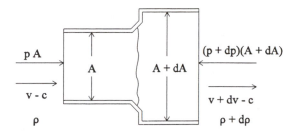

Figure 9.4. Control volume moving with wave speed c for water hammer.

hydraulic systems, particularly when solenoid valves are used that open and close suddenly.[1] In large hydraulic installations such as dams, the opening and closing of large gates can cause severe water hammer pressures because of the extremely large size of the gates. The standard practice for dealing with these situations is to use standpipes or accumulators, or to program the rate of opening and closing valves. Without such measures damage to the installation can occur, including rupture of pipes and damage to fluid machinery.

Pulsating pumps are another source of water hammer. One example of this is the circulatory system of humans. The heart is a pulsating pump that sends acoustic waves down the main arteries. The elasticity and soft tissue support of the arteries accommodate these pulses and dampens them. However, as vessels become more rigid or diseased the pulses may no longer be dampened, and damage can occur.

The main difference in the analysis of water hammer from our previous study of compressible flows is the recognition that the elasticity of the conduit walls and the support of the conduit both play major roles in the propagation of the waves. Thus area changes that depend on the internal pressure must be taken into account.

We start our analysis by considering the control volume of Figure 9.4. Since the wave phenomenon is unsteady, we take a control volume moving with the wave speed c. The continuity and momentum equations are then

$$\rho(v - c)A = (\rho + d\rho)(v + dv - c)(A + dA) \tag{9.3.1}$$

and

$$-A\,dp = \rho A(v - c)[(v + dv - c) - (v - c)] = \rho A(v - c)\,dv. \tag{9.3.2}$$

By dividing both sides of the continuity equation by $\rho A(v - c)$ we have

$$d\rho/\rho + dv/(v - c) + dA/A = 0. \tag{9.3.3}$$

Solving for dv from the momentum equation and using this to eliminate dv from the continuity equation we have

$$d\rho/\rho - dp/\rho(c - v)^2 + dA/A = 0. \tag{9.3.4}$$

[1]The liquid-fuel rockets used in the space program use such solenoid valves. Sudden closures at high pressures have resulted in the fracture of hydraulic control lines.

Solving for the wave speed we have

$$(c - v)^2 = \frac{dp/\rho}{dp/\rho + dA/A}. \tag{9.3.5}$$

From the definition of the bulk modulus $dp = K\, d\rho/\rho$. Eliminating $d\rho$ in the above expression we have

$$(c - v)^2 = \frac{dp/\rho}{dp/K + dA/A} = \frac{K/\rho}{1 + K\, dA/A\, dp}$$

$$= \frac{c^2_{\text{liquid}}}{1 + K\, dA/A\, dp}. \tag{9.3.6}$$

To proceed further we must evaluate dA/dp and use some elementary mechanics of materials relations for the pipe. We will consider a circular pipe of internal diameter D, wall thickness b, and $A = \pi D^2/4$. Consideration of other cross sections involve only minor changes.

From elementary differentiation of the area we find that

$$dA/A = 2dD/D. \tag{9.3.7}$$

With ε denoting strain (change of length per unit length), σ denoting stress, an l subscript denoting the direction along the pipe length, and a t subscript denoting the direction tangent to the pipe circumference, Hooke's law for the pipe material is

$$\varepsilon_t = \Delta D/D = (\sigma_t - v\sigma_l)/E, \tag{9.3.8}$$

$$\varepsilon_l = (\sigma_l - v\sigma_t)/E, \tag{9.3.9}$$

where E is the Young's modulus of the pipe and v is the Poisson ratio. From summation of forces on a pipe half as shown in Figure 9.5 we have

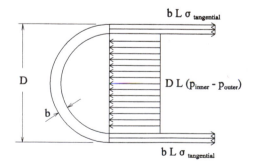

Figure 9.5. Free-body diagram for a pipe section of length L.

$$\sigma_t = 0.5D \, \Delta p / b. \tag{9.3.10}$$

To proceed further we consider three special cases.

a. Pipe closed-ended

In this case the pressure acts against the end of the pipe with a force of $\Delta p \pi D^2 / 4$ and is resisted by the force in the pipe wall that is $\sigma_l \pi D b$. Thus from sum of the forces we find

$$\sigma_l = D \, \Delta p / 4b. \tag{9.3.11}$$

Substituting the stresses into Hooke's law for the circumferential strain we have

$$\varepsilon_t = \Delta D / D = (\sigma_t - \nu \sigma_l)/E = D \, \Delta p (0.5 - 0.25\nu)/bE, \tag{9.3.12}$$

so that

$$\Delta A / A \, \Delta p = 2 \, \Delta D / D \, \Delta p = D(1 - 0.5\nu)/bE. \tag{9.3.13}$$

The wave speed is then given by inserting (9.3.13) into (9.3.6), obtaining

$$(c - v)^2 = \frac{K/\rho}{1 + KD(1 - 0.5\nu)/bE} = \frac{c_{\text{liquid}}^2}{1 + KD(1 - 0.5\nu)/bE}. \tag{9.3.14}$$

b. Pipe constrained from changing length

Here the strain in the direction of the pipe axis must be zero since the length cannot change. Thus from Hooke's law for the axial strain we have

$$\sigma_l = \nu \sigma_t = \nu D \, \Delta p / 2b. \tag{9.3.15}$$

Substituting the stresses into Hooke's law for the tangential strain, we have

$$\varepsilon_t = \Delta D / D = (\sigma_t - \nu \sigma_l)/E = D \, \Delta p (0.5 - 0.5\nu^2)/bE, \tag{9.3.16}$$

so that

$$\Delta A / A \, \Delta p = 2 \, \Delta D / D \, \Delta p = D(1 - \nu^2)/bE. \tag{9.3.17}$$

The wave speed is then given by inserting (9.3.17) into (9.3.6), obtaining

$$(c - v)^2 = \frac{K/\rho}{1 + KD(1 - \nu^2)/bE} = \frac{c_{\text{liquid}}^2}{1 + KD(1 - \nu^2)/bE}. \tag{9.3.18}$$

c. Pipe open-ended

Here the stress in the direction of the pipe axis must be zero since there is nothing for the stress to act on at the end of the pipe. Thus we have

$$\sigma_l = 0. \tag{9.3.19}$$

Substituting the stresses into Hooke's law for the tangential strain we have

$$\varepsilon_t = \Delta D/D = (\sigma_t - \nu\sigma_l)/E = 0.5D\,\Delta p/bE, \tag{9.3.20}$$

so that

$$\Delta A/A\,\Delta p = 2\,\Delta D/D\,\Delta p = D/bE. \tag{9.3.21}$$

The wave speed is then given by inserting (9.3.21) into (9.3.6), obtaining

$$(c - v)^2 = \frac{K/\rho}{1 + KD/bE} = \frac{c^2_{\text{liquid}}}{1 + KD/bE}. \tag{9.3.22}$$

Example 9.3.1. Wave speed in a steel pipe containing water

Water at 20°C flows in a steel pipe at a speed of 3 m/s. The pipe inner diameter is 75 mm and the wall thickness is 3 mm. What is the wave speed for the 3 types of pipe support considered above?

Sought: Wave speed c of water in a steel pipe.

Given: Pipe size and water velocity. The sonic speed of water is 1,481 m/s. For steel $E = 210$ GPa, Poisson's ratio = 0.29.

Assumptions: The fluid and pipe properties are uniform.

Solution: $K = \rho c^2_{\text{liquid}} = 998 \times 1{,}481^2 = 2.189 \times 10^9$.

a. Pipe closed-ended

$$\Delta A/A\,\Delta p = D(1 - 0.5\nu)/bE = 0.075(1 - 0.5 \times 0.29)/0.003 \times 210 \times 10^9$$

$$= 0.1017 \times 10^{-9}.$$

The wave speed is then given by

$$(c - v)^2 = \frac{1{,}481^2}{1 + 0.1017 \times 10^{-9} \times K} = \frac{1{,}481^2}{1.223}.$$

Therefore

$$c = 3 + 1{,}339.2 = 1{,}342.2 \text{ m/s}.$$

b. Pipe constrained from changing length

$$\Delta A/A\,\Delta p = D(1 - \nu^2)/bE = 0.075(1 - 0.29^2)/0.003 \times 210 \times 10^9 = 0.1090 \times 10^{-9}.$$

The wave speed is then given by

$$(c - v)^2 = \frac{1{,}481^2}{1 + 0.109 \times 10^{-9} \times K} = \frac{1{,}481^2}{1.239}.$$

Therefore

$c = 3 + 1{,}330.7 = 1{,}333.7$ m/s.

c. Pipe open-ended

$\Delta A / A\ \Delta p = D/bE = 0.075/0.003 \times 210 \times 10^{-9} = 0.1190 \times 10^{-9}$.

The wave speed is then given by

$$(c - v)^2 = \frac{1{,}481^2}{1 + 0.119 \times 10^{-9} \times K} = \frac{1{,}481^2}{1.26}.$$

Therefore

$c = 3 + 1{,}319.1 = 1{,}322.1$ m/s.

Notice that the three wave speeds differ by less than 2% for these conditions.

4. Ideal Gas Thermodynamics

An ideal, or thermally perfect, gas is defined as one that obeys the state equation

$$p = \rho R T, \tag{9.4.1}$$

and whose internal energy depends only on temperature. In equation (9.4.1), both pressure and temperature must be measured on an absolute scale. Gage pressure, or the use of relative temperature scales such as Celsius or Fahrenheit, is not acceptable.

The constant R is the gas constant for a given gas. It is defined as the universal gas constant $\bar{R}$, where

$\bar{R} = 8.31434$ J/mol-K $= 49{,}708$ ft-lb/slug-mol-°R

divided by the molecular weight of the gas. Thus for air with molecular weight 28.97,

$R = 8.31434/28.97 = 0.287$ kJ/kg-K

or equivalently in British gravitational units,

$R = 49{,}708/28.97 = 1{,}716$ ft-lb/slug-°R.

Example 9.4.1. Ideal gas law
Find the mass density of air at 15 psia and 70°F, and at 103 kPa and 20°C.

Sought: Air mass density at the given conditions.
Given: The conditions of 15 psia and 70°F for the first case, and 103 kPa and 20°C for the second case.

Assumptions: Air is an ideal gas under these conditions.

Solution: For air at 15 psia and 70°F, the mass density is

$$\rho = p/RT = 15 \times 144/1{,}716 \times (460 + 70) = 0.00237 \text{ slug/ft.}$$

For air at 103 kPa and 20°C, the mass density is

$$\rho = p/RT = 103/0.287 \times (273 + 20) = 1.23 \text{ kg/m}^3.$$

A combination of terms that appears in the energy law of thermodynamics is internal energy and flow work (p/ρ). This combination appears so often in calculations that it has been given its own special name, ***specific enthalpy***, and symbol h, with

$$h \equiv u + p/\rho. \tag{9.4.2}$$

Enthalpy is tabulated separately in thermodynamic tables as a state function (i.e., as a function of the thermodynamic variables such as pressure and temperature).

Other useful terms we will need in our study of compressible flow are the specific heats, or the amount of heat needed to raise a unit mass of fluid a degree in temperature. Particularly for gases, specific heats depend on the thermodynamic process that the gas undergoes. We define two specific heats. The first is defined for a constant pressure process by

$$c_p \equiv (\partial h/\partial T)_p. \tag{9.4.3}$$

The second specific heat is defined for a constant specific volume, or density, process by

$$c_v \equiv (\partial u/\partial T)_p. \tag{9.4.4}$$

For an ideal gas, since u depends only on temperature, it follows that

$$c_v = du/dT.$$

We can thus write (9.4.2) as $h = u + RT$. Taking the differential of this, we have

$$dh = du + R\, dT = c_v\, dT + R\, dT = (c_v + R)\, dT. \tag{9.4.5}$$

We see then that for an ideal gas, h as well as u will depend only upon temperature. Since

$$dh = (\partial h/\partial T)_p\, dT = c_p\, dT$$

from (9.4.3), comparison of this and (9.4.5) shows that for an ideal gas,

$$R = c_p - c_v. \tag{9.4.6}$$

We can integrate (9.4.3) and (9.4.4) to obtain the internal energy and enthalpy, namely,

$$u = \int c_v \, dT, \qquad h = \int c_p \, dT. \tag{9.4.7}$$

Often in compressible flow the temperature range is small enough that the specific heats are to a good approximation constant over the range. In that case

$$u = c_v T, \qquad h = c_p T, \tag{9.4.7a}$$

where we have not included an integration constant since we will deal only with changes in u and h. Equation (9.4.7a) will be used exclusively in our analyses.

The ratio of specific heats comes up often in our calculations. We designate it by k, defined by

$$k \equiv c_p / c_v. \tag{9.4.8}$$

Thus from the collection of four quantities k, R, c_p, and c_v, we need define only two, for between (9.4.7) and (9.4.8) it is possible to develop the relations

$$c_p = kR/(k-1), \qquad c_v = R/(k-1). \tag{9.4.9}$$

A last concept we need is that of ***specific entropy***. Entropy is a property of the flow that tells us the thermodynamic direction in which a flow can take place. It is particularly useful in flows where we find that the equations of continuity, energy, and momentum leave us with a multiplicity of solutions: entropy and the second law tell us which of these can be realized as a solution.

Specific entropy (entropy per unit mass) is defined by the equation

$$ds = (dq/T)_{\text{rev}}, \tag{9.4.10}$$

where q is the heat addition per unit mass rate of flow. Here s is a function of the thermodynamic state. The subscript rev indicates that the integration is to be performed along a reversible path connecting the initial and final states, and not along the thermodynamic path used by the actual process. To find s, the momentum equation written in terms of differentials is

$$-dp = \rho v \, dv. \tag{9.4.11}$$

We note that the energy equation in differential form is

$$dq = du + d(p/\rho) + v \, dv, \tag{9.4.12}$$

where we have neglected viscous effects and changes in potential energy. Eliminating the velocity v between these two equations, and using the product rule of calculus, we reduce the above expression for ds to

$$ds = [du + p \, d(1/\rho)]/T. \tag{9.4.13}$$

We have omitted writing the reversibility subscript in (9.4.13), since this is implied through the neglect of friction and the assumption that only continuous gradual changes occur.

Substituting for du from our ideal gas law (9.4.1) into (9.4.13) and using the state equation, we have

$$ds = (c_v \, dT - p \, d\rho/\rho^2)/T = c_v \, dT/T - R \, d\rho/\rho. \tag{9.4.14}$$

In general we need to know how c_v varies with temperature to be able to integrate equation (9.4.14). If however the temperature and density of the flow do not vary greatly from point to point in the flow, and if over the temperature range of interest the variation of c_v with temperature is small enough that c_v can be considered to be constant, then (9.4.14) can be integrated to give

$$s_2 - s_1 = c_v \ln T_2/T_1 - R \ln \rho_2/\rho_1$$

or using the state equation and the second of (9.4.9), we get

$$s_2 - s_1 = c_v[\ln p_2\rho_1/p_1\rho_2 - (k - 1) \ln \rho_2/\rho_1]$$

$$= c_v \ln [(p_2/p_1)(\rho_1/\rho_2)^k]. \tag{9.4.15}$$

It is seen from (9.4.15) that for a process involving an ideal gas to undergo no change in entropy, that is, for the process to be *isentropic*, it is sufficient that during that process

$$p\rho^{-k} = \text{constant}. \tag{9.4.16}$$

Such a process is both adiabatic (no heat transfer to or from the surroundings) and reversible.

5. Isentropic Flow of an Ideal Gas

The flow of an ideal gas through a variable area conduit such as a nozzle illustrates many of the flow phenomena found in compressible flow. Consider the control volume shown in Figure 9.6. By conservation of mass, we have

$$\dot{m} = \rho A v = (\rho + d\rho)(A + dA)(v + dv).$$

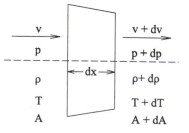

Figure 9.6. Control volume for variable area compressible flow analysis.

Expanding and omitting terms of order higher than first in the infinitesimals, this becomes

$$\rho A\,dv + \rho v\,dA + Av\,d\rho = 0$$

or after dividing by the mass flow rate ρAv,

$$dA/A + dv/v + d\rho/\rho = 0. \tag{9.5.1}$$

Summing forces on the control volume for use in the momentum equation, we see that on the back face of the control volume we have a force pA. On the front face there is a force $(p + dp)(A + dA)$. We must also consider that the pressure acts on the edge of our control volume. To lowest order, the horizontal component of this force is $p\,dA$. The net sum of these three forces on the control volume then is $-A\,dp$, which must be equal to the rate of change of momentum. Thus we get

$$-A\,dp = \dot{m}\,(v + dv - v)$$

or after dividing by the area and using $\dot{m} = \rho Av$,

$$-dp = \rho v\,dv. \tag{9.5.2}$$

We will assume that there is no heat loss through the wall of the control volume, so the flow will be adiabatic. Also, by using infinitesimal changes in the various quantities, we are saying that over a small distance there is only a small change in the thermodynamic variables. We have implied by this that the flow is reversible. Our first law then states that, since the heat flow is zero, the enthalpy plus the kinetic energy is constant, i.e.,

$$c_p T + v^2/2 = h_o = \text{constant}, \tag{9.5.3a}$$

where h_o is the enthalpy at a stagnation point (a point in the flow where the velocity has decelerated to zero). The constant h_o can be eliminated by differentiation of equation (9.5.3a), so that it can be written in differential form as

$$c_p\,dT + v\,dv = 0. \tag{9.5.3b}$$

In order to find how the various flow quantities vary, we will first combine the momentum equation (9.5.2) and the energy equation (9.5.3b) in such a way as to relate all flow properties to the Mach number. Along the way we will use the state equation and the definition of the Mach number. We will next relate the Mach number to the area through the continuity equation (9.5.1). Then we will be in a position to relate all quantities to the local area.

In working with these equations, we will soon find that we need to evaluate certain constants of integration as we go along. There are two choices for these constants that have been found useful. One choice is a point of zero velocity, i.e., either a reservoir or a stagnation point. Both these points have zero Mach number. This reference point will be denoted by an o subscript. The second choice is the sonic point, or location where the Mach number is unity. These reference quantities will be denoted by a superscript (*).

The reason why two reference points are necessary is that each works well for most of our flow variables p, ρ, T, v, c, M, A, but neither works well for all of our flow variables. For instance, the reservoir area must be infinite, so that A_o is not a convenient reference quantity for area. (Interpreting this as a stagnation point "area" does not help the situation.) Alternatively, we may not have a sonic location in a given problem, in which case (*) quantities are actually fictitious—or virtual—quantities for our flow field. In any case, if we develop our equations using either of these reference points, we can always refer them to a different reference point appropriate to our problem by a suitable ratio of quantities.

We first evaluate the constant in equation (9.5.4a) by referring it to a stagnation point. Since the velocity vanishes at the stagnation point, we have

$$c_p T + v^2/2 = c_p T_o. \tag{9.5.3c}$$

Dividing by the stagnation enthalpy $c_p T_o$, recognizing that $c_p = kR/(k-1)$, and bringing in the definitions of c and M gives, after a little algebra,

$$T_o/T = 1 + 0.5(k-1)\mathrm{M}^2. \tag{9.5.4}$$

Since by (9.1.5) $c_o/c = (T_o/T)^{1/2}$, (9.5.4) can also be written as

$$c_o/c = [1 + 0.5(k-1)\mathrm{M}^2]^{1/2}. \tag{9.5.5}$$

Next, using the state equation for an ideal gas as well as the condition of isentropy, we have

$$T_o/T = p_o\rho/p\rho_o = (\rho/\rho_o)^{k-1} = (p_o/p)^{(k-1)/k}. \tag{9.5.6}$$

Combining this with (9.5.4) and thereby eliminating the temperature,

$$p_o/p = [1 + 0.5(k-1)\mathrm{M}^2]^{k/k-1}. \tag{9.5.7}$$

Using the isentropy condition, we have from (9.5.7)

$$\rho_o/\rho = [1 + 0.5(k-1)\mathrm{M}^2]^{1/k-1}. \tag{9.5.8}$$

To illustrate the effect of compressibility, plots of equations (9.5.4), (9.5.5), (9.5.7), and (9.5.8) are shown as functions of Mach number in Figure 9.7. They are also tabulated in Appendix D for the case $k = 1.4$ (air).

Example 9.5.1. Pitot tube in a compressible flow

Find the flow past a pitot tube in air, taking into account the fact that air is compressible. Compare the result with the corresponding result for incompressible flow found in Chapter 3.

Sought: Compressibility effects in the flow past a pitot tube.

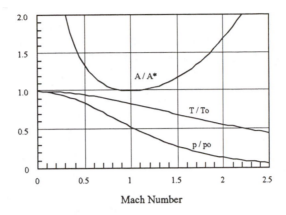

Figure 9.7. Variation of pressure, temperature, and area with Mach number. The fluid is air ($k = 1.4$).

Given: Local and stagnation pressures, and local density. (*Note*: Local density would usually be determined from the local pressure and temperature, and the ideal gas law.)
Assumptions: Air behaves as an ideal gas.
Solution: We will use equation (9.5.7), which says that

$$p_o/p = [1 + 0.5(k-1)M^2]^{k/k-1}.$$

For comparison purposes, recall that for incompressible flow we found in Chapter 3 that

$$v = \sqrt{2(p_o - p)/\rho}.$$

First rearrange equation (9.5.7) to solve for M, obtaining

$$M^2 = \frac{(p_o/p)^{(k-1)/k} - 1}{(k-1)/2}.$$

Next use $v = Mc$ to solve for v, obtaining

$$v^2 = \frac{(p_o/p)^{(k-1/k)} - 1}{c^2(k-1)/2}.$$

Using next the expression

$$c^2 = kp/\rho,$$

we find the relation between velocity, pressure, and density for isentropic compressible flow to be

$$v = \sqrt{[2pk/\rho(k-1)][(p_o/p)^{k-1/k} - 1]}.$$

While the compressible results for velocity look quite different from the incompressible results, for sufficiently small Mach number they are in very close agreement. To see this, expand equation (9.5.7) using the binomial theorem. This gives

$$p_o/p = 1 + 0.5k\text{M}^2 + 0.125k\text{M}^4 + \cdots$$

or

$$p_o = p + 0.5pk\text{M}^2(1 + 0.25\text{M}^2 + \cdots) = p + 0.5\rho v^2(1 + 0.25\text{M}^2 + \cdots).$$

The first two terms in the expansion thus give the incompressible result. We see that for a Mach number of 0.3, 0.25M^2 is equal to 0.0225, and so we can say that for values of the Mach number below about 0.3, the error in considering a gas to be incompressible introduces an error of the order of 2% or less.

To complete our discussion of isentropic flow, we need to know the effect of area variation. Area is brought into the problem through consideration of the continuity equation (9.5.1). Differentiating first the isentropic relation p/ρ^k = constant and then dividing by kp, we have

$$d\rho/\rho = dp/kp.$$

Introducing the momentum equation (9.5.2) to eliminate dp, we find that

$$d\rho/\rho = -\rho v \, dv/kp = -v \, dv/kRT = -v \, dv/c^2.$$

Putting this into (9.5.1), we find that

$$dA/A = (\text{M}^2 - 1) \, dv/v. \tag{9.5.9}$$

We next find the variation of velocity with Mach number. Since $\text{M} = v/c$, by differentiation of this we have

$$d\text{M} = dv/c - v \, dc/c^2.$$

From the definition of c for isentropic flow we know that $c^2 = kRT$. Differentiating this and dividing by c, we have

$$dc/c = dT/2T.$$

Combining this with (9.5.3b) we have

$$dc/c = -(k - 1)v \, dv/2c^2, \tag{9.5.10}$$

and so

$$d\text{M} = \text{M}[1 + 0.5(k - 1)\text{M}^2] \, dv/v. \tag{9.5.11}$$

Inserting these results into the continuity equation (9.5.1) yields

$$dA/A = (\text{M}^2 - 1) \, d\text{M}/\text{M}[1 + 0.5(k - 1)\text{M}^2] = \{(k + 1)/[2 + (k - 1)\text{M}^2] - 1/\text{M}^2\} \, d\text{M}^2/2.$$

The above result can be integrated (at least with the help of tables) to find A as a function of M. Using the sonic point to evaluate the constant of integration, we have finally

$$\frac{A}{A^*} = \frac{1}{M}\left[\frac{2 + (k-1)M^2}{k+1}\right]^{(k+1)/2(k-1)}. \tag{9.5.12}$$

The set of equations (9.5.3), (9.5.4), (9.5.5), (9.5.6), (9.5.7), and (9.5.12) enables us to relate the various quantities to the local area. Appendix D is a tabulation of results calculated from these equations for $k = 1.4$ (air).

We can use these results also to change our reference point from a stagnation point to a sonic point. Inserting M = 1 into the above set of equations, we have

$$T_o/T^* = (k+1)/2, \tag{9.5.13}$$

$$c_o/c^* = [(k+1)/2]^{1/2}, \tag{9.5.14}$$

$$p_o/p^* = [(k+1)/2]^{k/k-1}, \tag{9.5.15}$$

$$\rho_o/\rho^* = [(k+1)/2]^{1/k-1}, \tag{9.5.16}$$

$$v^* = c^*. \tag{9.5.17}$$

Thus the relations for T_o/T, c_o/c, p_o/p, ρ_o/ρ can all be changed to expressions involving the sonic point as a reference by proper multiplication by (9.5.13) through (9.5.17). We summarize these along with the useful equations we have developed so far in Table 9.1 for an ideal gas with arbitrary k, and also for air ($k = 1.4$).

Table 9.1. Isentropic equations for a general value of k and for $k = 1.4$ (air)

Equation (general k)	Source	Equation ($k = 1.4$)	Source
$v = Mc$	(9.1.6)		
$\dot{m} = \rho Av$	(9.5.1)		
$T_o/T = 1 + 0.5(k-1)M^2$	(9.5.4)	$T_o/T = 1 + 0.2M^2$	(9.5.4a)
$c_o/c = [1 + 0.5(k-1)M^2]^{1/2}$	(9.5.5)	$c_o/c = (1 + 0.2M^2)^{1/2}$	(9.5.5a)
$p_o/p = [1 + 0.5(k-1)M^2]^{k/k-1}$	(9.5.7)	$p_o/p = (1 + 0.2M^2)^{3.5}$	(9.5.7a)
$\rho_o/\rho = [1 + 0.5(k-1)M^2]^{1/k-1}$	(9.5.8)	$\rho_o/\rho = (1 + 0.2M^2)^{2.5}$	(9.5.8a)
$\dfrac{A}{A^*} = \dfrac{1}{M}\left[\dfrac{1 + 0.5(k-1)M^2}{0.5(k+1)}\right]^{(k+1)/2(k-1)}$	(9.5.12)	$\dfrac{A}{A^*} = \dfrac{1}{M}\left[\dfrac{1 + 0.2M^2}{1.2}\right]^3$	(9.5.9a)
$T_o/T^* = 0.5(k+1)$	(9.5.13)	$T_o/T^* = 1.09554$	(9.5.10a)
$c_o/c^* = [0.5(k+1)]^{1/2}$	(9.5.14)	$c_o/c^* = 1.44$	(9.5.11a)
$p_o/p^* = [0.5(k+1)]^{k/k-1}$	(9.5.15)	$p_o/p^* = (1.2)^{3.5} = 1.893$	(9.5.12a)
$\rho_o/\rho^* = [0.5(k+1)]^{1/k-1}$	(9.5.16)	$\rho_o/\rho^* = (1.2)^{2.5} = 1.577$	(9.5.13a)
$v^* = c^*$	(9.5.17)		

While the equations we have just derived and their tabulated values present a complete description of variable area isentropic flows, the possibilities and details of the flow are obscured by the mathematics. A clearer picture can be obtained by considering how changes in area influence the signs of the changes in the other variables.

Suppose that a flow starts out subsonic (M < 1) in a nozzle whose area first decreases (converges) and then increases (diverges). In the converging portion, dA is negative, as is $M^2 - 1$. Thus (9.5.9) tells us that dv is positive, or in other words, the velocity will increase with distance along the nozzle. Referring back to the momentum equation (9.5.2), we see that this means the pressure will decrease. The changes of the other variables can be found from our other equations in differential form. The results are summarized in Table 9.2.

At the narrowest part of the nozzle (the **throat**), dA will be zero. According to equations (9.5.9), (9.5.2), (9.5.10), and (9.5.11), there are then two possibilities. If the Mach number has not yet reached unity, then dv, dp, dc, dM, and in fact all the rest of the differential changes must be zero. Alternatively, if the Mach number just reaches the value of unity, then dv can be whatever value it wants to be: dp and the other quantities will follow it with the appropriate sign. Which of these possibilities is realized depends on the exit pressure.

In the diverging portion of the nozzle (the **diffuser**), where dA is positive, if the Mach number has not reached unity at the throat, dv must be negative. This means that the velocity, and by (9.5.11) the Mach number as well, will decrease as the flow moves downstream. The signs of the changes in all flow quantities are in fact the opposite of

Table 9.2. Signs of the changes in flow quantities in a converging-diverging nozzle

Converging region ($dA < 0$), M < 1					
$dv > 0$	$dp < 0$	$dT < 0$	$d\rho < 0$	$dc < 0$	$dM > 0$

Throat region ($dA = 0$)					

Case 1, M < 1

$dv = 0$	$dp = 0$	$dT = 0$	$d\rho = 0$	$dc = 0$	$dM = 0$

Case 2, M = 1

All infinitesimal quantities can have the signs that occur for either case 1 or case 2 for the diverging region listed below.

Diverging region ($dA > 0$)					

Case 1, M < 1

$dv < 0$	$dp > 0$	$dT > 0$	$d\rho > 0$	$dc > 0$	$dM < 0$

Case 2, M > 1

$dv > 0$	$dp < 0$	$dT < 0$	$d\rho < 0$	$dc < 0$	$dM > 0$

the signs in the converging portion of the flow. The flow then in the diffuser must be everywhere subsonic.

If, however, the Mach number at the throat does reach unity, there are two possibilities in the diverging portion, depending on the exit pressure. The first possibility is that the Mach number will decrease, and the flow will go back to being subsonic as discussed in the previous paragraph. This occurs if the exit pressure is only a little lower than the entrance pressure. The second possibility is that, if the exit pressure is low enough, the Mach number can continue increasing, and the flow will become supersonic. Thus if we start from a subsonic flow, supersonic flow in a nozzle can be reached only if the Mach number at the throat is unity. Note that the throat is the only place in an isentropic flow where the Mach number can be unity.

The mass rate of flow can be found by multiplying the mass density, area, and velocity at any point in the flow. A convenient place to do this is at the throat. (Obviously a stagnation point would not be a good choice for a reference point since we would be multiplying the zero velocity by the infinite area.) If we think of an experiment where the exit pressure and entrance pressure start out equal, and the exit pressure is gradually decreased with the entrance pressure held fixed, we would find the mass flow rate slowly increasing until the flow at the throat first becomes sonic. Decreasing the exit pressure beyond this point leaves the conditions at the throat unchanged, and hence the mass flow rate through the nozzle is unchanged. In such a case, we say the nozzle is ***choked***. Choking is defined as follows: For steady subsonic flows with a specified mass rate of flow and Mach number at a given section, there is a maximum contraction in cross-sectional area that is possible to pass the given mass flow rate. That area occurs where the flow is sonic.

Example 9.5.2. Isentropic flow in a converging-diverging nozzle

Air flows isentropically from a reservoir at an absolute pressure of 200 kPa and a temperature of 400 K through a nozzle with a throat area of 1 cm^2 and an exit area of 2 cm^2. At what exit pressures will a maximum mass flow rate be achieved, and what is this rate?

Sought: Maximum flow rate from a nozzle as a function of exit pressure.
Given: The reservoir is at an absolute pressure of 200 kPa and a temperature of 400 K. The nozzle has a throat area of 1 cm^2 and an exit area of 2 cm^2.
Assumptions: The gas is ideal, and the flow is isentropic.
Solution: To achieve a maximum flow rate, the nozzle must be choked, and so $A^* = A_{throat}$. From Appendix D we see that for $A_{exit}/A^* = 2$, there are two possibilities:

$$M_{exit} = 0.306, \qquad T_{exit} = 0.9816T_o = 0.9816 \times 400 = 392.64 \text{ K},$$

$$p_{exit} = 0.9370p_o = 0.9370 \times 200{,}000 = 187.4 \text{ kPa},$$

$$\rho_{exit} = 0.9546\rho_o,$$

and

$M_{exit} = 2.2, \qquad T_{exit} = 0.5075T_o = 0.5075 \times 400 = 203$ K,

$p_{exit} = 0.0931p_o = 0.0931 \times 200{,}000 = 18.62$ kPa,

$\rho_{exit} = 0.1835\rho_o.$

Thus for the given reservoir conditions a maximum flow rate is achieved under isentropic conditions when the exit pressure takes on either of the values

$p_{exit} = 187.4$ kPa $\qquad$ or $\qquad$ 18.62 kPa.

From the state equation, we have

$\rho_o = p_o/RT_o = 200{,}000/287 \times 400 = 1.742$ kg/m^3,

and so for the exit pressure of 187.4 kPa we find

$\rho_{exit} = 0.9546 \times 1.742 = 1.663$ kg/m^3.

The speed of sound for this exit pressure is

$c_{exit} = \sqrt{kRT_{exit}} = \sqrt{1.4 \times 287 \times 392.64} = 397.2$ m/s.

Thus the exit velocity is

$v_{exit} = M_{exit}c_{exit} = 121.5$ m/s.

The mass rate of flow is then

$\dot{m} = (\rho Av)_{exit} = 1.632 \times 121.5 \times 0.0002 = 0.0397$ kg/s.

The mass rate for the exit pressure of 18.62 kPa must be the same.

Example 9.5.3. Subsonic flow in a converging-diverging nozzle
Repeat the previous example with an exit pressure of 190 kPa.

Sought: Flow rate from a nozzle at a given exit pressure.
Given: The reservoir is at an absolute pressure of 200 kPa and a temperature of 400 K. The nozzle has a throat area of 1 cm^2 and an exit area of 2 cm^2. Exit pressure is 190 kPa.
Assumptions: Adiabatic compressible flow, ideal gas.
Solution: In this case the flow is subsonic throughout, and we must first calculate A^*, which will be somewhat less than A_{throat}. Since

$p_{exit}/p_o = 190/200 = 0.95,$

from Appendix D we see that

$A_{exit}/A^* = 2.229,$

and so

$A^* = 2/2.229 = 0.897$ cm^2.

Since from Appendix D

$M_{exit} = 0.2716$

and

$T_{exit} = 0.9855T_o = 394.2$ K,

we compute

$c_{exit} = \sqrt{kRT_{exit}} = 398.0$ m/s,

$\rho_{exit} = 0.9640\rho_o = 1.6793$ kg/m^3,

and finally

$\dot{m} = \rho_{exit}v_{exit}A = 1.6793 \times (398.0 \times 0.2716) \times (2 \times 10^{-4}) = 0.0363$ kg/s.

---

6. Normal Shock Waves

In discussing isentropic flow in a nozzle, we saw in the previous section (e.g., Example 9.5.2) that when the Mach number at the throat is unity, there can be only two values of the exit pressure for which the flow will be isentropic everywhere between the throat and the exit. If there is a different exit pressure, our equations as they stand will not describe the situation completely. One of the assumptions used in our derivation must somewhere have been violated.

What happens in this situation is that there is a very narrow region in the flow in which the assumption of flow reversibility breaks down. This sudden change is called a *normal shock wave*, normal in the sense that the shock will be perpendicular to the flow direction. Across this normal shock wave, which we will assume to have zero thickness (its thickness can be shown to depend on the viscosity of the gas and is of the order of a few mean free path lengths), there will be abrupt changes in all our variables (pressure, density, temperature, velocity, sonic velocity, Mach number). The situation is analogous to the hydraulic jump in open channel flow, where the elevation suddenly changes over a very short distance.

To analyze the normal shock wave, we use our three basic equations much as we did in the previous isentropic case, but because we have said that the shock wave is very thin, there is no change in area. The remaining flow quantities can have finite, rather than infinitesimal, changes. Using a u subscript to denote quantities immediately upstream of the shock wave, and a d subscript for the quantities downstream from the shock, the continuity, momentum, and energy equations become

$$\rho_u v_u = \rho_d v_d, \tag{9.6.1}$$

$$p_u - p_d = \rho_u v_u (v_d - v_u), \tag{9.6.2}$$

$$c_p T_u + v_u^2/2 = c_p T_d + v_d^2/2. \tag{9.6.3}$$

Note that in writing (9.6.3), we have kept the assumption that the flow is adiabatic, but by considering finite abrupt changes in flow quantities, we have dropped the assumption of reversibility.

Starting with the momentum equation (9.6.2), we first divide it by the mass rate of flow per unit area, and then use the ideal gas law and the definition of the speed of sound and the Mach number. After some straightforward but tedious algebraic manipulation, the result is

$$\frac{v_d}{v_u} = \frac{k + 1/M_u^2}{k + 1/M_d^2}.$$ (9.6.4)

Dividing the energy equation (9.6.3) by the specific enthalpy $c_p T_u$, using (9.4.9) to eliminate c_p and $c = v/M$ to eliminate c, after some manipulation (9.6.4) becomes

$$v_d/v_u = \sqrt{[(k-1) + 2/M_u^2]/[(k-1) + 2/M_d^2]}.$$ (9.6.5)

Eliminating the velocity ratio v_d/v_u between (9.6.4) and (9.6.5) we have, after dividing out the possible root $M_d = M_u$ (no shock),

$$M_d^2 = \frac{2 + (k-1)M_u^2}{2kM_u^2 - (k-1)}.$$ (9.6.6)

The consequences of (9.6.6) can be more easily seen by rearranging it into the form

$$M_d^2 = 1 + \frac{(k+1)(1 - M_u^2)}{2kM_u^2 - (k-1)}.$$ (9.6.6a)

The second term on the right-hand side of (9.6.6a) is always negative if M_u is greater than 1, and so M_d is always less than 1.

We can use (9.6.6) to find the ratios of the other flow quantities across the shock. We insert (9.6.6) into (9.6.5), and the result is

$$\frac{v_d}{v_u} = \frac{\rho_u}{\rho_d} = \frac{2 + (k-1)M_u^2}{(k+1)M_u^2},$$ (9.6.7)

the density ratio following from the continuity equation (9.6.1). From (9.6.6) and (9.6.7), the temperature ratio is

$$T_d/T_u = (c_d/c_u)^2 = v_u^2 M_d^2/v_d^2$$

$$= \frac{[2 + (k-1)M_u^2][2kM_u^2 - (k-1)]}{M_u^2(k+1)^2},$$ (9.6.8)

and from the ideal gas law the pressure is

$$\frac{p_d}{p_u} = \frac{T_d \rho_d}{T_u \rho_u} = \frac{2k M_u^2 - (k-1)}{k+1}. \tag{9.6.9}$$

Since the flow through the shock wave is irreversible, there is a change of entropy across the shock. Using (9.4.12), we see that

$$s_d - s_u = c_v \ln\left[(p_d/p_u)(\rho_u/\rho_d)^k\right]. \tag{9.6.10}$$

This can be expressed in terms of the upstream Mach number by use of (9.6.7) and (9.6.9), giving

$$s_d - s_u = c_v \ln\left\{\frac{2k M_u^2 - (k-1)}{k+1} \frac{[2 + (k-1)M_u^2]^k}{(k+1)M_u^2}\right\}. \tag{9.6.11}$$

When $M_u = 1$, then $s_d = s_u$. That is, at a Mach number of unity there is no change of entropy, and of course no shock. When $M_u > 1$, then $s_d > s_u$, which is an allowed process according to the second law. If M_u were less than 1, this would give $s_d < s_u$, which, according to the second law, is not a possible process. Thus shock waves can occur only if the upstream flow is supersonic.

The equations for a normal shock wave are summarized in Table 9.3. Plots of the shock relations as a function of upstream Mach number are given in Figure 9.8.

In our equations for isentropic flow it was useful to refer quantities to their stagnation values, as long as the area did not appear in the expression. Since the shock wave isolates the upstream flow from the downstream flow, it is necessary to see if the two flows have the same stagnation values.

Consider first the temperature. From (9.5.4) we have

$$T_{ou}/T_u = 1 + 0.5(k-1)M_u^2$$

and

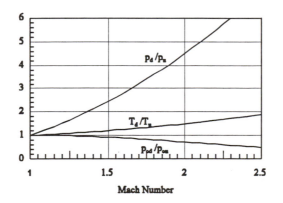

Figure 9.8. Shock-wave ratios. The fluid is air ($k = 1.4$).

Table 9.3. Equations for a normal shock wave

Equation	Source
$$\frac{v_d}{v_u} = \frac{k + 1/M_u^2}{k + 1/M_d^2}$$	(9.6.4)
$$v_d/v_u = \sqrt{[0.5(k-1) + 1/M_u^2]/[0.5(k-1) + 1/M_d^2]}$$	(9.6.5)
$$M_d^2 = \frac{2 + (k-1)M_u^2}{2kM_u^2 - (k-1)}$$	(9.6.6)
$$\frac{v_d}{v_u} = \frac{\rho_u}{\rho_d} = \frac{2 + (k-1)M_u^2}{(k+1)M_u^2}$$	(9.6.7)
$$T_d/T_u = (c_d/c_u)^2 = v_u^2 M_d^2/v_d^2 = \frac{[2 + (k-1)M_u^2][2kM_u^2 - (k-1)]}{M_u^2(k+1)^2}$$	(9.6.8)
$$\frac{p_d}{p_u} = \frac{T_d\rho_d}{T_u\rho_u} = \frac{2kM_u^2 - (k-1)]}{k+1}$$	(9.6.9)
$$s_d - s_u = c_v \ln\left\{\frac{2kM_u^{2-(k-1)}[2 + (k-1)M_u^2]^k}{(k+1)[(k+1)M_u^2]}\right\}$$	(9.6.11)
$$T_{ou}/T_{od} = h_{ou}/h_{od} = 1$$	(9.6.12)
$$\frac{p_{ou}}{p_{od}} = \left[\frac{2 + (k-1)M_u^2}{(k+1)M_u^2}\right]^{k/k-1}\left[\frac{2kM_u^2 - (k-1)}{k+1}\right]^{1/k-1} = \rho_{od}/\rho_{ou}$$	(9.6.13)
$$\frac{A_d^*}{A_u^*} = \left\{\left[\frac{2 + (k-1)M_u^2}{(k+1)M_u^2}\right]^{k/k-1}\left[\frac{2kM_u^2 - (k-1)}{k+1}\right]^{1/k-1}\right\} = p_{ou}/p_{od}$$	(9.6.14)

$T_{od}/T_d = 1 + 0.5(k-1)M_d^2$.

Using (9.6.8), we get

$$\frac{T_{ou}}{T_{od}} = \frac{M_u^2(k+1)^2}{[2 + (k-1)M_d^2][2kM_u^2 - (k-1)]}.$$

Eliminating M_d by (9.6.6), we end up with

$$T_{ou}/T_{od} = c_{ou}/c_{od} = h_{ou}/h_{od} = 1. \qquad (9.6.12)$$

That is, (9.6.12) tells us that the stagnation temperature, stagnation sonic speed, and the stagnation enthalpy all do not change across the shock wave. This could also have been arrived at by use of the energy equation as a consequence of the adiabatic assumption.

The stagnation values of pressure and density do, however, change across the shock. Upstream from the shock, from (9.5.7) we have

$$p_{ou}/p_u = [1 + 0.5(k-1)M_u^2]^{k/k-1}$$

and downstream we have

$$p_{od}/p_d = [1 + 0.5(k-1)M_d^2]^{k/k-1}.$$

Taking the ratio, and eliminating p_d/p_u by (9.6.9) and M_d by (9.6.6), the result is

$$\frac{p_{ou}}{p_{od}} = \left[\frac{2 + (k-1)M_u^2}{(k+1)\,M_u^2}\right]^{k/k-1} \left[\frac{2kM_u^2 - (k-1)}{k+1}\right]^{1/k-1} = \frac{\rho_{od}}{\rho_{ou}}, \qquad (9.6.13)$$

the density relation coming from the ideal gas law and (9.6.7).

The value of A^* will also change across the shock wave. From (9.5.12) written before and after the shock, we have

$$\frac{A_u^*}{A_u} = M_u \left[\frac{1 + 0.5(k-1)M_u^2}{0.5(k+1)}\right]^{-(k+1)/2(k-1)}$$

and

$$\frac{A_d^*}{A_d} = M_d \left[\frac{1 + 0.5(k-1)M_d^2}{0.5(k+1)}\right]^{-(k+1)/2(k-1)}.$$

Since the shock is assumed to be of zero thickness, $A_u = A_d$. Taking the ratio of the above two equations and eliminating M_d by using (9.6.6), we have the result

$$\frac{A_d^*}{A_u^*} = \left[\frac{2 + (k-1)M_u^2}{(k+1)M_u^2}\right]^{k/k-1} \left[\frac{2kM_u^2 - (k-1)}{k+1}\right]^{1/k-1} = \frac{p_{ou}}{p_{od}}. \qquad (9.6.14)$$

Example 9.6.1. Normal shock wave

A shock wave occurs in air at a place where the Mach number is 2. Find the downstream Mach number and the pressure, stagnation pressures, temperature, and velocity ratios, across the shock.

Sought: Flow conditions downstream of a normal shock wave where the upstream Mach number is 2.

Given: The upstream Mach number is 2, $k = 1.4$.

Assumptions: The flow across the shock is irreversible adiabatic. Air can be considered an ideal gas.

Solution: From (9.6.6), $M_d^2 = \dfrac{2 + (k-1)M_u^2}{2kM_u^2 - (k-1)} = \dfrac{2 + (1.4-1)2^2}{2 \times 1.4 \times 2^2 - (1.4-1)} = 0.3333;$

hence $M_u = 0.5774$. From (9.6.9), (9.6.8), and (9.6.7),

$$p_d = \frac{2kM_u^2 - (k-1)}{k+1} p_u = \frac{2 \times 1.4 \times 2^2 - (1.4-1)}{1.4+1} p_u = 4.5 p_u,$$

$$T_d = \frac{[2 + (k-1)M_u^2][2kM_u^2 - (k-1)]}{M_u^2(k+1)^2} T_u = \frac{[2 + (1.4-1)2^2][2 \times 1.4 \times 2^2 - (1.4-1)}{2^2(1.4+1)^2} T_u$$

$$= 1.688 T_u,$$

$$v_d = \frac{2 + (k-1)M_u^2}{(k+1)M_u^2} v_u = \frac{2 + (1.4-1)2^2}{(1.4+1)2^2} v_u = 0.375 v_u.$$

From (9.6.13),

$$p_{ou} = \left[\frac{2 + (k-1)M_u^2}{(k+1)M_u^2} \right]^{k/k-1} \left[\frac{2kM_u^2 - (k-1)}{k+1} \right]^{1/k-1} p_{od}$$

$$= \left[\frac{2 + (1.4-1)2^2}{(1.4+1)2^2} \right]^{(1.4/(1.4-1))} \left[\frac{2 \times 1.4}{1.4+1} \times 2^2 - (1.4-1) \right]^{1/(1.4-1)} p_{od}$$

$$= (0.375)^{3.5}(4.5)^{2.5} p_{od} = 1.3872 p_{od},$$

and from (9.6.14),

$$A_d^* = (p_{ou}/p_{od}) A_u^* = 1.3872 A_u^*.$$

The absolute pressure thus suddenly increases by 450% across the shock, causing a rapid deceleration so that the velocity is reduced to 37.5% of its upstream value.

Note that we could have avoided doing the above calculations by using the tables in Appendix D. The computations were carried out by the formulas directly to illustrate their use.

Fanno and Rayleigh lines

A graphical interpretation of the preceding is helpful in order to understand some of the physics of what is happening when a shock wave occurs. From (9.6.1) we see that the mass rate of flow per unit area, $G = \rho v$, remains constant across the shock, as does (from (9.6.12)) stagnation enthalpy $h_o = h + v^2/2$. The entropy (9.4.15) can then be written in terms of the enthalpy by use of continuity and the definition of h in the forms

$$v = \sqrt{2(h_o - h)}, \tag{9.6.15}$$

$$\rho = G/\sqrt{2(h_o - h)}, \tag{9.6.16}$$

and

$$p = \rho(h - u) = \rho(h - c_v h/c_p) = \rho h R/c_p = h(k-1)G/k\sqrt{2(h_o - h)}. \tag{9.6.17}$$

Putting these into (9.4.15) gives

$$s - s_1 = c_v \ln\left[(p/p_1)(\rho_1/\rho)^k\right]$$

$$= c_v \ln\left\{\frac{h(k-1)\,G^{1-k}\rho_1^k[2(h_o - h)]^{0.5(k-1)}}{kp_1}\right\}.$$

This can be further simplified by the use of (9.6.16) and (9.6.17) evaluated at the reference point 1, obtaining

$$s - s_1 = c_v \ln\left[\frac{h(h_o - h)^{0.5(k-1)}}{h_1(h_o - h_1)^{0.5(k-1)}}\right]. \tag{9.6.18}$$

Differentiating s in (9.6.18) with respect to h and setting ds/dh to zero, we find that s is a maximum when $h = 2h_o/(k+1)$. Putting this into (9.6.15), we see that

$$M_{@\max s} = v/\sqrt{kRT_{@\max s}} = v/\sqrt{kRh_{@\max s}/c_p} = \sqrt{2[h_o - 2h_o/(k+1)]}/\sqrt{kR2h_o/c_p(k+1)}$$

$$= 1.$$

This entropy maximum thus occurs where $M = 1$. A plot of (9.6.18) is given in Figure 9.9, where it is labeled the ***Fanno curve***. On the lower portion of the curve, the flow is supersonic. On the upper portion the flow is subsonic.

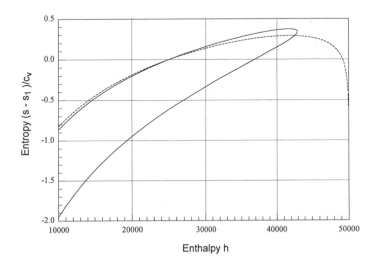

Figure 9.9. Fanno (dashed) and Rayleigh (solid) lines. A shock can occur between the intersections. $h_{\text{stagnation}} = 50,000$.

In developing (9.6.18) we used the continuity and energy equations, but not the momentum equation. Thus (9.6.18) gives a continuum of flows, of which only the two that satisfy momentum are realizable. We can complete the solution by looking at the momentum equation (9.6.2), and noting with the help of (9.6.1) that the quantity $F = p + \rho v^2$ is constant across the shock. Writing the entropy in terms of enthalpy again, but this time using continuity and momentum, but not the energy equation, we have

$$p = F - G^2/\rho \qquad (9.6.19)$$

and

$$h = c_p T = c_p p / R\rho = kp/\rho(k-1). \qquad (9.6.20)$$

From this we see that

$$p/\rho = (k-1)h/k.$$

Using (9.6.19) and (9.6.20) to solve for the density, we find that

$$(G/\rho)^2 - F/\rho + h(k-1)/k = 0.$$

This equation is quadratic in the mass density. There are then two possible values for ρ, given by

$$1/\rho = F(1 \pm \sqrt{1 - 4G^2(k-1)h/kF^2})/2G^2.$$

This expression can be simplified and its understanding enhanced if we introduce conditions at a point where sonic conditions are reached (i.e., where $M = 1$). At that point, denoted by an asterisk superscript, we have

$$v^* = c^*, \qquad h^* = c_p T^* = c_p c^{*2}/kR = c^{*2}/(k-1) = 2h_o/(k+1),$$

$$p^* = \rho^* R T^* = \rho^* c^{*2}/k, \qquad G = \rho^* v^*, \qquad F = p^* + \rho^* v^{*2} = c^{*2}\rho^*(k+1)/k,$$

$$1/\rho^* = F[1 \pm \sqrt{1 - 4G^2(k-1)h^*/kF^2}]/2G^2.$$

Then

$$G/F = \rho^* v^* k/\rho^* c^{*2}(k+1) = k/c^*(k+1),$$

and

$$4G^2(k-1)/kF^2 = 4k(k-1)/c^{*2}(k+1)^2 = 4k/h^*(k+1)^2 = 2k/h_o(k+1).$$

Putting all this together and using a reference point 2 to evaluate the constants, the entropy expression becomes

$$s - s_2 = c_v \ln[(p/p_2)(\rho_2/\rho)^k]$$

$$= c_v \ln\left\{ \frac{h \, [1 \pm \sqrt{1 - 4kh/h^*(k+1)^2}]^{k-1}}{h_2[1 \pm \sqrt{1 - 4kh_2/h^*(k+1)^2}]^{k-1}} \right\}. \qquad (9.6.21)$$

The above equation with the plus/minus signs represents two different curves, one being subsonic and the other supersonic. They join at the sonic point $h = h^*$, where $ds/dh = 0$. To represent the two curves, it is convenient to use the same reference point, the sonic point, for both. Denoting the two curves by p and m to correspond to the plus and minus signs, we have

$$s_p - s^* = c_v \ln \left\{ \frac{h[1 + \sqrt{1 - 4kh/h^*(k + 1)^2}]^{k-1}}{h^*[2k/(k + 1)]^{k-1}} \right\} \tag{9.6.22}$$

on the plus curve, and

$$s_m - s^* = c_v \ln \left\{ \frac{h[1 - \sqrt{1 - 4kh/h^*(k + 1)^2}]^{k-1}}{h^*[2/(k + 1)]^{k-1}} \right\} \tag{9.6.23}$$

on the minus curve.

Plots of (9.6.22) and (9.6.23) are shown on Figure 9.9, where they are labeled the *Rayleigh curve*.

Recall that the Fanno curve was derived using continuity and energy, but not the momentum equation. Thus the Fanno curve is the locus of all points satisfying continuity and energy. In developing (9.6.22) and (9.6.23) we did not use the energy equation; therefore the Rayleigh curve is the locus of all flows satisfying continuity and momentum, but not the energy equation. Only the two points where the Rayleigh curve intersects the Fanno curve satisfy continuity, momentum, and energy equations. The gap between these intersection points indicate the increase in enthalpy and entropy that occur across a shock wave. The Fanno and Rayleigh curves are analogs of the specific energy and momentum curves used in Chapter 5 while studying the hydraulic jump.

7. Flow in a Nozzle

With the theory of the normal shock wave now available to us, we can complete our discussion of flow in a nozzle. For the sake of definiteness, we will pick the specific ratio for the exit and throat areas as being 2. Comparable results would hold for other area ratios. The results for pressure, Mach number, and area are summarized in Figures 9.10a through 9.10c.

First we find the exit pressures for which the flow is everywhere isentropic, and for which the throat Mach number is 1. For this case, A^* equals the throat area. Looking in Appendix D for pressure values where $A_{ex}/A^* = 2$, we find $p_1 = 0.937 p_o$ at an exit Mach number of 0.305, and $p_3 = 0.0942 p_o$ at an exit Mach number of 2.195. We see that there is the possibility of a shock occurring at the exit in the second case if the exit pressure is $p_2 = 0.0942 \times 5.455 p_o = 0.514 p_o$ at an exit Mach number of 0.548. These three values of the exit pressure are the dividing values that bracket our results.

The situation can be summed up in the following special ranges for the exit pressure:

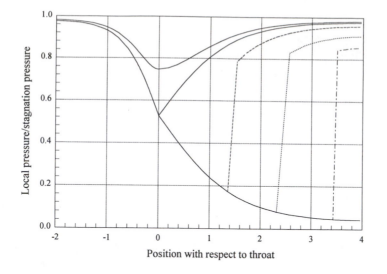

Figure 9.10a. Pressure variation in a nozzle. Isentropic flows are shown as solid lines, shocks as broken lines.

$p_1 < p_{\text{exit}} \leq p_o$.   The flow within the nozzle is everywhere isentropic and subsonic. The Mach number is a maximum at the throat, but its value is less than one. The pressure is a minimum at the throat, but the throat pressure value is greater than $0.528 p_o$. A^* is less than A_{throat}, and does not correspond to any area actually occurring in the nozzle.

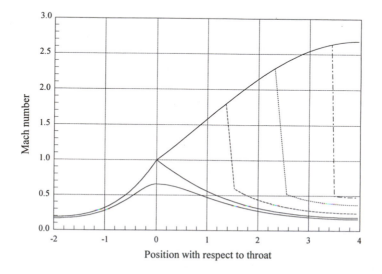

Figure 9.10b. Mach number variation in a nozzle. Isentropic flows are shown as solid lines, shocks as broken lines.

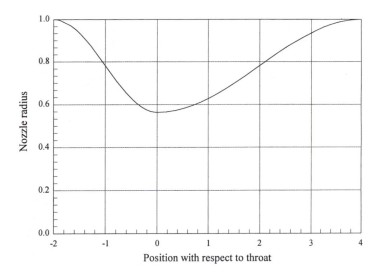

Figure 9.10c. Nozzle shape.

$p_{exit} = p_1$.   The same comments hold as in case a, except the flow reaches sonic conditions at the throat, $p_{throat} = 0.528p_o$, and $A^* = A_{throat}$. When the exit pressure is reduced below the value p_1, flow conditions at the throat will remain unchanged.

$p_2 < p_{exit} < p_1$.   A normal shock occurs in the diverging part of the nozzle. Upstream and downstream of the shock the flow is isentropic.

The location of the shock wave in the diffuser introduces an additional unknown. To determine this location, we first note that the mass rate of flow must be the same upstream and downstream of the shock. In particular, this implies

$$\dot{m} = \rho_u^* A_u^* c_u^* = \rho_d^* A_d^* c_d^*. \tag{9.7.1}$$

Since $c_u^* = c_d^*$, and since sonic densities are proportional to stagnation densities, with the constant of proportionality depending only on the fluid property k, (9.7.1) can be written as

$$\rho_{ou} A_u^* = \rho_{od} A_d^*. \tag{9.7.2}$$

Since from (9.6.12) $T_{ou} = T_{od}$, use of the ideal gas law allows us to rewrite (9.7.2) as

$$p_{ou} A_u^* = p_{od} A_d^* \tag{9.7.3a}$$

or

$$p_{ou}/p_{od} = A_d^*/A_u^*. \tag{9.7.3b}$$

These relations allow us to compute the flow downstream of the shock (either through the equations or the table of Appendix D) to check to see that the correct exit pressure is achieved.

$p < p_{exit} < p_2$.   The flow is isentropic everywhere within the nozzle, but a shock wave occurs at the nozzle exit. If the exit pressure is p_2, this shock will be a normal shock wave. For smaller exit pressures, an oblique set of shock waves, three-dimensional and so not described by our equations, will exist outside of the nozzle. Since changes in index of refraction (due to sudden changes in the density) occur across shock waves, these shock waves can be seen with suitable back-lighting.

$p_{exit} = p_3$.   The flow is everywhere isentropic in the nozzle (no shocks) and supersonic in the diverging portion of the nozzle.

$0 < p_{exit} < p_o$.   A series of expansion waves, first analyzed by Prandtl and Meyer and named after them, exists outside of the nozzle. The thermodynamic variables change continuously across the Prandtl-Meyer expansion waves, rather than abruptly as in a shock wave.

For all exit pressures less than or equal to p_1, the flow is choked, and the mass rate of flow will be the same as if the exit pressure were p_1.

Thus we see that the procedure used for computing a nozzle flow problem depends strongly on the exit pressure. If the exit pressure is p_1, or lies in the range less than or equal to p_3, the flow inside the nozzle is isentropic and the computations are straight-forward. If the exit pressure is between p_1 and p_o, A^* must first be calculated from given data. If the exit pressure lies between p_1 and p_2, the location of the shock must first be determined. This generally requires a trial-and-error procedure. The techniques for the various cases will be demonstrated by the following examples.

Example 9.7.1. Shockless flow in a nozzle

For a nozzle with $A_{exit} = 2A_{throat}$, the exit pressure is 0.93 times the reservoir pressure. Find A^* and the pressure and Mach number at the throat.

Sought: Throat pressure and Mach number and A^* for a given nozzle pressure ratio.

Given: $A_{exit} = 2A_{throat}$, the exit pressure is 0.93 times the reservoir pressure.

Assumptions: The fluid is an ideal gas. The flow is isentropic.

Solution: From Appendix D, these conditions correspond to an area ratio of $A_{exit}/A^* = 1.91$ at a Mach number of 0.322. Thus

$$A^* = A_{exit}/1.91 = 1.047A_{throat}$$

and

$$M_{throat} = 0.78, \qquad p_{throat} = 0.669p_o.$$

Example 9.7.2. Normal shock wave in a nozzle

A normal shock occurs at a position in the nozzle where $A = 1.5A_{throat}$. Find the exit area and Mach number.

Sought: Exit area and Mach number for a nozzle in which a shock wave occurs.

Given: A normal shock wave occurs where $A = 1.5A_{throat}$.

Assumptions: The fluid is an ideal gas, and the shock wave is normal to the axis of the nozzle and is of zero thickness. Except at the shock wave the flow can be considered one-dimensional and isentropic.

Solution: At the point where the shock wave occurs, from Appendix D we have

$$M_u = 1.855, \qquad p_u = 0.16 p_{ou},$$

$$M_d = 0.604, \qquad p_d = 3.85 p_u = 0.616\, p_{ou}, \qquad p_{od} = 0.788 p_{ou}.$$

From (9.6.11)

$$A_d^*/A_u^* = 1.269$$

and so

$$A_{exit}/A_d^* = 2A_u^*/A_d^* = 2.536.$$

Thus

$$p_{exit} = 0.962 p_{od} = 0.758 p_{ou}$$

and

$$M_{exit} = 0.235.$$

---

8. Oblique and Curved Shock Waves

Geometrical conditions are sometimes such that a shock wave is called for, but a normal shock will not perform the necessary correction in the flow. In that case the shock wave can still be a plane wave, but it will be inclined with respect to the incoming flow direction. This is termed an ***oblique shock wave***. Conversely, neither a normal nor an oblique shock wave may be able to perform the needed correction in flow direction. In such a case, a ***curved shock wave*** can result.

As an example of an oblique shock wave, consider the flow past a slender wedge, as shown in Figure 9.11a. If the incoming flow is supersonic, small disturbances cannot propagate upstream, and so the flow will arrive at the wedge and still be parallel to the incoming direction. If the flow were to continue this way, the flow would penetrate the wedge. Instead, an oblique shock wave, attached to the nose of the wedge, forms. This oblique shock acts to turn the flow through an angle δ so that downstream from the shock, the flow is parallel to the wedge.

In dealing with the oblique shock, we fortunately have to derive only a minimum number of new equations. Looking at Figure 9.11b, if we were to write the momentum equation tangent to the shock wave, we would notice that, since there is no pressure change along the shock wave, there is no force acting on the fluid in this direction. The result is that

$$v_{td} = v_{tu} \equiv v_t. \tag{9.8.1}$$

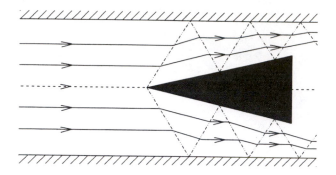

Figure 9.11a. Flow past a wedge in a channel showing multiple shock waves.

Writing the continuity and energy equations together with the momentum equation in the direction normal to the shock, we would find that they are exactly like the ones for the normal shock wave, with the one exception that the tangential velocity components do not appear. In fact, the equations are the same as for the normal shock wave, but with the total velocities everywhere replaced by the normal velocity components only. Thus, if we used the normal shock wave equations with the Mach numbers based on only the normal component of velocity, our shock equations together with (9.8.1) specify the flow completely.

One extra quantity we are interested in is the angle β, the inclination of the shock wave. From Figure 9.11b, the vector diagrams for the upstream and downstream velocities in terms of the shock inclination angle β and the velocity deflection angle δ give

$$v_t = v_u \cos \beta = v_d \cos (\beta - \delta), \tag{9.8.2}$$

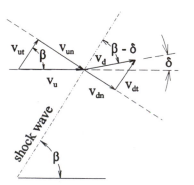

Figure 9.11b. Detail of velocities and velocity components at an oblique shock wave.

$$v_{un} = v_u \sin \beta, \tag{9.8.3}$$

$$v_{dn} = v_d \sin (\beta - \delta). \tag{9.8.4}$$

From these it is seen by taking the ratio of (9.8.3) and (9.8.2) that

$$v_{un}/v_t = \tan \beta \tag{9.8.5}$$

and from the ratio of (9.8.4) and (9.8.2) that

$$v_{dn}/v_t = \tan (\beta - \delta). \tag{9.8.6}$$

Eliminating the tangential velocity component, and using the continuity equation together with (9.6.7), we get the result

$$v_{dn}/v_{un} = \tan (\beta - \delta)/\tan \beta = \rho_u/\rho_d = 2\frac{[1 + 0.5(k - 1)\, M_u^2 \sin^2 \beta]}{(k + 1)M_u^2 \sin^2 \beta}. \tag{9.8.7}$$

The angle β enters into this result since, for the oblique shock, we use a Mach number based on upstream velocity normal to the shock. Here M_u is the Mach number based on total speed.

Solving this equation for the angle δ (we would prefer to solve for β, but that is not possible in a simple form), the result after splitting the tangent of the differences of the two angles by a familiar trigonometric formula and some manipulation is

$$\tan \delta = \frac{(M_u^2 \sin^2 \beta - 1)\cot \beta}{1 + 0.5M_u^2(k + \cos 2\beta)}. \tag{9.8.8}$$

It is seen that δ will be zero for β equal to either $\pi/2$ or $\sin^{-1}(1/M_u)$. Thus, somewhere within this range of angles, there is a maximum angle of δ, say, δ_{max}, whose value depends on M_u. In principle, δ_{max} can be found by differentiating (9.8.2) with respect to theta, and then setting $d\delta/d\beta$ to zero. The result is very messy, and we instead content ourselves with a plot of equation (9.8.2) as seen in Figure 9.12. This δ_{max} curve is shown in this figure as a dashed line. If the flow is turned through an angle δ that is less than δ_{max}, there are thus two possible solutions for β. The solution for the larger shock wave angle has the strongest discontinuity and is called a *strong* shock wave. The solution for the smaller shock wave angle has a weaker discontinuity and is called a *weak* shock.

If δ is greater than δ_{max}, neither an oblique shock nor a normal shock is possible. Instead the shock becomes ***detached***. That is, it is not attached to (touching) the body, but instead stands upstream of the body. This shock will also be curved. Detached shocks always occur with blunt bodies. A wedge becomes a "blunt body" if δ is greater than δ_{max}.

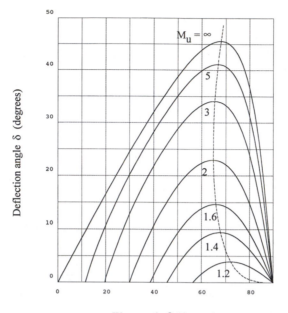

Figure 9.12. Oblique shock solutions for various Mach numbers. Curves to the left of the dashed line represent weak shock waves; those to the right, strong shock waves.

Example 9.8.1. Oblique shock wave

A 10° wedge is placed in a wind tunnel with an upstream Mach number of 1.5 and a temperature of 270 K. What angle does the shock wave at the wedge apex make with the horizontal? Is more than one shock wave possible?

Sought: Shock wave angle for an oblique shock.

Given: The wedge angle is 10°, and the wedge is placed in a wind tunnel with an upstream Mach number of 1.5 and a temperature of 270 K.

Assumptions: The incoming flow is parallel to the wind tunnel walls, the shock wave is of zero thickness, and the fluid behaves as an ideal gas except at the shock, where the flow is isentropic.

Solution: The flow must be deflected 5° so as to be parallel to the wedge once it has passed through the shock. Thus we have

$$M_u = 1.5, \qquad k = 1.4, \qquad \delta = 5°.$$

Equation (9.8.8) tells us that

$$\tan 5° = (2.25 \sin^2 \beta - 1) \cot \beta / (2.575 + 1.125 \cos 2\beta).$$

According to Figure 9.12, β is about 48°. As a check, substituting this into the right-hand side, we have 0.0889. The tangent of 5° is 0.0875, showing sufficient agreement.

The upstream sonic speed is $\sqrt{1.4 \times 287 \times 270} = 329$ m/s, so the upstream velocity is $1.5 \times 329 = 493.5$ m/s. The upstream normal and tangential velocity components are then $493.5 \sin 48° = 366.7$ m/s and $493.5 \cos 48° = 330.2$ m/s. The normal and tangential stream velocity components are $330.2 \tan (48° - 5°) = 307.9$ and 330.2, giving a total downstream speed of 451.5 m/s. The upstream Mach number based on the normal velocity component is $366.7/329 = 1.11$. Using this in (9.6.6), we find that the downstream Mach number based on the normal velocity component is

$$[(1 + 0.5 \times 0.4 \times 1.11^2)/1.4 \times 1.11^2 - 0.5 \times 0.4)]^{1/2} = 0.904.$$

This tells us that the sonic speed downstream is $307.9/0.904 = 340.6$ m/s. The downstream Mach number based on total speed is $451.5/340.6 = 1.33$. Thus further shock waves are possible.

9. Adiabatic Pipe Flow with Friction

Flow through constant area ducts can show some of the same trends as variable area flow if wall friction is present. We will consider the infinitesimally thin control volume shown in Figure 9.13. (An infinitesimal control volume is chosen since the wall shear will vary along the length of the control volume.) A is the duct area, and the control volume length is dx. The continuity, momentum, and energy equations for steady uniform flow are

$$\rho v = \dot{m}/A \qquad \text{a constant,} \tag{9.9.1}$$

$$-A \, dp - \tau_o C \, dx = \rho v A \, dv, \tag{9.9.2}$$

$$c_p T + 0.5 v^2 = h_o, \qquad \text{a constant,} \tag{9.9.3}$$

where τ_o is the wall shear stress and C is the pipe circumference. The stagnation enthalpy h_o is constant since the flow is adiabatic. Introducing the hydraulic diameter

$$D = 4A/C,$$

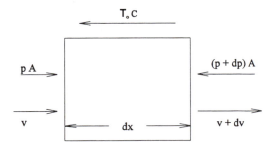

Figure 9.13. Control volume for compressible flow in a pipe with friction.

and writing the shear stress in the Darcy-Weisbach form

$$\tau_o = \rho f v^2 / 8,$$

the momentum equation becomes

$$dp + \rho f v^2 \, dx/2D + \rho v \, dv = 0. \tag{9.9.4}$$

The energy equation in differential form is obtained by rewriting (9.9.3) in differential form, obtaining

$$c_p \, dT + v \, dv = 0.$$

After division by $c_p T$ and introducing the Mach number by letting $c_p T = c/(k-1) = v M/(k-1)$, this becomes

$$dT/T = -(k-1)M^2 \, dv/v. \tag{9.9.5}$$

Differentiating the expression $v^2 = M^2 k R T$, we get the result

$$2v \, dv = 2M k R T \, dM + M^2 k R \, dT,$$

or upon division by $2v^2$,

$$dv/v = dM/M + dT/2T. \tag{9.9.6}$$

Eliminating the temperature between (9.9.5) and (9.9.6) and solving for dv/v gives

$$dv/v = dM/M[1 + 0.5(k-1)M^2]. \tag{9.9.7}$$

This can be integrated to give

$$v = K_v M/\sqrt{[1 + 0.5(k-1)M^2]}, \tag{9.9.8}$$

where K_v is a constant of integration. Letting an r subscript denote an arbitrary reference point and evaluating the above expression there, we find the constant to be

$$K_v = v_r \sqrt{[1 + 0.5(k-1)M_r^2]}/M_r, \tag{9.9.8a}$$

so that

$$v = v_r(M/M_r)\sqrt{[1 + 0.5(k-1)M_r^2]}/\sqrt{[1 + 0.5(k-1)M^2]}. \tag{9.9.8b}$$

We can also find the temperature as a function of the Mach number from the above. Combining (9.9.5) and (9.9.6) and solving for dT/T gives

$$dT/T = -(k-1)dM^2/2[1 + 0.5(k-1)\, M^2].$$

Integration of this gives

$$T = K_T/[1 + 0.5(k-1)\, M_r^2]. \tag{9.9.9}$$

Evaluation of this at the reference point r gives

$$K_T = T_r \left[1 + 0.5(k-1)M^2\right] \qquad (9.9.9a)$$

or

$$T = T_r \left[1 + 0.5(k-1)M^2\right] / \left[1 + 0.5(k-1)M_r^2\right]. \qquad (9.9.9b)$$

From the ideal gas law and continuity

$$p = \rho RT = \dot{m}RT/Av.$$

Differentiation of this and division by p gives

$$dp/p = dT/T - dv/v = -\left[1 + (k-1)M^2\right] dv/v. \qquad (9.9.10)$$

Combining (9.9.4), (9.9.7), and (9.9.10) and solving for the dx term, gives the result

$$f\, dx/D = (1 - M^2)\, dM^2 / kM^4 \left[1 + 0.5(k-1)M^2\right]. \qquad (9.9.11)$$

Integration of (9.9.11) from x_r to x gives

$$f(x - x_r)/D = (1/M_r^2 - 1/M^2)/k + \left[(k+1)/2k\right] \ln \left\{ \frac{M_r^2 \left[1 + 0.5(k-1)M^2\right]}{M^2 \left[1 + 0.5(k-1)M_r^2\right]} \right\}. \qquad (9.9.12)$$

Since (9.9.12) is fairly complicated in appearance, it is convenient at this point to introduce the function

$$F(M) = -\frac{1}{kM^2} + \frac{k+1}{2k} \ln \left\{1/M^2 + (k-1)/2\right\}. \qquad (9.9.13)$$

Then (9.9.12) can be rewritten in the form

$$f(x - x_r)/D = F(M) - F(M_r). \qquad (9.9.14)$$

Values of $F(M)$ are given for air ($k = 1.4$) in Appendix D and shown as a plot in Figure 9.14.

For a positive dp, the left side of (9.9.10) is positive, and the right side must be positive also. This tells us that, if M is initially less than unity, dM^2 must be positive, and so the Mach number will increase toward unity. Similarly, if M is initially greater than 1, the Mach number must decrease toward unity. A maximum length can thus be found from (9.9.10) by setting M_L to 1. That gives

$$fL_{\max}/D = F(1) - F(M_r).$$

If this maximum length is exceeded, the flow will act to readjust M_o if the flow is initially subsonic, and will produce shocks if the flow is initially supersonic. In both cases the effect is to reduce the flow rate by choking due to friction.

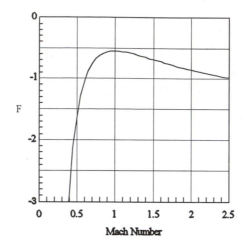

Figure 9.14. The function $F = -1/k\mathrm{M} + 0.5(k+1)\ln[\mathrm{M}^2 + 0.5(k-1)]$ plotted versus M for air $(k = 1.4)$.

It may initially seem surprising that, for subsonic flow, friction will cause the Mach number to increase. Combining (9.9.7) and (9.9.8) and solving for dp/p, we find

$$\frac{dp}{p} = -\frac{[1 + (k-1)\mathrm{M}^2]\,d\mathrm{M}^2}{\mathrm{M}^2[2 + (k-1)\mathrm{M}^2]}, \tag{9.9.15}$$

which after integration becomes

$$p = K_p/\mathrm{M}\sqrt{[1 + 0.5(k-1)\mathrm{M}^2]}, \tag{9.9.16}$$

where K_p is a constant of integration. Evaluation of (9.9.16) at our arbitrary reference point r gives

$$K_p = p_r\mathrm{M}_r\sqrt{[1 + 0.5(k-1)\mathrm{M}_r^2]}, \tag{9.9.16a}$$

so that

$$p = p_r\,(\mathrm{M}_r/\mathrm{M})\sqrt{[1 + 0.5(k-1)\mathrm{M}_r^2]}/\sqrt{[1 + 0.5(k-1)\mathrm{M}^2]}. \tag{9.9.16b}$$

Equation (9.9.16) is plotted in Figure 9.15 for $k = 1.4$ (air).

From equation (9.9.15) we see that dp has a sign that is opposite to that of $d\mathrm{M}^2$. From (9.9.16) we see that if M is less than M_r, p will be greater than p_r. Alternatively, if M is greater than M_r, p is less than p_r. In words, if the reference flow is supersonic, the Mach number downstream decreases and friction acts to decrease the pressure. Similarly, if the reference flow is subsonic, the Mach number downstream increases

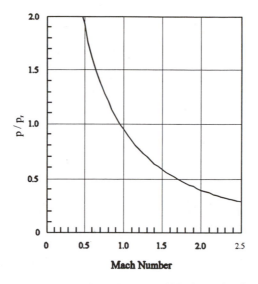

Figure 9.15. The function p/p_r plotted as a function of Mach number for air ($k = 1.4$).

and friction acts to increase the pressure. Generally, we can say that friction acts to increase the entropy and reduce the stagnation pressure and stagnation density.

Example 9.9.1. Adiabatic pipe flow with friction

Air flows in a well-insulated 5-cm-diameter duct. The friction factor is 0.03, and the Mach number at one point is 0.7. How far from this point will sonic velocity be achieved?

Sought: Sonic point in adiabatic pipe flow with friction.

Given: The duct diameter is 5 cm, the friction factor is 0.03, and the Mach number at the reference point is 0.7.

Assumptions: The flow is adiabatic and irreversible throughout, and air behaves as an ideal gas.

Solution: From (9.9.13),

$$F(0.7) = -1/k\mathrm{M}^2 + \frac{(k+1)}{2k} \ln\left[1/\mathrm{M}^2 + (k-1)/2\right]$$

$$= -1/1.4 \times 0.7^2 + \frac{(1.4+1)}{2 \times 1.4} \ln\left[1/0.7^2 + (1.4-1)/2\right] = -0.766,$$

$$F(1.0) = -1/k\mathrm{M}^2 + \frac{(k+1)}{2k} \ln\left[1/\mathrm{M}^2 + (k-1)/2\right]$$

$$= -1/1.4 \times 1^2 + \frac{(1.4 \times 1)}{2 \times 1.4} \ln\left[1/1^2 + (1.4-1)/2\right] = -0.558.$$

Thus from (9.9.14),

$$L_{max} = (D/f)[F(1.0) - F(0.7)] = (5/0.03) \times [-0.558 + 0.766] = 34.7 \text{ cm}.$$

Example 9.9.2. Adiabatic pipe flow with friction

In the preceding example, if the Mach number at one point was 1.4, how far from this point is sonic velocity achieved?

Sought: Location of sonic point in adiabatic pipe flow.

Given: The duct diameter is 5 cm, the friction factor is 0.03, and the Mach number at the reference point is 1.4.

Assumptions: The flow is adiabatic and irreversible, and air behaves as an ideal gas.

Solution: Again from (9.9.11),

$$F(1.4) = -1/k\text{M}^2 + \frac{(k+1)}{2k} \ln [1/\text{M}^2 + (k-1)/2]$$

$$= -1/1.4 \times 1.4^2 + \frac{(1.4+1)}{2 \times 1.4} \ln [1/1.4^2 + (1.4-1)/2] = -0.658,$$

and $F(1.0) = -0.558$ as in the previous example. Thus

$$L_{max} = (D/f)[F(1.4) - F(1.0)] = (5/0.03) \times [-0.558 + 0.658] = 16.6 \text{ cm}.$$

---

10. Frictionless Pipe Flow with Heat Transfer

Heating also can introduce Mach number effects similar to those caused by friction or area changes. If we consider constant area flow in a pipe with no friction and use the finite control volume of Figure 9.16, the continuity, momentum, and energy equations for steady uniform flow are

$$\dot{m} = \rho A v \qquad \text{(constant)}, \tag{9.10.1}$$

$$p_1 - p_2 = \dot{m}(v_2 - v_1)/A, \tag{9.10.2}$$

$$qL = c_p(T_2 - T_1) + (v_2^2 - v_1^2)/2 = c_p(T_{o2} - T_{o1}), \tag{9.10.3}$$

where here q is the constant rate of heat addition per unit mass rate over the length L of pipe between points 1 and 2. The subscript o on the temperature denotes stagnation values. We consider only the case of uniform heat addition (q = constant) so as to be able to obtain an analytic solution.

In the previous section we considered compressible pipe flow with friction and because the wall friction varied with x used an infinitesimal control volume. Here, since we are considering only constant rate of heat addition, integration can be avoided by considering a control volume of finite length.

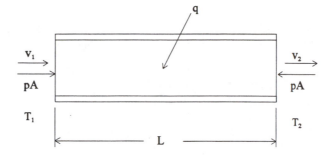

Figure 9.16. Control volume for frictionless pipe flow with heat addition.

Using the ideal gas equation of state in the form $\rho = p/RT$ to eliminate the density in equation (9.10.2), we can rewrite (9.10.2) in the form

$$p_1(1 + k\mathrm{M}_1^2) = p_2(1 + k\mathrm{M}_2^2).$$

To simplify the notation, we let 1 be the reference point r and 2 an arbitrary unsubscripted point. Then

$$p = K_p/(1 + k\mathrm{M}^2), \tag{9.10.4}$$

where the constant K_p can be determined at a reference point r as

$$K_p = p_r\,(1 + k\mathrm{M}_r^2). \tag{9.10.4a}$$

Then with the help of the state equation, continuity, and (9.1.5), (9.10.4) and (9.10.4a) become

$$p/p_r = \rho T/\rho_r T_r = v_r c^2/v c_r^2.$$

With the help of (9.10.4) and continuity, we can now write

$$\frac{v}{v_r} = \frac{\rho_r}{\rho} = \frac{\mathrm{M}^2\,(1 + k\mathrm{M}_r^2)}{\mathrm{M}_r^2\,(1 + k\mathrm{M}^2)}, \tag{9.10.5}$$

$$\frac{T}{T_r} = \frac{[\mathrm{M}\,(1 + k\mathrm{M}_r^2)]^2}{[\mathrm{M}_r\,(1 + k\mathrm{M}^2)]^2}. \tag{9.10.6}$$

From the definition of stagnation enthalpy and from (9.1.5), we have

$$c_p T_o = c_p T + v^2/2 = c^2[1 + 0.5(k - 1)\mathrm{M}^2]/(k - 1). \tag{9.10.7}$$

Thus the heat transfer needed to change the Mach number from M_1 to M_2 is obtained from (9.10.3) in the form

$$(x - x_r)\,q = c_p\,(T_{o2} - T_{o1})$$

$$= \frac{c^2[1 + 0.5(k - 1)\mathrm{M}^2] - c_r^2[1 + 0.5(k - 1)\mathrm{M}_r^2]}{k - 1}, \tag{9.10.8}$$

where c and c_r are the sonic speeds at their respected locations. Evaluation of equation (9.10.8) shows that if heat is added to the flow, the Mach number will increase (M > M_r), while if heat is subtracted from the flow Mach number will decrease (M < M_r).

Example 9.10.1. Heat addition in pipe flow

Air has a velocity of 100 m/s at a place where the pressure is 500 kPa and the temperature is 20°C. The pipe area is 0.1 m². At what rate must heat be added to have the flow just become sonic?

Sought: Rate of heat addition needed to obtain sonic flow.

Given: The air velocity is 100 m/s at a place where the pressure is 500 kPa and the temperature is 20°C. The pipe area is 0.1 m².

Assumptions: The flow is reversible but not adiabatic. The air behaves as an ideal gas.

Solution: Computing first the conditions at point 1,

$$c_1 = \sqrt{kRT_1} = \sqrt{1.4 \times 287 \times (273 + 20)} = 343.1 \text{ m/s},$$

$$M_1 = v_1/c_1 = 100/343.1 = 0.291.$$

Therefore from (9.10.7) we have

$$c_pT_{o1} = c_1^2[1 + 0.5(k-1)M_1^2]/(k-1) = 343.1^2 \times (1 + 0.5 \times (1 - 1.4) \times 0.291^2)/0.4$$

$$= 299{,}103.8 (\text{m/s})^2,$$

and from (9.10.6),

$$T_2 = T_1[M_2(1 + kM_1^2)/M_1(1 + kM_2^2)^2]^2 = 293[(1 + 1.4 \times 0.291^2)/0.291 \times (1 + 1.4)]^2$$

$$= 751.6°\text{R}.$$

Since

$$c_2 = \sqrt{kRT_2} = 549.5 \text{ m/s},$$

(9.10.7) gives

$$c_pT_{o2} = c_2^2[1 + 0.5(k-1)M_2^2]/(k-1) = 549.5^2[1 + 0.5 \times (1.4 - 1)]/(1.4 - 1)$$

$$= 905{,}948.0 \ (\text{m/s})^2.$$

Then from (9.10.3),

$$qL - c_p(T_{o2} - T_{o1}) = 905{,}948.0 - 299{,}103.8 = 606{,}844 \ (\text{m/s})^2.$$

The initial mass density is

$$\rho_1 = p_1/RT_1 = 500{,}000/287 \times 293 = 5.95 \text{ kg/m}^3,$$

and so the mass rate of flow is

$\dot{m} = \rho_1 v_1 A_1 = 5.95 \times 100 \times 0.1 = 59.5$ kg/s.

Thus the total rate of heat addition over the length of the pipe must be

$qL\dot{m} = 606{,}844 \times 59.5 = 36{,}107{,}218$ (kg/s)(m/s)^2 = 36.108 MJ/s.

Suggestions for Further Reading

Anderson, J. D., Jr., *Modern Compressible Flow with Historical Perspective*, McGraw Hill, New York, 1990.

Kuethe, A. M., and J. D. Schetzer, *Foundations of Aerodynamics*, Wiley, New York, 1950.

Liepman, H. W., and A. Roshko, *Elements of Gasdynamics*, Wiley, New York, 1957.

NACA, "Equations, tables, and charts for compressible flow," National Advisory Committee for Aeronatics Report 1135, 1953.

Parmakian, J., *Waterhammer Analysis*, Prentice Hall, New York, 1955.

Shapiro, A. H., *The Dynamics and Thermodynamics of Compressible Flow*, vols. I and II, Ronald Press, New York, 1953.

Wiley, E. B., and V. L. Streeter, *Fluid Transients in Systems*, Prentice Hall, New York, 1993.

Problems for Chapter 9

Speed of sound—liquids

9.1. A column filled with linseed oil ($\rho = 933$ kg/m^3, bulk modulus = 1.723 GPa) is to be used as an acoustic delay line in a recording studio to impart an echo effect to a recording. A portion of the electrical sound wave is converted to an acoustic wave, sent through the oil, then converted back to an electrical wave and mixed with the undelayed signal. If a delay of 1 ms is desired, how long should the oil column be?

9.2. The sudden closing of a water faucet causes a pressure wave to be transmitted through the connecting steel ($E = 210$ GPa, Poisson ratio = 0.29) piping system. An elbow 3 m from the faucet reflects 40% of the pressure wave back to the faucet, transmitting the remainder. The reflected wave is in turn totally reflected back to the elbow by the faucet. How long does it take for the tenth reflection to reach the elbow? What is the ratio of the amplitude of the tenth reflection to the pressure first reaching the elbow? Consider the pipe to be closed ended with inner diameter = 40 mm and wall thickness 3 mm.

9.3. Sonar uses pressure pulses to detect the distance between the signal source and its reflection. For seawater, take $\rho = 1{,}025$ kg/m^3, bulk modulus = 2.28 GPa. What is the relation between the distance and the time from transmission to receiving of the signal?

9.4. In the testing of tires to the point of bursting, it is much safer to perform the testing by inflating the tire with water rather than air. Using the energy equation, explain the reasoning for this.

Speed of sound—gases

9.5. The speed of sound measured in a soft drink at 15°C is found to be 250 m/s. Assuming that the soft drink consists mainly of water and dissolved carbon dioxide ($\rho_{CO_2} = 1.83$ kg/m^3), estimate the percentage of dissolved gas.

9.6. The Concorde supersonic airplane travels at Mach 2 at an altitude of 15 km in a standard atmosphere. How far past a ground observer will the Concorde be before it is first heard? If the temperature were 40°C warmer than standard and the plane flies at the same speed, what is the new Mach number?

9.7. What is the reading on a water-air differential manometer connected to a pitot-static head tube attached to an airplane flying at a Mach number of 0.8 through air at 20°F and 13 psia? Assume the correction coefficient for the pitot for tube is 1, and use the nominal specific weight for water.

9.8. If nitrogen at 60°F is flowing in a pipe and the temperature at the stagnation point of a body in the pipe is measured at 100°F, what is the velocity in the undisturbed region of the pipe?

9.9. Air flows isentropically around a submerged object. Far away from the object, $p_1 = 14.7$ psia, $\rho_1 = 0.0024$ slug/ft^3, and $v_1 = 450$ ft/s. At point 2, a stagnation point on the object, $p_2 = 5.5$ psia. Calculate the stagnation temperature.

9.10. The NASA X-33, an experimental space vehicle, will be tested at altitudes of 50 miles and at speeds up to Mach 15. At this altitude, Mach 15 corresponds to 11,000 mph. Estimate the temperature at this altitude, assuming that the air behaves as an ideal gas.

Isentropic flow

9.11. A nozzle having a throat area of 2 in^2 and an exit area of 2.4 in^2 is connected to an air reservoir at 50 psia and 40°F. What is the density of air and the speed of sound in the tank? What is the maximum mass discharge possible from the tank?

9.12. A reservoir contains air ($k = 1.4$) at 100 psi and 500°R. It discharges through a converging-diverging nozzle with an exit area of 1 in^2, and an exit Mach number of 2. Find the exit p and T. Find the Mach number and the area at the throat of the nozzle.

9.13. Air passes from a 4-in-diameter pipe into a convergent nozzle with a throat diameter of 2 in. If the throat Mach number is 0.9 and the pressure in the pipe just at the entrance to the nozzle is 30 psia at 500°R, assuming isentropic conditions, find the Mach number in the pipe. What are the velocity, temperature, density, and pressure at the throat?

9.14. Air flowing at a Mach number of 0.7 in a 2-in-diameter pipe passes through a venturi meter in the pipe. Sonic velocity is reached in the venturi meter. What is the throat diameter of the venturi meter?

9.15. A supersonic wind tunnel (reservoir plus nozzle) is to be designed such that at the test section $p = 14.7$ psia, $T = 70°F$, M = 3.0. If the test section has a diameter of 3 in, what should be the diameter of the throat? At what temperature and pressure must the reservoir air be kept? What is the mass rate of flow?

9.16. Air in a large tank flows through a 6-in-diameter pipe and discharges into the atmosphere through a convergent nozzle of 4-in tip diameter. The tank temperature and pressure are 100°F and 100 psia, and atmospheric pressure is 15 psia. Determine the pressure, temperature, Mach number, and velocity in the 6-in pipe, and the Mach number at the nozzle exit.

9.17. A nozzle with a throat area of 2 in^2 is connected between air at atmospheric pressure and a tank at reduced pressure. It is desired to have the air leave the nozzle at a Mach number of 1.5, with no shocks occurring in the nozzle. What should be the exit area and pressure of the nozzle?

9.18. Oxygen flows from one tank to another through a converging tube with a 1/2-in-diameter throat. The pressure at the throat is 30 psia and the temperature there is 80°F. What must be the pressure in the reservoir to obtain a Mach number of 0.9 at the throat?

9.19. If a Mach number of 3 is desired at the exit of a converging-diverging nozzle with air flowing, what must be the ratio of exit area to throat area and the ratio of exit pressure to reservoir pressure to obtain this Mach number?

9.20. For air flow through a converging nozzle from a reservoir into standard atmospheric conditions, if the exit (throat) area is 1 in², what should be the minimum reservoir pressure (in atmospheres) for maximum mass discharge? What would be this mass discharge, given exit conditions $T = 70°F$?

9.21. A tank of air is pressurized to 22.5 psia at 60°F. A converging nozzle with a throat area of 1 in² is attached to the side of the tank. If the nozzle discharges into an atmospheric pressure of 15 psia, what is the mass rate of discharge? To obtain a greater rate of mass discharge, the tank pressure is increased further. Is it possible to obtain twice the mass discharge you found in this manner without changing the exit conditions?

9.22. Air in a reservoir at 350 psia and 290°F flows out through a 2-in-diameter throat. What is the maximum mass rate of flow possible? What is the pressure at the throat for this condition?

9.23. Air at 100°C and 1 MPa in a large tank flows into a 45-mm-diameter pipe, from which it discharges into the atmosphere (100 kPa) through a convergent nozzle of 20-mm tip diameter. Determine the temperature, Mach number, pressure, and velocity in the pipe, assuming isentropic flow.

9.24. Carbon dioxide discharges from a tank through a convergent nozzle into the atmosphere at 100 kPa. If the tank temperature and pressure are 300 K and 150 kPa, respectively, what exit temperature, Mach number, and velocity can be expected?

9.25. If the nozzle in the previous problem has the given reservoir conditions, but is followed by a diffuser, and $M_{throat} = 1$, what should be the exit area to throat area ratio and exit pressure to have a Mach number of 2.2 at the exit?

9.26. Air passes from a 2-in-diameter pipe into a convergent nozzle with a throat diameter of 1 in. The pressure in the pipe just at the entrance to the nozzle is 30 psia, the temperature there is 40°F, and the flow in the pipe is subsonic. If the maximum discharge to occur for this geometry, what is the Mach number in the pipe upstream of the nozzle? What is the stagnation temperature?

9.27. A tank contains air at 60°F and 100 psia. It discharges through a converging nozzle. The ambient pressure $p_{ambient} = 60$ psia is such that the flow is not sonic at the exit of the converging nozzle. Calculate the exit Mach number and temperature.

9.28. A converging-diverging nozzle has a throat area of 1 cm² and an exit area of 2.5 cm². The reservoir temperature and pressure are 400 K and 500 kPa (absolute), respectively. The gas is air. For what exit pressures will the flow be everywhere isentropic? For what exit pressure will a normal shock wave occur at the exit? Given that the flow is choked, find the mass rate of flow.

9.29. For the nozzle in the previous problem, with $M_{throat} = 1$ and a diffuser attached, what is the temperature, pressure, and velocity at a point where M = 1.3?

9.30. Air discharges from a 40-mm-diameter pipe into a convergent nozzle with a throat diameter of 20 mm. The exit Mach number is 0.8 and the exit pressure is atmospheric pressure at 20°C. Find the pressure, temperature, and mass rate of flow of the air in the pipe.

9.31. A nozzle having a throat area of 1 in² and an exit area of 1.2 in² is connected to a tank containing air at 100 psia and 40°F. What is the mass density of the air in the tank?

What is the speed of sound in the tank? What is the maximum mass discharge possible from the tank? For what range of exit pressures does this maximum mass discharge occur?

9.32. Isentropic flow occurs in a nozzle with $A_{exit}/A_{throat} = 2$ and $M_{throat} = 1$. For what two values of p_{exit}/p_{throat} is this possible? Plot p/p_{exit} and M versus A/A_{throat} for the following values of A/A_{throat}: 2, 1.75, 1.5, 1.25, 1, 1.25, 1.5, 1.75, 2.

9.33. A nozzle with throat area of 0.5 in^2 is connected to an air reservoir at 200 psi and 400°R. If the exit pressure is 100 psi and the exit area is 1 in^2, what are the Mach number, speed of sound, and pressure at the throat?

9.34. Air flows from a large reservoir where $p = 200$ psia and $T = 40$°F through a converging-diverging nozzle with a throat area of 0.2 in^2 and an exit area of 0.3 in^2. Assuming that the flow is isentropic everywhere with the nozzle and that sonic conditions hold at the throat, what are the Mach number, temperature, and velocity at the throat? What are the Mach number temperature, and pressure at the exit?

Normal shock waves

9.35. A supersonic flow past a pitot tube will have a shock wave standing in front of the probe. At a stagnation temperature of 250 K, a pitot tube on the front of an airplane reads $p_{stagnation} = 160$ kPa, and $p_o = 120$ kPa. How fast is the plane traveling?

9.36. A nozzle has a throat area of 2 in^2 and an exit area of 2.8 in^2. It is connected to a reservoir of air with a pressure and temperature of 100 psia and temperature of 500°R. A shock wave occurs just at the exit. What is the mass density in the reservoir? What is the Mach number at the throat? What is the Mach number and the pressure at the exit just before the shock? What is the pressure at the exit just downstream of the shock?

9.37. A normal shock wave is observed to occur in a nozzle where the area is 1.3 cm^2 and the throat area is 1 cm^2. Air is flowing. What are the values of the Mach number upstream and downstream from the shock? What are the values of A^* upstream and downstream from the shock?

9.38. A shock wave occurs at $A/A_{throat} = 1.5$ in the diffuser of a converging/diverging nozzle. What is $p_{exit}/p_{entrance}$ if $A_{exit}/A^* = 1.7$ and $A_{entrance}/A^* = 1.59$?

9.39. A normal shock wave occurs at a place where the Mach number is 3.6, the pressure is 30 psia, and the temperature is 400°R. What are the Mach number, pressure, stagnation pressure, and stagnation temperature on the downstream side of the shock wave? What are the stagnation (or reservoir) pressure and temperature on the upstream side of the shock?

9.40. A converging-diverging nozzle is used to discharge air from a large tank ($p = 1$ MPa, $T = 20$°C) into the atmosphere ($p = 100$ kPa, $T = 20$°C). The area of the throat is 1 cm^2 and the exit area is 1.8 cm^2. What is A^*? What is the Mach number at the nozzle exit? What is the mass rate of flow through the nozzle?

9.41. Air in a reservoir at 14.7 psia, 60°F, $\rho = 2.4 \times 10^{-2}$ slug/ft^3, is drawn into a wind tunnel. A shock wave occurs in the test section when the Mach number there is 1.41. Find the pressure, velocity, and Mach number upstream and downstream from the shock.

9.42. The X-34 is an experimental space vehicle under development by NASA that would be launched from another aircraft and would be capable of flying at speeds of Mach 8. Assuming that the X-34 is flying at Mach 8 in a standard atmosphere at 30,000 m, and that a normal shock is present in the vicinity of the nose of the vehicle, what are the temperatures and pressures before and after the shock?

Oblique shock waves

9.43. A wedge of angle 10° sits on the bottom of a wind tunnel. Air at 350 K and a Mach number of 2 flows past it, forming an oblique shock wave. What angle does the shock wave make with the tunnel floor? What is the Mach number after the shock?

9.44. What is the maximum angle the wedge in the previous problem can have and still have a plane shock attached at the tip of the wedge?

9.45. For a wedge half-angle δ of 10° and a flow situation as shown in Figure 9.11a, find the angles of inclination β of the first four shock waves on the top of the wedge. What are the Mach numbers between the shocks, given that the incoming flow is at Mach 2.2 with a velocity of 500 m/s. The wedge is placed symmetrically within the wind tunnel of height 200 mm.

Pipe—friction

9.46. Air flows in a 3-in-diameter insulated pipe with a friction factor of 0.025. At one point the Mach number is found to be 0.6. How far downstream is the Mach number equal to 1?

9.47. Helium gas flows in a 3-in-diameter insulated pipe with a friction factor of 0.025. At one point the Mach number is found to be 0.6. How far downstream is the Mach number equal to 1?

9.48. Air flows in a 3-in-diameter insulated pipe with a friction factor of 0.025. At one point the Mach number is found to be 1.6. How far downstream is the Mach number equal to 1?

Pipe—heating

9.49. Air at a Mach number of 0.5, pressure 400 kPa, and temperature 20°C flows in a frictionless duct with a cross-sectional area of 20 cm^2. What uniform rate of heat addition is needed to bring the Mach number up to 1 within 1,000 m?

9.50. Helium gas at a Mach number of 0.5, pressure 400 kPa, and temperature 20°C flows in a frictionless duct with a cross-sectional area of 20 cm^2. What uniform rate of heat addition is needed to bring the Mach number up to 1 within 1,000 m?

9.51. Air at a Mach number of 0.5, pressure 400 kPa, and temperature 20°C flows in a frictionless duct with a cross-sectional area of 20 cm^2. What uniform rate of cooling is needed to bring the Mach number down to 0.1 within 1,000 m?

chapter 10

Measurement of Flow and Fluid Properties

Chapter Overview and Goals

In previous chapters we have looked briefly at several flow-measuring devices such as the venturi meter and the pitot tube. In our study of laminar viscous flows we also studied several flow cases that were suited to the measurement of fluid properties. In this chapter we elaborate on some of the details of these devices to better understand their use, and also introduce other types of measuring instruments. The devices are divided into the following categories: measurement of velocity, measurement of mass rate or volume rate, measurement of pressure, measurement of viscosity.

In this chapter we give a representative sampling of some of the measuring instrumentation currently in use. The list is by no means complete, since the list of commercially available devices is continually changing with advances in technology. The theory developed in previous chapters finds considerable use in giving simple theoretical results that are then refined through calibration procedures. The list of references at the end of the chapter should be consulted for further details and methods.

1. Velocity-measuring Devices

a. Pitot tubes

Pitot tubes, also called *pitot-static tubes*, can be used for measurements of steady velocities at a point. They are named after Henri de Pitot, a French engineer of the eighteenth century. Pitot's idea was an important contribution to flow measurement, even though his analysis of the theory of his device was in error.

Pitot tubes are designed with either rounded or blunt noses. (See Figure 10.1.) An opening at the nose of the pitot tube measures stagnation pressure, and openings on the circumference of the tube downstream of the stagnation point read the static

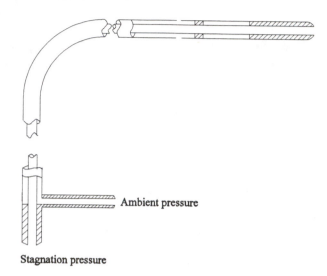

Ambient pressure

Stagnation pressure

Figure 10.1. Cross section of a pitot-static tube.

pressure. Typically the static pressure openings are 10 or more diameters downstream from the stagnation point, and the bends in a (typical) L-shaped tube are at least 18 diameters downstream from the nose. The static pressure holes can consist of a series of holes equally spaced around the circumference of the tube, or can be a circumferential slot. Because the flow along the tube is essentially a thin boundary layer, the side holes measure static pressure. Thus our Chapter 3 formula with velocity equal to the square root of twice $(p_{\text{stagnation point}} - p_{\text{ambient}})/\rho$ is valid, providing the Reynolds number based on tube diameter and fluid velocity is greater than 10^5. Below this value the formula overpredicts the flow velocity.

It is important that standard pitot tubes be in general alignment with the flow. They are relatively insensitive to misalignment as long as they are aligned within $\pm 15°$ of the flow direction. For cases where the flow direction is not known a priori and it is still desired to use a pitot tube, special pitot tubes have been constructed with a number of pressure taps on the rounded nose. By reading pressures at various locations it is possible to deduce the flow direction and find the desired velocity.

Pitot tubes generally operate satisfactorily when they can be aligned easily with the flow and when the flow has no swirl. Viscosity effects are not a concern provided the Reynolds number based on the nose radius is greater than 500. Manufacturers provide commercial pitot tubes in diameters of 1/8 in and up. Smaller pitot tubes for use in the laboratory and that measure only stagnation pressure can be made by squaring off the end of a hypodermic needle and bending the (heated) needle to get the L-shape. Pitot tubes generally have a low frequency response, the response being determined by the length of the pressure lines connecting the pitot tube to the pressure-measuring device and the density of the fluid. They will normally give only time-averaged values of the velocity.

Pitot tubes introduce little if any energy loss in the flow. They can have difficulties with fouling in flows with suspended solids, which can plug the openings. In cases where the pitot tube is used in a liquid, great care must be exercised to eliminate all traces of gas bubbles in the connecting lines attached to the pitot tube.

b. Hot-wire and hot-film anemometers

Most mechanical velocity-measuring devices have too low a frequency response to be useful in measuring turbulent velocity components. One of the first devices to be developed for such purposes was the ***hot-wire anemometer***. This instrument consists of a short (typically 1 cm or less), fine (typically 2.5 microns), wire, usually platinum, that is heated by passing a current through it. (See Figure 10.2.) The small wire size gives a low thermal inertia and thus a rapid response to temperature changes.

If the wire length is L, its resistance is R, and the wire is heated to a temperature T above the ambient temperature T_0 in an air stream with mean velocity V, then

$$\text{Rate of heat loss} = L\,(T - T_0)(A + BV) = RI^2. \tag{10.1.1}$$

Here the A term represents the part of the heat loss due to free convection and radiation, and the B term that part of the heat loss due to forced convection. Rearranging (10.1.1) somewhat, and letting the resistance of the wire be given by

$$R = R_0[1 + \alpha(T - T_0)], \tag{10.1.2}$$

where α is the coefficient of thermal resistance for the wire and R_0 the resistance at temperature T_0, we have

$$RI^2/(R - R_0) = (A + B\sqrt{V})L/\alpha R_0. \tag{10.1.3}$$

There are two modes of operating a hot-wire anemometer.

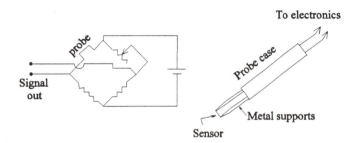

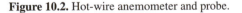

Figure 10.2. Hot-wire anemometer and probe.

Constant temperature mode

In this mode we set a desired operating temperature somewhat above ambient, and provide whatever current is necessary to keep R constant. Then we get

$$I^2 \simeq A + B\sqrt{V}. \tag{10.1.4}$$

By measuring I and calibrating the instrument, we can thus determine the flow velocity.

Constant current mode

In the constant current mode we set a desired operating current and keep its value constant. By measuring the resistance, we have

$$1/(1 - R_0/R) \simeq A + B\sqrt{V}, \tag{10.1.5}$$

and the velocity can be found.

Both operating modes are useful in practice, and their implementation can be carried out fairly cheaply with modern electronics. The output (current or resistance, depending on the mode of operation) varies nonlinearly with velocity, thus "linearizing" circuits are normally incorporated into the designs to provide an electrical signal that is linearly proportional to the local fluid velocity.

The disadvantages of hot-wire anemometers are that the probes are fairly delicate and need great care in their handling. Thus they are more often used in laboratory measurements than for field measurements.

Hot-wire probes have been constructed that have two or three wires, placed at right angles to one another, so that velocities in two or three orthogonal directions at the same small flow region can be measured simultaneously.

Hot wires can be used only in gases, since liquids would short the electrical circuitry. An alternative to the hot-wire anemometer is the ***hot-film anemometer***, which is a film of resistive material deposited on a glass wedge. Such films have been used successfully in measuring turbulent flow in liquids, although care has to be taken that suspended dirt does not deposit on the film and thus thermally insulate the probe.

c. Laser Doppler velocimeter

Laser Doppler velocimeters (LDV), also called laser Doppler anemometers (LDA), have been used successfully as measuring devices in optically transparent fluids. The simplest configuration is shown in Figure 10.3. Light from the laser is passed through a prism combination called a beamsplitter that divides the light into two beams, usually of equal intensities. By means of mirrors or prisms these beams are made parallel to one another, and then passed through a lens. After leaving the lens, the focused beams intersect at the lens focal point. The focusing power of the lens separates the two intersecting beams by an angle 2θ, where the angle θ is given by

$$\sin \theta = \text{(lens focal length)}/\text{(one-half the spacing between the two focused beams)}.$$

$$\tag{10.1.6}$$

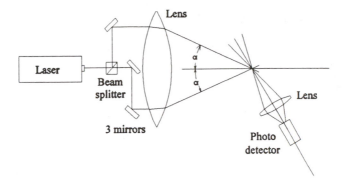

Figure 10.3. Laser Doppler anemometer.

At the point of beam intersection, the beams set up an interference pattern as shown in Figure 10.4. The laser light wave fronts are spaced a distance λ apart, where λ is the wavelength of the laser light.[1] By simple geometry, the fringes in the interference pattern formed at the lens focal are spaced a distance $\lambda/\sin\theta$ apart. The fluid will normally contain some suspended particles in the micron range: if there are not a sufficient number of these particles present, it is necessary to seed the flow with additional particles. As these suspended particles pass through this interference pattern in the flow, a changing light intensity is seen by a photodetector (photodiode or photomultiplier tube) focused on the beam intersection. The frequency at which this intensity varies is equal to the fluid velocity component lying in the plane of the pattern and perpendicular to the fringes, divided by the fringe spacing. From the geometry this is readily seen to be

$$f_D = V/(\lambda/\sin\theta) = V\sin\theta/\lambda, \tag{10.1.7}$$

where V is the instantaneous velocity component perpendicular to the fringes. The frequency shift f_D, called the Doppler frequency, is then converted by the electronics into a voltage proportional to velocity.

Equation (10.1.7) tells us the frequency shift, which is the information needed in processing the signal to determine velocity. Before this processing takes place, however, there is a certain amount of preprocessing that is necessary. First, the signal is going to be received from a test volume that is roughly ellipsoidal in shape. The minimum diameter of a focused laser beam (called the "Airy diameter") is

$$d = 4f\lambda\pi D, \tag{10.1.8}$$

where λ is the wavelength of the light, f is the focal length of the lens, and D is the diameter of the beam incident on the lens. Considering the two intersecting light beams

---

[1]The geometric description we use here for a laser Doppler velocimeter is the simplest model of its operation. It does not, however, explain how a Doppler shift occurs, as the fringe pattern formed is a virtual, rather than a real, image. A more detailed model using scattered wave theory is possible. The conclusions of the two theories are, however, identical, and the geometric model does provide necessary information for the design of the measuring volume.

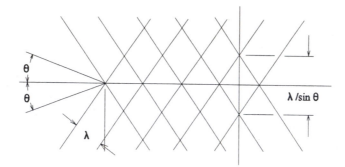

Figure 10.4. Wave front pattern showing the formation of fringers in a laser Doppler velocimeter.

to be circular cylinders of diameter d given by (10.1.8), simple geometrical considerations (Figure 10.5) show that the illuminated ellipsoid will have a length of $d/\sin \theta$ and a circular cross section of diameter $d/\cos \theta$. Thus a particle passing through the maximum diameter of the illuminated ellipsoid will pass through a maximum of N fringes, where N is given by

$$N = \text{volume diameter/fringe spacing} = (d/\cos \theta)/(\lambda/\sin \theta)$$

$$= d \tan \theta / \lambda = 4f \tan \theta / \pi D. \qquad (10.1.9)$$

The light intensity that the particle sees will vary from a minimum at the edges of the ellipsoid to a maximum at the ellipsoid center in an approximately Gaussian distribution. This means that the light seen by the photodetector will have the useful signal superimposed upon a DC Gaussian pedestal. This pedestal contains no velocity information and must first be removed, usually by means of a low-pass electrical filter.

Similarly, if there are particles in the flow that are either much larger or much smaller than the fringe spacing, the signals they give to the photodetector will

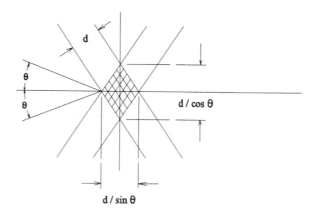

Figure 10.5. Measuring volume dimensions in a laser Doppler anemometer.

contribute to signal noise. Thus signals are usually bandpass-filtered, the width and center frequency of the filter being selected by consideration of the flow conditions being measured.

The Doppler frequency of (10.1.7) is measured by counting the number of fringes crossed by a particle in a given time. To ensure validity of the signal, modern LDV signal processors make counts at two different time intervals. If the two counts do not give velocities that agree within a certain selectable percentage, the processor rejects that measurement. Thus the output signal of an LDV processor consists of a series of steps of variable length. The number of particles in the flow must be controlled so that the step length is short enough to give a sufficiently accurate record.

The LDV or LDA has both pluses and minuses to its credit when compared to the hot-wire anemometer. It is linear in its response to velocity and works in both liquids and gases, as long as the fluid is optically transparent. The suspended particles needed to provide the signal should be roughly of the size of the fringe spacing, typically of the order of 1 micron. Such small particles will follow the flow very well. Since, however, the LDV follows any one particle for only a very short time, as time goes on the signal consists of data from many particles, with a signal "dropout" when no particles are present in the beam intersection. Thus the signal is intermittent. Also, the LDV needs a good optical path. Curved conduit walls can cause problems in focusing of the two beams, as can temperature gradients in the flow. Both the LDV and the hot-wire anemometer require a high degree of understanding of the principles of their operation and of turbulent flow to ensure successful operation.

Both hot-wire anemometers and LDVs are relatively nonintrusive measuring devices. In the case of the hot-wire anemometer, the hot wire itself is extremely small so as to keep thermal inertia small. Thus the main intrusion comes through the mechanical support of this wire. It might at first be assumed that the LDV is completely nonintrusive. However, the laser light can induce thermal gradients in the flow, thereby altering both the flow and the index of refraction of the fluid. Both these factors can contribute to measuring errors.

2. Volume Rate-measuring Devices

a. Venturi meter

The *venturi meter* was introduced in Section 3.8 of Chapter 3. (See Figure 10.6.) While named after Venturi, it was actually invented by Herschel in 1887. Venturi meters are designed primarily for measuring flow in a pipe or tube, but they can also be mounted on the side of a moving vehicle to measure velocity. These meters have the advantage that, providing the meter is properly designed, no permanent energy loss is introduced by inserting it into a flow. Proper design in this case means that changes in diameter occur smoothly, and that the diffuser angle is sufficiently small that flow separation does not take place. Venturi meters are also suitable for use when suspended solids are present.

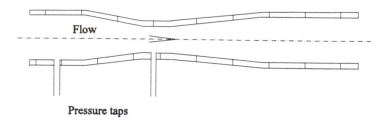

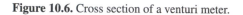

Pressure taps

Figure 10.6. Cross section of a venturi meter.

The meter consists of a short inlet section, a short throat (minimum area) section, and a diverging exit section. The inlet section continues the pipe diameter, and smoothly blends it into a short throat. At the throat exit the converging section consisting of a cone with an included angle of $21°\pm2°$. This converging section reduces the diameter to somewhere between one-third and three-fourths of the entrance diameter. At the point where the area reaches a minimum (the throat), the meter continues as a short cylindrical section of constant diameter. At the end of this constant diameter throat section, the diffuser cone with an included angle of 5–$15°$ brings the meter diameter back up to the entrance diameter. Typically the first pressure tap is 0.25–0.75 pipe diameters upstream from the converging section, and the second tap is 0.5 throat diameter downstream from the throat.

Venturi meters should change cross section smoothly in order to minimize energy losses. For the divergent part of the meter, angles are kept around $7°$ to avoid flow separation. In practice, manufacture is made much easier by using what basically are truncated cones for both the converging and diverging portions, with little degradation in performance of the meter. Considerable study has gone into the design of these meters, and engineering organizations such as ASME (American Society of Mechanical Engineers) regularly publish standard specifications for their construction.

The theoretical formula we developed in Chapter 3 for discharge as measured by the venturi meter is good to within 1–3%, provided that the Reynolds number based on inlet diameter is greater than 10^5. For values of Reynolds number below 10^5, the flow is not fully turbulent, and the actual velocity will lie below the value we predicted. (Remember that we assumed the velocity profile to be uniform across a section, which is more likely to be true for turbulent flow.) The meter should always be placed a minimum of 15 pipe diameters downstream from any bends, valves, or the like, since such departures from a straight pipe can introduce secondary flows with swirl velocity components that could affect the pressure readings. Consideration should be made of the pressure levels expected and the diameter ratio selected, so that cavitation does not occur at the throat of the meter. For best results the pressure should be read at four or more points around the periphery of the meter. Great care must be taken to remove any gas bubbles in gage connections.

b. Orifice plate

An ***orifice plate*** is basically a disk with a hole in it placed inside a pipe or tube. (See Figure 10.7.) Most commonly the hole is circular and centered in the plate, but it can be placed eccentrically in the plate, or it may even consist of just a baffle across part of the pipe. Typically the hole is sharp-edged and beveled, with the sharp edge placed on the upstream side of the meter. Assuming that the flow is turbulent so that head loss $\Delta p/\rho g$ is proportional to velocity squared, the discharge is given by

$$Q = KA\sqrt{2\,\Delta p/\rho}, \tag{10.2.1}$$

where Δp is the pressure drop across the orifice and A is the orifice area. For centered orifices, the discharge coefficient K is typically of the order of 0.62 for Reynolds numbers greater than 10^5.

The location of the pressure taps in an orifice is not critical. For some orifice plates, the taps are placed as close to the plate as possible. Generally it is a good idea not to have the downstream tap farther than one-half pipe diameter from the plate. This is approximately where the jet passing through the orifice has a minimum diameter (the ***vena contracta***). Further away, the pressure is likely to fluctuate and provide unreliable readings.

The orifice plate has the advantage that it is easy to construct and takes up a minimum amount of room. It has the disadvantages of introducing a nonrecoverable head loss and of being unreliable in dirty fluids.

c. Nozzles

A flow nozzle is another device for measuring flow in a pipe. Nozzles introduce a somewhat smaller energy loss than do orifice plates, and are both cheaper and shorter

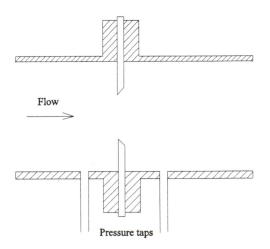

Figure 10.7. Sharp-edge orifice plate meter.

than a venturi meter. They are basically the same as the venturi meter up to the throat, with the divergent nozzle omitted. (See Figure 10.8.) The usual nozzle design used is the ASME long-radius nozzle, where the nozzle shape upstream of the throat is that of a quadrant of an ellipse. Pressure taps are usually located one pipe diameter upstream of the nozzle inlet and at the nozzle throat. Again the discharge will be given by an equation of the type

$$Q = KA\sqrt{2\,\Delta p/\rho}, \tag{10.2.2}$$

where Δp is the pressure drop across the nozzle and A is the orifice area. The discharge coefficient K is typically in the range of 0.99 to 1.11 for Reynolds numbers based on a pipe diameter greater than 10^5.

The nozzle has many of the same disadvantages as does the orifice plate.

d. Elbow meter

An elbow in a pipeline can be easily converted into a flow meter by placing pressure taps at the inside and outside bends of the nozzle. (See Figure 10.9.) The momentum of the flow is such that the flow on the inside of the bend will have a faster velocity than on the outside of the bend, creating a pressure difference across the elbow cross section. The analysis of the flow in an elbow meter is more complicated than for many of the other meters we consider here. An example of such an analysis can be found in the ASME publication (1971) referenced at the end of this chapter. The result is that the pressure difference developed by this momentum change will be related to the discharge by a formula very much of the same form as (10.2.1). The performance of an elbow meter is much like that of a venturi meter, with many of the same advantages. Since elbows are frequently occurring elements of flow systems, these meters are inexpensive alternates to venturi meters and orifice plates.

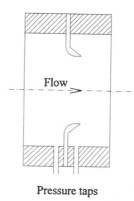

Pressure taps

Figure 10.8. Flow-measuring nozzle.

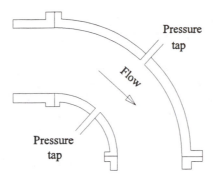

Figure 10.9. Elbow meter.

e. Positive displacement meter

A positive displacement meter has one or more moving parts that channel portions of the flow into one or more calibrated volumes. The flow is then measured by counting how many times the volumes are filled.

One example of a positive displacement meter is the reciprocating piston meter, frequently used in measuring the natural gas delivered to a home. This meter is essentially a piston pump running backward. The interior of the meter is separated into three chambers, the middle chamber being a double bellows that can intrude into the other two fixed chambers. (See Figure 10.10.) The incoming gas first pushes the middle piston to the end of its stroke. During this pushing phase, the intake valves open,

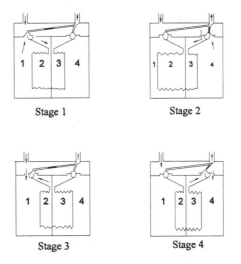

Figure 10.10. Positive displacement meter. Flow is controlled by valves driven by bellows pressure. Stage 1: Gas enters 2, leaves 3. Stage 2: Gas enters 3, leaves 4. Stage 3: Gas enters 1, leaves 2. Stage 4: Gas enters 4, leaves 3.

allowing gas to enter the piston. When the piston reaches the end of the stroke, the intake valves close and the discharge valves open. The piston returns to its original point, and the cycle repeats. A simple mechanical counter driven by a crank attached to the bellows usually is used to count the number of cycles.

A similar principle is utilized in measuring water delivered to the home. The interior of these meters consists of a chamber with spherical sidewalls and conically shaped top and bottom. (See Figure 10.11.) This chamber is divided by a radial partition. A slotted disk supported by a sphere in the center of the chamber fits over this partition. The partition restricts the disk from rotating. As entering water strikes the disk, the disk nutates and allows the water to pass through the meter. As the disk nutates, a pin in the supporting sphere traces out a circular path and drives a set of gears connected to a counter. More recent models of these meters have transmitters that output an electrical signal that can be read by small handheld dedicated computers.

The accuracy of positive displacement meters generally varies with flow rate, and also is affected by temperature changes. They are used in many applications, since they are cheap and reliable, and can run without maintenance in adverse conditions for long periods of time.

f. Rotameter

The **rotameter**, or variable area meter, is frequently seen in laboratories, hospitals, and manufacturing facilities. It consists of slightly tapered, usually transparent, vertical tubes with a float inside. (See Figure 10.12.) The fluid enters the tube at the bottom, causing the float to rise until the drag force on the float is equal to its weight minus the buoyancy force. The walls of the tube are calibrated to give a relation between height and flow rate for a given fluid. Some special float designs allow for a range of fluid densities to be used with no recalibration. In some cases all the flow passes through the rotameter. In other designs, only a fraction of the flow is diverted through the device, with the remainder bypassing the rotameter.

A rotameter is simple to use. It introduces a negligible pressure drop in a system, and is accurate to use. By changing the float, a wide range of flows can be measured with one meter body. Typical problems include the float sticking to the wall of the tube, and dirt blocking openings and coating the walls. Special designs can sometimes compensate for these problems.

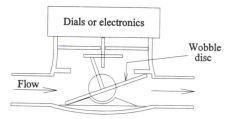

Figure 10.11. Home water meter.

Figure 10.12. Rotameter.

g. Turbine meter

A turbine meter consists of a housing supporting an impeller through bearings. The impeller can be coaxial with the flow direction or perpendicular to it. (See Figure 10.13.) Often a magnet is placed in one of the impeller vanes that, as the impeller rotates, activates a magnetic pickup in the housing. With well-designed vanes on the impeller, the impeller will rotate at a speed proportional to the flow velocity. The magnetic pickup relays the turbine speed to a counter.

Variations of these designs consist of propellers, sometimes in a housing, that can be hand-held for insertion in air or water currents. A turbine meter can be made quite accurate and in a variety of sizes. Its electrical output permits it to be used in control circuits. It can have a problem with bearings being contaminated or worn, giving false readings.

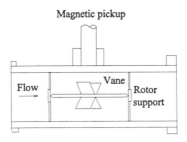

Figure 10.13. Turbine meter.

h. Doppler-acoustic flow meter

In flows containing typically 25 parts per million of suspended particles, or bubbles 30 microns or larger in diameter, a doppler-acoustic flow meter can be used. It consists of piezoelectric transducers that send a pressure pulse into the fluid. The pulse is reflected by the particles or bubbles, and the reflected pulse is then measured by another transducer. (See Figure 10.14.) Sometimes the configuration of the transducers is such that the same transducer can be used as both transmitter and receiver. The change in frequency between the transmitted and reflected pulses is due to the Doppler frequency shift in the signal induced by the suspended particles, and is proportional to the flow speed. These meters operate much on the same principle as radar detectors used by police departments for speed checks.

To get a good performance from this meter, field calibration is generally necessary, since its performance is sensitive to the form of the contaminants present. If the suspended particles are too large, the particles will not move with the flow velocity. This must be accommodated for by calibration of the meter.

i. Vortex-shedding flow meter

For Reynolds numbers greater than several thousand, the wake behind most bodies is characterized by an axisymmetric pattern of alternating eddies, or vortices. For a broad range of Reynolds numbers above the critical one (the Reynolds number at which shedding starts), typically from 2,000 to 2×10^5, the Strouhal number (frequency of vortex shedding times the body width divided by the stream velocity) remains fairly constant with Reynolds number.

The velocity disturbances associated with the shed vortices is accompanied by a corresponding pressure disturbance. By placing a blunt body across a pipe with a pressure transducer in its downstream side, a fluctuating pressure field with frequency proportional to velocity is measured. (See Figure 10.15.) The wide range of flow rates, linearity of response, and low induced pressure drop are some of the advantages associated with this meter.

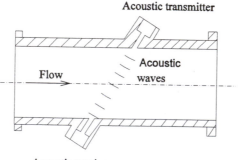

Figure 10.14. Acoustic flow meter.

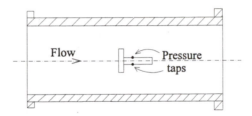

Figure 10.15. Vortex-shedding flow meter.

j. Magnetic flow meter

For electrically conductive fluids, Faraday's law states that the voltage induced across a conducting fluid as it moves at right angles through a magnetic field is proportional to the velocity of the fluid. A magnetic flow meter has a device for applying a magnetic field to the fluid, and another for measuring the voltage induced. (See Figure 10.16.) The magnetic flow meter does not obstruct the flow system in any way, which can be important in flows with low pressure drops. It can also be used where abrasive, toxic, or corrosive fluids are being metered. It does require periodic maintenance and cleaning, and is affected by temperature changes. If alternating current fields are used, the power requirements can be high. Again, linearity of response is a plus for this meter.

k. Weirs

Weirs are simple and reliable devices for measuring steady flows in open channel flows. A weir is an obstruction across a channel, usually a plate with a cut-out area, over which the liquid flows.

We assume that the approach kinetic energies can be neglected, that the flow over the weir is at atmospheric pressure, and that there are no energy losses. H is the upstream depth measured with respect to the lowest depth of the weir, and h, the depression of the free surface just above the weir with respect to the upstream conditions, is small compared to H. See Figure 10.17.

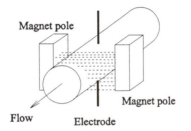

Figure 10.16. Magnetic flow meter.

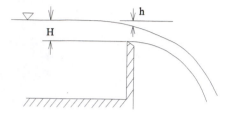

Figure 10.17. Flow over a weir.

With these assumptions, and if we assume that only approximately half of the potential energy head H is lost, the energy equations tells us that just above the weir the velocity is given by

$$v = \sqrt{2g(H/2)} = \sqrt{gH}.$$

For a rectangular weir with an opening across the entire channel (Figure 10.18) the area is approximately wH. Then

$$Q = vA = wH\sqrt{gH} = w\sqrt{gH^3}. \tag{10.2.3}$$

For a triangular weir with an approximate area $H^2 \cot \alpha$ (Figure 10.19), a similar analysis gives

$$Q = H^2 \cot \alpha \sqrt{gH} = \cot \alpha \sqrt{gH^5}. \tag{10.2.4}$$

Thus by measuring the depth upstream from the weir, a reliable indication of the discharge is obtained.

In practice, because of the many assumptions made the above formulas are not particularly accurate. Complicated and numerous experimental tests are necessary for more accurate formulas. Presently accepted empirical results for a rectangular weir to replace (10.2.2) are

$$Q = (2/3)C_R\, w\sqrt{2g(h')^3}, \tag{10.2.3a}$$

with $h' = H +$ either 0.003 ft or 0.9 mm, and for a triangular weir to replace (10.2.4),

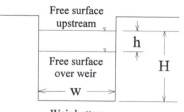

Weir bottom

Figure 10.18. Rectangular weir.

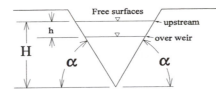

Figure 10.19. Triangular weir.

$$Q = (8/15)C_T \cot \alpha \sqrt{2g(h')^5}, \qquad\qquad (10.2.4a)$$

with $h' = H + \Delta$. The terms C_R, C_T, and Δ are empirical correction coefficients that depend on various geometric factors. Details are given in the ASME reference at the end of this chapter. Even with these modifications, in practice a weir should be calibrated in situ.

The presence of a weir means that suspended solids will settle out on the upstream side of the weir, thus the channel must be periodically cleaned. Weirs do introduce an energy loss into the system.

I. Integration of velocity measurements

Any device that measures velocity can also be used for measuring discharge, albeit with some additional work. For a circular pipe, the usual approach is to divide the area into a number of concentric equal-area rings, and to measure the velocity at some representative point inside each ring.

For example, suppose we divide a circle of radius R into n rings, each with equal areas. The outer radius of the ith ring is thus given by

$$r_i = R\sqrt{i/n}. \qquad\qquad (10.2.5)$$

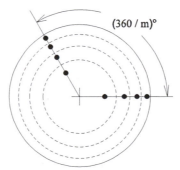

Figure 10.20. Equal area segments and measuring points for determining discharge from velocity measurements.

We might decide to take measurements at a point inside the ring so that the areas on either side of the ring containing the measuring point are equal. (See Figure 10.20.) Thus we would take measurements at

$$R_i = R\sqrt{(2i - 1)/2n}. \tag{10.2.6}$$

If the velocity readings at the R_i are denoted by V_i, then

$$Q = \pi R^2 \sum_{i=1}^{n} V_i/n. \tag{10.2.7}$$

The accuracy of this procedure can be improved in several ways:

1. Take measurements of V_i at several equally spaced points on the circle of radius R_i, and in (10.2.6) use the average of these for each V_i.
2. Take measurements using increasing values of n, until the value found for Q using (10.2.6) does not change.
3. The choice of the radius R_i for the measuring circle is arbitrary, and may not ensure great accuracy in the determination of discharge. Choices of other points as given in the references will improve accuracy.

3. Mass Rate-measuring Devices

If the mass density is virtually constant, any of the previous methods used for measuring volume rate of flow can be converted to a mass-measuring device simply by multiplying Q by the mass density. For liquid flow at low Mach numbers (less than 0.25), it may be sufficient to measure the local temperature of the liquid by means of a thermocouple or similar device, and use this together with tables for density as a function of temperature to compute mass flow. In many problems, however, particularly in manufacturing processes, this is not a sufficiently accurate or convenient procedure. Several commercial schemes follow.

a. Hastings mass flow meter

A flow meter manufactured by the Teledyne-Hastings company uses a device for measuring mass flow rate that relies on heat conduction due to a laminar flow. The main body of the flow meter contains a flow-straightening device, which is used to induce a pressure gradient across the meter length. This in turn causes flow to occur through a small capillary bypass of about 1 mm diameter. (See Figure 10.21.) A temperature gradient is imposed along the capillary tube by an electric coil that heats the capillary with a constant amount of heat flow. Thermocouples at either end of the capillary measure the temperature gradient in the capillary. This temperature gradient is proportional to the rate of mass flow.

The device has to be calibrated for a given fluid, given temperature rate, and given mass flow. These calibrations are carried out in the factory, although they should be

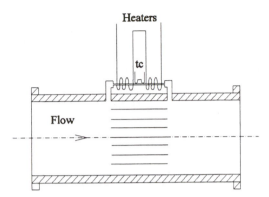

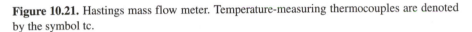

Figure 10.21. Hastings mass flow meter. Temperature-measuring thermocouples are denoted by the symbol tc.

checked locally before installation. The output is a voltage that can conveniently be read by a voltmeter or sent to a computer for further processing. The meter will respond to only slow changes in mass flow rate due to the thermal inertias involved. Care must be taken that the gas is free of dirt that might plug the bypass capillary or the flow straightener.

b. Coriolis force mass flow meter

A mass flow meter manufactured by the Micro Motion Corporation utilizes the Coriolis forces that are induced in an oscillating U-tube containing a flowing fluid. The meter operates by causing some of the flow to pass through a small capillary sensor tube. The U-tube is forced to vibrate at its natural frequency by electromagnets. This vibration is in a direction perpendicular to the plane containing the U-tube. (See Figure 10.22.) This motion accelerates the fluid in the inlet side of the U-tube, and decelerates the fluid in the exit side. Momentarily it appears as if the fluid is being

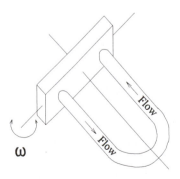

Figure 10.22. The Micro Motion Coriolis flow meter.

rotated at an angular velocity equal to the exciting frequency Ω. This induces a small twist force in the U-tube that can be measured electronically. The magnitude of this flow-induced Coriolis force is

$$\mathbf{F} = m2\Omega \times \mathbf{V}, \tag{10.3.1}$$

where m is the mass in the U-tube and $\mathbf{V}$ is the fluid velocity. If the two sides of the U-tube are separated by a distance $2r$, the twisting moment will be

$$M = 2rF = 4rm\Omega V. \tag{10.3.2}$$

Since the mass flow rate is equal to mV/L, where L is the length of the U-tube, we have

$$M = 4r\dot{m}\Omega L. \tag{10.3.3}$$

Since the U-tube is made of an elastic material, we expect that the moment will cause the U-tube to twist, with the angle of twist being linearly proportional to the moment. Thus

$$\theta = M/k, \tag{10.3.4}$$

where θ is the angle of twist and k is a torsional spring constant for the U-tube. Thus combination of (10.3.3) and (10.3.4) gives

$$\dot{m} = k\theta/krL\Omega. \tag{10.3.5}$$

The mass flow rate for a given U-tube is therefore linearly proportional to the angle of twist, which can be measured by two magnetic position detectors.

Again, the meter must be factory-calibrated and the calibration checked before use. The fluid must be free of dirt particles to avoid clogging the U-tube.

4. Pressure-measuring Devices

The manometer was introduced early in Chapter 2 as a simple and reliable pressure-measuring device. It is the standard by which other devices are usually calibrated. It, however, has the disadvantages that (1) it is not always easy to read if the flow is fluctuating, (2) its length can become very long if the pressure is high, and (3) it is not easily adapted to give remote readings. Thus alternative pressure-measuring means have been devised.

a. Bourdon-type gage

The common dial-type pressure gages seen in everyday use are variations of a Bourdon gage. This gage consists of a curved hollow tube, roughly elliptical in cross section and closed at one end. The open end of the tube is connected to the pressure that is being read. (See Figure 10.23.) The closed end has a mechanical linkage attached to

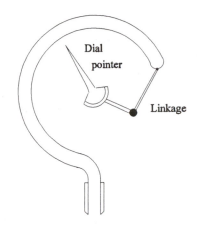

Figure 10.23. Bourdon tube pressure gage.

the dial needle. As the pressure is increased, the tube straightens out more and more. This moves the linkage, and the dial needle follows accordingly. These gages are simple to read, inexpensive, and can be made quite accurate. Friction in the linkage requires that they be "tapped" before every reading is taken.

Modifications of these gages mainly amount to replacing the Bourdon tube with a bellows or other expandable device. The means of operation of these modified devices is basically the same.

b. Strain gage and capacitance gage pressure cells

A device frequently used for pressure measurement that gives an electrical output is the pressure cell. It consists of a chamber divided in two parts by a thin diaphragm. Pressure is introduced on one or both sides of the chamber, depending on whether gage pressure or pressure difference is desired. (See Figure 10.24.) The pressure difference across the diaphragm causes the diaphragm to stretch. A resistance strain

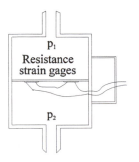

Figure 10.24. Strain gage pressure cell.

gage bridge bonded to the diaphragm records the amount of strain produced on the diaphragm. This strain is proportional to the pressure difference across the diaphragm. Such cells are rugged and are designed for wide ranges of operating pressures. They are compact in size and the instrumentation is well-developed.

A variation of the strain gage pressure cell is the capacitance gage pressure cell. The basic operation is much the same as with a strain gage cell, but the diaphragm this time serves as one plate of a capacitor. As the pressure varies, so too does the distance between this capacitor plate and a fixed plate, varying the capacitance. A bridge circuit in the cell measures the change in capacitance, which is proportional to the pressure.

c. Piezoelectric crystals and semiconductors

Inexpensive phonographs (pre-CDs) used piezoelectric crystals to convert the mechanical forces produced by needle movement to an electrical signal. Such crystals can clearly also be used as pressure-measuring devices. Semiconductor devices that develop voltages proportional to applied pressures have also recently been developed. These devices have the advantage that they can be readily miniaturized and used in applications where long pressure lines would mean the filtering out of the high-frequency portions of pressure changes. Applications include medical ones, where they can be inserted into arteries and the heart for direct measurement of blood flow pressure and heart behavior. These devices are more sensitive to temperature and electrical drift and stability than pressure cells.

5. Viscosity-measuring Devices

Laminar pipe flow and the velocity of falling spheres at Reynolds numbers less than 1 have already been pointed out as suitable means for measuring viscosity. These and other solutions have been incorporated into a wide range of instruments in the field of viscometry.

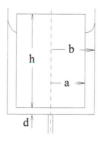

Figure 10.25. Rotating cylinder viscometer.

a. Rotating cylinder viscometer

The fluid to be measured is placed between two concentric cylinders. The outer cylinder is rotated at a constant speed Ω, and the inner cylinder is attached to a torsion wire or other torque-measuring device, so that the torque exerted on the inner cylinder can be measured. (See Figure 10.25.) The gap between the cylinders is small, so that the velocity profile is close to that of plane Couette flow, with a constant velocity gradient. The shear stress on the side of the inner cylinder is then

$$\tau_s = \mu b\Omega/(b-a), \tag{10.5.1}$$

and the torque on the side of this cylinder is

$$T_s = a(\tau_s 2\pi a h) = 2\pi a^2 h b\Omega\mu/(b-a). \tag{10.5.2}$$

The flow at the bottom of the inner cylinder will also have a constant velocity gradient in the vertical direction, although this will vary linearly with the radius because of the rigid-body rotation of the cylinder. The shear stress on the bottom is then

$$\tau_b = \mu r\Omega/d, \tag{10.5.3}$$

resulting in a torque of

$$T_b = \int_0^a r\tau_b 2\pi r \, dr = \pi a^4 \mu\Omega/2d. \tag{10.5.4}$$

The total torque $T = T_s + T_b$ is then given by

$$T = 2\pi a^2 h b\Omega\mu/(b-a) + \pi a^4\Omega\mu/2d. \tag{10.5.5}$$

The viscosity μ can then be solved for from (10.5.5), giving

$$\mu + T/\pi a^2\Omega[2hb/(b-a) + a^2/2d]. \tag{10.5.6}$$

Even though the instrument geometry is known for a given device, the various approximations using in obtaining (10.5.6) require that the viscometer be calibrated with a fluid of known viscosity.

b. Oswald-Cannon-Fenske viscometer

This viscometer is frequently used in laboratories because of its high accuracy. It is less commonly used in the field because of its fragility. It consists of a glass tube bent roughly in the shape of a U, with two spherical chambers in one leg connected by a capillary tube. Two lines are marked on either side of the capillary tube. (See Figure 10.26.) A given charge of fluid is added to the right-hand leg of the viscometer, usually around 7 mL. The time it takes for the upper free surface to pass between the two lines is linearly related to the viscosity according to a formula provided by the manufacturer. These viscometers come in a variety of capillary diameters suitable for measurements in different ranges of viscosity.

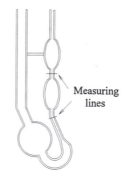

Figure 10.26. Oswald-Cannon-Fenske viscometer.

A version of this viscometer with a slightly different geometry is the Ubbelohde viscometer.

c. Saybolt viscometer

Another version of the capillary tube viscometer is the Saybolt viscometer. It consists of a vertical tube with a rounded bottom containing a short capillary tube. (See Figure 10.27.) A charge greater than 60 cm^3 of a fluid is introduced into the viscometer and a plug at the bottom then removed, allowing the fluid to flow out under a falling head. The time for 60 cm^3 of a fluid to flow out is related to the viscosity by a formula of the type

Figure 10.27. Saybolt viscometer.

$\nu = At + B/t.$

Because of the frequent use of this type of viscometer in industrial situations, viscosity is often reported in Saybolt seconds, also abbreviated as SUS for Saybolt universal seconds. SUS then is the time it takes for 60 cm^3 of a fluid to pass through the viscometer orifice. The relationship between this viscosity unit and kinematic viscosity in stokes is given by

$$\nu \text{ (stokes)} = 0.0022t - 1.80/t. \tag{10.5.7}$$

d. Falling-body viscometer

The drag on a sphere falling in a very large container of otherwise stationary fluid is given by

$$F_{\text{drag}} = C_D(\pi D^2/4)\rho V^2/2. \tag{10.5.8}$$

For very low Reynolds number (less than 2 or 3), Stokes found that the drag coefficient is given by

$$C_D = 24/\text{Re}, \tag{10.5.9}$$

where Re is the Reynolds number based on the sphere diameter. Thus, combining (10.5.8) and (10.5.9) we find

$$F_{\text{drag}} = 3\mu\pi DV. \tag{10.5.10}$$

The net force on a falling sphere is

$$F_{\text{drag}} + F_{\text{buoyancy}} - F_{\text{gravity}} = 0$$

for a constant (terminal) velocity V. The gravity and buoyancy forces are simply the sphere volume times the specific weight of the body and fluid, respectively. Thus

$$3\mu\pi DV + \rho_{\text{fluid}}g\pi D^3/6 - \rho_{\text{sphere}}g\pi D^3/6 = 0. \tag{10.5.11}$$

Dividing by pi and the sphere diameter D and solving for the viscosity we have

$$\mu = (\rho_{\text{sphere}} - \rho_{\text{fluid}})gD^2/18V. \tag{10.5.12}$$

Thus by measuring the sphere velocity (typically by determining the time it takes the sphere to traverse a marked distance) the viscosity of the fluid is easily determined.

Because the flow is so slow, the velocity field caused by the sphere motion dies out slowly with distance from the sphere. Thus the container walls can appreciably affect the Stokes drag relation. Equation (10.5.12) is valid providing that the ratio of container diameter to sphere diameter is of the order of 100 or more. Care must be taken to release the sphere along the tube centerline.

6. Surface-Tension-measuring Device

Surface tension can be measured in several ways. One device consists of a thin platinum ring supported by a stirrup. This is connected to a lever arm attached to a torsion wire. The ring is first cleaned in a solvent, then passed through a hot flame to remove the solvent, and finally placed in a cup of the fluid. The torsion wire is then twisted until the ring raises above the surface of the fluid and is held in place by a thin film of the liquid. Twisting continues until the film breaks, and this torque is recorded. Surface tension can be deduced from this with empirical formulas. Use of more complicated rings allows measurement of the surface tension at an interface of two liquids.

Another technique for measuring surface tension is the pendent drop technique. A drop of the liquid whose surface tension is to be found is created at the outlet of a small tube. A free-body diagram of the suspended drop is shown in Figure 10.28. Summing forces in the vertical direction gives

$$\Sigma F_{\text{up}} = 2\pi R\sigma \cos\theta - \gamma V - (p_0 - \gamma Z)\,\pi R^2 = 0. \tag{10.6.1}$$

Here Z is the length of the drop, V is the volume of the drop given by

$$V = \int_0^R 2\pi r^2\,dz,$$

R is the drop radius at the top of the drop, θ is the angle between the vertical and the tangent to the drop at its top, and p_0 is the pressure at the bottom interior of the drop. This pressure can be found by measuring the pressure somewhere in the line and using the hydrostatic law to transfer this pressure to the drop bottom. Solving (10.6.1) for the surface tension gives

$$\sigma = [\gamma V + (p_0 - \gamma Z)\,\pi R^2]/R\cos\theta. \tag{10.6.2}$$

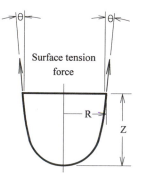

Figure 10.28. Free-body diagram of suspended drop for surface tension measurement.

To determine Z, V, and the other geometric quantities, the drop may either be photographed, or its silhouette enlarged by projection onto a surface, where the shape may be traced. Breaking up the silhouette into N equally spaced sections of length Δz and measuring the $N + 1$ radii at the ends of each section, we have the approximations

$$V \approx 2\pi \sum_{1}^{N} 0.5(r_{i+1} + r_i)\, \Delta z = \pi(r_1 + 2r_2 + 2r_3 + \cdots + 2r_N + r_{N+1})\, \Delta z,$$

where $r_1 = 0$ and $r_{N+1} = R$. The angle θ is found from

$$\theta = \pi - \tan^{-1}(dr/dz)_{z=Z},$$

where the derivative is computed with a backward difference formula such as

$$(dr/dz)_{z=Z} \approx (3r_{N+1} - 4r_N + r_{N-1})/8\, \Delta z.$$

Thus all the needed data can be computed. The advantage of this method is that the needed instrumentation is inexpensive.

7. Calibration

We have seen in this chapter a wide range of devices for measuring flow and fluid properties. Any analytic results we have relating fluid properties to measurable flow quantities are fair game for developing an instrument, and many other instruments do exist for special purposes. The American Society of Mechanical Engineers (ASME), the American Society for Testing Materials (ASTM), the National Bureau of Standards (NBS), and similar organizations provide detailed specifications for designing and calibrating these devices, as well as for producing fluids of known properties for use in calibration. Many manufacturers also provide extensive literature for applying their

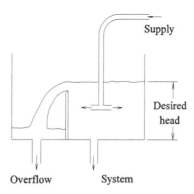

Figure 10.29. Constant head tank.

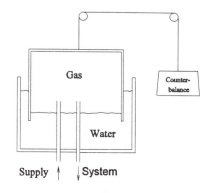

Figure 10.30. Gasometer.

products, as well as technical personnel to aid in the choice of a system. These should all be consulted when deciding on instrumentation for a given task.

Along with all the assistance that is available from these various sources, it is preferable if a measuring system can be tested and calibrated in place, and under the conditions in which it is to be used. If a venturi meter, orifice plate, or the like has already been calibrated, it can be placed in the line in series with the device to be tested and the two sets of readings compared.

A more reliable calibration in the case of liquids is to use a constant head tank for a supply source, and a weight tank as a receiver. A constant head tank is a tank with a large cross section (Figure 10.29). The liquid to be used is supplied to the constant head tank at a rate greater than the fluid will be used in the system. An overflow weir in the tank ensures that the level of fluid is kept constant, or very nearly so, at all times. Thus the supply head of the system will remain constant during a calibration run.

A weigh tank is simply a receiver tank mounted on a scale. Usually it is used by timing the interval it takes to add a given weight of fluid to the tank.

In the case where the working fluid is a gas, the constant head and weigh tanks are replaced by gasometers. These consist of an inverted moving upper cylinder closed at the top and inserted into closely fitting lower cylinder with a liquid in between acting to seal the gas inside the two cylinders (Figure 10.30). Weights and counterweights are used to maintain a constant pressure inside the gasometer. The change of height of the gasometer is then proportional to the volume of gas added or subtracted from the gasometer. Before natural gas replaced coal gas as a heating source in many cities, gasometers were used for the storage of coal gas. The outline of large gasometers were then a normal and prominent sight on a city skyline. Large gasometers can still be found on many industrial sites, where they are used for gas storage. Smaller ones can be seen in medical diagnostic laboratories, where they are used to measure lung capacity.

Suggestions for Further Reading

ASHRAE, *ASHRAE Fundamentals*, American Society of Heating, Refrigeration, and Air Conditioning Engineers, New York, 1989.

ASME, *Fluid Meters, Their Theory and Application*, 6th edition, American Society of Mechanical Engineers, New York, 1971.

ASME Fluid Meters Research Committee, "The ISO-ASME Orifice Coefficient Equation," *Mech. Eng.*, pp. 44–45, July 1981.

Baker, W. C., and J. F. Pouchet, "The measurement of gas flow, parts 1 and 2," *J. Air Pollut. Contr. Assoc.*, vol. **33**, pages 66–72 and 156–162, 1983.

Bean, H. S., editor, *Fluid Meters: Their Theory and Application*, 6th edition, American Society of Mechanical Engineers, New York, 1971.

Beckwith, T. G., and R. D. Marangoni, *Mechanical Measurements*, 4th edition, Addison-Wesley, Reading, MA, 1990.

Benedict, R. P., *Fundamentals of Temperature, Pressure, and Flow Measurement*, 3rd edition, Wiley, New York, 1984.

Buckwith, T. G., and N. Lewis Buck, *Mechanical Measurements*, 2nd edition, Addison-Wesley, Reading, MA, 1973.

Dally, J. W., W. F. Riley, and K. G. McConnell, *Instrumentation for Engineering Measurements*, Wiley, New York, 1984.

Durst, F., A. Melling, and J. H. Whitelaw, *Principles and Practice of Laser Doppler Anemometry*, Academic, New York, 1976.

Figliola, R. S., and D. E. Beasley, *Theory and Design for Mechanical Measurements*, Wiley, New York, 1991.

Goldstein, R. J., editor, *Fluid Mechanics Measurements*, Hemisphere, New York, 1983.

Graber, Jr., J. C., "Ultrasonic Flow," *Meas. Contr.*, pp. 258–266, October 1983.

Granger, R. A., *Experiments in Fluid Mechanics*, Holt, New York, 1988.

Hatschek, E., *The Viscosity of Liquids,* Van Nostrand, New York, 1928.

Hayward, A. T. J., *Flowmeters; a Basic Guide and Source-book for Users,* Macmillan, New York, 1979.

Hele-Shaw, H. J. S., "Investigation of the Nature of the Surface Resistance of Water and of Streamline Motion under Certain Experimental Conditions," *Trans. Soc. Inst. Nav. Archit.*, vol. **40**, p. 25, 1898.

Holman, J. P., *Experimental Methods for Engineers*, 5th edition, McGraw-Hill, New York, 1989.

Kopp, J. G., "Vortex Flowmeters," *Meas. Contr.*, pp. 280–284, June 1983.

Moore, A. D., "Fields from Fluid Flow Mappers," *J. Appl. Phys.*, vol. **20**, pp. 790–804, 1949.

Partington, J. R., *An Advanced Treatise on Physical Chemistry:* vol. II of *Properties of Liquids*, Longmans, Green, New York, 1951.

Shercliff, J. A., *Electromagnetic Flow Measurement*, Cambridge University, New York, 1962.

Streeter, V. L., editor, *Handbook of Fluid Dynamics*, McGraw-Hill, New York, 1961.

van Water, J. R., *Viscosity and Flow Measurement*, Interscience, New York, 1963.

Problems for Chapter 10

Density, specific weight

10.1. An object weighs 5 lb in air and 3.6 lb fully submerged in water. What is its volume and specific weight?

10.2. A hydrometer consists of a hollow 5-mm-diameter, 300-mm-long hollow glass cylinder and a weighted sphere of 20-mm diameter. Its mass is 5 g. What range of specific gravities can be read, assuming that the scale starts and ends 15 mm from each end of the cylinder?

Viscometry

10.3. A concentric cylinder viscometer consists of a stationary inner cylinder, closed at the bottom with outer diameter 2 in, and a rotating outer cylinder, also closed on the bottom, that has an inner diameter of 2.1 in. The bottom of each cylinder is separated by a distance of 0.15 in. The space between the cylinders are filled to a depth of 9 in. At a speed of 30 revolutions every 4 s, a torque of 3 in-lb is measured on the inner cylinder. What is the viscosity of the fluid? What proportion of the measured torque is due to the fluid between the bottoms of the cylinders?

10.4. Many viscometers have falling heads during their operation. An example is the Saybolt viscometer, which consists of a larger diameter cylinder with a short open tube on the bottom. For a tube of inner diameter 0.040 in and length 3 in, while withdrawing a 50-cm^3 sample the depth of liquid is seen to fall from 10 to 9.35 in in 142 s. Which viscosity, absolute or kinematic, can be found from this experiment? Find this viscosity, assuming that the flow in the tube is fully established, that there are no entrance losses, and that the average head can be used.

10.5. A rotameter (also called a variable area meter) consists of a disk-shaped float in a transparent tapered tube. A theoretical formula for discharge through a rotameter is:

$$Q_{ideal} = A_{annulus} [2gV_{float} (\rho_{float}/\rho_{fluid} - 1)/A_{float}]^{1/2},$$

where $A_{annulus} = \pi [(D + by)^2 - d^2]/4$, D is the diameter of the rotameter tube at its bottom, b is the rate of change of tube diameter with height, d is the maximum float diameter, V_{float} is the float volume, and $A_{float} = \pi d^2/4$. For a tube whose diameter varies from 1 in at $y = 0$ to 1.4 in at $y = 4$ in, a maximum float diameter of 0.9 in, a float volume of 2 in, and a float density of 3 lb/in^3, what range of flow rates of water can be measured?

10.6. A viscometer sometimes used for very viscous fluids consists of two parallel plates, each of diameter D and separated by a gap d. The upper plate is fixed, and the lower plate is rotated at a fixed speed Ω. The torque T needed to turn the lower plate is measured. Derive an expression relating the fluid viscosity to the various moment and geometric parameters.

10.7. A variation of the viscometer of Problem 10.6 is the cone and plate viscometer. The lower plate of the previous viscometer is replaced by a cone. The gap between the plate and cone at the center is essentially zero, and the angle between the cone and plate is θ_{cone}. The upper plate may contain manometer tubes to measure the pressure along the plate. Derive an expression relating the fluid viscosity to the various moment and geometric parameters. (This device is often used for measuring the properties of non-Newtonian fluids. It is often referred to as the "cone and plate rheogoniometer," since it can measure fluid properties other than viscosity.)

Falling spheres

10.8. Stokes' law states that the drag coefficient for a sphere falling in a quiescent fluid is 24/Re. Knowing that the forces are due to gravity, buoyancy, and viscosity, derive an expression for the viscosity in terms of the densities, sphere diameter, and terminal velocity.

10.9. For a 4-mm-diameter glass sphere (SG = 2.6) falling in a stagnant fluid with SG = 0.8, the sphere is seen to move 80 mm in 20 s. What is the viscosity of the fluid? Check to see that the Reynolds number is less than 1 for Stokes' law to be valid for this problem.

10.10. Wall effects can have pronounced effects on the drag of particles. For a sphere of radius a falling along the axis of a cylinder of radius R, the drag force is given by $6\pi\mu aVK$, in which

$$K = \frac{1 - 0.75857r^5}{1 - 2.1050r + 2.0865r^3 - 1.7068r^5 + 0.72603r^6},$$

where $r = a/R$. Compute the correction factor K for $r = 0.2$ and 0.6. (This expression for K was obtained by Haberman and Sayre, David Taylor Model Basin Report No. 1143, Washington, D.C., U.S. Navy Dept., 1961. It is valid to within 5% up to $r = 0.6$. For larger r's it underestimates wall effects.)

Venturi meters

10.11. The performance of a venturi meter is within about 1.5% of the ideal formula developed in Chapter 3, providing that the Reynolds number based on the diameter of the throat is greater than about 2.5×10^5. The following measurements were taken in a laboratory for water at ($\rho = 1,000$ kg/m^3) and a venturi meter with a throat diameter of 100 mm and an inlet diameter of 150 mm:

Δp (Pa)	65.2	101	201	404	684	910	2,528	3,610
Q(m^3/s)	0.003	0.0038	0.0053	0.0077	0.0101	0.0116	0.0194	0.0233

Plot the ratio of $V_{actual}/V_{theoretical}$ versus Reynolds number. Use a log scale on the Reynolds number.

Pulsatile flows

10.12. Frequently it is desired to measure the flow into or out of pulsatile pumps (e.g., fuel pumps, hearts, engines). Orifice meters and similar devices can have considerable errors under such conditions. A method to avoid measurement problems is to place a chamber of volume V between the source of the pulsations and the meter. Suitable values of V can be obtained from use of the Hodgson number $H = fV\delta p/Qp$, where f is the frequency of the pulsations, Q the flow rate, δp the magnitude of the pressure pulsations, and p absolute pressure at the flow meter. Experience shows that values of H greater than 2 give meter errors less than 1%. For a pump operating at 1,800 rpm, where the pressure pulsations are 3% of the absolute pressure and the discharge is 0.005 ft^3/s, what is the minimum volume needed to achieve this accuracy?

Velocity-to-discharge measurements

10.13. A 5-in-inner-diameter pipe is divided into six rings of equal area. The velocities are measured at the half-radius of each ring by means of a pitot tube. Pressure measurements obtained from the tube are

Ring #	1	2	3	4	5	6
Δp (mmHg)	33	32	30.5	28	25.5	23

Find the velocities, their locations, and the total discharge through the pipe. The density is 1.94 slugs/ft^3.

Weirs

10.14.   A triangular weir with $\alpha = 90°$ discharges 19 m^3 of water in 1,000 s. The depth measured over the apex of the triangle is 180 mm. What is the ratio of this discharge to the theoretical discharge?

10.15.   Derive the theoretical relation between discharge and flow depth for a trapezoidal weir. The bottom width of the trapezoid is L, and the angle the side wall makes with the horizontal is α. Your results should reduce to those for the triangle and rectangular weirs as limiting cases.

10.16.   Broad-crested weirs are frequently used for measurement of flow rates in channels. They consist of a large rectangular block placed across a channel. If the weir is broad enough and if its front edge is rounded, very little energy loss occurs on the upstream side of the weir. Using this, derive an expression for the discharge per unit width over the weir in terms of the given parameters and the local depth y. Let H be the upstream depth measured from the bottom of the weir, and h be the depression of the free surface above the weir. Neglect the upstream kinetic energy. Show that this discharge is maximized when the depth above the weir is the critical depth, i.e., when the Froude number based on this depth is unity.

Hydrocyclones

10.17.   Hydrocyclones are frequently used to separate contaminants from fluid flows for metering and other applications. They can be seen on the roofs and sides of factories as the familiar silver cylinder/cone sheet metal assemblages. The principle of operation is that the contaminated fluid is injected tangentially at the top of the cylinder, giving a swirl velocity component to the flow. As the fluid moves down into the conical portion, the swirl velocity increases proportionally to the reciprocal of the radius squared. Heavier particles are thrown to the boundary, where they slide down and are collected. The cleaned fluid exits through a pipe in the top. The behavior of these devices is controlled by the Reynolds number based on cyclone diameter D and discharge Q, the pressure coefficient C_p based on pressure drop across the cyclone, and the Stokes number $\mathrm{St} = 2Qx^2/9\pi D^3\mu$, where x is the particle size. For a family of similar cyclones, $\mathrm{St}_{50} \times C_p = K_1 = $ constant and $C_p/\mathrm{Re}^n = K_2 = $ constant, where the subscript 50 is based on the particle size that has a 50–50 chance of being separated from the flow (the "cut size"). A particular hydrocyclone has $K_1 = 0.333$, $K_2 = 1,200$, and $n = 0.15$. If the cut size is to be reduced by a factor of 2, what must be the percent change in Q and the pressure drop across the cyclone?

chapter 11

Hydraulic Machinery

Chapter Overview and Goals

Hydraulic machinery is an important engineering application of the material we have covered in earlier chapters. In studying the behavior of hydraulic machines, we first present the suitable dimensionless parameters needed to describe the machines, and select those suitable for a given application. Descriptions of several types of pumps are then presented, along with a discussion of turbines used for electric power generation. Other issues involved in the design of a pumping system are discussed in Appendix G.

1. Pump Classification and Selection

The use and design of hydraulic machinery is important to many engineering tasks. Hydraulic machinery can be broadly divided into two classes: pumps and turbines. A pump converts mechanical or electrical energy from an outside source into hydraulic energy, often in the form of a pressure rise. Many hydraulic machines used in engineering applications are centrifugal machines, where an essential part of the machine is a rotating member. This member is called an impeller, or a rotor, or a runner, depending on the type of machine. The pressure rise across the pump is due to the kinetic energy imparted to the fluid by the rotation. Positive displacement pumps on the other hand use pistons or rotary vanes to increase the pressure by compressing the fluid.

In a sense a turbine is a pump running backward, in that it converts hydraulic energy into mechanical energy. Most turbines are centrifugal machines.

There are many varieties of pumps available, each satisfying different needs or a different range of operating conditions. They may be divided two main classifications of pumps, the positive displacement pump and the turbomachine pump.

a. Positive displacement pumps

A positive displacement pump is used when it is necessary to develop high heads or create a suction lift. It can also be used as a metering device. Positive displacement pumps are built either in the reciprocating configuration (or alternatively the piston configuration) or the rotary configuration.

The reciprocating or piston pump is usually a low flow rate pump with a given delivery capacity to any required head. They may be single or multiple piston devices. They are used with low to moderate viscosity fluids and will pull a vacuum (suction), i.e. they can be considered self-priming. Some examples of these types of pumps are automobile fuel pumps, reactor charging pumps that deliver 100 gpm at 2,000 psi, ship bilge pumps that deliver 700 gpm at 125 psi with a 15-ft suction lift, and multipiston pumps for use with large die-casting machines that deliver 10 gpm at 2,000 psi.

Rotary-type positive displacement pumps are available in various mechanical designs. They can be the wobble plate piston type, the gear type, the sliding vane type, the lobe type (which is a modified gear type), and the screw type. All these pumps are for low flow rates at any given head. They can handle fluids of any reasonable viscosity, such as fuel oils, lubricating oils, hydraulic oils, barnyard sewage, etc. An example of a lobe pump is the oil pump in an automobile.

The wobble plate piston type and the sliding vane type are usually built as variable volume-pressure compensated pumps. These operate at a constant speed. When the system's volume requirements are changed, the position of the pump's internal mechanical components automatically change, altering the delivery volume.

b. Turbomachine pumps

The turbomachine class consists of two subtypes, the centrifugal machine (sometimes called the "radial machine") and the axial machine (sometimes called the "propeller machine"). The centrifugal (radial) pump is the most commonly found pump in industrial operations. In single-stage configurations these pumps are used to provide low to moderate heads, and low to high flow rates. Multiple-stage pumps are used when higher pressure heads are needed. An example would be a large boiler feed pump that can deliver 450 gpm at 1,100 psi. These pumps are built with either single or double suction lines, and will not operate with a dry suction line—i.e. they are not self-priming. The prime is usually prevented from being lost through the use of a foot (check) valve on the end of the suction pipe inlet.

The axial- (propeller-) type pump is usually used as a circulating pump. It is used where a high flow rate is required with a low head. An example would be a circulating pump for a condenser in a power plant. This type of pump requires a flooded suction.

An important engineering task is to choose the type of hydraulic machine best suited to a given application. Each type has certain advantages and disadvantages. The engineer must decide what the ranking of the performance factors are for the application. For example, in selecting a pump, is the main purpose of the pump to provide a large pressure head, or is discharge more important? Are suction characteristics important to the application, or are they immaterial? Is efficiency of major importance? Size, availability, ease of service, expected life, and first cost are other major nontechnical factors that play a role in the decision.

As a guide to pump selection, analysis and experience give the following rules:

- Centrifugal machines have small inlet sizes compared to the size of the rotor. Thus flow rates must be kept small in order not to have large inlet velocities. The exit diameters are usually larger than the inlet diameters to accommodate the increase in speed as the fluid passes through the machine. Centrifugal machines are thus suited for low flow rates and large pressure increases. Streamlines and velocity vectors lie in planes.

- Axial flow machines have unrestricted inlet areas, allowing large flow rates. Each row of blades, however, can give only a limited change of pressure across each stage of blades. Thus axial flow machines are suited to large flow rates with low pressure increase. Streamlines are helices of constant radii.

- Mixed-flow machines that combine radial and axial flow have medium flow rates and medium pressure increases. They thus are a compromise between the centrifugal and axial machines. The streamlines lie on nonconstant radii helices. The flow is complicated and highly three-dimensional.

2. Centrifugal Machines

When we deal with rotating fluid machinery, the following dimensional quantities are of prime interest and provide a general description of the machine:

Q—flow discharge through the machine,

N—angular speed of the rotating member,

D—outside diameter of the rotating member,

P—power needed to drive the pump or provided by the turbine,

H—energy head (pressure change/specific weight) provided by a pump or lost by a turbine.

This list describes the rotating machine itself. In addition, we have the fluid properties of mass density (representing momentum forces) and viscosity (representing viscous forces) to add to our description of the complete system. Counting these variables, we see that to perform a dimensional analysis of our centrifugal machine we have seven dimensional quantities $(Q, N, D, gH,$[1] $P, \rho, \mu)$ involving three dimensions. Thus four independent dimensionless parameters can be formed. The customary dimensionless parameters formed from this list are:

$$C_Q\text{—capacity coefficient} = Q/ND^3, \tag{11.2.1}$$

$$C_H\text{—head coefficient} = gH/N^2D^2, \tag{11.2.2}$$

$$C_P\text{—power coefficient} = P/\rho N^3 D^5, \tag{11.2.3}$$

$$\text{Re—Reynolds number} = 4Q/\pi D^2 \nu. \tag{11.2.4}$$

---

[1]Recall that gH is used rather than just H, because the quantity of interest is the pressure change. Since we already have density in our list, we use $gH = \Delta p/\rho$ as the fundamental quantity.

Besides the dimensionless parameters given in this list, there are two other parameters that are frequently used. These are called "specific speeds." The specific speed used for pumps is C_{NQ}, defined on a flow rate basis, and is given by

$$C_{NQ} = C_Q^{1/2}/C_H^{3/4} = NQ^{1/2}/(gH)^{3/4}. \tag{11.2.5}$$

Since C_{NQ} does not have any pump dimensions in it, it is useful in selecting the type and size of a pump from a given design family to be used for a given application. For a given family of pumps, the specific speed will be approximately constant.

Similarly, for turbines the useful speed parameter is C_{NP}, the specific speed defined on a power basis, and is given by

$$C_{NP} = C_P^{1/2}/C_H^{5/4} = N\sqrt{P}/\rho(gH)^{5/4}. \tag{11.2.6}$$

Again, since C_{NP} does not contain the pump dimensions, it serves the same purpose for turbines as C_{NQ} does for pumps.

In using these dimensionless parameters, a consistent set of units should, in principle, be used. That is, Q should be feet cubed per second or meters cubed per second, N should be radians per second, D should be feet or meters, P should be foot-pounds per second or watts, and H should be either feet or meters. In American industrial practice, Q is usually given in gallons per minute, N in revolutions per minute, and P in horsepower. Frequently g is omitted (set to unity). While these units do not allow ready conversion to other systems of units, nevertheless they are convenient for the uses to which they are put. In this text we will use the consistent set of dimensionless numbers except for N, where we will follow practice and use revolutions per second. In these units we would have

$$N_s = NQ^{1/2}/H^{3/4}$$

for pumps and

$$N_s = N\sqrt{P}/H^{5/4}$$

for turbines.

As an example of the use of these parameters, the following gives general guidelines in determining the type of centrifugal pump needed for a given application:

N_s in the range 400–1,100 rpm: a radial flow pump is indicated.

N_s in the range 1,100–10,000 rpm: a mixed flow pump is indicated.

N_s in the range 10,000 rpm and up: an axial flow pump is indicated.

We would expect that plots for a given machine would look something like the three-dimensional plots seen in Figures 11.1 and 11.2. Unfortunately, such complete data as are depicted usually do not exist for any given machine! Manufacturers generally test their machinery for only one or two fluids, and so the complete Reynolds number data does not exist. Typical dimensionless results that might be provided for a centrifugal pump are shown in Figure 11.3.

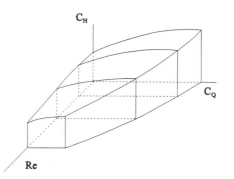

Figure 11.1. Typical C_H curves for centrifugal pumps.

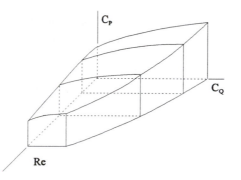

Figure 11.2. Typical C_P curves for centrifugal pumps.

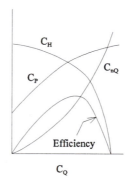

Figure 11.3. Representative centrifugal pump curves.

3. Centrifugal Pumps

A centrifugal pump consists of an impeller with blades or vanes inside a casing. The blades spiral radially outward and against the direction of rotation, and the casing can be either concentric or spiral in shape (see Figure 11.4). The fluid enters the casing axially and encounters the rotating impeller, which then throws the fluid radially outward into the diffuser case. The **shaft power** SP (also called the "brake horsepower") is the power delivered to the pump shaft by the motor. This can be usually measured by a dynamometer. The **water power** WP is $\rho g H Q$, the rate at which energy has been gained in passing through the pump. The pump hydraulic efficiency η is defined as

$$\eta = WP/SP. \tag{11.3.1}$$

Peak pump efficiencies typically range from 0.5–0.9 for centrifugal pumps. These pumps are used when relatively large heads and low flow rates are desired.

Centrifugal pumps are frequently used in a location where they are placed above the level of the supply fluid. In that case a **check valve** (also called a **foot valve**) is usually installed upstream of the pump to ensure that the flow is unidirectional and does not drain out of the pump when it is stopped. Sometimes filter screens are also installed upstream of the pump to remove suspended debris. A centrifugal pump must generally first be **primed**. That is, the pump casing and supply line are first filled with the fluid to be used, to remove all air. The check valve ensures that this fluid stays in the pump. In some installations a given pump can be self-priming.

Theoretically, a pump can be placed at a height

$$\text{NPSH} = h_{\text{suction}} = (p_{\text{atmos}} + p_{\text{supply}} - p_{\text{vapor}} - \rho V^2/2)/\rho g \tag{11.3.2}$$

above the supply, where NPSH is the net positive suction pressure, p_{supply} is the supply source pressure, and p_{vapor} is the vapor pressure. (See Figure 11.5.) In practice, this is a head of about 45–60% of an atmosphere above the supply pressure. If the height of the **suction line** (the inlet line to the pump) is made larger than this, the flow will cavitate and the pump cease to function satisfactorily.

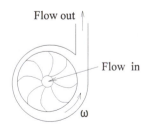

Figure 11.4. Centrifugal pump.

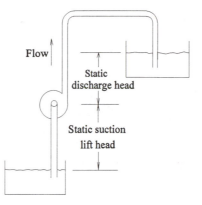

Figure 11.5. Definition of suction terms for centrifugal pumps.

A pump manufacturer will usually provide the NPSH for pumps, but not always, or perhaps not always at the desired conditions. In that case an estimate can be made using Thoma's cavitation parameter. This is defined as

$$\sigma = \text{NPSH}/H,$$

where H is the head in the first stage of a pump expressed in the same units as NPSH. The Hydraulic Institute suggests the following empirical results:

$$\sigma = 6.3 \times 10^{-6} \times N_s^{4/3} \qquad \text{for a single suction pump,}$$

$$= 4.0 \times 10^{-6} \times N_s^{4/3} \qquad \text{for a double suction pump.}$$

Here N is in rpm. Using the previous definition for N_s gives

$$\text{NPSH} = 6.3 \times 10^{-6} \times H(NQ^{1/2}/H^{3/4})^{4/3} = 6.3 \times 10^{-6}(NQ^{1/2})^{4/3}$$

$$\text{for a single suction pump,}$$

$$= 4.0 \times 10^{-6} (NQ^{1/2})^{4/3} \quad \text{for a double suction pump,}$$

again with N in rpm and Q in gpm. This should be used only when manufacturer information is not provided.

An expression for the theoretical torque needed to drive a centrifugal pump can be found by considering the velocity diagrams and control volume of Figure 11.6. Let 1 and 2 denote inlet and outlet conditions, respectively, where the fluid enters at the smaller of the two radii, r_1. The various velocity components are as given in Table 11.1.

With an impeller having a passage of uniform height w, the normal velocity components are related to discharge according to

$$Q = 2\pi r_1 w V_1 \sin \alpha_1 = 2\pi r_2 w V_2 \sin \alpha_2. \tag{11.3.3}$$

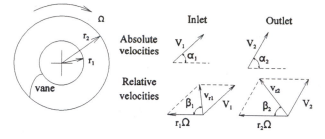

Figure 11.6. Centrifugal pump velocities.

We will consider only the case where the flow enters and exits tangent to each vane of the impeller. For this case the vane slope is the same as the slope of the velocity component. Thus the ideal entering and exiting vane angles β_1 and β_2 are given by

$$\tan \beta_1 = V_1 \cos \alpha_1 / (r_1 \Omega - V_1 \cos \alpha_1)$$

and

$$\tan \beta_2 = V_2 \cos \alpha_2 / (r_2 \Omega - V_2 \cos \alpha_2). \tag{11.3.4}$$

Having the flow enter tangentially to the vane ensures that there is no momentum loss normal to the impeller. Other angles will result in inlet losses.

The head delivered to the fluid in passing through the pump is, by the energy equation,

$$H = (p_2/\gamma + V_2^2/2g + z_2) - (p_1/\gamma + V_1^2/2g + z_1). \tag{11.3.5}$$

Normally the greater portion of head delivered is due to the pressure rise, since the entry and outlet areas are nearly equal.

The ideal torque applied to the pump (the torque needed if there were no losses) is the mass rate of flow times the moment of momentum change. This involves only the tangential components of the entering and exiting velocities. Thus

$$T_{\text{ideal}} = \rho Q (r_2 V_2 \cos \alpha_2 - r_1 V_1 \cos \alpha_1). \tag{11.3.6}$$

Notice that the ideal torque can be maximized by having the fluid enter the impeller normally ($\alpha_1 = 90°$).

Centrifugal pumps usually have their vanes curved backward against the direction of curvature, so β lies somewhere between 0 and 90°. This is done in order to reduce the velocity of discharge, since only a portion of the discharge energy can be recovered, the rest going into losses. It also reduces the tendency of the pump to cavitate, which

Table 11.1. Centrifugal pump velocity components

	Inlet velocity components		Outlet velocity components	
	Tangential	Normal	Tangential	Normal
Absolute	$V_1 \cos \alpha_1$	$V_1 \sin \alpha_1$	$V_2 \cos \alpha_2$	$V_2 \sin \alpha_2$
Relative	$V_1 \cos \alpha_1 - r_1 \Omega$	$V_1 \sin \alpha_1$	$V_2 \cos \alpha_2 - r_2 \Omega$	$V_2 \sin \alpha_2$

would cause it to cease operation. Thus with a typical centrifugal pump, the head gain increases with increasing the impeller speed, but reduces with increasing discharge. Values of β greater than 90° can result in pump surging, where the pump operation point keeps changing and no steady state is reached.

To determine how a centrifugal pump operates in a given situation, we have to consider the entire operating system characteristics along with the pump characteristics. The operating system includes the pump, all attached plumbing lines and connections, and any other devices that can affect flow velocities and pressure. In Figure 11.7, we show the pump curve plotted together with a typical head loss curve for a system. The intersection of the two curves is the ***operating point*** for the system. Operation at $Q = 0$ is called ***deadheading***, and conditions at that point are called ***deadhead conditions***.

Generally we want the pump and operating curves to intersect at a sufficiently large angle, as close to 90° as is practical. If the curves intersect at a small angle, so that the curves are nearly tangent, the operating point will shift considerably with small changes in the characteristics of the system, and the pump may "hunt" in an attempt to find a stable operating point.

Example 11.3.1. Centrifugal pump
A centrifugal water pump has an impeller of width $h = 5$ cm, an inner radius of 7 cm, and an outer radius of 30 cm. It turns at 1,800 rpm. The inlet velocity is 6 m/s and the exit velocity is 7 m/s. What theoretical head and power are developed, and what is the pressure rise across the impeller?

Sought: Discharge, head, power, and pressure rise in a pump.
Given: The impeller width is 5 cm, its inner radius $r_1 = 7$ cm, its outer radius $r_2 = 30$ cm, and the angular speed $\Omega = 1{,}800$ rpm $= 60\,\pi$ rad/s. The inlet velocity $V_1 = 6$ m/s and the exit velocity $V_2 = 7$ m/s. The fluid is water.
Assumptions: Ideal (theoretical) behavior of a pump. Assume an optimum inlet angle $\alpha_1 = 90°$.
Solution: According to (11.3.1)

$$Q = 2\pi r_1 w V_1 = 2\pi \times 0.07 \times 0.05 \times 6 = 0.1319 \text{ m}^3\text{s} = 2\pi r_2 w V_2 \sin \alpha_2.$$

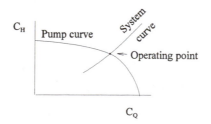

Figure 11.7. Pump and system curves superimposed.

Thus $\sin \alpha_2 = 0.1319/2\pi r_2 w V_2 = 0.1319/2\pi \times 0.3 \times 0.05 \times 7 = 0.2$. Therefore $\alpha_2 = 11.54°$.

From (11.3.4)

$$\tan \beta_1 = V_1/r_1\Omega = 6/0.7 \times 60\pi = 0.4547, \quad \text{or } \beta_1 = 24.45°,$$

and

$$\tan \beta_2 = V_2 \sin \alpha_2/(r_2\Omega - V_2 \cos \alpha_2)$$

$$= 7 \times 0.2/(0.3 \times 60\pi - 7 \times \cos 11.54) = 0.0282, \quad \text{giving } \beta_2 = 1.614°.$$

From (11.3.6) the theoretical torque is

$$T_{\text{ideal}} = \rho Q(r_2 V_2 \cos \alpha_2 - r_1 V_1 \cos \alpha_1) = 998 \times 0.1319(0.3 \times 7 \times \cos 11.54° - 0)$$

$$= 270.8 \text{ W}.$$

The head gain is

$$H_{\text{gain}} = T_{\text{ideal}} \Omega/\rho g Q = 270.8 \times 60\pi/998 \times 9.81 \times 0.1319 = 39.53 \text{ m}.$$

The pressure rise is found from the energy equation with the head gain term included. Then

$$gH_g + V_2^2/2 + p/\rho = V_1^2/2 + p_1/\rho$$

or

$$p_1 - p = \rho[gH_g + (V_1^2 - V_2^2)/2] = 998[9.81 \times 39.53 + (6^2 - 7^2)/2]$$

$$= 998[387.8 - 6.5] = 380.5 \text{ kPa}.$$

Example 11.3.2. Centrifugal pump

For a centrifugal pump with impeller diameter of 1 ft and belonging to the pump family of Figure 11.3, Figure 11.4 can be converted to Figures 11.8 and 11.9 using water properties. If the pump is to be operated at 1,725 rpm and is connected to a system where $h_L = 10^{-5}Q^2$, what pressure difference will the pump develop and what power must be supplied to the pump?

Sought: Pressure difference across the pump and needed supply power.
Given: Impeller diameter of 1 ft rotates at 1,725 rpm. The head loss equation is given.
Assumptions: Ideal (theoretical) behavior of a pump.
Solution: The above head loss–flow relationship is shown as a dotted line on Figure 11.9. Interpolating, the flow rate and head for 1,725 rpm are 2,300 gpm and 52 ft. The pressure difference across the pump is thus

$$\Delta p = \rho g Q H = 62.4 \times (2,300/60) \times 7.48 \times 52 = 16,629 \text{ psf} = 115 \text{ psi}.$$

From Figure 11.9 the power needed is 47 hp. The efficiency is approximately 40%.

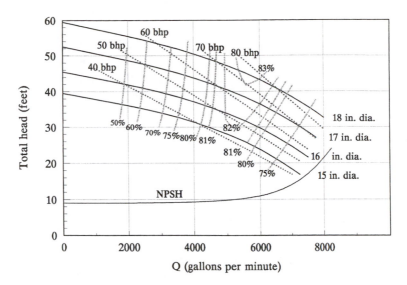

Figure 11.8. Typical head versus discharge curves for a family of centrifugal pumps. Also shown are NPSH, pump power input, and efficiencies.

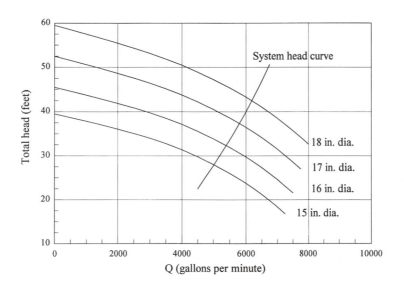

Figure 11.9. System head curve plotted together with head versus discharge curves for a family of centrifugal pumps.

Example 11.3.3. Centrifugal pump

A centrifugal pump of the family of Example 11.3.2 has an impeller diameter of 8 in. At what speed should the pump be run to give the same discharge as the 1-ft-impeller-diameter pump of Example 11.3.2 that runs at 1,725 rpm? What will be the pressure rise across the pump, and what power is needed?

Sought: Pressure rise across a pump and needed power supply.

Given: Impeller diameter is 8 in. The discharge is known.

Assumptions: Ideal (theoretical) behavior of a pump.

Solution: The C_Q must be the same for the two pumps. Since Q is the same, the requirement is that ND^3 be the same. Thus

$$N_2 = N_1(D_2/D_1)^3 = 1,725(12/8)^3 = 5,822 \text{ rpm}.$$

Since

$$C_Q = Q/ND^3 = (2,300/60 \times 7.48)/(5,822/60) \times (0.6667)^3 = 0.178,$$

from Figure 11.3, $C_H = 5.5$ and $C_P = 0.65$, and so

$$H = C_H(ND)^2/g = 5.5 \times (0.6667 \times 5,822/60)^2/32.17 = 722.6 \text{ ft}.$$

Then

$$\Delta p = \rho g H = 62.4 \times 722.6 = 45,091 \text{ psf} = 313 \text{ psi},$$

$$P = \rho C_P N^3 D^5 = 1.935 \times 0.65 \times (5,822/60)^3 \times (0.6667)^5 = 1.513 \times 10^5 \text{ ft-lb/s} = 275 \text{ hp}.$$

Example 11.3.4. Centrifugal pump

The characteristics of a centrifugal pump are given in a catalog as follows:

H (ft)	100	95	80	40
Q (gpm)	0	100	200	300

A horizontal galvanized pipe of 3-in diameter and 1,000-ft length is connected to the pump and discharges water into the atmosphere. What will be the discharge rate? The water temperature is 75°F.

Sought: Discharge flow rate from a pump.

Given: Pump characteristics as follows:

H (ft)	100	95	80	40
$Q(\text{ft}^3/\text{s})$	0	0.223	0.446	0.668
V(ft/s)	0	4.54	9.09	13.61 (for 3-in-diameter pipe)

From Appendix B we have $\nu = 10^{-5}$ ft^2/s. For the pipeline

$$D = 0.25 \text{ ft}, \quad L = 1,000 \text{ ft}, \quad \text{roughness} = 0.0005 \text{ ft}.$$

The dimensionless roughness is then $0.0005/0.25 = 0.002$.

Assumptions: Stated characteristic behavior of a pump is valid.

Solution: Recalling our work with the Moody diagram in Chapter 7, the expression for head loss in a pipeline is

$$h_L = f \frac{L}{D} \frac{V^2}{2g} = f \frac{L}{D^5} \frac{16Q^2}{2g\pi^2}.$$

In the fully developed turbulent flow regime (according to the Moody diagram the region where Re > 60,000), the friction factor is $f = 0.0238$. Then

$$h_L = f \frac{L}{D^5} \frac{16Q^2}{2g\pi^2} = \frac{0.0238 \times 1000 \times 16Q^2}{0.25^5 \times 2 \times 32.2 \times \pi^2} = 613.5 \, Q^2.$$

Since the Reynolds number must be greater than 60,000 and Re $= 4Q/\pi\nu D$, this is valid for

$$Q = 60,000 \times \pi \times 10^{-5} \times 0.25/4 = 0.118 \text{ ft}^3/\text{s},$$

or very likely for the entire flow range we are interested in.

One approach for completing the determination of Q would be to plot the pump characteristics and the piping system characteristics on the same graph with head on the vertical axis and discharge on the horizontal axis. The two curves would intersect at the operating point, giving the Q and head.

An alternative approach is to fit the pump curve data with a polynomial. Since we are given four points on the curve, a cubic polynomial is suggested. Letting

$$H = a + Q \times [b + Q \times (c + dQ)],$$

and substituting the four data pairs of H and Q, we find

$$H = 100 + Q \times [-22.81 + Q \times (52.88 - 229.34Q)].$$

Equating this to the piping system curve

$$h_L = 613.5Q^2,$$

we find from numerical evaluation of the two expressions that they intersect at

$$Q = 0.376 \text{ ft}^3/\text{s} \qquad \text{and} \qquad h_L = 86.71 \text{ ft}.$$

When manufacturer specifications are not available for your desired operating conditions, the available data can be scaled by using the dimensionless parameters previously described in this chapter. As an example, you may wish to operate at a different motor speed, or perhaps the pump impeller is available in other diameters. From (11.3.1), (11.3.2), and (11.3.3), within a design family of pumps we would expect dimensional similarity. Thus for two pumps within the same family,

$$(Q/ND^3)_1 = (Q/ND^3)_2,$$

$$(gH/N^2D^2)_1 = (gH/N^2D^2)_2,$$

and

$$(P/\rho N^3 D^5)_1 = (P/\rho N^3 D^5)_2.$$

Providing that both are pumping the same fluid, this means that

$$Q_2/Q_1 = N_2 D_2^3/N_1 D_1^3,$$

$$H_2/H_1 = N_2^2 D_2^2/N_1^2 D_1^2,$$

and

$$P_2/P_1 = N_2^3 D_2^5/N_1^3 D_1^5.$$

Thus discharge, head, and power needed from the pump can be simply scaled from the speed and diameter ratios.

Scaling of the NPSH ratio is more difficult, but an estimate can be made from the empirical use of Thoma's cavitation parameter. Our development there showed that NPSH was proportional to $[NQ^{1/2}]^{4/3}$. Inserting the above scaling for Q gives

$$\text{NPSH}_2/\text{NPSH}_1 = [N_2 Q_2^{1/2}/N_1 Q_1^{1/2}]^{4/3} = [N_2^2 Q_2/N_1^2 Q_1]^{2/3} = [N_2^2 N_2 D_2^3/N_1^2 N_1 D_1^3]^{2/3}$$

$$= N_2^2 D_2^2/N_1^2 D_1^2,$$

that is, NPSH can be scaled approximately the same way as the head H.

4. Positive Displacement Pumps

A positive displacement pump generally is used where high heads with low flow rates are needed, or to create a suction lift. They can also be used as metering devices. These pumps can be of a reciprocating or piston configuration, of the rotary type such as a gear pump, or can have rotary vanes. Most positive displacement pumps have the advantage of being self-priming.

The rotary-type positive displacement pump can be of the wobble plate type, a sliding vane type, a gear type, or the screw type. The wobble plate type is similar to the wobble plate flow meter described in Chapter 7. The sliding vane type of pump has the rotor blades free to move in radial slots in the rotor, or in some cases, the vanes are made out of a flexible material such as neoprene. The blade ends follow the housing as the rotor turns and trap the fluid into compartments, moving the fluid from the pump inlet to the pump exit. These pumps are sometimes used as electrically driven automotive fuel pumps.

The *gear*-, or *lobe*-, type of positive displacement pump consists of two counterrotating gears, or lobes, which trap the fluid and carry it through the pumps. These pumps have the advantage of being self-priming, but are easily injured by suspended grit.

The *screw*-type positive displacement pump, frequently referred to as an Archimedes screw, consists of a helical screw inclined at an angle with the horizontal,

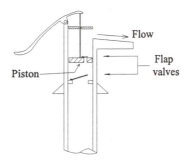

Figure 11.10. Bucket, or pitcher, pump.

usually between 30 and 40°. The angle of the inclination is less than the helix angle of the screw, so admitted fluid always is running downhill. These pumps typically run at very low speeds, but are capable of very high efficiencies, and can be made in very large diameters. A typical head would be less than 10 ft.

A simple example of a reciprocating pump is the manually operated ***bucket pump***, seen in Figure 11.10. On the upstroke, the piston reduces the chamber pressure closing the piston check valve, and fluid is drawn in through the check valve at the bottom of the pump. On the downstroke, the bottom check valve closes, and the fluid is forced through the piston check valve. Such pumps are not self-priming, but can lift fluids approximately 2/3 of an atmosphere. A variation of this pump is the diaphragm pump, sometimes used as a mechanically driven fuel pump in an automobile.

5. Axial Flow Fans and Pumps

An axial flow fan or pump (the usual distinction is that a fan handles gas, while a pump handles liquid) is fundamentally one or more propellers in a duct, casing, or shroud. The term ***axial*** means that the flow is substantially parallel to the axis of the impeller. It can be used as a pump to transport fluid, or it can provide a thrust for propulsive purposes. The casing allows a static pressure to develop. There are usually fixed or moveable (variable pitch) vanes that keep the flow direction axial. They improve the pumping performance by producing relatively high pressures, and also reduce losses.

The analysis used for an axial flow fan depends on the ratio of the cord length L to the blade spacing B, where at any radius r the blade spacing is given by $2\pi r/n$, where n is the number of blades. If L/B is less than unity, a blade-element, or air foil, analysis is used. In this case the blades are relatively far apart, and analysis as a single blade is possible. When L/B approaches or exceeds unity, there is a good deal of mutual interaction between the blades, and a lattice, or cascade, type of analysis is appropriate. We will describe only the blade-element approach.

Blade-element analysis.   We will consider the case of a rotor between stationary inlet and exit guide vanes, and the fluid density to be constant. (By taking either the inlet or exit turning angle to be zero, the case of only one stationary set of guide vanes

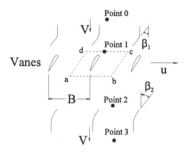

Figure 11.11. Axial fan pump velocities.

is easily obtained.) The flow will be taken to be axial both before the inlet guide vanes, and after the exit guide vanes. The fan area will be taken constant, so that the axial velocity V is the same upstream and downstream of the fan. The local rotor velocity is $u = r\Omega$, where Ω is the rotor angular velocity. Figure 11.11 explains the geometry.

Following the approach used in Chapter 3, the vane velocities are shown in Figure 11.12. By continuity, the axial velocities are V at points 0 through 3. From trigonometry, we have the following relations for velocities relative to the vane from considering Figure 11.12:

$$w_1 = \sqrt{V^2 + (u + V \tan \beta_1)^2},$$ (11.5.1)

$$w_2 = \sqrt{V^2 + (u + V \tan \beta_2)^2},$$ (11.5.2)

$$w_3 = \sqrt{V^2 + (\tan \beta - u)^2}.$$ (11.5.3)

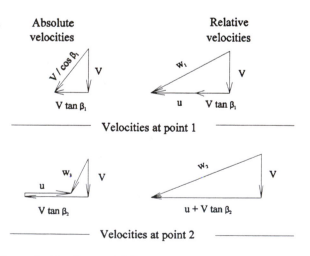

Figure 11.12. Vane velocities for an axial fan pump.

Writing the Bernoulli equation between the various points and assuming no energy losses, we have

$$p_1 - p_0 = 0.5\rho(V^2/\cos^2\beta_1 - V^2 = 0.5\rho V^2 \tan^2\beta_1, \tag{11.5.4}$$

$$p_2 - p_1 = 0.5\rho(w_2^2 - w_1^2)$$

$$= 0.5\rho V(\tan\beta_2 - \tan\beta_1)[V(\tan\beta_2 + \tan\beta_1) + 2u], \tag{11.5.5}$$

$$p_3 - p_2 = 0.5\rho(V^2 - w_3^2) = -0.5\rho V^2 \tan^2\beta_2. \tag{11.5.6}$$

The overall pressure rise is then found by adding the above three equations, giving

$$p_3 - p_0 = (p_3 - p_2) + (p_2 - p_1) + (p_1 - p_0) = \rho u V(\tan\beta_2 - \tan\beta_1)$$

$$= \rho r \Omega V(\tan\beta_2 - \tan\beta_1). \tag{11.5.7}$$

Generally it is desirable to have both p_0 and p_3 uniform over their respective areas. Thus equation (11.5.7) indicates that the stationary blade angles should vary as the radius changes [i.e., that $r(\tan\beta_2 - \tan\beta_1)$ is constant or that the local angular momentum increases by the same amount], in order to keep the right-hand side of (11.5.7) constant.

To investigate the forces on the fan, we perform a momentum analysis on the control volume with cross section *abcd* (Figure 11.11) and height *dr*. The total force *dF* has a thrust component *dT* and a circumferential component *dC* as shown in Figure 11.13. (These are the components of force that the fluid exerts on the blade.) The sides *bc* and *da* are streamlines, and occupy similar locations between their adjacent blades. Since the flow rate through the control volume is *VB dr*, application of the momentum equations to the control volume gives

$$-dC = \rho VB\, dr(u + V\tan\beta_2 - u - V\tan\beta_1)$$

$$= \rho V^2 B\, dr(\tan\beta_2 - \tan\beta_1) \tag{11.5.8}$$

and

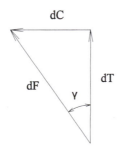

Figure 11.13. Force components for an axial fan pump.

$$dT + (p_1 - p)B\,dr = 0,$$

or

$$dT = (p - p_1)B\,dr. \tag{11.5.9}$$

Taking the ratio of dC and dT, we have

$$\tan\theta = dC/dT = \rho V^2(\tan\beta_2 - \tan\beta_1)/(p_2 - p_1)$$

$$= V/[u + 0.5V(\tan\beta_2 + \tan\beta_1)]. \tag{11.5.10}$$

Since dF is the total incremental force and dC is its circumferential component, we have

$$dF = dC/\sin\theta,$$

where θ is the angle that dF makes with the radial direction. Then

$$dF = -\rho UVB\,dr(\tan\beta_2 - \tan\beta_1), \tag{11.5.11}$$

where

$$U = V/\sin\theta \tag{11.5.12}$$

is a velocity lying between w_1 and w_3 (Figure 11.14), and represents the mean velocity for the blade.

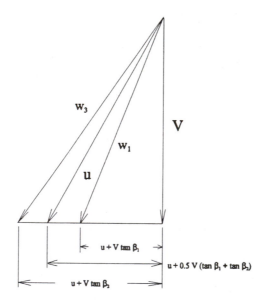

Figure 11.14. Velocities in an axial fan pump.

From the previous local results we can deduce some global results. If the blades extend from an inner radius r_i to an outer radius r_o, then the total flow rate through the fan is

$$Q = Vn \int B \, dr = V \int_{r_i}^{r_o} 2\pi \, dr = \pi V(r_o^2 - r_i^2), \tag{11.5.13}$$

and the net axial (thrust) force for n blades is

$$F_{\text{axial}} = n \int dT = \int_{r_i}^{r_o} (p - p_1) 2\pi r \, dr. \tag{11.5.14}$$

The moment needed to drive the fan is then

$$M = n \int r \, dC = n\rho V^2 \int_{r_i}^{r_o} Br \, dr (\tan \beta_2 - \tan \beta_1)$$

$$= nV(p_3 - p_0) \int_{r_i}^{r_o} B \, dr / \Omega$$

$$= \pi V(r_o^2 - r_i^2)(p_3 - p_0)/\Omega$$

$$= Q(p_3 - p_0)/\Omega. \tag{11.5.15}$$

Expression (11.5.15) for the applied moment M states that the net power put into the impeller, ΩM, is equal to the power gained by the fluid, $Q(p_3 - p_0)$. This is a consequence of our assumption that no losses take place, and represents an ideal case. More realistically, we could include a drag force D acting on the vane as well. Introducing lift and drag coefficients C_L and C_D, then improved versions of the circumferential and thrust forces including drag are

$$dC = dF \sin \theta + dD \cos \theta = 0.5\rho U^2 B \, (C_L \sin \theta + C_D \cos \theta) \, dr, \tag{11.5.16}$$

and

$$dT = dF \cos \theta - dD \sin \theta = 0.5\rho U^2 B (C_L \cos \theta - C_D \sin \theta) \, dr. \tag{11.5.17}$$

Loss coefficients for the pressure could also be included in the stationary guide sections in the standard form where the pressure loss can be assumed to be proportional to a kinetic energy, in the manner which we included them in Chapter 3. Losses that are more difficult to predict are those due to the gap that exists between the rotating blade tips and the outer casing. These can be appreciable, and minimizing such gaps is an important task for the designer.

Example 11.5.1. Axial fan pump

An axial flow fan is to deliver 7.5 m^3/s of air with a static pressure rise of 0.25 kPa across the nine-bladed rotor. There is a set of guide vanes upstream of the rotor that turn the air flow. The guide vanes and blades have a hub (inner) radius of 0.25 m and a tip (outer) radius of 0.4 m. The rotor turns at a rate of 90 rad/s. Air density is 1.1 kg/m^3. What is the rotor blade inclination and speed, and what is the ideal power needed to turn the rotor?

Sought: Rotor blade inclination and ideal needed power to be supplied.

Given: The nine-bladed rotor has an inner radius of 0.25 m and an outer radius of 0.4 m, and turns 90 rad/s. Air density is 1.1 kg/m^3, discharge is 7.5 m^3/s, and the static pressure rise is 0.25 kPa.

Assumptions: Ideal (theoretical) behavior of a pump.

Solution: Since $u = 90\,r$ and $\beta_2 = 0°$, from (11.5.3) we have

$$V = Q/\pi(r_o^2 - r_i^2) = 7.5/\pi(0.4^2 - 0.25^2) = 24.49 \text{ m/s}.$$

From (11.5.7), since $p_3 - p = 250$ Pa,

$$p_3 - p_0 = 250 = 0.5\rho u V(\tan \beta_2 - \tan \beta_1) = 1.1 \times 90 \times r \times 24.49 \times (0 - \tan \beta_1);$$

therefore solving for β we have

$$\beta_1 = -\tan^{-1}(2.5/1.1 \times 24.49r) = \tan^{-1}(0.0928/r).$$

From (11.5.10) and (11.5.12), the rotor blade inclination and speed are

$$\theta = \tan^{-1}[\rho V^2(\tan \beta_2 - \tan \beta_1)/(p_2 - p_1)]$$

$$= \tan^{-1}\{24.49/[90r + 0.5 \times 24.49(-2.5/1.1 \times 24.49r)]\} = \tan^{-1}\{24.49/[90r - 1.136/r]\},$$

$$U = V/\sin \theta = 24.49/\sin \theta.$$

From (11.5.15)

$$M = Q(p_3 - p_0)/\Omega = 7.5 \times 250/90 = 20.83 \text{ N-m}$$

and

$$M\Omega = 7.5 \times 250 = 1,875 \text{ N-m/s} = 1.875 \text{ kW}.$$

6. Other Pumps

A pump developed in the eighteenth century that uses a flowing river as a power source is the **hydraulic ram**, seen in Figure 11.15. The operation is started by opening the spring loaded waste valve V_w, thereby allowing the water to flow through the ram. When this valve is released, and if the flow rate is large enough to allow the spring to close, the valve V_w will close suddenly. This gives a high transient pressure in the valve box, which opens the delivery valve V_d. Water flowing into the air chamber (also called

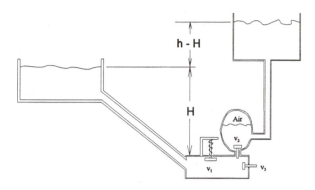

Figure 11.15. Hydraulic ram.

an **_accumulator_**) partially flows into the supply pipe. The remaining flowing water acts to compress the air in the air chamber. The transient pressure in the valve box reduces as a result of this air compression, allowing the cycle to repeat. Water continues to flow up the supply pipe even when V_s is closed, because the compressed air in the accumulator forces the water up. Air is retained in the accumulator either by enclosing it in a flexible bag, or by adding a valve V_{air} that allows air to enter the valve box when its pressure falls below atmospheric.

The hydraulic ram tends to be noisy, since the opening and closing of the valves is very rapid, and the pressure transients upon which the pump is based are the same as the water hammer that causes the familiar banging of pipes in the home. A good percentage of the water is lost through the waste valve V_w. The efficiency of the pump is tied very much to the length of the supply pipe and the head H, which should be tuned to the ram. However, when there is a supply of flowing water and the situation allows installation at about a meter or so below the stream surface, this is a good choice for a free-energy-cost pump.

The **_jet pump_** (Figure 11.16) was already discussed in Chapter 3. These pumps are sometimes used at the inlets of centrifugal pumps in order to locally lower the pressure. They can also be used as an inexpensive vacuum pump. A good application for these

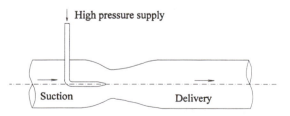

Figure 11.16. Jet pump.

pumps is in mixing dry powders with solvents, where it is necessary to add powder slowly to the fluid in the mixing process to avoid forming large clumps of the powder. The powder can be blown in slowly through the suction line to allow gradual mixing. Jet pumps are not particularly efficient, but their inherent simplicity, low first cost, and freedom from maintenance makes them attractive for many applications.

The *air lift pump* (Figure 11.17) is another very old pump that has important modern applications. Compressed air is introduced into the system at a low point. As the air bubbles rise in the liquid, they expand due to the lower hydrostatic pressure, reducing the pressure further and giving a reduced specific gravity to the fluid. The higher pressure at the bottom then lifts the fluid. The air bubbles should be distributed as evenly as possible across the cross section of the liquid pipe, and they should be kept small in diameter, typically 1 cm or so. Efficiencies of the order of 65% or more have been obtained with these pumps. The advantages of this type of pump is that it can transport corrosive liquids or liquids with suspended solids, and it has no moving mechanical parts in the fluid. Also, these pumps work in deep wells without the need of putting moving mechanical parts below the surface. It may be necessary to pass the raised fluid through a chamber to remove the introduced air.

The height to which the fluid is lifted must be less than twice the static head at the point where the air is introduced. In some cases, this may mean a well has to be deepened, increasing the initial cost of the system. These pumps have been used to raise water from depths as great as 500 ft.

The air lift pump can be reversed and used as a hydraulic air compressor by introducing an air chamber at the bottom and a means of introducing the air at the top. In Figure 11.18 air is induced into the liquid at the top by a simple jet pump. When the combination of air and liquid reaches the air chamber, the air separates from the liquid and is at nearly the hydrostatic pressure of the liquid at that depth.

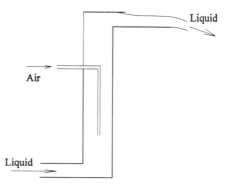

Figure 11.17. Air lift pump.

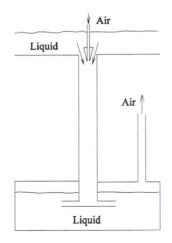

Figure 11.18. Air compressor.

7. Hydraulic Turbines

The old-fashioned water wheel and the steam turbine of Hero are examples of crude turbines that have been around for many centuries. Modern design of these turbines started in the late eighteenth century, and many of the designs now in use first appeared in the nineteenth century. The general terminology used in describing turbines thus has evolved over a long period of time, and owes its ancestry to the water wheel and its accompanying mill.

For a hydraulic turbine such as is used in power generation, the water is typically carried from its dammed reservoir through a ***headrace*** (an open channel) to a ***penstock tube***, which is a conduit leading the water to the turbine. After passing through the turbine, the water generally enters a ***draft tube*** that conducts the water to the ***tailrace***, another open channel that leads the water usually to the river bed at the base of the dam. Most of the elevation change occurs in the penstock tube, allowing the turbine to be placed at or near the tailrace elevation, where it is able to utilize most of the available head.

Hydraulic turbines are of two types, the ***impulse turbine*** and the ***reaction turbine***. In the impulse turbine the entire energy head is converted into kinetic energy at essentially atmospheric pressure. The water meets the turbine in the form of one or more high-velocity jets, which are made to impinge on a series of vanes or buckets mounted on a wheel, causing the wheel to turn. The jets pass through the turbine at a constant pressure. The casing volume contains a high percentage of air, and a draft tube thus cannot be used to cause pressure adjustments. The impulse turbine is usually placed at or slightly above the highest level of the tailrace. The moment exerted on the turbine by the water is due primarily to the change of momentum of the jets.

The impulse turbine is usually run fully submerged, and the flow is made to impinge on all vanes simultaneously. The moment produced is due to a combination of momentum and pressure changes. Most modern turbines use inward flow runners, where the flow is introduced at the outer circumference of the runner, usually through guide vanes, and leaves on the inner circumference of the runners. This arrangement gives a better mechanical construction, and better efficiencies than the older outward-flow runners.

a. Impulse Turbines

The ***Pelton turbine***, or Pelton wheel, is an example of an impulse turbine. It was originally designed by Lester A. Pelton (1829–1908) for use in the California gold fields to operate stamping machines for crushing ore. It consists of a disk, or wheel, with buckets on its circumference. One or more jets impinge on the buckets to produce the motion. The buckets are usually designed with a splitting vane (Pelton's principal contribution to an older design) that turns the jet through an angle of nearly 180°. Best efficiency is achieved when the bucket speed is in the range 44–48% of the jet speed. Today Pelton turbines are used where the available head is large, generally of the order of more than 500 ft, although in early applications they were used at much smaller heads. Since there are no regions of narrow clearances, they run well with silt-laden water and there is little or no risk of cavitation damage. The efficiency curves tend to be flat over a wide range of flow rates. Control of speed and/or load is easily carried out either by deflecting the jets by means of a mechanically controlled obstacle, or by reducing the flow in the jets by means of a needle valve. Pelton turbines are usually constructed without a draft tube and can be very simple to manufacture.

A typical bucket of a Pelton wheel is shown in Figure 11.19. Fluid with a flow rate Q leaves a nozzle with exit area A_n, and impinges on a series of buckets that turn the fluid through an angle $180° - \alpha$, measured relative to the bucket. If Ω is the angular velocity at which the water wheel turns, then the velocity at which the jet approaches a bucket as seen from the rotating wheel is $V_{jet} - r\Omega$, where $V_{jet} = Q/A_n$. We use equation (3.10.3) again and the counterclockwise torque T exerted about the wheel axis is

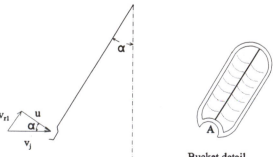

Figure 11.19. Pelton turbine bucket and arm.

$$T = r\rho Q(1 - \cos \beta)(V_{\text{jet}} - r\Omega). \tag{11.7.1}$$

Since all of the jet will strike one or the other of the buckets, the total discharge of the jet is used. Some kinetic energy is lost in the process, and the $\cos \beta$ term is often divided by a factor $(1 + k)^{1/2}$ to incorporate this loss.

The minimum number of buckets that a Pelton turbine should have can be found from Figure 11.20 and a simple calculation. Require that at the instant all of the jet of diameter d has just encountered bucket 2, it should just start to leave bucket 1. Denote the angle between two consecutive buckets by 2θ. The minimum number of buckets needed is then given by

$$n = 2\pi/2\theta. \tag{11.7.2}$$

If R is the extreme outer radius of the receiving edge of the buckets and r is the pitch circle radius (the distance from the center of the jet to the wheel circle when the jet is farthest up the bucket), we see from Figure 11.20 that

$$\cos \theta = (r - 0.5d)/R, \tag{11.7.3}$$

so

$$\theta \cong \sin \theta = \sqrt{1 - \cos^2 \theta} = \sqrt{1 - (r - 0.5d)^2/R^2}, \tag{11.7.4}$$

and the number of blades is given by

$$n = \pi/\sqrt{1 - (r - 0.5d)^2/R^2}. \tag{11.7.5}$$

Typically, $(R - r)/d$ ranges from 0.6 for a small wheel of 3-ft diameter to 0.565 for a wheel of 6-ft diameter or greater.

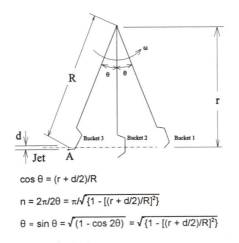

$\cos \theta = (r + d/2)/R$

$n = 2\pi/2\theta = \pi/\sqrt{\{1 - [(r + d/2)/R]^2\}}$

$\theta \approx \sin \theta = \sqrt{(1 - \cos 2\theta)} = \sqrt{\{1 - [(r + d/2)/R]^2\}}$

Figure 11.20. Pelton turbine arm analysis.

The efficiency of a Pelton turbine can be found as follows. Refer to Figure 11.20 to find that the velocity u of the bucket is $r\Omega$. From (11.7.11) the power produced by the turbine is

$$P = \Omega T = \rho Q r \Omega (v_j - r\Omega)(1 - \cos \beta). \tag{11.7.6}$$

Note from this that P is a maximum when $r\Omega = v_j/2$ and $\beta = 180°$.

If we assume that no kinetic energy is left in the fluid exiting the turbine, the efficiency is defined as the ratio of the power produced divided by the power available, i.e., the incoming kinetic energy. Thus

$$\eta = P/0.5\rho Q v_j^2 = 2(r\Omega/v_j)(1 - r\Omega/v_j)(1 - \cos \beta). \tag{11.7.7}$$

Example 11.7.1. Pelton turbine

A Pelton turbine is to be constructed where the maximum available head is 400 ft. If the 4-ft-pitch-diameter wheel is to turn at 300 rpm and the power output is to be 400 hp, what is the efficiency, jet diameter, minimum number of buckets, and distance the buckets project past the pitch circle?

Sought: Turbine efficiency, number of needed buckets, location of buckets with respect to the pitch circle radius.

Given: Wheel diameter is 4 ft. The wheel turns at 300 rpm, with a power output of 400 hp. The fluid is water.

Assumptions: Ideal (theoretical) behavior of a turbine.

Solution: For the given head, the maximum jet velocity is

$$v_j = \sqrt{2gh} = \sqrt{2 \times 32.2 \times 400} = 160.5 \text{ ft/s}.$$

The wheel speed is

$$u = r\Omega = 2 \times (300 \times 2\pi/60) = 62.8 \text{ ft/s},$$

or about 39% that of the jet. For closely spaced buckets the angle α will be small, and we can approximate the cosine of α as 1. The relative exit velocity will be some fraction of the input velocity depending on the bucket design, so we write

$$v_r \cos \theta = -a(v_j - u),$$

where a is less than 1. (The minus sign is introduced since θ will lie in the second quadrant.) Plotting efficiency as a function of a (Figure 11.21), we see a linear relationship. With proper bucket design, a value of a close to 0.8 can be reached. For our purposes, let us assume that we can at least achieve $a = 0.75$. Then $\eta = 83.4\%$. The force divided by the area is

$$F/A = \rho u(v_j - u)(1 + a) = 17,808 \text{ lb/ft}^2.$$

The jet area is then found from

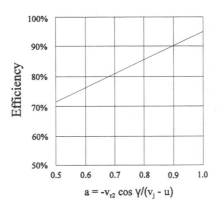

Figure 11.21. Pelton turbine efficiency.

$A = P/(Fu/A) = 550 \times 400/17{,}808 \times 62.8 = 0.1967 \text{ ft}^2,$

corresponding to a jet diameter of 0.5 ft or 6 in.

The minimum number of buckets depends on the quantity $R - r$. If we select this at 0.58 of the jet diameter, since $r = 2$ ft,

$R = 2 + 0.58 \times 0.5 = 2.29 \text{ ft}.$

Then

$n = \pi/\sqrt{1 - (2 + 0.25)^2/(2.29)^2} = 17,$

or, better for the purposes of balancing the turbine, 18 or 20.

b. Reaction turbines

Francis Turbine.   The Francis turbine (James Bicheno Francis, 1815–1892) is an example of a reaction turbine, and is useful for moderate heads (typically 100–500 ft). It has a set of outward adjustable guide vanes (called the *wicket gates*) that turn the flow inward in a vortex flow toward the runner. The runner consists of numerous closely spaced fixed vanes. The flow can either be introduced through the penstocks into an outer spiral chamber surrounding the guide vanes, or the entire turbine can be placed in a flume without any external casing. The guide vanes are designed to give separation-free flow to the runner, and the runner vanes to give a minimum exit velocity. The latter is accomplished by reducing the inner diameter and having the vanes turn the flow so that it leaves the runner with a velocity that is largely axial. The vanes of the runner always run fully submerged.

The analysis of the reaction turbine is very similar to that of the centrifugal pump, the difference mainly being an interchange of the subscripts. We repeat the results from that section, commenting on the differences. We let 1 and 2 denote inlet and outlet conditions, respectively. The inlet radius r_1 is now the greater of the two radii.

Table 11.2. Reaction turbine velocity components

	Inlet velocity components		Outlet velocity components	
	Tangential	Normal	Tangential	Normal
Absolute	$V_1 \cos \alpha_1$	$V_1 \sin \alpha_1$	$V_2 \cos \alpha_2$	$V_2 \sin \alpha_2$
Relative	$V_1 \cos \alpha_1 - r_1\Omega$	$V_1 \sin \alpha_1$	$V_2 \cos \alpha_2 - r_2\Omega$	$V_2 \sin \alpha_2$

The various velocity components are as given in Table 11.2, and are shown in Figure 11.22.

With blades of uniform height w, the normal velocity components are related to discharge according to

$$Q = 2\pi r_1 w V_1 \sin \alpha_1 = 2\pi r_2 w V_2 \sin \alpha_2. \tag{11.7.8}$$

Again we consider only the case where the flow enters and exits each blade of the runner tangentially. This is called **_shockless flow_**. For shockless flow the blade slope is again the same as the slope of the velocity component. Thus the ideal entering and exiting blade angles β_1 and β_2 are given by

$$\tan \beta_1 = V_1 \cos \alpha_1 / (r_1\Omega - V_1 \cos \alpha_1)$$

$$\tan \beta_2 = V_2 \cos \alpha_2 / (r_2\Omega - V_2 \cos \alpha_2). \tag{11.7.9}$$

Having the flow enter tangentially to the vane ensures that there is no momentum loss normal to the impeller. Other angles will result in inlet losses.

The head lost by the fluid in passing through the turbine is, by the energy equation,

$$H = (p_2/\gamma + V_2^2/2g + z_2) - (p_1/\gamma + V_1^2/2g + z_1). \tag{11.7.10}$$

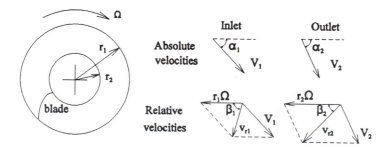

Figure 11.22. Reaction turbine velocities.

The ideal torque produced by the turbine (the torque produced if there were no losses) is the mass rate of flow times the moment of momentum change. This involves only the tangential components of the entering and exiting velocities. Thus

$$T_{ideal} = \rho Q (r_1 V_1 \cos \alpha_1 - r_2 V_2 \cos \alpha_2). \tag{11.7.11}$$

Notice that the ideal torque can be maximized by having the fluid exit the turbine runner normally ($\alpha_2 = 90°$).

Since the blade angles α are fixed, to accommodate changing discharges the wicket gates adjust (i.e., α_1 changes with changes in Q) so that the fluid always enters tangent to the runner blades.

Kaplan turbines.   With heads around 100 ft or less the Kaplan turbine (Victor Kaplan, 1876–1934) is appropriate. The runner vanes look like a large ship's propeller and are movable. The guide vanes are also adjustable, and placed further from the runner than in a Francis turbine.

Draft tubes are important components of a successful reaction turbine design. The pressure at the entrance to the draft tube is negative, and the draft tube thus allows the reaction turbine to be set above the level of the tailwater without loss in head. By having a draft tube that is gradually diverging, the head loss and the exit kinetic energy can be reduced. They are generally designed to turn the flow through the requisite number of 90° bends in a manner so as to minimize losses.

Suggestions for Further Reading

Balje, O. E., *Turbomachines: A Guide to Design, Selection, and Theory*, Wiley, New York, 1981.

Bathe, W. W., *Fundamentals of Gas Turbines*, Wiley, New York, 1984.

Beny, C. H., *Flow and Fan: Principles of Moving Air Through Ducts*, 2d edition, Industrial Press, New York, 1963.

Betz, A., *Introduction to the Theory of Flow Machines*, Pergamon, New York, 1966.

Brown, R. N., *Compressors*, Gulf Pub., Houston, 1986.

Church, A. H., *Centrifugal Pumps and Blowers*, Wiley, New York, 1944.

Cooper, P., editor, "Pumping Machinery—1989," *Proc. ASME Symp. FED*, vol. **81**, American Society of Mechanical Engineers, New York, 1989.

Csanady, G. T., *Theory of Turbomachines*, McGraw-Hill, New York, 1964.

Cumpsty, N. A., *Compressor Aerodynamics*, Longmans, London, 1989.

Eggleston, D. M., and F. S. Stoddard, *Wind Turbine Engineering Design*, Van Nostrand, New York, 1987.

Eldridge, F. R., *Wind Machines*, 2d edition, Van Nostrand Reinhold, New York, 1980.

Fritz, J. J., *Small and Mini Hydropower Systems*, McGraw-Hill, New York, 1984.

Greitzer, E. M., "The Stability of Pumping Systems: The 1980 Freeman Scholar Lecture," *J. Fluids Eng.*, vol. **103**, pp. 193–242, June 1981.

Gulliver, J. S., and R. E. A. Arndt, *Hydropower Engineering Handbook*, Haimerl, L. A., "The Crossflow Turbine," *Waterpower*, January 1960, pp. 5–13; see also ASME Symp. Small Hydropower Fluid Mach., vol. **1**, 1980, and vol. 2, 1982. McGraw-Hill, New York, 1990.

Hawthorne, W. R., and R. A. Novak, "The Aerodynamics of Turbo-Machinery," *Ann. Rev. Fluid Mech.*, vol. **1**, 1969.

Henshaw, T. L., *Reciprocating Pumps*, Van Nostrand Reinhold, New York, 1987.

Hicks, T. G., and T. W. Edwards. *Pump Application Engineering*, McGraw-Hill, New York, 1971.

Horlock, J. H., *Axial Flow Compressors*, Butterworth, London, 1958.

Horlock, J. H., *Axial Flow Turbines*, Butterworth, London, 1966.

Hydraulic Institute, *Hydraulic Institute Standards for Centrifugal, Rotating, and Reciprocal Pumps*, New York, 1983.

Inglis, D. R., *Wind Power*, The University of Michigan Press, Ann Arbor, 1978.

Jarass, L., *Wind Energy: An Assessment of the Technical and Economic Potential*, Springer-Verlag, Berlin, 1981.

Jorgensen, R., editor, *Fan Engineering*, Buffalo Forge, Buffalo, NY, 1983.

Karassik, I. J., W. C. Krutzsch, W. H. Fraser, and J. P. Messina, *Pump Handbook*, 2nd edition. McGraw-Hill, New York, 1985.

Karassik, I. J., and R. Carter, *Centrifugal Pumps*, Dodge, New York, 1960.

Koeppl, G. W., *Putnam's Power from the Wind*, 2nd edition, Van Nostrand Reinhold, New York, 1982.

Lakshminarayana, B., and P. Runstadler, Jr., editors, "Measurement Methods in Rotating Components of Turbomachinery," *ASME Symp. Proc.*, New Orleans, vol. **I00130**, 1980.

Lakshminarayana, B., "An Assessment of Computational Fluid Dynamic Techniques in the Analysis and Design of Turbomachinery: The 1990 Freeman Scholar Lecture," *J. Fluids Eng.*, vol. **113**, pp. 315–352, September 1991.

Lambeck, R. P., *Hydraulic Pumps and Motors: Selection and Application for Hydraulic Power Control Systems*, Marcel Dekker, New York, 1983.

Lobanoff, V. L., and R. R. Ross, *Centrifugal Pumps, Design and Application*, Gulf Pub., Houston, 1985.

Logan, Jr., E. S., *Turbomachinery: Basic Theory and Applications*, Marcel Dekker, New York, 1981.

McGuigan, D., *Small Scale Water Power*, Prism Press, Dorchester, MA, 1978.

Miller, J. L., *The Reciprocating Pump: Theory, Design and Use*, Wiley, New York, 1987.

Moody, L. F., "The Propeller Type Turbine," *ASCE Trans.*, vol. **89**, p. 628, 1926.

Morris, H. M., *Applied Hydraulics in Engineering*, Ronald, New York, 1963.

Osborne, W. C., *Fans*, 2nd edition, Pergamon, London, 1977.

Poynton, J. P., *Metering Pumps*, Marcel Dekker, New York, 1983.

Robinson, M. L., "The Darrieus Wind Turbine for Electrical Power Generation," *Aeronaut. J.*, pp. 244–255, June 1981.

Roco, M. C., P. Hamelin, T. Cader, and G. Davidson, "Animation of LDV Measurements in a Centrifugal Pump," in *Fluid Machinery Forum—1990*. U.S. Rohatgi, editor, FED, vol. **96**, American Society of Mechanical Engineers, New York, 1990.

Shepherd, D. G., *Principles of Turbomachinery*, Macmillan, New York, 1956.

Stepanoff, A. J., *Centrifugal and Axial Flow Pumps*, 2nd edition, Wiley, New York, 1957.

Stepanoff, A. J., *Pumps and Blowers: Two-Phase Flow*, Wiley, New York, 1965.

Stewart, H. L., *Pumps*, 5th edition, Macmillan, New York, 1991.

Streeter, V. L., editor, *Handbook of Fluid Dynamics*, McGraw-Hill, New York, 1961.

Walker, R., *Pump Selection*, 2nd edition, Butterworth, London, 1979.

Wallis, R. A., *Axial Flow Fans and Ducts*, Wiley, New York, 1983.

Warne, D. F., and P. G. Calnan, "Generation of Electricity from the Wind," *IEE Rev.*, vol. **124**, no. 11R, pp. 963–985, November 1977.

Warnick, C. C., *Hydropower Engineering*, Prentice-Hall, Englewood Cliffs, NJ, 1984.

Warring, R. H., *Pumps: Selection, Systems, and Applications*, 2nd edition, Gulf Pub., Houston, 1984.

Wilson, D. G., "Turbomachinery—From Paddle Wheels to Turbojets," *Mech. Eng.*, vol. **104**, pp. 28–40, October 1982.

Wislicenus, G. F., *Fluid Mechanics of Turbomachinery*, 2nd edition, McGraw-Hill, New York, 1965.

Wortman, A. J., *Introduction to Wind Turbine Engineering*, Butterworth, Woburn, MA, 1983.

Worthington Co., *Pump Selector for Industry*, Worthington Pump, Mountainside, NJ, 1977.

Problems for Chapter 11

Centrifugal machine

11.1. A 24-in-diameter fan operating at 1,750 rpm produces an air flow of 150 ft^3/s and a pressure head of 7 in of water. At a speed of 1,650 rpm the efficiency and air flow pattern remain the same. What is the flow rate and head at 1,650 rpm?

11.2. A 24-in-diameter fan operating at 1,750 rpm produces an air flow of 150 ft^3/s and a pressure head of 7 in of water. It is desired to build a larger fan that will deliver the same head but at the lower speed of 1,650 rpm. What should be the diameter of this larger fan, and what will be the delivered flow rate?

11.3. A pump manufacturer wishes to design a new series of pumps. A prototype with a 12-in-diameter impeller produces 8 ft of head at a flow rate of 900 gpm and a rotation speed of 1,200 rpm. The pump requires 10 hp and is 80% efficient. The largest pump in this new family will have an impeller diameter of 20 in. What rotational speed will give the same discharge? What will be the head and required horsepower?

11.4. A pump manufacturer states that at a flow rate of 20 ft^3/s the NPSH of a centrifugal pump is 17 ft. For a water temperature of 68°F and standard atmospheric pressure, is the pump able to safely draw water from a reservoir 5 ft below the pump without fear of cavitation? The head loss in the suction line is 3.5 ft. The inlet diameter is 3 in.

11.5. What is the specific speed (rpm) of a pump operating at 1,000 rpm that delivers 900 gpm at a head of 30 ft? On the basis of this number, which type of pump should be selected: radial flow, mixed flow, axial flow. $[N_S \text{ (rpm)} = NQ^{1/2}/H^{3/4}.]$

11.6. A speed used for selecting turbine type is ω_P, where in American practice $\omega_P = \Omega P^{1/2}/H^{5/3}$, where Ω is the rotational speed in rpm, P is the power in horsepower ($P = \gamma QH/550$ for 100% efficiency), and H is the working head. If ω_P is less than about 80, a Francis turbine is indicated. If ω_P is greater than 80, a propeller turbine is indicated. For water flowing into the turbine at a rate of 1,500 ft³/s with an average head of 220 ft, which type of turbine should be selected if the angular velocity is to be kept below 2,000 rpm?

11.7. A turbine rotating at 180 rpm at a head of 130 m and a flow rate of 11 m³/s of water develops a torque of 610 kN-m. What is the efficiency of the turbine?

11.8. A turbine rotating at 180 rpm at a head of 105 ft develops 1,900 hp at a flow of 170 ft³/s of water. What is the efficiency of the turbine?

11.9. If the turbine in Problem 11.8 were run at a head of 210 ft, what should be the new operating speed, Q, and P to ensure dimensional similitude?

11.10. If the diameter of the turbine in Problem 11.8 were doubled in size but the head was kept at 105 ft, what should be the new operating speed, Q, and P to ensure dimensional similitude?

11.11. A 1/15 scale model of a pump is built and tested with water. It is found that at 3,000 rpm a flow rate of 0.08 m³/s is obtained at a head of 35 m. The efficiency is 86%. If the prototype is to be used at a 50-m head and the same efficiency is desired, what should be the speed and discharge for the prototype?

Centrifugal pumps

11.12. A centrifugal pump manufacturer gives the following data for a family of pumps operating at 1,000 rpm:

H	40	38	30	12	(feet of head)
Q	0	1	2	3	(thousands of gallons per minute)
η	0	60.8	83.6	68.4	(%)

This data is for a pump with 16-in-diameter impeller and with inlet and outlet diameters equal to 6 in. Fit the curve $H = H_{DH} + AQ + BQ^2 + CQ^3$ to the H-Q data. (That is, insert each data point such as $H = 40$, $Q = 0$, obtaining four linear algebraic equations in the four unknowns H_{DH}, A, B, and C. Solve for H_{DH}, A, B, and C.)

11.13. Fit the curve $\eta = aQ + bQ^2 + cQ^3$ to the data of Problem 11.12. At what value of Q is the efficiency maximized?

11.14. Compute the power added to the water by the pump in Problem 11.12. Power is found from $Q \, \Delta p$.

11.15. Use your results from Problems 11.13 and 11.14 to find the motor power that must be supplied to the pump for operation.

11.16. What would be the data for H and Q corresponding to that given in Problem 11.12 if the same pump were operated at 1,750 rpm?

11.17. What would be the data for H and Q corresponding to that given in Problem 11.12 if the pump was geometrically scaled up to a 20-in-diameter impeller, but still run at 1,000 rpm?

11.18. A 6-in-diameter pipe with roughness of 0.005 in and length 280 ft is carrying water at 68°F. The pump described in Problem 11.12 is used to drive the flow. What is the flow rate?

11.19. A model for flow in a centrifugal pump is a radial flow combined with a swirl, giving radial and tangential velocities in the form of A/r and Ωr, respectively. For a pump rotating at a speed of 1,000 rpm and with a discharge of 500 gpm of water, what is the change of kinetic energy head as the fluid moves from a radius of 1 in to a radius of 10 in? The inlet diameter is 1.5 in and the interior distance between the pump casing walls is 1 in.

11.20. Water enters a centrifugal pump impeller at the ideal angle with a flow rate of 4 ft³/s. The impeller is rotating at 1,200 rpm and has an inner diameter of 7 in and an outer diameter of 20 in. The exit velocity from the impeller is 11 ft/s. The pump is 80% efficient under these conditions. Neglecting any losses due to shaft bearings, seals, and the like, what is the needed power input to the fluid? The impeller width is 1 in.

11.21. A pipeline delivers 3,000 gpm to a point in the pipeline where there are two pumps in parallel. The larger pump has an H-Q relation given by $H = 40 - 0.8Q^2$, while the smaller has an H-Q relation given by $H = 35 - 0.6Q^2$. (Q is in 1,000 gpm, H in ft.) The discharges from the two pumps join on the high-pressure (downstream) side of the pumps. What is the flow through each pump? What is the head produced?

11.22. A centrifugal pump has an inlet diameter of 6 in and an outlet diameter of 4 in. The inlet pressure read on a vacuum gage is 10 in of mercury. The outlet pressure is 24 psi, with a discharge of 2.5 ft³/s of water. If 40 hp must be delivered to the pump, what is the pump efficiency? What percentage of the power added to the water is provided by the kinetic energy change as the water passes through the pump?

Turbine

11.23. The classic American farm windmill is an example of a horizontal-axis wind turbine. An approximation to its performance is given by

$$C_P = \text{power}/0.5\rho A V_{\text{wind}}^3 = 0.957s(1 - 0.937s + 0.212s^2),$$

where $s = r\Omega/V_{\text{wind}}$ is the ratio of the blade tip speed to wind speed. For a 12-ft-diameter fan and a wind speed of 20 mph, what is the fan angular velocity and power produced at optimum C_P when the air temperature is 86°F? Using the actuator disk theory of Chapter 3, what is the air velocity in the wake?

Pitcher pump

11.24. What is the maximum "lift" (the height the water can be raised) of a bucket, or pitcher, pump raising water from a well that is open to the atmosphere? Assume perfect seals between the piston and the pump walls.

Pelton turbine

11.25. A Pelton impulse turbine has a wheel diameter of 10 ft and a jet diameter of 6 in. The water jet velocity is 300 ft/s, the bucket turns the flow through an angle of 150°, and the loss factor k is 0.2. Compute the torque and horsepower output at the following angular speeds: 0, 100, 200, 300 rad/s.

Reaction turbine

11.26. The runner of a radial-flow reaction turbine has a water flow rate of 2,500 gpm, an entrance radius of 14 in, and an exit radius of 6 in. The entrance and exit blade angles are 40° and 15°, respectively. The entering and exiting flows are shockless (tangent to the blade), the runner width is 4 in, and the runner velocity is 100 rpm. Overall efficiency is 80%. Find the absolute entry and exit angles of the jet and the ideal torque.

chapter 12

Conclusion

In this book we have covered material that is standard in most introductory courses in fluid dynamics. You may wonder what further topics in fluid mechanics exist that you might wish to study further.

The material that you have already covered has given you an introduction to the basic concepts of fluid mechanics and how to apply them in simple situations. You may well find in fact that it covers many if not most of the fluids-related problems you will encounter as you pursue your profession as an engineer. And the fundamental concept of the control volume and its application in accounting for mass, momentum, and energy transport is a tool that can be applied in many other engineering situations. However, there are many other topics that require a more detailed knowledge that can be acquired by additional study. A few of the topics you might wish to pursue through further course work follow.

Flow in porous media. Civil engineers are concerned with the flow of groundwater and sand filters to purify water. Chemical and mechanical engineers use porous beds to enhance chemical processes and heat transfer, and petroleum engineers are concerned with the flow of oil and gases in the ground. A fundamental tool for treating these topics, in addition to the topics covered in this book, is *Fick's law*, which governs the flow in such situations. It states that in situations where the media dominates momentum transfer and the usual viscous interactions between fluid particles, the velocity is directly proportional to the negative of the pressure gradient. This is an empirical law, but at least in steady-flow situations it has proved quite reliable.

Hydraulic transients. This topic has been briefly touched on in Chapters 8 and 9. Knowledge of how to control them is important in dealing with large hydraulic machinery, hydraulic lines where pressure levels are very high, in rivers, estuaries, and tidal bays, and in the flow in irrigation canals and dam channels. Timing of gate and valve openings and closures is often critical to the safety of systems and humans.

Vibrations. Vibrations in mechanical structures caused by flows over and past pipelines and underwater and aerial structures, pump-caused pressure oscillations, and the like are important in such diverse fields as offshore oil technology, power line technology, and bioengineering. The heart and lung are two examples of mechanical systems that produce oscillating flow fields. Understanding the mechanisms that can interfere with their proper operation is necessary to providing the safest methodology for treating the patient. The strong currents and waves encountered by offshore

567

oil-drilling platforms and the lines connecting them to the shore can cause devastating damage to personnel as well as to the environment.

Dispersion.   Distribution of pollutants in the atmosphere, lakes, rivers, and seas is an important environmental problem that still requires additional study. Early engineering practice was to build tall chimneys, with the hope that the winds would carry pollutants far enough away and in a sufficiently reduced concentration that there was no health problem. More recently it was understood that the old maxim "the solution to pollution is dilution" is not satisfactory. While ideally no pollutants should be released, until this standard can be reached it is necessary to understand what happens when pollutants are released in the air and waters, where they go, and how their effects can be rendered harmless.

Combustion.   Combustion of fuels can take place either in open areas such as water heaters or in confined areas such as the cylinders of engines. In both cases important considerations are: control of the combustion for safety reasons, pollution reduction, and fuel and cost efficiency. Too early ignition or improper mixture of gases can result in higher noise levels, excessive wear, lowered efficiency, and increased pollution. Propagation of explosion waves in industrial situations, explosion of dust-laden air, and even the burning in confined areas of such diverse materials as coal, peat, buried garbage, and iron scrap are other technological problems that often arise through the lack of proper technical attention.

Fluidics.   In 1800, Thomas Young noted that at moderate Reynolds numbers a fluid jet issuing in a wedge-shaped container had a tendency to attach to a wall. This effect was rediscovered by a number of investigators over the years, particularly by Henri Coanda in the 1930s. The effect has been named the Coanda effect in his honor. In 1959 the Harry Diamond Laboratories started extensive development of this and related effects, and the speciality of fluidics was born. Very simple flow geometries have been set up that act as amplifiers, switches, and other computer elements. These can be connected to form a small dedicated computer, complete with a fluidic readout, and made to perform duties such as detection and counting of items on an assembly line. As power amplifiers they are used to switch the direction of jets in rockets and aircraft. They can operate using either gas or liquid. One of their principal advantages is that a jet of moderate power can control a power source such as a high-momentum jet without the use of moving mechanical parts or hardware.

Wave motion.   Waves generated by the bow of a ship have long been of interest to the naval architect, since they are a major factor in ship drag and in determination of the marine power plant. A spectacular technological advance was made in the late 1950s when, based on theoretical considerations, it was shown that adding a bulbous bow to the bow of a ship below the waterline produced a wave that interfered with the basic bow wave, thereby canceling a good portion of the wave drag. (The same principle is now being used in noise reduction technology.) This reduction in drag was so significant that, as a result, many bulbous bows were added to existing ships, which at the same time were lengthened so that increased cargo could be carried using the same power plant. In other applications of wave theory, waves on the surfaces of lakes and oceans are studied to see why they concentrate at certain points due to depth variations, and to understand why, no matter what the direction of the wind, waves always seem to be coming into the shore and in the process increasing in height. Another use of waves, the soldering of pc boards by generating waves on the surface of liquid solder, is now standard commercial practice.

High-speed jets.   High-pressure water jets have been used to cut materials such as coal in mines and timber. The jets are of high speed, typically a quarter or more of the sonic speed of water, and have the advantage of removing coal dust or sawdust in the process. Russian engineers have been the chief developers of these high-pressure (around 500,000 psi) jets. They can be used to etch or cut glass and other materials. (They can also cut soft tissue, so care must be taken in their use.[1])

Density stratification.   Stratified flows occur in valleys (e.g., the Los Angeles basin, where it contains the smog), rivers, and lakes. Interfacial waves exist where freshwater rivers discharge into saltwater, resulting in high drag on ships proceeding through the channel. Saltwater ponds have also been found to be useful in storing solar energy for heating and electrical energy production.

Physical meteorology and oceanography.   The study of currents in oceans and the atmosphere has shown the cause of jet streams, the Gulf currents, el Niño, and many other similar but lesser known phenomena. These affect our weather strongly as well as the ground speed of aircraft. They also act as distributors of pollutants, carrying them into the highest region of the atmosphere.

Lubrication.   Lubrication of machinery by liquids or gases is important so that dry friction can be avoided. Ball and roller bearings can run only if sufficient lubricant is supplied. Different support applications and needs require different types of bearings; e.g., propeller shafts on ships use water pumped along the shaft, high-precision instrumentation may require the low drag provided by gas lubricants, while knee joints require biocompatible materials with long life and that can safely be attached to neighboring bones.

Heat transfer.   Fluids used in heat transfer must have a high capacity for carrying heat, withstand system temperatures and environment, and be nontoxic. There is a continuing need for ever-smaller and more efficient heat exchangers, and many innovative technologies directed to the production of power depend on the success of this search.

Bioengineering.   Biofluid engineers originally attacked problems involving circulation in the blood vessels and air passages. However, there are other fluid transport problems in the human body (e.g., lymph transport, transport of oxygen and other nutrients through the eyes) that have received less attention and are still outstanding. Their study requires knowledge of chemistry and diffusion processes as well as the more usually studied fluid transport cases.

An important aspect of fluid flow in the body is the diffusion of mass. This was neglected in our derivations of the continuity equation, but it can be important when

---

[1]A story is often recounted of an operator of a diesel engine who suspected that there was a small hole developing in the exhaust manifold of the engine. To locate it he moved his finger along the manifold, with unfortunate results when the hole was found.

the fluid contains different chemical constituents that move in the flow under concentration gradients as well as convection. Other situations besides bioengineering also involve diffusion phenomena, including diffusion of pollutants in lakes, rivers, and the atmosphere, chemical processing, and many mixing and separation processes.

Aerodynamics and compressible flows.   The development of the airplane has been one of the spectacular technical achievements of the twentieth century. At the start of the century, manned powered flight was still a dream. The Wright brothers were the first to achieve this goal, but it wasn't until the 1930s that scheduled commercial flights were a reality. Now supersonic commercial flight is possible on a limited basis, and certainly it will be standard procedure on longer flights as the twenty-first century proceeds. Rockets and manned spacecraft have launched space platforms and satellites that have revolutionized our communications technology and hold promise for manufacturing processes not possible on earth.

While these are some of the more dramatic aspects of compressible flow, more mundane applications are vitally important in our daily life. The transport of natural gas in pipelines revolutionized the heating of homes in the last half of the twentieth century. Industries manufacture gases useful in water purification, medical applications, heat treating, and many other industrial applications must transport these gases safely and economically to the end user. The foundations of compressible flow technology have been presented in Chapter 9, but more detailed information is useful as new or more complex applications arise.

Oil spills.   Oil spills have unfortunately become commonplace in our time. A standard procedure for controlling them is to surround the spill with a barrier, thus buying time and allowing the oil to be picked up at leisure. However, it is known that if a current is present in the water, oil will be pulled off the contained slick in little droplets if the currents are slow, and in strong jets if the current is fast. (A rough estimate of the transition between "slow" and "fast" currents can be made using the Froude number based on current speed and barrier depth. Values less than unity mean that the current can be considered "slow," while as this Froude number approaches or exceeds unity the current is "fast.") No acceptable technique has yet been found that can contain/retrieve large oil spills in fast currents.[2]

These examples are but a few of the various areas of fluid mechanics that may interest you and knowledge of which can aid your career as an engineer. Many of these topics can be studied using a one-dimensional approach and only a moderate amount of mathematics. But, to approach two- and three-space dimensional problems, or problems involving time, it becomes necessary to extend this analysis to higher dimensions. The basic equations that result from this are the *Euler equations* if viscosity is not included, and the *Navier-Stokes equations* if viscosity is included. Both of these equations are partial differential equations presented in Chapter 4.

---

[2]We tend to think of the more spectacular oil spills such as the Exxon Valdez. However, parking lots and streets tend to accumulate oil from vehicles that is eventually washed into the storm sewers and waterways by rain. This is a more dilute example, but because it is steady and widely dispersed it produces a substantial amount of pollution daily.

The Euler equations state that the mass density times the acceleration equals the negative of the pressure gradient plus the specific weight vector. While these equations have been known for more than a century, their nonlinearity in the convective acceleration terms makes them very difficult to solve analytically unless there is no vorticity present and the velocity potential can be used. In that case the nonlinearity can be avoided, and many solutions exist. However, the classical solution techniques are limited to relatively simple geometry, and it wasn't until the widespread availability of digital computers that the theory emerged as a design tool. The aircraft industry in particular has used it to design commercial aircraft, even dealing with such complicated situations as a space shuttle perched atop of a Boeing 747. Euler-based models that include vorticity and compressible effects are also used extensively in studies of compressible gas flows (including the prediction of shock waves), meteorological flows (including weather forecasting), flow around vehicles, and combustion of gases.

The more general Navier-Stokes equations include the Euler equations as a special case. The Navier-Stokes equations state that the mass density times the acceleration equals the negative of the pressure gradient, plus the specific weight vector, plus the gradient of the viscous stress terms. While the viscous terms do not increase the mathematical nonlinearity beyond that of the Euler equations, they do raise the mathematical order of the equations, thereby increasing both the mathematical and physical complexity. Something of the order of 50 exact solutions of the Navier-Stokes equations have been found, all in the form of similarity solutions in unbounded regions. However, the boundary layer equations that can be deduced from the Navier-Stokes equations can be used in conjunction with solutions of the Euler equations when the Reynolds number is sufficiently high. Solutions for lower values of the Reynolds numbers can be obtained by computational means.

The rapid growth in the power and availability of computers has had a pronounced impact on developments in the applications of fluid mechanics. New computational methods have been developed for the Navier-Stokes equations, and the classical Green's function approach of the nineteenth century developed to a point where it can be used on present-day personal computers for the solution of the Euler equations. There has in fact been an interaction between fluid mechanics, numerical methods, and computer designers, which has led to new ideas that better match the capabilities and needs of all.

Commercial computer codes are now available that claim to be able to tackle most any fluids problem imaginable. It might be thought that we have arrived at the stage where it is no longer necessary to learn the theoretical foundations of fluid mechanics—all we need to do is understand how to run the code! While many engineering problems do lend themselves to this approach, unfortunately others do not. Failure to fully understand the physical nature of a flow, the complexities it may have, and the physical basis of the computer model can lead to the use of a computer code that is totally unsuitable. It is still true that a little knowledge can be a dangerous thing, and nothing can replace a thorough knowledge of the field in which one practices.

References for Further Study

The following books are natural continuations of the topics that have been considered in this book. Many of them require some familiarity with differential equations and advanced mathematical topics such as complex variable theory. Others are just fun to read.

> Some books are to be tasted, others to be swallowed, and some few to be chewed and digested.
>
> —Francis Bacon

General topics

Batchelor, G. K., *An Introduction to Fluid Dynamics*, Cambridge University Press, Cambridge, 1967.

Goldstein, S., *Modern Developments in Fluid Dynamics*, vols. **1** and **2**, Dover, New York, 1965. (Originally published in 1938.)

Goldstein, S., *Lectures on Fluid Mechanics*, Interscience, New York, 1960.

Lamb, H., *Hydrodynamics*, 6th edition, Dover, New York, 1932.

Milne-Thomson, L. M., *Theoretical Hydrodynamics*, 4th edition, Macmillan, New York, 1960.

Prandtl, L., *Essentials of Fluid Dynamics*, Hafner, New York, 1952.

Prandtl, L., and O. G. Tietjens, *Fundamentals of Hydro- and Aeromechanics*, Dover, New York, 1957.

Robertson, J. M., *Hydrodynamics in Theory and Application*, Prentice Hall, Englewood Cliffs, NJ, 1965.

Rosenhead, L., editor, *Laminar Boundary Layers*, Oxford University Press, Oxford, 1963.

Schlichting, H., *Boundary Layer Theory*, 7th edition, McGraw-Hill, New York, 1979.

Vallentine, H. R., *Applied Hydrodynamics*, 4th edition, Macmillan, New York, 1967.

Yih, C.-S., *Fluid Mechanics*, West River Press, Ann Arbor, 1988.

Aerodynamics

Abbott, I. H., and A. E. von Doenhoff, *Theory of Wing Sections*, Dover, New York, 1959.

Anderson, J. D., *Fundamentals of Aerodynamics*, McGraw-Hill, New York, 1984.

Bertin, J. J., and M. L. Smith, *Aerodynamics for Engineers*, 2nd edition, Prentice Hall, New York, 1989.

Kuethe, A. M., and C.-Y. Chow, *Foundations of Aerodynamics*, 4th edition, Wiley, New York, 1986.

Kuethe, A., and J. D. Schetzer, *Foundations of Aerodynamics*, Wiley, New York, 1950.

Liepmann, H. W., and A. Roshko, *Elements of Gasdynamics*, Wiley, New York, 1957.

Prandtl, L., "Applications of Modern Hydrodynamics to Aeronautics," NACA Rep. 116, 1921.

Thwaites, B., editor, *Incompressible Aerodynamics*, Clarendon Press, Oxford, England, 1960.

Tucker, V., and G. C. Parrott, "Aerodynamics of Gliding Flight of Falcons and Other Birds," *J. Exp. Biol.*, vol. **52**, pp. 345–368, 1970.

Yu, T. Y., et al., editors, *Swimming and Flying in Nature: Proceedings of a Symposium*, Plenum, New York, 1975.

Compressible flow

Anderson, J. D., Jr., *Modern Compressible Flow, with Historical Perspective*, 2nd edition, McGraw-Hill, New York, 1990.

Anderson, J. D., Jr., *Hypersonic and High Temperature Gas Dynamics*, McGraw-Hill, New York, 1989.

Cambel, A. B., and B. H. Jennings, *Gas Dynamics*, McGraw-Hill, New York, 1958.

Chapman, A. J., and W. F. Walker, *Introductory Gas Dynamics*, Holt, New York, 1971.

Cheers, F., *Elements of Compressible Flow*, Wiley, New York, 1963.

Courant, R., and K. O. Friedrichs, *Supersonic Flow and Shock Waves*, Interscience, New York, 1948; reprinted by Springer-Verlag, New York, 1977.

Husain, Z., *Gas Dynamics through Problems*, Halstead, New York, 1989.

Imric, B. W., *Compressible Fluid Flow*, Halstead, New York, 1974.

John, J. E., *Gas Dynamics*, 2nd edition, Allyn & Bacon, Boston, 1984.

Liepmann, H. W., and A. Roshko, *Elements of Gas Dynamics*, Wiley, New York, 1957.

Moran, J., *An Introduction to Theoretical and Computational Aerodynamics*, Wiley, New York, 1984.

Owczarek, J. A., *Gas Dynamics*, International Textbook, Scranton, PA, 1964.

Pope, A. Y., *Aerodynamics of Supersonic Flight*, 2nd edition, Pitman, New York, 1958.

Pope, A. Y., and K. L. Coin, *High Speed Wind Tunnel Testing*, Wiley, New York, 1965.

Saad, M. A., *Compressible Fluid Flow*, Prentice Hall, Englewood Cliffs, NJ, 1985.

Schreier, S., *Compressible Flow*, Wiley, New York, 1982.

Shapiro, A. H., *The Dynamics and Thermodynamics of Compressible Fluid Flow*, Ronald, New York, 1953.

Thompson, P. A., *Compressible Fluid Dynamics*, McGraw-Hill, New York, 1972.

Vincenti, W. G., and C. Kruger, *Introduction to Physical Gas Dynamics*, Wiley, New York, 1965.

von Mises, R., *Mathematical Theory of Compressible Fluid Flow*, Academic, New York, 1958.

Zucrow, M. J., and J. D. Hoffman, *Gas Dynamics*, Wiley, New York, 1976.

Computational fluid mechanics

Anderson, D. A., J. C. Tannehill, and R. H. Pletcher, *Computational Fluid Mechanics and Heat Transfer*, McGraw-Hill, New York, 1984.

Brebbia, C. A., and J. Dominquez, *Boundary Elements—An Introductory Course*, 2nd edition, Computational Mechanics Publications, Southampton, England, and McGraw-Hill, New York, 1991.

Carey, G. F., and J. T. Oden, *Finite Elements: Fluid Mechanics*, vol. 6, Prentice Hall, Englewood Cliffs, NJ, 1986.

Hess, J. L., and A. M. O. Smith, "Calculation of Nonlifting Potential Flow about Arbitrary Three-Dimensional Bodies," *J. Ship Res.*, vol. **8**, pp. 22–44, 1964.

Huebner, K. H., and E. A. Thornton, *The Finite Element Method for Engineers*, 2nd edition, Wiley, New York, 1983.

Patankar, S. V., *Numerical Heat Transfer and Fluid Flow*, McGraw-Hill, New York, 1980.

Roache, P. J., *Computational Fluid Dynamics*, Hermosa Press, Albuquerque, 1976.

Flow visualization

National Committee for Fluid Mechanics Films, *Illustrated Experiments in Fluid Mechanics*, M.I.T. Press, Cambridge, 1972.

Merzkirch, W., *Flow Visualization*, 2nd edition, Academic, New York, 1987.

Nakayama, Y., editor, *Visualized Flow*, Pergamon, Oxford, 1988.

Van Dyke, M., *An Album of Fluid Motion*, Parabolic Press, Stanford, 1982.

Yang, W. J., editor, *Handbook of Flow Visualization*, Hemisphere, New York, 1989.

Turbulence

Batchelor, G. K., *Homogeneous Turbulence*, Cambridge University Press, Cambridge, 1967.

Hinze, J. O., *Turbulence*, 2nd edition, McGraw-Hill, 1975.

Rodi, W., *Turbulence Models and Their Application in Hydraulics*, Brookfield Pub., Brookfield, VT, 1984.

Rodi, W., *Turbulence Models and Their Application in Hydraulics*, Brookfield Pub., Brookfield, VT, 1984.

Waves and oceanography

Defant, A., *Ebb and Flow, the Tides of Earth, Air, and Water*, The University of Michigan Press, Ann Arbor, 1958.

Ippen, A. T., *Estuary and Coastline Hydrodynamics*, McGraw-Hill, New York, 1966.

Kinsman, B., *Wind Waves: Their Generation and Propagation on the Ocean Surface*, Prentice Hall, Englewood Cliffs, NJ, 1964; Dover, New York, 1984.

Knauss, J. A., *Introduction to Physical Oceanography*, Prentice Hall, Englewood Cliffs, NJ, 1978.

Lighthill, M. J., *Waves in Fluids*. Cambridge University Press, London, 1978.

Mei, C. C., *The Applied Dynamics of Ocean Surface Waves*, Wiley, New York, 1983.

Neumann, G., and W. J. Pierson, Jr., *Principles of Physical Oceanography*, Prentice Hall, Englewood Cliffs, NJ, 1966.

Sorenson, R. M., *Basic Coastal Engineering*, Wiley, New York, 1978.

Stoker, J. J., *Water Waves*, Interscience, New York, 1957.

U.S. Department of Commerce, *Tidal Current Tables*, National Oceanographic and Atmospheric Administration, Washington, 1971.

Warren, B., and C. Wunsch, editors, *Evolution of Physical Oceanography*, M.I.T. Press, Cambridge, 1981.

Other specialized topics

Chang, P. K., *Control of Flow Separation*. McGraw-Hill, New York, 1976. See also P. K. Chang, *Recent Development in Flow Separation*, Pang Han Pub. Co., Seoul, Korea, 1983.

Comstock, J. P., editor, *Principles of Naval Architecture*, Society of Naval Architects and Marine Engineers, New York, 1967.

Drazin, P. G., and W. H. Reid, *Hydrodynamic Stability*, Cambridge University Press, London 1981.

Gillmer, T. C., *Modern Ship Design*, United States Naval Institute, Annapolis, 1970.

Griffin, O. M., and S. E. Ramberg, "The Vortex Street Wakes of Vibrating Cylinders", *J. Fluid Mech.*, vol. **66**, pt. 3, pp. 553–576, 1974.

Jones, G. W., Jr., "Unsteady Lift Forces Generated by Vortex Shedding about a Large, Stationary, Oscillating Cylinder at High Reynolds Numbers," *ASME Symp. Unsteady Flow*, 1968.

Kirshner, J. M., and S. Katz, *Design Theory of Fluidic Components*, Academic, New York, 1975.

Knapp, R. T., J. W. Daily, and F. G. Hammitt, *Cavitation*, McGraw-Hill, New York, 1970.

Koschmieder, E. L., "Turbulent Taylor Vortex Row," *J. Fluid Mech.*, vol. **93**, pt. 3, pp. 515–527, 1979.

Lin, C. C., *The Theory of Hydrodynamic Stability*, Cambridge University Press, Cambridge, 1967.

Newman, J. N., *Marine Hydrodynamics*, M.I.T. Press, Cambridge, 1977.

Pletcher, R. H., and C. L. Dancey, "A Direct Method of Calculating through Separated Regions in Boundary Layer Flow," *J. Fluids Eng.*, pp. 568–572, September 1976.

Pletcher, R. H., "Calculation of Separated Flows by Viscous-Inviscid Interaction," *Recent Advances in Numerical Methods for Fluids,* vol. **3**, pp. 383–414, 1984.

Roshko, A., "On the Development of Turbulent Wakes from Vortex Streets," NACA Report 1191, 1954.

Saffman, P. G., *Vortex Dynamics*, Cambridge University Press, Cambridge, 1992.

Schubauer, G. B., and H. K. Skramstad, "Laminar Boundary Layer Oscillations and Stability of Laminar Flow," Natl. Bur. Stand. Res. Pap. 1772, April 1943 (see also *J. Aero. Sci.*, vol. **14**, pp. 69–78, 1947, and NACA Rep. 909, 1947).

Taylor, G. I. "Stability of a Viscous Liquid Contained between Two Rotating Cylinders," *Phil. Trans. Roy. Soc. London Ser. A*, vol. **223**, pp. 289–343, 1923.

Telionis, D. P., *Unsteady Viscous Flows*, Springer-Verlag, New York, 1981.

Thwaites, B., "Approximate Calculation of the Laminar Boundary Layer," *Aeronaut. Q.*, vol. **1**, pp. 245–280, 1949.

Truesdell, C., *The Kinematics of Vorticity*, Indiana University Press, Bloomington, 1954.

Van Dyke, M., *Perturbation Methods in Fluid Mechanics*, Parabolic Press, Stanford, 1964.

von Kármán, T., "On Laminar and Turbulent Friction," *Z. Angew. Math. Mech.*, vol. **1**, pp. 235–236, 1921.

Yih, C.-S., *Dynamics of Nonhomogeneous Fluids*, Macmillan, New York, 1965.

General references

ASHRAE Handbook of Fundamentals, ASHRAE, Atlanta, 1996.

Blevins, R. D., *Applied Fluid Dynamics Handbook*, Van Nostrand Reinhold, New York, 1984.

CRC Handbook of Tables for Applied Engineering Science, 2nd edition, CRC Press, Boca Raton, FL, 1973.

Encyclopedia of Science and Technology, 7th edition, McGraw-Hill, New York, 1993.

Handbuch der Physik, Springer-Verlag, Berlin, 1963.

Selby, S. M., *CRC Handbook of Tables for Mathematics*, 4th edition, CRC Press, Cleveland, 1976.

Conversion of Units and Useful Constants

Classified list of units

To convert from	to	multiply by
Acceleration		
foot/second^2	meter/second^2	3.048 000E-01
inch/second^2	meter/second^2	2.540 000E-02
Area		
acre	meter^2	4.046 873E+03
centimeter^2	meter^2	1.000 000E-04
foot^2	meter^2	9.290 304E-02
hectare	meter^2	1.000 000E+04
inch^2	meter^2	6.451 600E-04
mile^2	meter^2	2.589 998E+06
yard^2	meter^2	8.361 274E-01
Bulk modulus		See force/area below.
Energy, moment, or work		
Btu (Int. Steam Table)	joule (N-m)	1.055 056E+03
calorie (Int. Steam Table)	joule (N-m)	4.186 800E+00
calorie (mean)	joule (N-m)	4.190 020E+00
erg (dyne/cm^2)	joule (N-m)	1.000 000E-07
foot-pound	joule (N-m)	1.355 818E+00
gram calorie (U.S. NBS)	kilowatt-hour	1.162 222E-03
kilogram calorie (U.S. NBS)	watt-hour	1.162 222E-03
watt-second	joule (N-m)	1.000 000E+00
Energy/mass, work/mass		
Btu/pound-mass	kilojoule/kilogram	2.326 000E+00
Energy/mass-degree (specific heat)		
Btu/pound-mass-°R	joules/kilogram-K	4.186 000E+00
Force		
kilogram-force	newton	9.806 650E+00
pound	newton	4.448 222E+00

Classified list of units (*continued*)

To convert from	to	multiply by
dyne	newton	1.000 000E-05
poundal	newton	1.382 550E-01

Force/area (bulk modulus, pressure, stress)

atmosphere	pascal	1.013 250E+05
bar	pascal	1.000 000E+05
centimeter Hg(0°C)	pascal	1.333 220E+03
centimeter $H_2O(4°C)$	pascal	9.806 380E+01
dyne/centimeter^2	pascal	1.000 000E-01
foot water (39.2°F)	pascal	2.988 980E+03
inch mercury (60°F)	pascal	3.376 850E+03
inch water (60°F)	pascal	2.488 400E+02
kilogram-force/cm^2	pascal	9.806 650E+04
kilogram-force/meter^2	pascal	9.806 659E+00
pound/inch^2 (psi)	pascal	6.894 757E+03
pound/foot^2 (psf)	pascal	4.788 026E+01
torr [mmHg (0°C)]	pascal	1.333 224E+02

Force/length (surface tension)

pound/inch	newton/meter	1.751 268E+02
pound/foot	newton/meter	1.459 390E+01

Force/volume (specific weight)

pound/foot^3	newton/meter^3	1.570 875E+02

Heat transfer rate

Btu/hour	watt	2.930 570E-01

Heat flux

Btu/hour-foot^2	watt/meter^2	3.153 580E+00

Ideal gas constant, heat capacity

calorie (mean)/gram-K	kilojoule/kilogram-K	4.184 000E+00
Btu/pound-mass-°R	kilojoule/kilogram-K	4.186 800E+00
pound-foot/pound-mass-°R	kilojoule/kilogram-K	5.380 320E-03

Length

angstrom	meter	1.000 000E-10
foot	meter	3.048 000E-01
inch	meter	2.540 000E-02
micron	meter	1.000 000E-06
mil (0.001 in)	meter	2.540 000E-05
yard	meter	9.144 000E-01

Classified list of units (*continued*)

To convert from	to	multiply by
Mass		
gram	kilogram	1.000 000E-03
ounce (avdp)	kilogram	2.834 952E-02
pound-mass	kilogram	4.535 924E-01
slug	kilogram	1.459 390E+01
Mass/time (mass flow rate)		
pound-mass/second	kilogram/second	4.535 924E-01
slug/second	kilogram/second	1.459 390E+01
Mass/volume (mass density)		
gram/centimeter^3	kilogram/meter^3	1.000 000E+03
pound-mass/foot^3	kilogram/meter^3	1.601 846E+01
slug/foot^3	kilogram/meter^3	5.153 788E+02
Moment		See energy above.
Power		
Btu/hour (Int. Table)	watt	2.930 711E-01
foot-pound/second	watt	1.355 818E+00
horsepower (550 foot-pound/second)	watt	7.456 999E+02
ton refrigeration (12,000 Btu/hour)	watt	3.316 853E+03
Power/length-degree, thermal conductivity		
Btu/hour-foot	watt/meter-K	4.188 300E-01
calorie/second-centimeter-°R	watt/meter-K	1.730 640E+00
Power/area-degree, heat transfer coefficient		
Btu/hour-foot^2	watt/meter^2-K	2.390 100E+03
calorie/second-centimeter^2-°R	watt/meter^2-K	5.677 947E+00
Power/volume, heat generation rate		
Btu/hour-foot^3	watt/meter^3	1.034 661E+01
Pressure or stress		See force/area above.
Specific weight		See force/volume above.
Surface tension		See force/length above.
Temperature		
degree Celsius	kelvin	$t_K = t_C + 273.15$
degree Fahrenheit	kelvin	$t_K = (t_F + 459.67)/1.8$
degree Rankine	kelvin	$t_K = t_R/1.8$
degree Fahrenheit	celsius	$t_C = (t_F - 32)/1.8$

Classified list of units (*continued*)

To convert from	to	multiply by
Velocity		
centimeter/second	meter/second	1.000 000E-02
foot/second	meter/second	3.048 000E-01
inch/second	meter/second	2.540 000E-02
kilometer/hour	meter/second	2.777 778E-01
knot (Int.)	meter/second	5.144 444E-01
mile/hour	meter/second	4.470 400E-01
Viscosity (absolute)		
kilogram/meter-second	pascal-second	1.000 000E+00
poise (gram/centimeter-second)	pascal-second	1.000 000E-01
centipoise	pascal-second	1.000 000E-03
micropoise	pascal-second	1.000 000E-07
pound-second/foot^2	pascal-second	4.788 026E+01
pound-mass/foot-second	pascal-second	1.488 162E+00
slug/foot-second	pascal-second	4.788 026E+01
Viscosity (kinematic)		
centistoke	meter^2/second	1.000 000E-06
foot^2/second	meter^2/second	9.290 304E-02
stoke	meter^2/second	1.000 000E-04
Volume		
fluid ounce (U.S.)	meter^3	2.975 353E-05
foot^3	meter^3	2.831 685E-02
gallon (U.S.)	meter^3	3.785 412E-03
inch^3	meter^3	1.638 706E-05
liter	meter^3	1.000 000E-03
yard^3	meter^3	7.645 549E-01
Volume/time (volumetric discharge)		
centimeter^3/second	meter^3/second	1.000 000E-06
foot^3/minute	meter^3/second	4.719 474E-04
foot^3/second	meter^3/second	2.831 685E-02
gallon (U.S.)/minute	meter^3/second	6.309 020E-05
liter/second	meter^3/second	1.000 000E-03
Work		See energy above.
Work/mass, energy/mass		
Btu/pound-mass	kilojoule/kilogram	2.326 000E+00

SI base units

ampere (A)	electric current
candela (cd)	luminous intensity
kelvin (K)	thermodynamic temperature
kilogram (kg)	mass
meter (m)	length
mole (mol)	amount of substance
second (s)	time

Supplementary SI units

radian (rad)	plane angle - meter/meter
steradian (sr)	solid angle - meter^2/meter^2

British derived units

Quantity	Type	Dimension
Fahrenheit (°F)	temperature	°Fahrenheit = °Rankine – 459.67
pound/foot^2	pressure	144 pound/foot^2 = 1 pound/inch^2
pound-mass	mass	32.174 pound-mass = 1 slug
poundal	force	32.174 poundals = 1 pound
horsepower (hp)	power	1 horsepower = 550 foot-pound/second

SI derived units

Quantity	Type	Dimensions
becquerel (Bq)	activity	$second^{-1}$
celsius (°C)	temperature	°Celsius = Kelvin − 273.15
coulomb (C)	quantity of electricity, electric charge	ampere-second
farad (F)	electric capacitance	ampere-second/volt
gray (Gy)	adsorbed dose	$meter^2/second^2$
henry (H)	electric inductance	volt-second/ampere
hertz (Hz)	frequency	$second^{-1}$
joule (J)	work, energy, quantity of heat	newton-meter
kilogram-force	force	1 kilogram-force = 9.807 newtons
lumen (lm)	luminous flux	candela steradian
lux (lx)	illuminance	$lumen/meter^2$
newton (N)	force	$kilogram\text{-}meter/second^2$
ohm (Ω)	electric resistance	volt/ampere
pascal (Pa)	pressure	$newton/meter^2$
siemens (S)	electric conductance	ampere/volt
tesla (T)	magnetic flux density	$weber/meter^2$
volt (V)	electric potential, potential difference, electromotive force	watt/ampere
watt (W)	power, radiant flux	joule/second
weber (Wb)	magnetic flux	volt-second

CGS units and their SI equivalents

Quantity	Type	Dimension
centimeter (cm)	length	10^{-2} meter
dyne (dyn)	force	$centimeter\text{-}gram/second^2 = 10^{-5}$ newton
erg (erg)	energy	$centimeter^2\text{-}gram/second = 10^{-7}$ joule
gram (g)	mass	10^{-3} kilogram
poise (P)	absolute viscosity	gram/centimeter-second = 0.1 pascal-second
stoke (St)	kinematic viscosity	$centimeter^2/second = 10^{-4}\ meter^2/second$

Abbreviations

	British		SI	
length	foot	ft	meter	m
	inch	in		
time	second	s	second	s
mass	slug	slug	kilogram	kg
	pound-mass	lbm		
force	pound	lb	newton	N
	poundal	poundal		
pressure	pound per square foot	psf	pascal	Pa
	pound per square inch	psi		
work	foot-pound	ft-lb	joule	J
power	horsepower	hp	watt	W

In all unit systems, pressure (gage) = pressure (absolute) − pressure (atmospheric).

Prefixes

10^{-18}	atto		
10^{-15}	femto		
10^{-12}	pico	10^{+12}	tera
10^{-9}	nano		
10^{-6}	micro	10^{+6}	mega
10^{-3}	milli	10^{+3}	kilo
10^{-2}	centi	10^{+2}	hecto
10^{-1}	deci	10^{+1}	deca

Note: The use of prefixes that are not an integer multiples of 3 is not recommended practice.

Useful constants

$g = 9.806\,650 \text{ m/s}^2 = 32.174\,048 \text{ ft/s}^2$

$\pi = 3.141\,593$

$\pi^{-1} = 0.318\,310$

$e = 2.718\,282$

$\gamma = 0.577\,216$ (Euler's constant)

$\log_a x = \log_b x / \log_b a$

$\log_{10} e = 0.434\,294$

$\log_e 10 = 2.302\,585$

$\log_{10} x = \log_e x / \log_e 10 = 0.434\,294 \log_e x$

$\log_e x = \log_{10} x / \log_{10} e = 2.302\,585 \log_{10} x$

Appendix B

Fluid Properties

Surface tensions for various fluids at 1 atm and 20°C (N/M)

Fluid	Surface tension	Fluid	Surface tension
Acetic acid-air	27.8E-3	Hexane-air	18.4E-3
Acetone-air	23.7E-3	Kerosene-air	26.8E-3
Ammonia-air	21.3E-3	Mercury-air	513E-3
Benzene-air	28.85E-3	Mercury-water	392E-3
Carbon tetrachloride–air	27.0E-3	Methanol-air	22.6E-3
Ethanol-air	22.8E-3	Octane-air	21.8E-3
Ether-air	16E-3	Soap solution–air (approximate)	25E-3
Ethyl alcohol–air	22.75E-3	Toluene-air	28.5E-3
Gasoline-air	21.6E-3	Turpentine-air	28.8E-3
Glycerine-air	63E-3	Turpentine-water	11.5E-3
Glycol-air	47.7E-3	Water-air	73.05E-3

Surface tensions of water and alcohol

Temperature (°C)	Surface tension (N/M)	
	Water	Alcohol
0	7.56E-2	2.4E-2
25	7.19E-2	2.18E-2
50	6.79E-2	1.98E-2
100	5.88E-2	

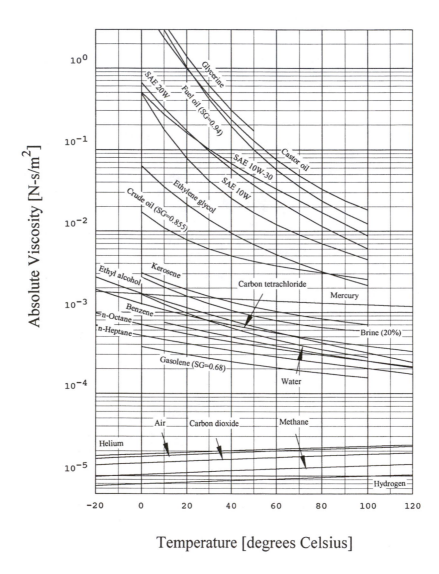

Figure B.1. Plot of absolute viscosity versus temperature in SI units.

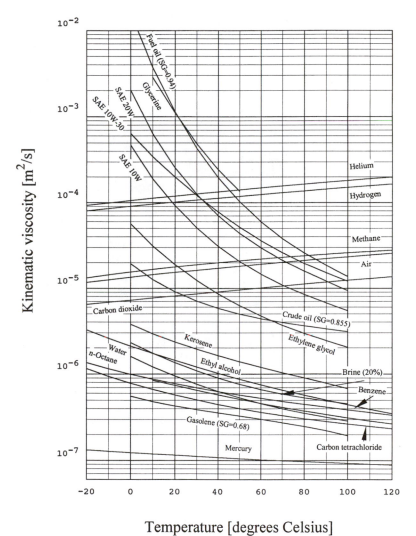

Figure B.2. Plot of kinematic viscosity versus temperature in SI units.

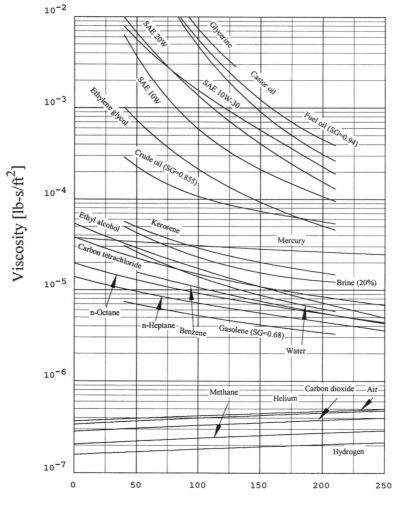

Temperature [degrees Fahrenheit]

Figure B.3. Plot of absolute viscosity versus temperature in British gravitational units.

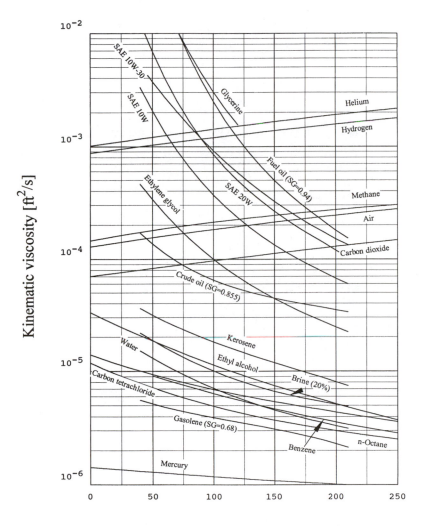

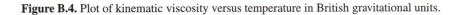

Temperature [degrees Fahrenheit]

Figure B.4. Plot of kinematic viscosity versus temperature in British gravitational units.

Contact angles for various fluids

Fluids	Contact angle α
Pure water–clean glass	≈ 0
Impure water–ordinary glass	$\approx 25°$
Pure mercury–clean glass	$140°$
Mercury–glass after exposure to air	$< 140°$
Turpentine–glass	$17°$
Kerosene–glass	$20°$

Approximate physical properties of water

T (°C)	γ (kN/m³)	ρ (kg/m³)	μ (Pa-s)	ν (m²/s)	p_{vapor} (kPa)	σ (N/m)	K (GPa)
0	9.81	1,000	1.75E-3	1.75E-6	0.611	0.0756	2.02
10	9.81	1,000	1.30E-3	1.30E-6	1.23	0.0742	2.10
20	9.79	998	1.02E-3	1.02E-6	2.34	0.0728	2.18
30	9.77	996	8.00E-4	8.03E-7	4.24	0.0712	2.25
40	9.73	992	6.51E-4	6.56E-7	7.38	0.0696	2.28
50	9.69	988	5.41E-4	5.48E-7	12.3	0.0679	2.29
60	9.65	984	4.60E-4	4.67E-7	19.9	0.0662	2.28
70	9.59	978	4.02E-4	4.11E-7	31.2	0.0644	2.25
80	9.53	971	3.50E-4	3.60E-7	47.4	0.0626	2.20
90	9.47	965	3.11E-4	3.22E-7	70.1	0.0608	2.14
100	9.40	958	2.82E-4	2.94E-7	101.3	0.0589	2.07

Approximate physical properties of some common liquids at 1 atm and 20°C

Liquid	γ (kN/m³)	ρ (kg/m³)	SG	μ (N-s/m²)	p_{vapor} (kPa)	σ (m/s)
Ammonia	8.13	829	0.83	2.20E-4	910	0.0213
Benzene	8.62	879	0.88	6.51E-4	10.1	0.0289
Carbon tetrachloride	15.57	1,588	1.59	9.67E-4	12.0	0.0270
Ethanol	7.73	7887	0.79	1.20E-3	5.75	0.0228
Gasoline	7.05	719	0.72	2.92E-4	55.1	—
Glycerine	12.34	1,258	1.26	1.49	1.4E-5	0.0633
Kerosene	8.03	819	0.82	1.92E-3	3.11	0.0277
Mercury	133.1	13,570	13.6	1.56E-3	1.1E-6	0.514
Methanol	7.73	788	0.79	5.98E-4	13.4	0.0226
SAE 10 oil	8.52	869	0.87	8.14E-2	—	0.0365
SAE 30 oil	8.71	888	0.89	4.40E-1	—	0.0350
Freshwater	9.79	998	1.00	1.02E-3	2.34	0.0728
Seawater	10.08	1,028	1.03	1.07E-3	2.34	0.0728

SAE motor oil viscosity allowable ranges

| SAE viscosity number | Viscosity range [(m²/s) × 10⁵] | | | |
| | At 0°F | | At 210°F | |
	Min	Max	Min	Max
5W crankcase	—	1,300	3.9	
10W crankcase	1,300	2,600	3.9	
20W crankcase	2,600	10,500	3.9	
20 crankcase			5.7	9.6
30 crankcase			9.6	12.9
40 crankcase			12.9	16.8
50 crankcase			16.8	22.7
75 transmission & axle		15,000		
80 transmission & axle	15,000	100,000		
90 transmission & axle			75	120
140 transmission & axle			120	200
250 transmission & axle			200	
Type A automatic transmission	39 @ 100°F	43 @ 100°F	7	8.5

Viscosity numbers ending in W (e.g., 10W-30) are classified according to their viscosities at 0°F. Numbers without a W (e.g., 20) are classified by their viscosities at 210°F. Multigrade oils such as 10W-30 are designed to minimize the variation of viscosity with temperature by the use of highly non-Newtonian additives.

Approximate physical properties of air at standard atmospheric pressure

T (°C)	γ (kN/m³)	ρ (kg/m³)	μ (Pa-s)	ν (m²/s)
0	12.7	1.29	1.72E-5	1.33E-5
10	12.2	1.25	1.77E-5	1.42E-5
20	11.8	1.2	1.81E-5	1.51E-5
30	11.4	1.16	1.86E-5	1.60E-5
40	11	1.13	1.91E-5	1.69E-5
50	10.7	1.09	1.95E-5	1.79E-5
60	10.4	1.06	1.99E-5	1.89E-5
70	10.1	1.03	2.04E-5	1.99E-5
80	9.8	1.00	2.09E-5	2.09E-5
90	9.53	0.972	2.13E-5	2.19E-5
100	9.28	0.946	2.17E-5	2.30E-5

Approximate physical properties of some common gases at 1 atm and 30°C

Gas	Mole. wt	ρ (kg/m³)	μ (N-s/m²)	R (N-m/kg-K)	c_p (N-m/kg-K)	c_v (N-m/kg-K)	k
Air	29.00	1.20	1.81E-5	287	1,003	716	1.40
Carbon dioxide	44.00	1.84	1.48E-5	189	858	670	1.30
Helium	4.00	0.166	1.97E-5	2,077	5,220	3,143	1.66
Hydrogen	2.02	0.0839	9.05E-6	4,124	14,450	10,330	1.41
Methane	16.00	0.666	1.34E-5	519	2,250	17,301	1.32
Nitrogen	28.00	1.16	1.76E-5	297	1,040	743	1.40
Oxygen	32.00	1.33	2.00E-5	260	909	649	1.40

Properties of the U.S. Standard Atmosphere

Altitude (m)	Temperature (K)	$p/p_{reference}$	$\rho/\rho_{reference}$	g (m/s²)	Viscosity (N-s/m² × 10⁵)	Sonic speed (m/s)
−500	291.4	1.0607	1.0489	9.808	1.805	342.208
0	288.2	1.0000	1.0000	9.807	1.789	340.294
500	284.9	0.9421	0.9529	9.805	1.774	338.370
1,000	281.7	0.8870	0.9075	9.804	1.758	336.435
1,500	278.4	0.8345	0.8638	9.802	1.742	334.489
2,000	275.2	0.7846	0.8217	9.801	1.726	332.532
2,500	271.9	0.7372	0.7812	9.799	1.710	330.563
3,000	268.7	0.6920	0.7423	9.797	1.694	328.583
3,500	265.4	0.6492	0.7048	9.796	1.678	326.592
4,000	262.2	0.6085	0.6689	9.794	1.661	324.589
4,500	258.9	0.5700	0.6343	9.793	1.645	322.573
5,000	255.7	0.5334	0.6012	9.791	1.628	320.545
6,000	249.2	0.4660	0.5389	9.788	1.595	316.452
7,000	242.7	0.4057	0.4817	9.785	1.561	312.306
8,000	236.2	0.3519	0.4292	9.782	1.527	308.105
9,000	229.7	0.3040	0.3813	9.779	1.493	303.848
10,000	223.3	0.2615	0.3376	9.776	1.458	299.532
11,000	216.8	0.2240	0.2978	9.773	1.422	295.154
12,000	216.7	0.1915	0.2546	9.770	1.422	295.069
13,000	216.7	0.1636	0.2176	9.767	1.422	295.069
14,000	216.7	0.1399	0.1860	9.764	1.422	295.069
15,000	216.7	0.1195	0.1590	9.760	1.422	295.069
16,000	216.7	0.1022	0.1359	9.757	1.422	295.069
17,000	216.7	0.08734	0.1162	9.754	1.422	295.069
18,000	216.7	0.07466	0.09930	9.751	1.422	295.069
19,000	216.7	0.06383	0.08489	9.748	1.422	295.069
20,000	216.7	0.05457	0.07258	9.745	1.422	295.069

Properties of the U.S. Standard Atmosphere (*continued*)

Altitude (m)	Temperature (K)	$p/p_{reference}$	$\rho/\rho_{reference}$	g (m/s^2)	Viscosity (N-s/m^2 $\times 10^5$)	Sonic speed (m/s)
22,000	218.6	0.03995	0.05266	9.739	1.432	296.377
24,000	220.6	0.02933	0.03832	9.733	1.443	297.720
26,000	222.5	0.02160	0.02797	9.727	1.454	299.056
28,000	224.5	0.01595	0.02047	9.721	1.465	300.386
30,000	226.5	0.01181	0.01503	9.715	1.475	301.709
40,000	250.4	2.834E-3	0.003262	9.684	1.601	317.189
50,000	270.7	7.874E-4	8.383E-4	9.654	1.704	329.799
60,000	255.8	2.217E-4	2.497E-4	9.624	1.629	320.606
70,000	219.7	5.448E-5	7.146E-5	9.594	1.438	297.139
80,000	180.7	1.023E-5	1.632E-5	9.564	1.216	269.44
90,000	180.7	1.622E-6	2.588E-6	9.535	1.216	269.44

$p_{reference} = 1.01325E+5$ N/m^2 abs = 14.696 psia.
$\rho_{reference} = 1.2250$ kg/m^3 = 0.002377 slug/ft^3.

References

ASHRAE, *ASHRAE Handbook of Fundamentals*, American Society of Heating, Refrigferation, and Air Conditioning Engineers, New York, 1993.

ASTM, "Viscosity-Temperature Charts for Liquid Petroleum Products," *ASTM Standard D341-77*, American Society for Testing and Materials, 1977.

Benedict, R. P., *Fundamentals of Temperature, Pressure, and Flow Measurement*, 3rd edition, Wiley, New York, 1984.

Blevins, R. D., *Applied Fluid Dynamics Handbook*, Van Nostrand Reinhold, New York, 1984.

Boltz, R., and G. L. Tuve, editors, *Handbook of Tables for Applied Engineering Science*, 2nd edition, CRC Press, Cleveland, 1979.

Eckert, E. R. G., and R. M. Drake, *Analysis of Heat and Mass Transfer*, McGraw-Hill, New York, 1972.

Hilsenrath, J., et al., "Tables of Thermodynamic and Transport Properties," *U.S. Natl. Bur. Stand. Circ. 564*, 1955; reprinted by Pergamon, New York, 1960.

Holman, J. P., *Experimental Methods for Engineers*, 5th edition, McGraw-Hill, New York, 1989.

Keenan, J. H., et al., *Steam Tables: SI Version*, 2 vols., Wiley, New York, 1985.

Meyer, C. A., et al., *ASME Steam Tables: With Mollier Chart*, 5th edition, American Society of Mechanical Engineers, New York, 1983.

Perry, J. H., *Chemical Engineers Handbook*, 4th edition, McGraw-Hill, 1963.

Raznjevic, K., *Handbook of Thermodynamic Tables and Charts*, Hemisphere, Washington, 1976.

Reid, R. C., J. M. Prausnitz, and T. K. Sherwood, *The Properties of Gases and Liquids*, 2nd edition, McGraw-Hill, New York, 1977.

SAE, "Crankcase Oil Viscosity Classification, Recommended Practice SAE J300b," *SAE Handbook*, 1976 ed., Society of Automotive Engineers, 1976.

Touloukian, Y. S., S. C. Saxena, and P. Hestermans, *Thermophysical Properties of Matter, the TPRC Data Series, Vol. 11—Viscosity*, Plenum, New York, 1975.

"The U.S. Standard Atmosphere (1962)," U.S. Government Printing Office, Washington D.C., 1962.

Vargaftik, N. B., *Tables of the Thermophysical Properties of Liquids and Gases*, 2nd edition, Hemisphere, Washington, 1975.

Viswanath, D. S., and G. Natarajan, *Data Book on the Viscosity of Liquids*, Taylor & Francis, New York, 1989.

Weast, R. C., editors, *Handbook of Chemistry and Physics*, 62nd edition, CRC Press, Boca Raton, FL, 1974.

Yaws, C. L., *Physical Properties—A Guide to the Physical, Thermodynamic, and Transport Property Data of Industrially Important Chemical Compounds*, McGraw-Hill, New York, 1977.

Yaws, C. L., *Handbook of Transport Property Data—Viscosity, Thermal Conductivity, and Diffusion Coefficients of Liquids and Gases*, Gulf Pub., Houston, 1995.

Mathematical Aids

1. Solution of Algebraic Equations—Descartes' Rule of Signs

An upper bound to the number of real positive and real negative roots of an algebraic equation $f(x) = 0$ with real coefficients can be obtained simply by counting the number of times the coefficients of the equation change sign. The number of real positive roots cannot be greater than this number. To find the number of real negative roots, replace x by $-x$ in the equation. Count the number of times the equation $f(-x) = 0$ changes sign; the number of real negative roots cannot be greater than this number.

For both positive and negative roots, if the number of roots is less than the number of sign changes, it will always be less by an even integer. When complex roots occur, they will always occur in complex conjugate pairs. In all cases, roots of multiplicity m are counted as m roots.

Example C.1.1. Descartes' rule of signs
How many real roots of the equation $f(x) = 2x^4 - 7x^3 - 3x^2 + 9x - 1 = 0$ are positive, and how many are negative?

The equation has three changes in sign of the coefficients. Thus there are at most three positive real roots. Since the coefficients of $f(-x) = 2x^4 + 7x^3 - 3x^2 - 9x - 1 = 0$ has only one change of sign, the equation $f(x) = 0$ has only one real negative root.

---

2. Cubic Equations

Several times in the text it was necessary to find the roots of cubic equations. While the formula for these roots is not as well known as that for a quadratic equation, it is very closely allied to it.

First write the equation in the form

$$f(x) = x^3 + a_1 x^2 + a_2 x + a_3 = 0 \tag{C.2.1}$$

by dividing through by the coefficient of the cubic term. Then eliminate the quadratic term by the transformation

$$x = y - a_1/3, \tag{C.2.2}$$

giving

$$f(y) = y^3 - 3Qy + 2R = 0, \tag{C.2.3}$$

where

$$Q = (a_1^2 - 3a_2)/9, \tag{C.2.4}$$

$$R = (2a_1^3 - 9a_1a_2 + 27a_3)/54. \tag{C.2.5}$$

The above transformation places $y = 0$ at the average of the roots, since the coefficient of the quadratic term is the sum of the three roots. Next, let $q = \text{sign}(Q)$, $r = \text{sign}(R)$, and scale the equation by the further transformation

$$y = r\sqrt{|Q|}\,z. \tag{C.2.6}$$

Then

$$z^3 - 3qz + 2P = 0, \tag{C.2.7}$$

where

$$P = |R|/|Q|^{3/2}. \tag{C.2.8}$$

One last transformation converts (C.2.7) into a quadratic equation. We let

$$z = t + q/t, \tag{C.2.9}$$

giving

$$t^6 + 2Pt^3 + q = 0, \tag{C.2.10}$$

which is quadratic in t^3. The roots are

$$t^3 = -(P + \sqrt{P^2 - q}), \qquad -q/(P + \sqrt{P^2 - q}). \tag{C.2.11}$$

The above is the computationally preferred form of the familiar quadratic formula. The usual plus/minus formula presented in algebra textbooks can give severe errors in one of the roots when P is large.

There are now two cases of interest, depending on whether the quantity inside of the square root sign is positive or negative.

Case 1. $P \geq 1$, $q = 1$, or $P \geq 0$, $q = -1$—one real root

The real root is then

$$z = - [P + \sqrt{P^2 - q} + q/(P + \sqrt{P^2 - q})]. \tag{C.2.12}$$

The other two roots are complex conjugates of one another, and can be found either by using (C.2.9) and (C.2.11), or by factoring (C.2.12) from (C.2.7).

Case 2. $P < 1$, $q = 1$—three real roots

Let

$$\theta = \cos^{-1} P, \tag{C.2.13}$$

so that (C.2.11) gives

$$t^3 = -r \exp(\pm i\theta).$$

The three roots are then given by

$$t_j = -r \exp \pm i\,[\theta + 2\pi(j-1)]/3, \qquad \text{where } j = 1, 2, \text{ or } 3.$$

Rewriting this solution in terms of z, we have

$$z_j = -2r \cos\{[\theta + 2\pi(j-1)]/3\}, \qquad \text{where } j = 1, 2, \text{ or } 3. \qquad (C.2.14)$$

Example C2.1. Solution of a cubic equation

Solve $x^3 - 3x^2 - 10x + 24 = 0$.

From Descartes' rule of signs, there can be no more than two real positive roots and one real negative root. First let

$$x = y + 1.$$

Then $Q = 13/3$, $q = +1$, $R = 6$, $r = +1$, and $y^3 - 13y + 12 = 0$.

Next let $y = \sqrt{3/13}z$.

Then $P = (18/3)\sqrt{3/13} = 0.66515$ and $z^3 - 3z + 2P = 0$.

Since $q = +1$ and $P < 1$, there are three real roots. First, finding $\theta = \cos^{-1} P = 0.8431$ rad from (C.2.13), the three roots are then

$$x_j = 1 - \sqrt{13/3}\,\cos\{[(\theta + 2\pi(j-1)]/3\} = -3, 4, 2.$$

---

3. Newton's Method for Finding the Roots of an Algebraic Equation

For transcendental equations, or polynomials of order higher than the cubic, exact formulas either do not exist or are too complicated to be practical. Thus a numerical solution is necessary, involving either iteration or repeated application of a simple algorithm. While there are many different iterative methods of finding the roots of an equation, Newton's method (also called "Newton-Rapheson") is probably the most frequently used because of its simplicity, its applicability to such a broad class of equations, and because its rate of convergence to a root is quite good, providing the slope of the function is not too small.

For the equation $f(x) = 0$, Newton's scheme is as follows. First guess a root x_0, using any physical understanding you might have from the problem. Then iterate according to the formula

$$x_{k+1} = x_k - f(x_k)/f'(x_k), \qquad (C.3.1)$$

where $f'(x)$ is the derivative of $f(x)$ with respect to x. This procedure amounts to constructing the local tangent to the curve $y = f(x)$ at the point x_k, and then locating where that tangent intersects the x axis. The process is repeated until a satisfactory accuracy is obtained. Once one root is obtained, other roots can be found by starting with a different starting guess x_0.

As a rule of thumb, the number of correct decimals roughly doubles with each application of Newton's formula. A useful feature of Newton's method is that it is self-correcting; that is, any errors made in the execution of the formula merely give a different point from which to draw the tangent, and do not affect the final result. If the coefficients and functions in the expression $f(x)$ are all real, and if the first guess is real, then Newton's method will find only real roots. Newton's method can be used for finding complex roots, as well as for equations with more than one independent variable, such as $f(x_1, x_2, \ldots x_N) = 0$, although the latter equation for even a quadratic equation can pose considerably more difficulties than the one-variable case does.

Example C.3.1. Newton's method

Find the real roots of $f(x) = x^3 - 9.2x^2 - 6.35x + 26 = 0$.

According to Descartes' rule of signs, since there are two changes of signs in the equation, there are either two or zero real positive roots, and one negative root. The recursive algorithm is

$$x_{k+1} = x_k - (x_k^3 - 9.2x_k^2 - 6.35x_k + 26)/(3x_k^2 - 18.4x_k - 6.35).$$

As a first guess, choose $x_0 = 1$. Then repeated use of the formula gives

$x_1 = 1.52644,$

$x_2 = 1.46915,$

$x_3 = 1.46858,$

$x_4 = 1.46858.$

We have thus converged on one of the roots. [As a check, $f(1.46858) = 1.907\text{E-}05$.]
To find the next root, start with $x_0 = 10$. Then

$x_1 = 9.6124,$

$x_2 = 9.57977,$

$x_3 = 9.57955,$

$x_4 = 9.57955.$

As a check, $f(9.57955) = -3.624\text{E-}05$.
For the negative root, start with $x_0 = -10$. Then

$x_1 = -6.1677,$

$x_2 = -3.82007,$

$x_3 = -2.52276,$

$x_4 = -1.97193,$

$x_5 = -1.85365,$

$x_6 = -1.84814,$

$x_7 = -1.84812,$

and $f(-1.84812) = 3.8147 \times 10^{-6}$.

4. Numerical Integration of Ordinary Differential Equations

Frequently a first-order ordinary differential equation (or set of equations if y and f are vectors) of the type

$$\frac{dy}{dx} = f(y, x)$$

cannot be integrated in closed form, and numerical methods must be resorted to if the solution is to be found. One of the simplest methods of doing this is **Euler's method**, where the solution is found by repeating the process

$$y(x + \Delta x) = y(x) + f[y(x), x] \, \Delta x. \tag{C.4.1}$$

This technique is easy to implement in a computer code and is readily extended to the case where f and F are arrays. It, however, is not particularly accurate, the error being of the order of Δx squared.

As an example of more accurate methods, we consider the family of solutions called **Runge-Kutta** methods. They require evaluation at intermediate points in the interval to achieve their accuracy. Two of them are listed here.

Second-order Runge-Kutta (The error is of order Δx cubed.)

$$y(x + \Delta x) = y(x) + [(1 - b)k_1 + bk_2] \, \Delta x, \tag{C.4.2}$$

with

$$k_1 = f[y(x), x], \tag{C.4.3}$$

$$k_2 = f[y(x) + pk_1, x + p], \tag{C.4.4}$$

$$p = \Delta x/2b. \tag{C.4.5}$$

The constant b can be chosen arbitrarily in the range 0 to 1. Common choices are 1/2 and 1. When b is chosen as 1 (modified Euler-Cauchy method), we have from (C.4.2) to (C.4.5)

$$y(x + \Delta x) = y(x) + k_2 \, \Delta x, \tag{C.4.6}$$

with

$$k_1 = f[y(x), x],$$

$$k_2 = f[y(x) + k_1 \, \Delta x/2, x + \Delta x/2].$$

When b is chosen as 1/2 (improved Euler-Cauchy method), (C.4.2) to (C.4.5) become

$$y(x + \Delta x) = y(x) + 0.5[k_1 + k_2] \, \Delta x, \tag{C.4.7}$$

with

$$k_1 = f[y(x), x],$$

$$k_2 = f[y(x) + k_1 \, \Delta x, x + \Delta x].$$

Both choices of b (1/2 and 1) give results with the same order of accuracy.

Fourth-order Runge-Kutta (The error is of order Δx to the fifth power.)

There are many different possibilities for a fourth-order accurate Runge-Kutta scheme. The general form is

$$y(x + \Delta x) = y(x) + \sum_{i=1}^{4} w_i k_i. \tag{C.4.8}$$

There are many different combinations of the w and k that give the same accuracy. All are easy to program. Three sets of commonly used formulas follow:

1. (Credited to Kutta.)

$$y(x + \Delta x) = y(x) + (k_1 + 2k_2 + 2k_3 + k_4)/6, \tag{C.4.9}$$

with

$$k_1 = f[(y(x), x] \, \Delta x,$$

$$k_2 = f[(y(x) + k_1 \, \Delta x/2, x + \Delta x/2] \, \Delta x,$$

$$k_3 = f[(y(x) + k_2 \, \Delta x/2, x + \Delta x/2] \, \Delta x,$$

$$k_4 = f[(y(x) + k_3 \, \Delta x, x + \Delta x] \, \Delta x.$$

2. (Credited also to Kutta.)

$$y(x + \Delta x) = y(x) + (k_1 + 3k_2 + 3k_3 + k_4)/8, \tag{C.4.10}$$

with

$$k_1 = f[(y(x), x] \, \Delta x,$$

$$k_2 = f[(y(x) + k_1 \, \Delta x/3, x + \Delta x/3] \, \Delta x,$$

$k_3 = f[(y(x) - k_1 \, \Delta x/3 + k_2 \, \Delta x, x + 2 \, \Delta x/3] \, \Delta x,$

$k_4 = f[(y(x) + k_1 \, \Delta x - k_2 \, \Delta x + k_3 \, \Delta x, x + \Delta x] \, \Delta x.$

3. (Credited to Gill.)

$$y(x + \Delta x) = y(x) + [k_1 + 2(1 - 1/\sqrt{2})k_2 + 2(1 + 1/\sqrt{2})k_3 + k_4]/6, \tag{C4.11}$$

with

$k_1 = f[y(x), x] \, \Delta x,$

$k_2 = f[y(x) + k_1 \, \Delta x/2, x + \Delta x/2] \, \Delta x,$

$k_3 = f[y(x) + (-1 + \sqrt{2})k_1 \, \Delta x/2 + (1 - \sqrt{2})k_1 \, \Delta x/2, x + \Delta x/2] \, \Delta x,$

$k_4 = F[y(x) - k_2 \, \Delta x/\sqrt{2} + (1 + 1/\sqrt{2})k_3 \, \Delta x, x + \Delta x] \, \Delta x.$

Again, all three of these choices give results with the same order of accuracy.

5. The Navier-Stokes Equations in Curvilinear Coordinates

The forms of our basic equations in curvilinear coordinates are much more complicated than in Cartesian coordinates. In the following, we restrict ourselves to orthogonal coordinate systems with coordinate axes designated by y_1, y_2, y_3. The distance between two infinitesimally distant points is given by

$$ds^2 = (h_1 \, dy_1)^2 + (h_2 \, dy_2)^2 + (h_3 \, dy_3)^2. \tag{C.5.1}$$

Then the gradient and Laplacian of a scalar quantity ϕ are given by

$$\text{grad } \phi = \nabla \phi = (\partial\phi/h_1 \, \partial y_1, \partial\phi/h_2 \, \partial y_2, \partial\phi/h_3 \, \partial y_3), \tag{C.5.2}$$

$$\nabla^2 \phi = \frac{1}{h_1 h_2 h_3} \left[\frac{\partial}{\partial y_1} \frac{(h_2 h_3 \partial \phi)}{\partial y_1 \, (h_1 \, \partial y_1)} + \frac{\partial}{\partial y_2} \frac{(h_3 h_1 \, \partial \phi)}{\partial y_2 \, (h_2 \, \partial y_2)} + \frac{\partial}{\partial y_3} \frac{(h_1 h_2 \, \partial \phi)}{\partial y_3 \, (h_3 \, \partial y_3)} \right]. \tag{C.5.3}$$

The divergence of a vector is given by

$$\text{div } \mathbf{v} = \nabla \cdot \mathbf{v} = \frac{1}{h_1 h_2 h_3} \left[\partial \frac{(h_2 h_3 v_1)}{\partial y_1} + \frac{\partial (h_3 h_1 v_2)}{\partial y_2} + \frac{\partial (h_1 h_2 v_3)}{\partial y_3} \right]. \tag{C.5.4}$$

The components of the curl of a vector are given by

$$\text{curl } \mathbf{v} = \nabla \times \mathbf{v} = \left\{ \frac{1}{h_2 h_3} \left[\frac{\partial(h_3 v_3)}{\partial y_2} - \frac{\partial (h_2 v_2)}{\partial y_3} \right], \frac{1}{h_3 h_1} \left[\frac{\partial(h_1 v_1)}{\partial y_3} - \frac{\partial(h_3 v_3)}{\partial y_1} \right], \right.$$

$$\frac{1}{h_1 h_2} \left[\frac{\partial(h_2 v_2)}{\partial y_1} - \frac{\partial(h_1 v_1)}{\partial y_2} \right] \right\} .$$

(C.5.5)

Equations (C.5.4) and (C.5.5) are useful in expressing the continuity equation and the vorticity vector in orthogonal curvilinear coordinates, and equation (C.5.3) for the continuity equation for an irrotational flow. For the rate of deformation tensor the expressions are

$$d_{11} = (1/h_1) [\partial v_1/\partial y_1 + (v_2/h_2) \, \partial h_1/\partial y_2 + (v_3/h_3) \, \partial h_1/\partial y_3],$$

$$d_{22} = (1/h_2) [\partial v_2/\partial y_2 + (v_3/h_3) \, \partial h_2/\partial y_3 + (v_1/h_1) \, \partial h_2/\partial y_1],$$

$$d_{33} = (1/h_3) [\partial v_3/\partial y_3 + (v_1/h_1) \, \partial h_3/\partial y_1 + (v_2/h_2) \, \partial h_3/\partial y_2],$$

(C.5.6)

$$2d_{12} = (h_3/h_2) \, \partial(v_3/h_3)/\partial y_2 + (h_2/h_3) \, \partial(v_2/h_2)/\partial y_3,$$

$$2d_{23} = (h_1/h_3) \, \partial(v_1/h_1)/\partial y_3 + (h_3/h_1) \, \partial(v_3/h_3)/\partial y_1,$$

$$2d_{31} = (h_2/h_1) \, \partial(v_2/h_2)/\partial y_1 + (h_1/h_2) \, \partial(v_1/h_1)/\partial y_2,$$

with the viscous stresses related to the rates of deformation by

$$\tau_{11} = \lambda \, \nabla \cdot \mathbf{v} + 2\mu d_{11}, \qquad \tau_{22} = \lambda \, \nabla \cdot \mathbf{v} + 2\mu d_{22}, \qquad \tau_{33} = \lambda \, \nabla \cdot \mathbf{v} + 2\mu d_{33},$$

$$\tau_{12} = 2\mu d_{12}, \qquad\qquad \tau_{13} = 2\mu d_{13}, \qquad\qquad \tau_{23} = 2\mu d_{23}.$$

(C.5.7)

The acceleration is arrived at by first writing the convective terms in a correct vector form. As can be verified by writing out the components in Cartesian coordinates and comparing, that form is

$$(\mathbf{v} \cdot \nabla)\mathbf{v} = -\mathbf{v} \times (\nabla \times \mathbf{v}) + 0.5\nabla \, (v^2) = -\mathbf{v} \times \text{curl } \mathbf{v} + 0.5 \text{ grad } (v^2).$$

(C.5.8)

The first term on the right is found in an arbitrary orthogonal system by applying (C.5.5), the second by use of (C.5.2).

The continuity equation is given by

$$\frac{\partial \rho}{\partial t} + \frac{1}{h_1 h_2 h_3} \left[\frac{\partial(h_2 h_3 \rho v_1)}{\partial y_1} + \frac{\partial(h_3 h_1 \rho v_2)}{\partial y2} + \frac{\partial(h_1 h_2 \rho v_3)}{\partial y_3} \right].$$

(C.5.9)

Since the Navier-Stokes equations involve taking the second derivative of a vector for the viscous terms, the procedure is much longer and tedious than for the continuity equation or Laplace's equation. The acceleration terms can be found from (C.5.8), and the pressure gradient from (C.5.2). The remaining terms involve the operation $\nabla^2 \mathbf{v}$. Writing this out is a long tedious process, and would require several pages of complicated equations. Instead, we note that

$$\nabla^2 \mathbf{v} = \nabla (\nabla \cdot \mathbf{v}) - \nabla \times (\nabla \times \mathbf{v}) \tag{C.5.10}$$

and (C.5.4) and (C.5.5) show us how to compute the divergence and curl. The Navier-Stokes equations then are of the form

$$\rho[\partial \mathbf{v}/\partial t - \mathbf{v} \times \boldsymbol{\omega} + \nabla(0.5\mathbf{v} \cdot \mathbf{v})] = -\nabla p + \rho \mathbf{g} + \mu[\nabla(\nabla \cdot \mathbf{v}) - \nabla \times (\nabla \times \mathbf{v})]. \tag{C.5.11}$$

To illustrate the previous, we next consider two familiar polar coordinate systems.

a. Cylindrical polar coordinates

The element of length in this coordinate system is

$$ds^2 = (dr)^2 + (rd\theta)^2 + (dz)^2, \tag{C.5.12}$$

with

$$(y_1, y_2, y_3) = (r, \theta, z), \tag{C.5.13}$$

and the polar coordinates are expressed in terms of the usual Cartesian coordinates by

$$r^2 = x^2 + y^2 + z^2, \quad \theta = \tan^{-1} y/x, \quad z = z. \tag{C.5.14}$$

Comparing (C.5.12) with (C.5.1), we see that

$$h_1 = h_3 = 1, \quad h_2 = y_1 = r. \tag{C.5.15}$$

We let

$$(u, v, w) = (v_r, v_\theta, v_z). \tag{C.5.16}$$

Then

$$\nabla \phi = [\partial \phi/\partial r, (1/r) \, \partial \phi/\partial \theta, \partial \phi/\partial z], \tag{C.5.17}$$

$$\nabla^2 \phi = (1/r)[(\partial/\partial r)(r\partial \phi/\partial r) + (1/r) \, \partial^2 \phi/\partial \theta^2 + r \, \partial^2 \phi/\partial z^2], \tag{C.5.18}$$

$$\nabla \cdot \mathbf{v} = (1/r)[\partial(ru)/\partial r + \partial v/\partial \theta + r \, \partial w/\partial z], \tag{C.5.19}$$

$$\nabla \times \mathbf{v} = \{(1/r)[\partial w/\partial \theta - r \, \partial v/\partial z], [\partial u/\partial z - \partial w/\partial r], (1/r)[\partial(rv)/\partial r - \partial u/\partial \theta)]\}, \tag{C.5.20}$$

$$d_{rr} = \partial u/\partial r, \quad d_\theta = (1/r)(\partial v/\partial \theta + u), \quad d_{zz} = \partial w/\partial z,$$

$$2d_{r\theta} = r \, \partial(v/r)/\partial r + (1/r)\partial u/\partial \theta), \quad 2d_{rz} = \partial u/\partial z + \partial w/\partial r,$$

$$2d_{z\theta} = (1/r)\partial w/\partial \theta + \partial v/\partial z. \tag{C.5.21}$$

The continuity equation is

$$\partial \rho/\partial t + (1/r)[\partial(r\rho u)/\partial r + \partial(\rho v)/\partial \theta + r \, \partial(\rho w)/\partial z] = 0. \tag{C.5.22}$$

The Navier-Stokes equations for constant density and viscosity are

$$\rho\left(\frac{\partial u}{\partial t} + u\frac{\partial u}{\partial r} + \frac{v}{r}\frac{\partial u}{\partial \theta} + w\frac{\partial u}{\partial z} - \frac{v^2}{r}\right) = -\frac{\partial p}{\partial r} + \rho g_r + \mu\left(\nabla^2 u - \frac{u}{r^2} - \frac{2}{r^2}\frac{\partial v}{\partial \theta}\right),$$

$$\rho\left(\frac{\partial v}{\partial t} + u\frac{\partial v}{\partial r} + \frac{v}{r}\frac{\partial v}{\partial \theta} + w\frac{\partial v}{\partial z} + \frac{uv}{r}\right) = -\frac{1}{r}\frac{\partial p}{\partial \theta} + \rho g_\theta + \mu\left(\nabla^2 v - \frac{v}{r^2} + \frac{2}{r^2}\frac{\partial u}{\partial \theta}\right),$$

$$\rho\left(\frac{\partial w}{\partial t} + u\frac{\partial w}{\partial r} + \frac{v}{r}\frac{\partial w}{\partial \theta} + w\frac{\partial w}{\partial z}\right) = -\frac{\partial p}{\partial z} + \rho g_z + \mu\,\nabla^2 w, \tag{C.5.23}$$

where the Laplacian ∇^2 is as given for a scalar in (C.5.18).

b. Spherical polar coordinates

The element of length in this coordinate system is

$$ds^2 = dR^2 + (R\,d\beta)^2 + (R\sin\beta\,d\theta)^2, \tag{C.5.24}$$

with $(y_1, y_2, y_3) = (R, \beta, \theta)$. Thus

$$h_1 = 1, \qquad h_2 = y_1 = R, \qquad h_3 = y_1\sin y_2 = R\sin\beta. \tag{C.5.25}$$

These polar coordinates are expressed in terms of Cartesian coordinates by

$$R^2 = x^2 + y^2 + z^2, \qquad \beta = \cos^{-1} z/R, \qquad \theta = \tan^{-1} y/x. \tag{C.5.26}$$

Let

$$(u, v, w) = (v_R, v_\beta, v_\theta). \tag{C.5.27}$$

Then

$$\nabla\phi = [\partial\phi/\partial R,\ (1/R)\,\partial\phi/\partial\beta,\ (1/R\sin\beta)\,\partial\phi/\partial\theta],$$

$$\nabla^2\phi = (1/R^2)[\partial(R^2\partial\phi/\partial R)/\partial R + (1/R^2\sin\beta)\,\partial(\sin\beta\,\partial\phi/\partial\beta)/\partial\beta$$

$$+ (1/R^2\sin^2\beta)\,\partial^2\phi/\partial\theta^2], \tag{C.5.28}$$

$$\nabla\cdot\mathbf{v} = (1/R^2)\,[\partial(R^2 u)/\partial R + (1/R\sin\beta)\,\partial(v\sin\beta)/\partial\beta + (1/R\sin\beta)\partial w/\partial\theta], \tag{C.5.29}$$

$$\nabla\times\mathbf{v} = \{(1/R\sin\beta)[\partial(w\sin\beta)/\partial\beta - \partial v/\partial\theta],\ (1/R\sin\beta)\,[\partial u/\partial\theta - \partial(Rw\sin\beta)/\partial R],$$

$$(1/R)[\partial(Rv)/\partial R - \partial u/\partial\beta)]\}, \tag{C.5.30}$$

$$d_{RR} = \partial u/\partial R, \qquad d_{\beta\beta} = (1/R)(\partial v/\partial\beta + u),$$

$$d_{\theta\theta} = [(1/\sin\beta)\partial w/\partial\theta + u + v\cot\beta\,]/R,$$

$$2d_{\beta\theta} = (\sin\beta/R)\,\partial(w/\sin\beta)/\partial\beta + (1/R\sin\beta)\,\partial v/\partial\theta,$$

$$2d_{\theta R} = (1/R \sin \beta) \, \partial u/\partial \theta + R \, \partial(w/R)/\partial R$$

$$2d_{R\beta} = R \, \partial(v/R)/\partial R + (1/R) \, \partial u/\partial \beta. \tag{C.5.31}$$

The continuity equation is

$$\partial \rho/\partial t + (1/R^2) \, [\partial(R^2\rho u)/\partial R + (1/R \sin \beta) \, \partial(\rho v \sin \beta)/\partial \beta + (1/R \sin \beta) \, \partial(\rho w)/\partial \theta] = 0. \tag{C.5.32}$$

The Navier-Stokes equations for constant density and viscosity are

$$\rho \left(\frac{\partial u}{\partial t} + u \frac{\partial u}{\partial R} + \frac{v}{R} \frac{\partial u}{\partial \beta} + \frac{w}{R \sin \beta} \frac{\partial u}{\partial \theta} - \frac{v^2+w^2}{R} \right)$$

$$= -\frac{\partial p}{\partial R} + \rho g_R + \mu \left(\nabla^2 u - \frac{2u}{R^2} - \frac{2}{R^2} \frac{\partial v}{\partial \beta} - \frac{2v \cot \beta}{R^2} - \frac{2}{R^2 \sin \beta} \frac{\partial w}{\partial \theta} \right),$$

$$\rho \left(\frac{\partial v}{\partial t} + u \frac{\partial v}{\partial R} + \frac{v}{R} \frac{\partial v}{\partial \beta} + \frac{w}{R \sin \beta} \frac{\partial v}{\partial \theta} + \frac{uv}{R} - \frac{w^2 \cot \beta}{R} \right)$$

$$= -\frac{1}{R} \frac{\partial p}{\partial \beta} + \rho g_\beta + \mu \left(\nabla^2 v - \frac{v}{R^2 \sin^2\beta} + \frac{2}{R^2} \frac{\partial u}{\partial \beta} - \frac{2 \cot \beta}{R^2 \sin \beta} \frac{\partial w}{\partial \theta} \right),$$

$$\rho \left(\frac{\partial w}{\partial t} + u \frac{\partial w}{\partial R} + \frac{v}{R} \frac{\partial w}{\partial \beta} + \frac{w}{R \sin \beta} \frac{\partial w}{\partial \theta} + \frac{wu}{R} + \frac{vw \cot \beta}{R} \right)$$

$$= -\frac{1}{R \sin \beta} \frac{\partial p}{\partial \theta} + \rho g_\theta + \mu \left(\nabla^2 w - \frac{w}{R^2 \sin^2\beta} + \frac{2}{R^2 \sin \beta} \frac{\partial u}{\partial \theta} + \frac{2 \cot \beta}{R^2 \sin \beta} \frac{\partial v}{\partial \theta} \right), \tag{C.5.33}$$

where the Laplacian ∇^2 is as given for a scalar in (C.5.28).

Appendix D

Compressible Flow Table for Air (k = 1.4)

Subsonic flow

M	p/p_0	ρ/ρ_0	T/T_0	A/A^*	V/c^*	M	p/p_0	ρ/ρ_0	T/T_0	A/A^*	V/c^*
0.00	1.0000	1.0000	1.0000	Infinite	0.0000	0.39	0.9004	0.9278	0.9705	1.6234	0.4209
0.01	0.9999	1.0000	1.0000	57.8738	0.0110	0.40	0.8956	0.9243	0.9690	1.5901	0.4313
0.02	0.9997	0.9998	0.9999	28.9421	0.0219	0.41	0.8907	0.9207	0.9675	1.5587	0.4418
0.03	0.9994	0.9996	0.9998	19.3005	0.0329	0.42	0.8857	0.9170	0.9659	1.5289	0.4522
0.04	0.9989	0.9992	0.9997	14.4815	0.0438	0.43	0.8807	0.9132	0.9643	1.5007	0.4626
0.05	0.9983	0.9988	0.9995	11.5914	0.0548	0.44	0.8755	0.9094	0.9627	1.4740	0.4729
0.06	0.9975	0.9982	0.9993	9.6659	0.0657	0.45	0.8703	0.9055	0.9611	1.4487	0.4833
0.07	0.9966	0.9976	0.9990	8.2915	0.0766	0.46	0.8650	0.9016	0.9594	1.4246	0.4936
0.08	0.9955	0.9968	0.9987	7.2616	0.0876	0.47	0.8596	0.8976	0.9577	1.4018	0.5038
0.09	0.9944	0.9960	0.9984	6.4613	0.0985	0.48	0.8541	0.8935	0.9559	1.3801	0.5141
0.10	0.9930	0.9950	0.9980	5.8218	0.1094	0.49	0.8486	0.8894	0.9542	1.3595	0.5243
0.11	0.9916	0.9940	0.9976	5.2992	0.1204	0.50	0.8430	0.8852	0.9524	1.3398	0.5345
0.12	0.9900	0.9928	0.9971	4.8643	0.1313	0.51	0.8374	0.8809	0.9506	1.3212	0.5447
0.13	0.9883	0.9916	0.9966	4.4969	0.1422	0.52	0.8317	0.8766	0.9487	1.3034	0.5548
0.14	0.9864	0.9903	0.9961	4.1824	0.1531	0.53	0.8259	0.8723	0.9468	1.2865	0.5649
0.15	0.9844	0.9888	0.9955	3.9103	0.1639	0.54	0.8201	0.8679	0.9449	1.2703	0.5750
0.16	0.9823	0.9873	0.9949	3.6727	0.1748	0.55	0.8142	0.8634	0.9430	1.2549	0.5851
0.17	0.9800	0.9857	0.9943	3.4635	0.1857	0.56	0.8082	0.8589	0.9410	1.2403	0.5951
0.18	0.9776	0.9840	0.9936	3.2779	0.1965	0.57	0.8022	0.8544	0.9390	1.2263	0.6051
0.19	0.9751	0.9822	0.9928	3.1123	0.2074	0.58	0.7962	0.8498	0.9370	1.2130	0.6150
0.20	0.9725	0.9803	0.9921	2.9635	0.2182	0.59	0.7901	0.8451	0.9349	1.2003	0.6249
0.21	0.9697	0.9783	0.9913	2.8293	0.2290	0.60	0.7840	0.8405	0.9328	1.1882	0.6348
0.22	0.9668	0.9762	0.9904	2.7076	0.2398	0.61	0.7778	0.8357	0.9307	1.1767	0.6447
0.23	0.9638	0.9740	0.9895	2.5968	0.2506	0.62	0.7716	0.8310	0.9286	1.1656	0.6545
0.24	0.9607	0.9718	0.9886	2.4956	0.2614	0.63	0.7654	0.8262	0.9265	1.1552	0.6643
0.25	0.9575	0.9694	0.9877	2.4027	0.2722	0.64	0.7591	0.8213	0.9243	1.1451	0.6740
0.26	0.9541	0.9670	0.9867	2.3173	0.2829	0.65	0.7528	0.8164	0.9221	1.1356	0.6837
0.27	0.9506	0.9645	0.9856	2.2385	0.2936	0.66	0.7465	0.8115	0.9199	1.1265	0.6934
0.28	0.9470	0.9619	0.9846	2.1656	0.3043	0.67	0.7401	0.8066	0.9176	1.1179	0.7031
0.29	0.9433	0.9592	0.9835	2.0979	0.3150	0.68	0.7338	0.8016	0.9153	1.1097	0.7127
0.30	0.9395	0.9564	0.9823	2.0351	0.3257	0.69	0.7274	0.7966	0.9131	1.1018	0.7223
0.31	0.9355	0.9535	0.9811	1.9765	0.3364	0.70	0.7209	0.7916	0.9107	1.0944	0.7318
0.32	0.9315	0.9506	0.9799	1.9219	0.3470	0.71	0.7145	0.7865	0.9084	1.0873	0.7413
0.33	0.9274	0.9476	0.9787	1.8707	0.3576	0.72	0.7080	0.7814	0.9061	1.0806	0.7508
0.34	0.9231	0.9445	0.9774	1.8229	0.3682	0.73	0.7016	0.7763	0.9037	1.0742	0.7602
0.35	0.9188	0.9413	0.9761	1.7780	0.3788	0.74	0.6951	0.7712	0.9013	1.0681	0.7696
0.36	0.9143	0.9380	0.9747	1.7358	0.3893	0.75	0.6886	0.7660	0.8989	1.0624	0.7789
0.37	0.9098	0.9347	0.9733	1.6961	0.3999	0.76	0.6821	0.7609	0.8964	1.0570	0.7883
0.38	0.9052	0.9313	0.9719	1.6587	0.4104	0.77	0.6756	0.7557	0.8940	1.0519	0.7975

Subsonic flow (*continued*)

M	p/p_0	ρ/ρ_0	T/T_0	A/A^*	V/c^*	M	p/p_0	ρ/ρ_0	T/T_0	A/A^*	V/c^*
0.78	0.6691	0.7505	0.8915	1.0471	0.8068	0.90	0.5913	0.6870	0.8606	1.0089	0.9146
0.79	0.6625	0.7452	0.8890	1.0425	0.8160	0.91	0.5849	0.6817	0.8579	1.0071	0.9233
0.80	0.6560	0.7400	0.8865	1.0382	0.8251	0.92	0.5785	0.6764	0.8552	1.0056	0.9320
0.81	0.6495	0.7347	0.8840	1.0342	0.8343	0.93	0.5721	0.6711	0.8525	1.0043	0.9407
0.82	0.6430	0.7295	0.8815	1.0305	0.8433	0.94	0.5658	0.6658	0.8498	1.0031	0.9493
0.83	0.6365	0.7242	0.8789	1.0270	0.8524	0.95	0.5595	0.6604	0.8471	1.0021	0.9578
0.84	0.6300	0.7189	0.8763	1.0237	0.8614	0.96	0.5532	0.6551	0.8444	1.0014	0.9663
0.85	0.6235	0.7136	0.8737	1.0207	0.8704	0.97	0.5469	0.6498	0.8416	1.0008	0.9748
0.86	0.6170	0.7083	0.8711	1.0179	0.8793	0.98	0.5407	0.6445	0.8389	1.0003	0.9833
0.87	0.6106	0.7030	0.8685	1.0153	0.8882	0.99	0.5345	0.6392	0.8361	1.0001	0.9916
0.88	0.6041	0.6977	0.8659	1.0129	0.8970	1.00	0.5283	0.6339	0.8333	1.0000	1.0000
0.89	0.5977	0.6924	0.8632	1.0108	0.9058						

Supersonic flow

M or M_{up}	Before shock					After shock				
	p/p_0	ρ/ρ_0	T/T_0	A/A^*	V/c^*	M_{down}	$\dfrac{p_{down}}{p_{up}}$	$\dfrac{\rho_{down}}{\rho_{up}}$	$\dfrac{T_{down}}{T_{up}}$	$\dfrac{p_{0\,down}}{p_{0\,up}}$
1.00	0.5283	0.6339	0.8333	1.0000	1.0000	1.0000	1.000	1.0000	1.0000	1.0000
1.01	0.5221	0.6287	0.8306	1.0001	1.0083	0.9901	1.023	1.0167	1.0066	1.0000
1.02	0.5160	0.6234	0.8278	1.0003	1.0166	0.9805	1.047	1.0334	1.0132	1.0000
1.03	0.5099	0.6181	0.8250	1.0007	1.0248	0.9712	1.071	1.0502	1.0198	1.0000
1.04	0.5039	0.6129	0.8222	1.0013	1.0330	0.9620	1.095	1.0671	1.0263	0.9999
1.05	0.4979	0.6077	0.8193	1.0020	1.0411	0.9531	1.120	1.0840	1.0328	0.9999
1.06	0.4919	0.6024	0.8165	1.0029	1.0492	0.9444	1.144	1.1009	1.0393	0.9998
1.07	0.4860	0.5972	0.8137	1.0039	1.0573	0.9360	1.169	1.1179	1.0458	0.9996
1.08	0.4800	0.5920	0.8108	1.0051	1.0653	0.9277	1.194	1.1349	1.0522	0.9994
1.09	0.4742	0.5869	0.8080	1.0064	1.0733	0.9196	1.219	1.1520	1.0586	0.9992
1.10	0.4684	0.5817	0.8052	1.0079	1.0812	0.9118	1.245	1.1691	1.0649	0.9989
1.11	0.4626	0.5766	0.8023	1.0095	1.0891	0.9041	1.271	1.1862	1.0713	0.9986
1.12	0.4568	0.5714	0.7994	1.0113	1.0970	0.8966	1.297	1.2034	1.0776	0.9982
1.13	0.4511	0.5663	0.7966	1.0132	1.1048	0.8892	1.323	1.2206	1.0840	0.9978
1.14	0.4455	0.5612	0.7937	1.0153	1.1126	0.8820	1.350	1.2378	1.0903	0.9973
1.15	0.4398	0.5562	0.7908	1.0175	1.1203	0.8750	1.376	1.2550	1.0966	0.9967
1.16	0.4343	0.5511	0.7879	1.0198	1.1280	0.8682	1.403	1.2723	1.1029	0.9961
1.17	0.4287	0.5461	0.7851	1.0222	1.1356	0.8615	1.430	1.2896	1.1092	0.9953
1.18	0.4232	0.5411	0.7822	1.0248	1.1432	0.8549	1.458	1.3069	1.1154	0.9946
1.19	0.4178	0.5361	0.7793	1.0276	1.1508	0.8485	1.485	1.3243	1.1217	0.9937
1.20	0.4124	0.5311	0.7764	1.0304	1.1583	0.8422	1.513	1.3416	1.1280	0.9928
1.21	0.4070	0.5262	0.7735	1.0334	1.1658	0.8360	1.541	1.3590	1.1343	0.9918
1.22	0.4017	0.5213	0.7706	1.0366	1.1732	0.8300	1.570	1.3764	1.1405	0.9907
1.23	0.3964	0.5164	0.7677	1.0398	1.1806	0.8241	1.598	1.3938	1.1468	0.9896
1.24	0.3912	0.5115	0.7648	1.0432	1.1879	0.8183	1.627	1.4112	1.1531	0.9884
1.25	0.3861	0.5067	0.7619	1.0468	1.1952	0.8126	1.656	1.4286	1.1594	0.9871
1.26	0.3809	0.5019	0.7590	1.0504	1.2025	0.8071	1.686	1.4460	1.1657	0.9857
1.27	0.3759	0.4971	0.7561	1.0542	1.2097	0.8016	1.715	1.4634	1.1720	0.9842
1.28	0.3708	0.4923	0.7532	1.0581	1.2169	0.7963	1.745	1.4808	1.1783	0.9827
1.29	0.3658	0.4876	0.7503	1.0621	1.2240	0.7911	1.775	1.4983	1.1846	0.9811
1.30	0.3609	0.4829	0.7474	1.0663	1.2311	0.7860	1.805	1.5157	1.1909	0.9794
1.31	0.3560	0.4782	0.7445	1.0706	1.2382	0.7809	1.835	1.5331	1.1972	0.9776
1.32	0.3512	0.4736	0.7416	1.0750	1.2452	0.7760	1.866	1.5505	1.2035	0.9758
1.33	0.3464	0.4690	0.7387	1.0796	1.2522	0.7712	1.897	1.5680	1.2099	0.9738
1.34	0.3417	0.4644	0.7358	1.0842	1.2591	0.7664	1.928	1.5854	1.2162	0.9718

Supersonic flow (*continued*)

	Before shock						After shock			
M or M_{up}	p/p_0	ρ/ρ_0	T/T_0	A/A^*	V/c^*	M_{down}	$\dfrac{p_{down}}{p_{up}}$	$\dfrac{\rho_{down}}{\rho_{up}}$	$\dfrac{T_{down}}{T_{up}}$	$\dfrac{p_{0\ down}}{p_{0\ up}}$
1.35	0.3370	0.4598	0.7329	1.0890	1.2660	0.7618	1.960	1.6028	1.2226	0.9697
1.36	0.3323	0.4553	0.7300	1.0940	1.2729	0.7572	1.991	1.6202	1.2290	0.9676
1.37	0.3277	0.4508	0.7271	1.0990	1.2797	0.7527	2.023	1.6376	1.2354	0.9653
1.38	0.3232	0.4463	0.7242	1.1042	1.2864	0.7483	2.055	1.6549	1.2418	0.9630
1.39	0.3187	0.4418	0.7213	1.1095	1.2932	0.7440	2.087	1.6723	1.2482	0.9607
1.40	0.3142	0.4374	0.7184	1.1149	1.2999	0.7397	2.120	1.6897	1.2547	0.9582
1.41	0.3098	0.4330	0.7155	1.1205	1.3065	0.7355	2.153	1.7070	1.2612	0.9557
1.42	0.3055	0.4287	0.7126	1.1262	1.3131	0.7314	2.186	1.7243	1.2676	0.9531
1.43	0.3012	0.4244	0.7097	1.1320	1.3197	0.7274	2.219	1.7416	1.2741	0.9504
1.44	0.2969	0.4201	0.7069	1.1379	1.3262	0.7235	2.253	1.7589	1.2807	0.9476
1.45	0.2927	0.4158	0.7040	1.1440	1.3327	0.7196	2.286	1.7761	1.2872	0.9448
1.46	0.2886	0.4116	0.7011	1.1501	1.3392	0.7157	2.320	1.7934	1.2938	0.9420
1.47	0.2845	0.4074	0.6982	1.1565	1.3456	0.7120	2.354	1.8106	1.3003	0.9390
1.48	0.2804	0.4032	0.6954	1.1629	1.3520	0.7083	2.389	1.8278	1.3069	0.9360
1.49	0.2764	0.3991	0.6925	1.1695	1.3583	0.7047	2.423	1.8449	1.3136	0.9329
1.50	0.2724	0.3950	0.6897	1.1762	1.3646	0.7011	2.458	1.8621	1.3202	0.9298
1.51	0.2685	0.3909	0.6868	1.1830	1.3708	0.6976	2.493	1.8792	1.3269	0.9266
1.52	0.2646	0.3869	0.6840	1.1899	1.3770	0.6941	2.529	1.8963	1.3336	0.9233
1.53	0.2608	0.3829	0.6811	1.1970	1.3832	0.6907	2.564	1.9133	1.3403	0.9200
1.54	0.2570	0.3789	0.6783	1.2042	1.3894	0.6874	2.600	1.9303	1.3470	0.9166
1.55	0.2533	0.3750	0.6754	1.2116	1.3955	0.6841	2.636	1.9473	1.3538	0.9132
1.56	0.2496	0.3710	0.6726	1.2190	1.4015	0.6809	2.673	1.9643	1.3606	0.9097
1.57	0.2459	0.3672	0.6698	1.2266	1.4075	0.6777	2.709	1.9812	1.3674	0.9062
1.58	0.2423	0.3633	0.6670	1.2344	1.4135	0.6746	2.746	1.9981	1.3742	0.9026
1.59	0.2388	0.3595	0.6642	1.2422	1.4195	0.6715	2.783	2.0149	1.3811	0.8989
1.60	0.2353	0.3557	0.6614	1.2502	1.4254	0.6684	2.820	2.0317	1.3880	0.8952
1.61	0.2318	0.3520	0.6586	1.2584	1.4313	0.6655	2.857	2.0485	1.3949	0.8915
1.62	0.2284	0.3483	0.6558	1.2666	1.4371	0.6625	2.895	2.0653	1.4018	0.8877
1.63	0.2250	0.3446	0.6530	1.2750	1.4429	0.6596	2.933	2.0820	1.4088	0.8838
1.64	0.2217	0.3409	0.6502	1.2836	1.4487	0.6568	2.971	2.0986	1.4158	0.8799
1.65	0.2184	0.3373	0.6475	1.2922	1.4544	0.6540	3.010	2.1152	1.4228	0.8760
1.66	0.2151	0.3337	0.6447	1.3010	1.4601	0.6512	3.048	2.1318	1.4299	0.8720
1.67	0.2119	0.3302	0.6419	1.3100	1.4657	0.6485	3.087	2.1484	1.4369	0.8680
1.68	0.2088	0.3266	0.6392	1.3190	1.4713	0.6458	3.126	2.1649	1.4440	0.8639
1.69	0.2057	0.3232	0.6364	1.3283	1.4769	0.6431	3.165	2.1813	1.4512	0.8599
1.70	0.2026	0.3197	0.6337	1.3376	1.4825	0.6405	3.205	2.1977	1.4583	0.8557
1.71	0.1996	0.3163	0.6310	1.3471	1.4880	0.6380	3.245	2.2141	1.4655	0.8516
1.72	0.1966	0.3129	0.6283	1.3567	1.4935	0.6355	3.285	2.2304	1.4727	0.8474
1.73	0.1936	0.3095	0.6256	1.3665	1.4989	0.6330	3.325	2.2467	1.4800	0.8431
1.74	0.1907	0.3062	0.6229	1.3764	1.5043	0.6305	3.366	2.2629	1.4873	0.8389
1.75	0.1878	0.3029	0.6202	1.3865	1.5097	0.6281	3.406	2.2791	1.4946	0.8346
1.76	0.1850	0.2996	0.6175	1.3967	1.5150	0.6257	3.447	2.2952	1.5019	0.8302
1.77	0.1822	0.2964	0.6148	1.4070	1.5203	0.6234	3.488	2.3113	1.5093	0.8259
1.78	0.1794	0.2931	0.6121	1.4175	1.5256	0.6210	3.530	2.3273	1.5167	0.8215
1.79	0.1767	0.2900	0.6095	1.4282	1.5308	0.6188	3.571	2.3433	1.5241	0.8171
1.80	0.1740	0.2868	0.6068	1.4390	1.5360	0.6165	3.613	2.3592	1.5316	0.8127
1.81	0.1714	0.2837	0.6041	1.4499	1.5411	0.6143	3.655	2.3751	1.5391	0.8082
1.82	0.1688	0.2806	0.6015	1.4610	1.5463	0.6121	3.698	2.3909	1.5466	0.8038
1.83	0.1662	0.2776	0.5989	1.4723	1.5514	0.6099	3.740	2.4067	1.5541	0.7993
1.84	0.1637	0.2745	0.5963	1.4836	1.5564	0.6078	3.783	2.4224	1.5617	0.7948
1.85	0.1612	0.2715	0.5936	1.4952	1.5614	0.6057	3.826	2.4381	1.5693	0.7902

Supersonic flow (*continued*)

M or M_{up}	p/p_0	ρ/ρ_0	T/T_0	A/A^*	V/c^*	M_{down}	$\dfrac{p_{down}}{p_{up}}$	$\dfrac{\rho_{down}}{\rho_{up}}$	$\dfrac{T_{down}}{T_{up}}$	$\dfrac{p_{0\,down}}{p_{0\,up}}$
				Before shock				After shock		
1.86	0.1587	0.2686	0.5910	1.5069	1.5664	0.6036	3.870	2.4537	1.5770	0.7857
1.87	0.1563	0.2656	0.5884	1.5187	1.5714	0.6016	3.913	2.4693	1.5847	0.7811
1.88	0.1539	0.2627	0.5859	1.5308	1.5763	0.5996	3.957	2.4848	1.5924	0.7765
1.89	0.1516	0.2598	0.5833	1.5429	1.5812	0.5976	4.001	2.5003	1.6001	0.7720
1.90	0.1492	0.2570	0.5807	1.5553	1.5861	0.5956	4.045	2.5157	1.6079	0.7674
1.91	0.1470	0.2542	0.5782	1.5677	1.5909	0.5937	4.089	2.5310	1.6157	0.7627
1.92	0.1447	0.2514	0.5756	1.5804	1.5957	0.5918	4.134	2.5463	1.6236	0.7581
1.93	0.1425	0.2486	0.5731	1.5932	1.6005	0.5899	4.179	2.5616	1.6314	0.7535
1.94	0.1403	0.2459	0.5705	1.6062	1.6052	0.5880	4.224	2.5767	1.6394	0.7488
1.95	0.1381	0.2432	0.5680	1.6193	1.6099	0.5862	4.270	2.5919	1.6473	0.7442
1.96	0.1360	0.2405	0.5655	1.6326	1.6146	0.5844	4.315	2.6069	1.6553	0.7395
1.97	0.1339	0.2378	0.5630	1.6461	1.6192	0.5826	4.361	2.6220	1.6633	0.7349
1.98	0.1318	0.2352	0.5605	1.6597	1.6239	0.5808	4.407	2.6369	1.6713	0.7302
1.99	0.1298	0.2326	0.5580	1.6735	1.6284	0.5791	4.453	2.6518	1.6794	0.7255
2.00	0.1278	0.2300	0.5556	1.6875	1.6330	0.5774	4.500	2.6667	1.6875	0.7209
2.1	0.1094	0.2058	0.5313	1.8369	1.6769	0.5613	4.978	2.8119	1.7705	0.6742
2.2	0.0935	0.1841	0.5081	2.0050	1.7179	0.5471	5.480	2.9512	1.8569	0.6281
2.3	0.0800	0.1646	0.4859	2.1931	1.7563	0.5344	6.005	3.0845	1.9468	0.5833
2.4	0.0684	0.1472	0.4647	2.4031	1.7922	0.5231	6.553	3.2119	2.0403	0.5401
2.5	0.0585	0.1317	0.4444	2.6367	1.8257	0.5130	7.125	3.3333	2.1375	0.4990
2.6	0.0501	0.1179	0.4252	2.8960	1.8571	0.5039	7.720	3.4490	2.2383	0.4601
2.7	0.0430	0.1056	0.4068	3.1830	1.8865	0.4956	8.338	3.5590	2.3429	0.4236
2.8	0.0368	0.0946	0.3894	3.5001	1.9140	0.4882	8.980	3.6636	2.4512	0.3895
2.9	0.0317	0.0849	0.3729	3.8498	1.9398	0.4814	9.645	3.7629	2.5632	0.3577
3.0	0.0272	0.0762	0.3571	4.2346	1.9640	0.4752	10.333	3.8571	2.6790	0.3283
3.1	0.0234	0.0685	0.3422	4.6573	1.9866	0.4695	11.045	3.9466	2.7986	0.3012
3.2	0.0202	0.0617	0.3281	5.1210	2.0079	0.4643	11.780	4.0315	2.9220	0.2762
3.3	0.0175	0.0555	0.3147	5.6286	2.0278	0.4596	12.538	4.1120	3.0492	0.2533
3.4	0.0151	0.0501	0.3019	6.1837	2.0466	0.4552	13.320	4.1884	3.1802	0.2322
3.5	0.0131	0.0452	0.2899	6.7896	2.0642	0.4512	14.125	4.2609	3.3151	0.2129
3.6	0.0114	0.0409	0.2784	7.4501	2.0808	0.4474	14.953	4.3296	3.4537	0.1953
3.7	9.903E–03	0.0370	0.2675	8.1691	2.0964	0.4439	15.805	4.3949	3.5962	0.1792
3.8	8.629E–03	0.0335	0.2572	8.9506	2.1111	0.4407	16.680	4.4568	3.7426	0.1645
3.9	7.532E–03	0.0304	0.2474	9.7990	2.1250	0.4377	17.578	4.5156	3.8928	0.1510
4.0	6.586E–03	0.0277	0.2381	10.7188	2.1381	0.4350	18.500	4.5714	4.0469	0.1388
4.1	5.769E–03	0.0252	0.2293	11.7147	2.1505	0.4324	19.445	4.6245	4.2048	0.1276
4.2	5.062E–03	0.0229	0.2208	12.7916	2.1622	0.4299	20.413	4.6749	4.3666	0.1173
4.3	4.449E–03	0.0209	0.2129	13.9549	2.1732	0.4277	21.405	4.7229	4.5322	0.1080
4.4	3.918E–03	0.0191	0.2053	15.2099	2.1837	0.4255	22.420	4.7685	4.7017	9.948E–02
4.5	3.455E–03	0.0174	0.1980	16.5622	2.1936	0.4236	23.458	4.8119	4.8751	9.170E–02
4.6	3.053E–03	0.0160	0.1911	18.0178	2.2030	0.4217	24.520	4.8532	5.0523	8.459E–02
4.7	2.701E–03	0.0146	0.1846	19.5828	2.2119	0.4199	25.605	4.8926	5.2334	7.809E–02
4.8	2.394E–03	0.0134	0.1783	21.2637	2.2204	0.4183	26.713	4.9301	5.4184	7.214E–02
4.9	2.126E–03	0.0123	0.1724	23.0671	2.2284	0.4167	27.845	4.9659	5.6073	6.670E–02
5.0	1.890E–03	0.0113	0.1667	25.0000	2.2361	0.4152	29.000	5.0000	5.8000	6.172E–02
5.5	1.075E–03	7.578E–03	0.1418	36.8690	2.2691	0.4090	35.125	5.1489	6.8218	4.236E–02
6.0	6.334E–04	5.194E–03	0.1220	53.1798	2.2953	0.4042	41.833	5.2683	7.9406	2.965E–02
6.5	3.855E–04	3.643E–03	0.1058	75.1343	2.3163	0.4004	49.125	5.3651	9.1564	2.115E–02
7.0	2.416E–04	2.609E–03	0.0926	104.1429	2.3333	0.3974	57.000	5.4444	10.47	1.535E–02
7.5	1.554E–04	1.904E–03	0.0816	141.8415	2.3474	0.3949	65.458	5.5102	11.88	1.133E–02
8.0	1.024E–04	1.414E–03	0.0725	190.1094	2.3591	0.3929	74.500	5.5652	13.39	8.488E–03
8.5	6.898E–05	1.066E–03	0.0647	251.0862	2.3689	0.3912	84.125	5.6117	14.99	6.449E–03
9.0	4.739E–05	8.150E–04	0.0581	327.1893	2.3772	0.3898	94.333	5.6512	16.69	4.964E–03

Supersonic flow (*continued*)

Before shock						After shock				
M or M_{up}	p/p_0	ρ/ρ_0	T/T_0	A/A^*	V/c^*	M_{down}	$\dfrac{p_{down}}{p_{up}}$	$\dfrac{\rho_{down}}{\rho_{up}}$	$\dfrac{T_{down}}{T_{up}}$	$\dfrac{p_{0\ down}}{p_{0\ up}}$
9.5	3.314E−05	6.313E−04	0.0525	421.1314	2.3843	0.3886	105.1	5.6850	18.49	3.866E−03
10.0	2.356E−05	4.948E−04	0.0476	535.9375	2.3905	0.3876	116.5	5.7143	20.39	3.045E−03
10.5	1.701E−05	3.920E−04	0.0434	674.9627	2.3958	0.3867	128.5	5.7397	22.38	2.422E−03
11.0	1.245E−05	3.137E−04	0.0397	841.9091	2.4004	0.3859	141.0	5.7619	24.47	1.945E−03
11.5	9.228E−06	2.533E−04	0.0364	1.041E+03	2.4045	0.3853	154.1	5.7814	26.66	1.576E−03
12.0	6.922E−06	2.063E−04	0.0336	1.276E+03	2.4080	0.3847	167.8	5.7987	28.94	1.287E−03
12.5	5.250E−06	1.693E−04	0.0310	1.553E+03	2.4112	0.3841	182.1	5.8140	31.33	1.059E−03
13.0	4.022E−06	1.400E−04	0.0287	1.876E+03	2.4140	0.3837	197.0	5.8276	33.80	8.771E−04
13.5	3.111E−06	1.165E−04	0.0267	2.252E+03	2.4166	0.3833	212.5	5.8398	36.38	7.315E−04
14.0	2.428E−06	9.760E−05	0.0249	2.685E+03	2.4188	0.3829	228.5	5.8507	39.05	6.138E−04
14.5	1.910E−06	8.224E−05	0.0232	3.184E+03	2.4209	0.3826	245.1	5.8606	41.83	5.180E−04
15.0	1.515E−06	6.968E−05	0.0217	3.755E+03	2.4227	0.3823	262.3	5.8696	44.69	4.395E−04
15.5	1.210E−06	5.935E−05	0.0204	4.406E+03	2.4244	0.3820	280.1	5.8777	47.66	3.748E−04
16.0	9.731E−07	5.080E−05	0.0192	5.145E+03	2.4259	0.3817	298.5	5.8851	50.72	3.212E−04
16.5	7.877E−07	4.368E−05	0.0180	5.980E+03	2.4273	0.3815	317.5	5.8918	53.88	2.765E−04
17.0	6.415E−07	3.772E−05	0.0170	6.921E+03	2.4286	0.3813	337.0	5.8980	57.14	2.390E−04
17.5	5.254E−07	3.271E−05	0.0161	7.977E+03	2.4297	0.3811	357.1	5.9036	60.49	2.074E−04
18.0	4.327E−07	2.847E−05	0.0152	9.159E+03	2.4308	0.3810	377.8	5.9088	63.94	1.807E−04
18.5	3.582E−07	2.488E−05	0.0144	1.048E+04	2.4318	0.3808	399.1	5.9136	67.49	1.580E−04
19.0	2.980E−07	2.181E−05	0.0137	1.195E+04	2.4327	0.3806	421.0	5.9180	71.14	1.386E−04
19.5	2.491E−07	1.919E−05	0.0130	1.357E+04	2.4335	0.3805	443.5	5.9221	74.88	1.221E−04
20	2.091E−07	1.694E−05	0.0123	1.538E+04	2.4343	0.3804	466.5	5.9259	78.72	1.078E−04
21	1.492E−07	1.331E−05	0.0112	1.956E+04	2.4357	0.3802	514.3	5.9327	86.69	8.478E−05
22	1.081E−07	1.057E−05	0.0102	2.461E+04	2.4369	0.3800	564.5	5.9387	95.06	6.741E−05
23	7.943E−08	8.483E−06	9.363E−03	3.065E+04	2.4380	0.3798	617.0	5.9438	103.8	5.414E−05
24	5.913E−08	6.870E−06	8.606E−03	3.783E+04	2.4389	0.3796	671.8	5.9484	112.9	4.388E−05
25	4.454E−08	5.611E−06	7.937E−03	4.631E+04	2.4398	0.3795	729.0	5.9524	122.5	3.586E−05
26	3.391E−08	4.619E−06	7.342E−03	5.624E+04	2.4405	0.3794	788.5	5.9559	132.4	2.953E−05
27	2.609E−08	3.830E−06	6.812E−03	6.781E+04	2.4411	0.3793	850.3	5.9591	142.7	2.450E−05
28	2.026E−08	3.197E−06	6.337E−03	8.121E+04	2.4417	0.3792	914.5	5.9620	153.4	2.046E−05
29	1.587E−08	2.685E−06	5.910E−03	9.666E+04	2.4422	0.3791	981.0	5.9645	164.5	1.719E−05
30	1.254E−08	2.269E−06	5.525E−03	1.144E+05	2.4427	0.3790	1050	5.9669	175.9	1.453E−05
31	9.976E−09	1.927E−06	5.176E−03	1.346E+05	2.4431	0.3790	1121	5.9689	187.8	1.235E−05
32	7.997E−09	1.646E−06	4.859E−03	1.576E+05	2.4435	0.3789	1195	5.9708	200.1	1.055E−05
33	6.454E−09	1.412E−06	4.570E−03	1.837E+05	2.4439	0.3789	1270	5.9726	212.7	9.053E−06
34	5.242E−09	1.217E−06	4.307E−03	2.131E+05	2.4442	0.3788	1349	5.9742	225.7	7.804E−06
35	4.283E−09	1.054E−06	4.065E−03	2.461E+05	2.4445	0.3788	1429	5.9756	239.1	6.757E−06
36	3.519E−09	9.157E−07	3.843E−03	2.832E+05	2.4448	0.3787	1512	5.9769	252.9	5.874E−06
37	2.907E−09	7.988E−07	3.639E−03	3.246E+05	2.4450	0.3787	1597	5.9782	267.1	5.125E−06
38	2.414E−09	6.994E−07	3.451E−03	3.707E+05	2.4453	0.3786	1685	5.9793	281.7	4.488E−06
39	2.014E−09	6.145E−07	3.277E−03	4.218E+05	2.4455	0.3786	1774	5.9803	296.7	3.944E−06
40	1.687E−09	5.417E−07	3.115E−03	4.785E+05	2.4457	0.3786	1867	5.9813	312.1	3.477E−06
41	1.420E−09	4.789E−07	2.966E−03	5.412E+05	2.4459	0.3785	1961	5.9822	327.8	3.075E−06
42	1.200E−09	4.247E−07	2.826E−03	6.102E+05	2.4460	0.3785	2058	5.9830	343.9	2.727E−06
43	1.019E−09	3.777E−07	2.697E−03	6.861E+05	2.4462	0.3785	2157	5.9838	360.5	2.425E−06
44	8.676E−10	3.368E−07	2.576E−03	7.694E+05	2.4463	0.3785	2259	5.9845	377.4	2.163E−06
45	7.416E−10	3.011E−07	2.463E−03	8.606E+05	2.4465	0.3784	2362	5.9852	394.7	1.934E−06
46	6.361E−10	2.698E−07	2.357E−03	9.603E+05	2.4466	0.3784	2469	5.9859	412.4	1.733E−06
47	5.474E−10	2.424E−07	2.258E−03	1.069E+06	2.4467	0.3784	2577	5.9864	430.5	1.557E−06
48	4.725E−10	2.182E−07	2.165E−03	1.187E+06	2.4468	0.3784	2688	5.9870	448.9	1.402E−06
49	4.091E−10	1.969E−07	2.078E−03	1.316E+06	2.4469	0.3784	2801	5.9875	467.8	1.265E−06
50	3.553E−10	1.780E−07	1.996E−03	1.455E+06	2.4470	0.3784	2917	5.9880	487.1	1.144E−06
55	1.825E−10	1.106E−07	1.650E−03	2.342E+06	2.4475	0.3783	3529	5.9901	589.1	7.110E−07

Supersonic flow (*continued*)

M or M_{up}	Before shock					After shock				
	p/p_0	ρ/ρ_0	T/T_0	A/A^*	V/c^*	M_{down}	$\dfrac{p_{down}}{p_{up}}$	$\dfrac{\rho_{down}}{\rho_{up}}$	$\dfrac{T_{down}}{T_{up}}$	$\dfrac{p_{0\ down}}{p_{0\ up}}$
60	9.936E–11	7.164E–08	1.387E–03	3.615E+06	2.4478	0.3782	4200	5.9917	700.9	4.606E–07
65	5.678E–11	4.804E–08	1.182E–03	5.391E+06	2.4480	0.3782	4929	5.9929	822.5	3.089E–07
70	3.382E–11	3.318E–08	1.019E–03	7.805E+06	2.4480	0.3782	5717	5.9939	953.7	2.134E–07
75	2.087E–11	2.350E–08	8.881E–04	1.102E+07	2.4484	0.3781	6562	5.9947	1095	1.512E–07
80	1.329E–11	1.703E–08	7.806E–04	1.521E+07	2.4485	0.3781	7467	5.9953	1245	1.095E–07
85	8.698E–12	1.258E–08	6.916E–04	2.058E+07	2.4486	0.3781	8429	5.9959	1406	8.092E–08
90	5.831E–12	9.452E–09	6.169E–04	2.739E+07	2.4487	0.3781	9450	5.9963	1576	6.082E–08
95	3.995E–12	7.214E–09	5.537E–04	3.588E+07	2.4488	0.3781	10529	5.9967	1756	4.642E–08
100	2.790E–12	5.583E–09	4.998E–04	4.637E+07	2.4489	0.3781	11667	5.9970	1945	3.593E–08

The function $F(M)$ as given by equation (9.9.3)

M	$F(M)$	M	$F(M)$	M	$F(M)$
0.40	–3.118	0.75	–0.6321	1.20	–0.8819
0.41	–2.898	0.76	–0.6255	1.22	–0.9022
0.42	–2.696	0.77	–0.6202	1.24	–0.9227
0.43	–2.512	0.78	–0.6159	1.26	–0.9432
0.44	–2.343	0.79	–0.6127	1.28	–0.9639
0.45	–2.189	0.80	–0.6104	1.30	–0.9841
0.46	–2.047	0.81	–0.6090	1.32	–1.005
0.47	–1.917	0.82	–0.6084	1.34	–1.025
0.48	–1.798	0.83	–0.6087	1.36	–1.045
0.49	–1.689	0.84	–0.6096	1.38	–1.065
0.50	–1.589	0.85	–0.6113	1.40	–1.085
0.51	–1.497	0.86	–0.6136	1.42	–1.105
0.52	–1.413	0.87	–0.6164	1.44	–1.125
0.53	–1.335	0.88	–0.6199	1.46	–1.144
0.54	–1.264	0.89	–0.6239	1.48	–1.163
0.55	–1.198	0.90	–0.6283	1.50	–1.183
0.56	–1.138	0.91	–0.6332	1.55	–1.230
0.57	–1.083	0.92	–0.6386	1.60	–1.275
0.58	–1.033	0.93	–0.6443	1.65	–1.320
0.59	–0.987	0.94	–0.6504	1.70	–1.363
0.60	–0.945	0.95	–0.6569	1.75	–1.404
0.61	–0.9060	0.96	–0.6637	1.80	–1.444
0.62	–0.8708	0.97	–0.6708	1.85	–1.483
0.63	–0.8388	0.98	–0.6782	1.90	–1.521
0.64	–0.8096	0.99	–0.6858	1.95	–1.557
0.65	–0.7832	1.00	–0.6937	2.00	–1.592
0.66	–0.7592	1.02	–0.7101	2.05	–1.625
0.67	–0.7376	1.04	–0.7273	2.10	–1.657
0.68	–0.7182	1.06	–0.7452	2.15	–1.689
0.69	–0.7008	1.08	–0.7637	2.20	–1.718
0.70	–0.6853	1.10	–0.7826	2.25	–1.747
0.71	–0.6716	1.12	–0.8019	2.30	–1.775
0.72	–0.6595	1.14	–0.8216	2.35	–1.802
0.73	–0.6490	1.16	–0.8415	2.40	–1.828
0.74	–0.6398	1.18	–0.8616	2.50	–1.876

A Brief History of Fluid Mechanics

When encountering a new subject in their education, students often get the impression that everything about the subject is already known, and in fact has always been known. Such is far from the truth. This section is designed to help you better understand how knowledge in a subject is built up over a period of time by many people in many places, and how the treatment of a subject such as fluid mechanics grows and changes with time.

The subject of mathematics has greatly influenced—and been influenced by—fluid mechanics. The invention of the calculus by Isaac Newton (1642–1727, Cambridge, England) and Gottfried Wilhelm von Leibniz (1646–1716, Germany) provided the first major tool for the codification of fluid mechanics. Newton's calculus used a superposed dot to indicate differentiation, and was closely tied to time changes. His notation has now been largely replaced by the notation and concepts developed by Leibniz, who along with improved notation also introduced the concept of infinitesimals. Leibniz also is credited with introducing the concept of *vis viva*, or living force, which when divided by two becomes our modern concept of kinetic energy.

Jean-Baptiste-Joseph Fourier's (1768–1830, France) invention of a method of solution of differential equations in the form of a series of trigonometric functions provided a major tool in the analytical implementation of solution techniques for the equations of fluid mechanics. Fourier's work was motivated by an interest in the heat developed by boring a cannon barrel. Karl Friedrich Gauss (1777–1855, Germany) showed that the solution of Laplace's equation, important in describing the behavior of flows in regions where viscosity is not important, could be represented by the superposition of sources and sinks. Jules Henri Poincaré (1854–1912, France) introduced the theory of asymptotic solutions of differential equations, of major importance in our understanding of how to find solutions that apply near the boundary layers at walls. Josiah Willard Gibbs (1839–1903, US) introduced modern vector notation that has simplified the interpretation and writing of equations describing physical phenomenon.

Calculation techniques have also been of great importance. When George Biddle Airy (1801–1892, Astronomer Royal of England) calculated what we now call the "Airy functions" (Bessel functions of order 1/3), a period of over 10 years of his time was required. By the early nineteenth century slide rules became available as well as mechanical adding machines. By the 1950s mechanical "adding" machines were developed that were able to multiply, and in some cases even divide, two numbers, the process taking several seconds. In 1955 the IBM 650, the first mass-produced computer with memory sufficient to store a program, appeared on the scene. Its use was mostly limited to the staff of computer and research centers. In the early 1970s the first pocket calculators appeared, and by the late 1970s the first desktop personal computers. By today's standards these early personal computers would seem expensive, awkward to use, and of very limited memory. But to those purchasing their first TRS 80 from Radio Shack, 64 kilobytes internal memory and 360 kilobyte floppy disks were important milestones. And hard drives for the personal computer soon made their appearance.

Today we measure computer memory in gigabytes and the capabilities of the modern home computer rival those of the mainframes of the 1970s. We have the capability of reproducing Airy's results in a matter of seconds. Such possibilities have changed what we are able to do as engineers, and are making major changes in how fluid mechanics is taught and practiced.

Early fluid mechanics was concerned with the supply of drinking and irrigation water, with transportation, and with the control of water—in other words, with much the same problems we deal with today. In Mesopotamia, Egypt, India, Pakistan, and China, records have been found dated as early as 4,000 B.C. describing wells, dams, irrigation canals, dikes, artificial conduits, bailers, crude pumps, and ships. Little is known about the engineers involved in their construction and how technological improvements were made and passed on to others, but certainly the first technology originated in prehistoric times and provided the first platform on which modern fluid mechanics is built.

The Greeks built upon existing technology, and a Greek philosopher, Archimedes (287–212 B.C.), is regarded as the founder of the subject of hydrostatics. Perhaps the most amazing part of the work of Archimedes is that it is reasoned from mathematical postulates and principles, rather than from the fuzzy philosophical principles common in his time. To this day little has been added to the subject of hydrostatics that Archimedes wouldn't understand. Besides his famous buoyancy concept, his name is also associated with a screw pump concept still in use. (A giant one can be seen at the Kinderdyke region of The Netherlands, pumping water to keep the land dry.) Other Greek inventions included fire pumps, water clocks, siphons, water organs, and steam turbines. The water organ refers to the use of water to compress air in a cylinder, the compressed air then being fed to the organ pipes to cause them to speak. The aeolian harp, consisting of strings and/or sticks placed in the wind, used vortices shed by the strings to provide pleasant sounds, as did small open vessels (Helmholtz resonators) placed in trees and later, windmills.

The Romans acquired the knowledge of the Greeks and became capable appliers of that technology. Rome, built on a small river (the Tiber) of uncertain and polluted flow, depended on eleven aqueducts to convey water from the neighboring mountains to the city. These aqueducts ended in control buildings in Rome that usually had as ornamentation elaborate fountains. The Trevi fountain is today possibly the most famous of these because of its many appearances in the cinema. Another famous aqueduct—or remnant of one—is the Pont du Gard near Nimes, France. This aqueduct, consisting of three tiers of arches, brought water from the mountains across the valley of the Gard river to the town of Nimes. (Cloth manufactured in Nimes bore the label "de Nimes," that is, "of Nimes." This gradually was corrupted to denim, the standard cloth of jeans manufacturers.) Remnants of other Roman aqueducts still exist in Africa, Italy, France, and Spain.

Besides aqueducts, the Romans provided indoor plumbing and central heating in some of their houses. Examples of this can be seen in the city ruins of Verulanium (near St. Albans, England) and Glanum (near St. Remy de Provence, France). The Romans also plumbed the springs at Bath, England, for medicinal waters. Unfortunately, for all of their construction and applied hydraulics, the Romans added little to our scientific study of fluid mechanics.

After the end of the Roman Empire in about A.D. 400, the next thousand or so years saw little scientific development of fluid mechanics. Water supplies continued to be built, and water wheels and other primitive pumping stations were developed. But indoor plumbing and heating disappeared along with the Roman Empire, not to reappear until late in the nineteenth century.

The next stage in the development of fluid mechanics was initiated by Leonardo da Vinci (1452–1519, Italy), who is now more famous as an architect and painter than as an engineer. Originally Leonardo trained as a military engineer and did design a canal in France. While his presentation of fluid resistance and the mechanisms of flight is not acceptable by today's standards, he did successfully describe the velocity distribution within a vortex, the profiles of free jets, the formation of eddies at a sudden expansion in a conduit, wake flow behind bodies, and the hydraulic jump. He also proposed the parachute, the anemometer, streamlining of bodies, the centrifugal pump, and several means of flow visualization. From a theoretical point of view, one of his most important contributions may have been the first clear statement of the principle of continuity. He did not make his work public during his lifetime, so that it had little immediate impact.

Leonardo and Galileo Galilei (1564–1642, Italy) were among the first to demonstrate the power of the approach of observation and experimentation, rather than of pure philosophical reasoning. Following them, Evangelista Torricelli (1608–1647, Italy) described the parabolic trajectories of water exiting an orifice and is considered to be the inventor of the barometer. A French prior, Edmé Mariotte (1620–1684, Burgundy, France), published studies on the impact of jets and devised crude balances for measuring the dynamic forces on vanes and surfaces.

Besides his invention of the calculus, Isaac Newton is known for his statement of the three laws of motion, although in fact all of these had been previously formulated by René Descartes (1596–1650, France) and used by others such as John Wallis (1616–1703, England), Christian Huygens (1629–1695, The Netherlands), and the famous architect Christopher Wren (1632–1723, England). Newton's contribution to mechanics was in his definitions of mass, momentum, inertia, and force, and in consolidating and clearly stating the laws of motion. He also was the first to state how viscous forces depend on velocity gradients, although at that point this work did not have strong experimental verification. He also obtained the correct form for discharge from a jet, and correctly stated that the velocity of a surface wave varied as the square root of the stream depth.

The Bernoulli family played a very important role in the founding of the subject of modern hydrodynamics. The Bernoulli family moved to Basel, Switzerland, from Antwerp, Belgium, in 1622 at a time of intense persecution of the Huguenots (Protestants). Jakob Bernoulli (1654–1705), a grandson of the original immigrant, became professor of physics and later rector of the University at Basel. He was succeeded by his brother Johann (1667–1748). Johann's son Daniel (1700–1782) wrote a treatise that introduced the term "hydrodynamics" and set forth a number of topics familiar to readers today, such as the use of openings in conduit walls for pressure taps, the kinetic theory of gases, surface elevations in accelerated and rotated containers of water, generalization of Newton's solution for water oscillating between two vessels, discussion of the establishment of flow in a conduit, origination of the concept of jet propulsion, and initiation of a discussion of the energy principle now bearing his name.

Daniel's friend Leonard Euler (1707–1783) was also a student of Johann Bernoulli. He developed the momentum equations we now refer to as the "Euler equations." By integration of his equations he arrived at what we now call the "Bernoulli equation," a relation between mechanical energy and work. He also generalized Jean Le Rond d'Alembert's (1717–1783, France) differential equation of continuity to the presently accepted form. He introduced the concepts of cavitation and centrifugal machinery. Even though blind in the last years of his life, Euler made valuable contributions to subjects such as the buckling of columns, convergence of infinite series, analytical geometry, and trigonometry, a field in which he introduced the presently used notation. He wrote the first comprehensive treatises on such fields as calculus, optics, mechanics, hydrodynamics, hydraulic machinery, and celestial mechanics. Fifty pages of his eulogy were needed just to list the titles of his writings!

The work of Euler and Bernoulli was further advanced by Joseph Louis Lagrange (1736–1813, France), who generalized the concepts we now call "Eularian" and "Lagrangian" of observing motion at a point (both concepts had in fact originally been introduced by Euler), and also introduced the concepts of stream and velocity potential functions. Pierre Simon Laplace (1749–1827, France) became a professor of mathematics at the age of eighteen and was subsequently made a marquis by Napoleon. He was given several administrative positions by Napoleon, who is said to have commented that Laplace was a mathematician of the first order who "rapidly revealed himself as a mediocre administrator; he carried into administration the spirit of the infinitely small." Nevertheless, he made contributions to our understanding of celestial mechanics, waves, the tides, and capillary forces. The formulas for stresses due to surface tension in a sphere are attributed to him, and the operator ∇^2 bearing his name is familiar to all students of advanced engineering and mathematics.

The final cornerstone of modern fluid mechanics, a differential equation form of Newton's momentum principle, was laid by Louis Marie Henri Navier (1785–1836, France), who was the first (1822) to set down what we now refer to as the Navier-Stokes equation. His analysis paralleled that of Euler, but Navier included the viscous stresses, using a complex model of molecular attraction to explain viscosity. Alternative derivations of these equations were made by Simeon Denis Poisson (1781–1840, France) in 1829[1] and Jean-Claude Barre de Saint-Venant (1797–1886, France). Saint-Venant's main interest was elasticity theory. Perhaps influenced by that he was the first of the three to also write down the expressions for the normal and shear stresses in terms of the velocity gradients, and to recognize that the equations held for both laminar and turbulent flows.

In the eighteenth century a number of engineers contributed to the advancement of experimental fluid mechanics. Henri de Pitot (1695–1771, France) developed the velocity-measuring device bearing his name, although his original device consisted of two separate tubes, one bent, one straight, rather than the concentric tubes we are now familiar with. Antoine Chezy (1718–1798, France) devised the formula for resistance in a sloped channel that is still in use. Jean Charles Borda (1733–1799, France) analyzed the orifice associated with his name, and also developed the concept that head loss was proportional to the square of the relative velocity divided by $2g$. Charles Bossut (1730–1814, France) developed the first large towing tank (approximately 100 ft long, 50 ft wide, and 7 ft deep) to investigate drag on bodies. (Benjamin Franklin had built a miniature towing tank earlier and came up with results similar to those of Bossut.) Giovanni Battista Venturi (1746–1822, Italy) demonstrated the eddies that form after a sudden expansion or contraction and showed that they could be eliminated by conical sections in the form of the flow meter we now associate with his name.

By the start of the nineteenth century the basics of inviscid flow theory had been set down and the fundamentals of applied hydraulics were off to a good start. These set the stage for the tremendous strides that were to take place. A paper published in 1800 by Charles Augustin de Coulomb (1736–1806, France) noted that in the drag acting on a body there was a term linearly dependent on velocity (due to viscous drag) as well as a term depending on velocity squared (form, or pressure, drag). This was perhaps the first clear statement of this behavior.

Coulomb's name, now used as the fundamental unit of charge, is more familiar today in the field of electricity than in mechanics. During the nineteenth century there was much fertile interaction between the fields of mechanics and electricity, with established concepts in one being used to advance knowledge in the other. The term *natural philosophy* was used to describe the combined fields of solid mechanics, fluid mechanics, and electricity.

Jean Louis Poiseuille (1799–1869, France), a physician interested in the movement of blood in veins and capillaries, showed experimentally that, when entrance effects were negligible, discharge through a tube was proportional to the viscosity times the head gradient times the tube diameter to the fourth power. Gotthilf Heinrich Ludwig Hagen (1797–1884, Germany) had predicted this earlier, but entrance effects were much larger in Hagen's experiments, and Poiseuille's results were more precise. Thus Poiseuille often is given the credit, although many still refer to laminar flow in a pipe as Hagen-Poiseuille flow. The theory governing these flows was not given for about 20 years after their experiments, when Franz Neumann (1798–1895, Germany) and

---

[1] Poisson had originally developed the equations of elasticity theory. He showed that much the same analysis applied to fluids.

Eduard Hagenbach (1833–1910, Basel, Switzerland) independently published analyses. Around 1888, Robert Manning (1816–1897, Ireland) published his famous formula for open channel flow resistance. It was identical to one proposed 20 years earlier by Phillippe Gaspard Gauckler (1826–1905, France). Manning recognized that the formula was not dimensionally proper, and also found the 2/3 power to be inconvenient for calculations. He later proposed a dimensionally homogeneous formula, which, however, received little attention. Vincenz Strouhal (1850–1922) in 1878 published a relationship between the current speed and the frequency of vortex-shedding behind a cylinder. The relationship holds for describing the sound of the wind through trees and the singing of power wires.

Drag and lift experiments were originally conducted by placing bodies on rotating arms. Otto Lillienthal (1848–1896, Germany) used such a device in 1866 to show that giving camber and body to a flat plate increased the lift. Using this, Otto and his brother Gustav constructed and flew the world's first glider. Horatio Frederick Phillips (1845–1912, England) in 1884 devised and built a small wind tunnel in 1884 and used it to arrive at much the same conclusions as Lillienthal. Frederick William Lanchester (1868–1946, England) sought to explain the lift effect in physical terms and came up with a preliminary form of modern circulation theory. He published his book *Aerodynamics* in 1907. Nicolai Egorovich Joukowski (1847–1921, Russia, also sometimes transliterated as Zhukovskii) in 1905 developed a general theorem for the side thrust on a rotating cylinder in a uniform stream with all viscous effects neglected. This, along with the results of Wilhelm Kutta (1867–1944, Germany), provided the two-dimensional Kutta-Joukowski theory still in use for the description of lift on an airfoil. While Lanchester's theory was not as rigorous as the Kutta-Joukowski theory, it had the advantage of being applicable to three-dimensional shapes. Another of Joukowski's achievements was the development of water hammer theory.

Osborne Reynolds' (1842–1912, England) many accomplishments were fundamental to the development of fluid mechanics. He was the first to demonstrate the phenomenon of cavitation, and attributed the noise that accompanies cavitation to the collapse of the vapor bubbles. He was the first to use the dimensionless number named after him to denote the dividing boundary between laminar and turbulent flow in a pipe. He used the Navier-Stokes equations to develop a theory of lubrication for plane and journal bearings, developed the concept of Reynolds stresses, and provided the framework for our present understanding of the mechanics of turbulent flows.

Towing tank tests to determine ship resistance were started by a father-and-son team, William (1810–1879, England) and Robert Edmund Froude (1846–1924, England). Their first efforts were privately funded and were derided in scathing terms by the leading naval architects of the day. Perseverance paid off, however, and the British Admiralty granted funds to build a 250-ft-long towing tank. Apparently Froude's models did not obey dimensional similitude and he did not in fact use what we now call the Froude number. He did, however, make important contributions to boundary layer research, now referred to mainly in the naval architecture literature.

The major U.S. contributors to fluid mechanics in the nineteenth century were Lester Allen Pelton (1829–1908), who developed an impulse turbine for work in the California gold fields, Clemens Herschel (1842–1930), who invented the venturi meter, and John Ripley Freeman (1855–1932), who conducted experiments on fire nozzles and the resistance of pipes, hoses, and fittings.

The final player in developing the differential momentum equations we now call the "Navier-Stokes equations" was George Gabriel Stokes (1819–1903, England). Stokes in 1845 was the first to use the Greek letter mu (μ) for the viscosity and to enunciate the no-slip theory for a fluid in contact with a solid boundary. (Saint-Venant, for instance, concluded that the tangential component of stress is zero in a direction in which there is no slipping.) Stokes' derivation of the equations did not rely on any assumptions as to how the various stress components arise, as did Navier's. His derivation is essentially the one used today.

Stokes was an interesting and prolific scientist, possibly England's greatest since Newton, and certainly one of the greatest the world had seen to that time. He was the first scientist since Newton to be granted the honors of the Lucasian professorship at Oxford University, and to be secretary and later president of the Royal Society. Besides his contribution to the derivation of the Navier-Stokes equation, he presented more than 100 papers before the Royal Society on a variety of topics. His accomplishments include a theory of low Reynolds number flow, an analysis of the second viscosity coefficient[2] for the effect of viscosity on mean normal stress, a study of waves at interfaces, and a formulation of theories for both deep-water and shallow-water small-amplitude waves.

During the time of Reynolds and Stokes, England had a number of other prominent fluid mechanicians. William Thomson—Lord Kelvin (1824–1907, Ireland-England) made numerous contributions to a broad range of technical subjects. His contributions to fluid mechanics included the analysis of irrotational flows, vortex motions, tides, open channel waves, capillary waves, and ship waves. He also introduced the word "turbulence" to the field of fluid mechanics. John William Strutt—Lord Rayleigh (1842–1919, England) contributed greatly to the study of acoustics (his book *The Theory of Sound* is in print today and still is considered a standard reference). He provided a model for the collapse of a bubble that is still used to estimate cavitation damage. He also popularized the principle of dynamic similitude, studied the instability of jets, and analyzed the onset of secondary flows between heated turbulent plates. The last was the first successful analysis of flow stability theory that included the effects of viscosity. It is known as the "Rayleigh-Bénard solution," Bénard having been the first to notice cellular structure in the drying of spermaceti. Variations of these cells are often observed in cloud formations and in the drying of metallic pigmented paints.

Horace Lamb's (1849–1934, England) lasting contributions were two books, *Hydrodynamics* and *The Dynamic Theory of Sound*, both of which are still in print. *Hydrodynamics* has gone through six editions and is considered a "must" reference for any serious student of fluid mechanics. While the majority of the book deals with

---

[2]Stokes originally decided that the second viscosity coefficient should be a function of the first, according to the relation $\mu^1 = 2\mu/3$. He later decided that this wasn't true in general. Later investigation has confirmed that for simple gases the above relation works well, but it is not correct for fluids of a more complicated molecular structure. This coefficient is particularly important for high-frequency oscillations in a fluid. Unfortunately, many books still give the simple formula $\mu^1 = 2\mu/3$ as holding for all fluids.

inviscid flow theory, the chapter on viscosity shows an understanding of the meaning of the approximate theories for low Reynolds number flows that was decades ahead of its time.

Lamb originally was named editor of what was to be a collaborative compendium of the status of developments in fluid mechanics, entitled *Modern Developments in Fluid Mechanics*. However, the work was actually done by his noted successor at the University of Manchester, Sydney Goldstein (1903–1989, England, USA). Among his many accomplishments, Goldstein demonstrated the form of the outer velocity for which similar solutions of the boundary layer equations are possible, matching conditions for higher order boundary layer solutions, and obtained a similarity solution for wake flow. Goldstein later moved to Harvard University. In between times he became rector at the Technion in Haifa. Not being one who enjoyed administrative duties, he soon decided he did not want to continue as rector and resigned. However, his resignation was rejected. He subsequently left Israel and said that he would not return until they accepted his resignation as rector. *Modern Developments in Fluid Mechanics* had a sequel some 25 years later, *Laminar Boundary Layers*, edited by L. Rosenhead.

Three more great minds of the nineteenth century deserve mention. Hermann Ludwig Ferdinand von Helmholtz (1821–1894, Germany) is certainly an ideal model for what we now call "renaissance man," or more colloquially, a "triple threat." He first taught physiology, then physics, and was also a medical doctor. His contributions to fluid mechanics include the following: a dimensional analysis of the Navier-Stokes equations along with the Froude, Reynolds, and Mach criteria for similitude; stability of interfacial waves; development of inviscid free streamline theory using conformal mapping; theory of water hammer; the role of free surfaces in cavitation; studies in acoustics, including an analysis of what is now known as the "Helmholtz resonator." His book, *On the Sensations of Tone as a Physiological Basis for the Theory of Music* is still in print today. It differs from Rayleigh's treatise, in that it deals largely with physiological acoustics, including the human formation and hearing of sound, and includes a survey of the esthetics of various musical styles. His book on ophthalmology and the eye is unfortunately out of print. It provides a rigorous background for the physics of the eye and vision that has not been much improved on to this day.

Gustav Robert Kirchhoff (1824–1887, Germany) was one of the first (as early as 1845) to grasp that irrotational flows could be superposed and that combinations of sources and sinks could be built up to solve more complex flows. He continued Helmholtz's work on free surface flows and advanced the theory of cavitation pockets behind bodies. Kirchhoff was also a student of elasticity, and his enunciation of the proper boundary condition to apply at a free boundary of a plate stands out as one of the earliest examples of a boundary layer phenomenon.

Ernst Mach (1838–1916, Germany) published the first picture of the shock wave ahead of a projectile traveling at supersonic speed. The picture was taken using a schlieren (streaks) method, now a standard technique for visualizing supersonic flows. Mach gave a theoretical explanation of this shock wave phenomenon, and showed the Mach cone to be an envelope of a series of spherical pulses. This has become the standard explanation of shock phenomena. He was the first to recognize the nature of shock waves traveling in air and to show how to generate shock waves by electric sparks. He also studied the reflection of shock waves, developed the Mach-Zehnder interferometer still widely used in aerodynamics, and gained acceptance of Christian Doppler's (1803–1853, Austria) explanation of the frequency shift of moving sound sources by conducting a series of laboratory experiments.

Thus at the start of the twentieth century the fundamentals of fluid mechanics were well established. The Navier-Stokes equations were accepted as a differential equation description of fluid flows, many of the experimental devices we use today had been established, similitude had been accepted, and the basics of hydraulics provided engineers with good working tools. Many solutions of the Euler equations had been published, most of them for irrotational flows, and a few exact solutions to the Navier-Stokes equations had been published by Stokes, albeit for flows for which the Navier-Stokes equations became linear. Yet there were still fundamental difficulties in our understanding of fluid mechanics.

These difficulties were often stated in the form of paradoxes. A formal definition of a paradox is that a plausible argument leads to conclusions that are not in agreement with experimental observations. An informal definition might be this—if you ask the wrong question, you get the wrong answer. We list a few of these paradoxes along with reasons why the wrong answer is obtained.

☞ *D'Alembert's paradox*: If irrotational flow theory is used, the drag on a stationary body in an infinite stream is zero. Two things join to give the wrong answer here. If viscosity is neglected, any effect of surface shear stress is naturally zero. Also, unless specific allowance is made in setting up the analytical problem, irrotational flow theory will not give either flow separation or a wake. Thus pressure drag is also lost.

☞ *Earnshaw's paradox*: Plane, finite-amplitude, stationary sound waves are impossible in a gas vibrating adiabatically. The problem here is that too many conditions have been imposed. The combination of adiabatic and finite amplitude is usually self-contradictory.

☞ *Eiffel paradox*: Near a critical Reynolds number of approximately 150,000 the drag on a sphere actually decreases as the velocity increases. This is a purely experimental result, and means that our expectation exceeds our understanding. The phenomenon is now usually referred to as the "drag crisis." It occurs because the flow within the boundary layer is changing from laminar to turbulent flow.

☞ *The Spin paradox*: Shooters, golfers, and baseball players knew for many years that a spinning shell drifted out of the vertical plane in which it was fired, and that the direction of the drift was in the direction of the velocity at the top of the bullet. Yet the Magnus effect (Heinrich Magnus, 1802–1870, German chemist and physicist) predicts a drift in the opposite direction. A partial explanation of this is suggested by boundary layer theory. Due to the rotation, a lift force is generated by the fact that the spin shifts the wake up and the stagnation point down, resulting in a lift force. Considering how frequently new experiments and analyses are reported in the quasi-scientific literature for baseballs, the last word has likely not been said on the matter.

☞ *Stokes paradox*: According to Stokes' approximation of the Navier-Stokes equations at very low Reynolds numbers, no flow is possible for flow past an infinitely long cylinder of any shape in an infinite stream. This is the most subtle of these paradoxes and was not satisfactorily explained until 1955. Stokes had published a linearized approximate form of the Navier-Stokes equations and used them to successfully determine the flow past a sphere in a uniform stream. Other three-dimensional shapes could also be computed, but applications to two-dimensional flows led to contradictions. (A leading early numerical analyst did in fact publish a "solution" in the 1920s, which shows that a little knowledge can sometimes be embarrassing.) In 1927 Carl Wilhelm Oseen (1879–1944, Upsala, Sweden) published a modified approximate form of the Navier-Stokes equations that linearized the convective acceleration terms, and in Lamb's later editions of *Hydrodynamics* a partial explanation is given based on Oseen's results. Nevertheless, the question as to why Stokes' equation was satisfactory for three-dimensional flows while Oseen's equation was necessary for two-dimensional flows remained a puzzle for decades, and was not satisfactorily settled until 1955.

The first major breakthrough of the new century was made by Ludwig Prandtl (1875–1953, Germany). In 1904, in an eight-page paper presented at the Third International Congress of Mathematicians, he presented his early work on what we now call "boundary layer theory." This short paper in fact contained in essence every major aspect of boundary layer theory. It was to a large degree ignored by his audience. However, the mathematician Felix Klein recognized the importance of Prandtl's contribution and appointed Prandtl to the University of Göttingen as professor and director of a small research institute. Klein did not realize it at the time, but he had not only just made possible a bold new direction in fluid mechanics research, but also in modern aviation. The outpouring of research from Prandtl and his students was closely followed throughout the scientific world, with NACA (National Center for Aviation Research, the forerunner of NASA) translating the results into English and rapidly disseminating them among the U.S. fluids mechanics scientists. In 1908 one of Prandtl's first students, Paul Richard Heinrich Blasius (1883–1970, Germany) published the solution for the flow past a flat plate using Prandtl's equations. This is the fundamental solution of boundary layer theory. In the next 5 years Blasius showed that the resistance coefficient for flow in a pipe should be a unique function of the Reynolds number, and presented the first diagram of drag coefficient versus Reynolds number for a flat plate. After decades of controversy, disputes between theory and experiment were being resolved.

Prandtl's institute produced many whose names resound in the fluid mechanics literature. A few of these follow: Johann Albert Betz (1885–1968, Germany), director of the Göttingen fluid machinery institute; Walter Gustav Johannes Tollmien (1900–1968, Germany), famous for studies in flow stability and turbulent diffusion; Jakob Ackeret (1898–1981, Germany), an authority on cavitation who later founded an aerodynamics laboratory in Zurich; Oscar Karl Gustav Tietjens (b. 1893, Germany), who turned Prandtl's lectures into book form; Hermann T. Schlichting (1907–1982, Germany), contributions to flow stability, boundary layer development, and author of a standard reference text; Johann Nikuradse (b. 1894, Germany), experiments in pipe resistance.

It is interesting to contrast Prandtl's contribution to science with that of Albert Einstein (1879–1955, Germany, U.S.), who published his paper on the special theory of relativity at about the same time (1905) as Prandtl's paper. To the world in general, Einstein is undoubtedly the more familiar name. He has been popularized in the news and entertainment media, his easily recognized portrait is found even on T-shirts and in advertisements, and if the man on the street were stopped and asked who was Einstein, the majority would at least recognize the name if not be able to expound on his theory. Yet Prandtl, who has been called the father of aviation, is a compete unknown outside of the fluid mechanics and heat transfer communities. Whose contribution has been used more by modern man?

Einstein did publish a paper on the viscosity of dilute suspensions in 1906. He used Stokes' low Reynolds number approximation and came up with a formula that is still in use. (A paper of his several years later corrected some calculational errors in the original paper.) Interestingly, another foray by Einstein into applied fields occurred in the late 1920s when he worked on refrigeration cycles, several of which he patented.

Perhaps the most influential student passing through Göttingen during Prandtl's time was Theodore von Kármán (1881–1963, Hungary, Germany, U.S.). Kármán completed his doctoral dissertation (dealing with the buckling of structures) in 1908, held several posts in Germany, and finally migrated to the U.S. in 1930. Kármán was a superb example of a scientist who could couple physical insight with mathematical analysis to get at the essence of a problem. His study of the stability of two counter-rotating lines of vortices is famous as the Kármán vortex street, and has been used to explain the oscillations of telephone and power lines in a cross wind, as well as the notorious collapse of "Galloping Gertie," the Tacoma Narrows bridge, in 1940. (Kármán was a member of the official investigating committee formed after the failure.) His other contributions include a model for turbulence, a model for the velocity profile in the boundary layer, work on gliders, Zeppelins, and monoplanes, early design of the Douglas DC-3, early design of the Northrup flying wing, impact forces on seaplane floats during landing, rockets, jet assisted takeoff, a blueprint for the then new U.S. Air Force, birth of the Jet Propulsion Laboratory, and other accomplishments far too numerous to mention here. During and before World War II, Kármán was a principle advisor to the U.S. Air Force (at that time still the Army Air Corps). Perhaps the most unusual role he played was at the end of World War II, when the military sent him to Europe to analyze the extent of German research in aerodynamics. This naturally brought him to a dramatic interview with Prandtl, whose institute had been a major research laboratory for the early Luftwaffe. (Later in the war Hitler thought that the work being done in Prandtl's institute was too far-out and

[3]Kármán met Einstein in 1911 and they became lifelong friends. Kármán recounts the story of being at a welcoming party for Einstein in Pasadena and being pulled aside by Einstein and asked to explain why a small ball dancing in the spray on the top of a fountain did not fall off. He was apparently impressed by Kármán's explanation and returned to the ceremony held in his honor, his curiosity satisfied.

long range to be of interest.) The account of the interview couldn't be more dramatic if it were fiction, and is described in Kármán's fascinating autobiography.[3]

The name of Karl Pohlhausen (1892–1980, Germany, U.S.) today is most often associated with the work he did using Kármán's integral form of the boundary layer equations in 1921. Yet that paper did much more. It gave the first clear proof that the boundary layer equations were in fact asymptotic forms of the Navier-Stokes equations; it derived Kármán's integral form of the boundary layer equations directly from Prandtl's boundary layer equations; it showed the importance of the conditions at the outer edge of the boundary layer; it introduced the concept of weighting various sublayers within the boundary layer; it introduced the simplifications possible when the external flow varied as a power of x, thus leading to the very important Falkner-Skan similarity solution of the Navier-Stokes equations; it obtained in closed form the flow in a converging channel. Thus this single paper led to decades of research in fluids mechanics as well as to important developments in aerodynamics.

Among Pohlhausen's other contributions are the extension of Prandtl's lifting line theory to biplanes and triplanes (then very much in vogue) and in showing that elliptical wing loading was optimal for form drag for finite span wings of elliptical planform. A story told concerning Pohlhausen, Prandtl, and Kármán is that when Kármán was at Göttingen in 1913 he was asked to teach the elementary course on applied mechanics. The students complained and asked for Kármán's removal on the grounds that (a) his German had such a strong Hungarian accent that effective communication was impossible, and (b) Kármán neither understand nor knew elementary mechanics. Prandtl responded that (a) was probably true and (b) was false. To correct the matter he assigned Pohlhausen the task of writing Kármán's lectures in proper German.

In 1923 Geoffrey Ingram Taylor (1886–1975, England) was made the Yarrow Research Professor at Cambridge University by the Royal Society. The position had no specified duties—not a bad job! The result was the outpouring of a steady stream of original work covering a wide variety of topics, many originating in his own mind. The collected works of G. I., as he came to be known, cover four thick volumes, three devoted mainly to fluid mechanics, and one to solid mechanics. His work includes studies of atmospheric eddies, a theory of turbulence introducing correlations that showed how viscosity dissipates energy in turbulent flows, the stability of Couette flow, a theory of the effects of moving a body in a rotating flow (the Taylor column), rotating atomizing nozzles, and countless others. Taylor's mother was the daughter of the mathematician George Boole of Boolean algebra fame, which algebra is the foundation of our modern computers, and his great-uncle was George Everest, the geodesist after whom the mountain made famous by climbers is named. G.I. spent his career at Cambridge University working with his assistant, and had no official teaching duties. Yet his work has influenced countless others through his publications, his many lectures, and his wonderful charming manner. His study of the stability of Couette flow was the first to include both viscosity and a nonstagnant primary flow. Lord Rayleigh, who had been the first to successfully include viscosity in a stability analysis, discouraged him in following this line of research. The problem Taylor was considering was more complicated mathematically than was Rayleigh's, and an exact solution was out of the question. G.I. persevered and provided an analysis that was an example of how to simplify a problem so as to include all the important details without including unnecessary ones, a model of elegance.

The study of the stability of parallel flows has a more extended history. The original equations were published by William McFarland Orr (1866–1934, Ireland) in 1906–1907. Orr was able to find an exact solution of his equation for plane Couette flow in terms of known functions (although considerable numerical effort is required to evaluate those functions), but was not able to find any unstable regions for that special case. In later attempts, the famed physicist Arnold Johannes Wilhelm Sommerfeld (1891–1951, Germany) independently derived the equations in 1908, but was not able to improve on Orr's results. In 1924, another famous physicist, Werner Karl Heisenberg (1901–1976, Germany), attempted a complicated asymptotic analysis that unfortunately ran into difficulties and led to an erroneous result. In 1929 Walter G. J. Tollmien (1900–1968, Germany) did a better job of treating the inviscid solution and developed a stability curve for boundary layer flows that was later confirmed by experiment. Hermann T. Schlichting (1907–1982, Germany) in 1933 and 1935 also calculated boundary layer stability curves that, however, had a poorer agreement with experiments than did Tollmien's results. In 1944 Chia Chiao Lin (b. 1916, China, U.S.) reproduced and extended Tollmien's results. In the process he clarified the nature of the behavior of the solution in the region where the disturbance wave speed equals the speed of the primary flow. In 1947 Galen B. Schubauer (1904–1995, U.S.) and Harold Kenneth Skramstad (b. 1908, U.S.) at the National Bureau of Standards completed a series of complex experiments on the effects of disturbances in the boundary layer that confirmed Lin's and Tollmien's analyses.

In the U.S. in the early twentieth century, Edgar Buckingham (1867–1940, U.S.) stimulated usage of dimensionless variables by the introduction of his famous pi theorem. Lewis Ferry Moody (1880–1953, U.S.) correlated pipe resistance data into his now ubiquitous diagram, and as well introduced a cavitation parameter for hydraulic machinery and improved draft-tube design.

By the 1950s many centers of excellence in fluid mechanics were developing throughout the world, and the number of engineers/scientists engaged in research and teaching of fluid mechanics was growing rapidly. The Space programs in the U.S. and the U.S.S.R. rapidly advanced the funding available for research. Along with the growing interest in fluid mechanics came new tools. Matched asymptotic theory was used to further clarify the asymptotic nature of boundary layer theory. It showed the regions of applicability of the Euler and Navier-Stokes equations at large Reynolds numbers, and the Stokes and Oseen approximations at small Reynolds numbers. It also showed how higher order approximations could be obtained when necessary, as well as the limits of asymptotic solutions near stagnation points and leading edges. The rising availability and sophistication of computers gradually changed the solution methodology for fluids problems. Prior to the computer, computations of practical flows almost always involved long and tedious algebraic manipulations. With the computer, the classical solutions could be used as starting points to build on, and complicated design situations became readily amenable. The singularity solutions of Gauss and Green could now be readily used to solve flows for such complex geometries as the Challenger space shuttle perched atop a Boeing 747. Flows that were in some regions supersonic and others subsonic could now be computed for the first time, and shock waves could be located. Details of compressible flow boundary layers with their triple-deck structure could be analyzed, as well as the surface wave structures due to moving ships, and the flows in the oceans and atmospheres.

Why then are you studying fluid mechanics? Hasn't it all been done? Why not just buy a software package and let the computer do it? Well, as they say, not quite. There is still more to be done and more understanding and research still needed. It is safe to say that at the end of the twenty-first century more things will known, and even more questions will have been raised. With weather predictions beyond 3 days unreliable and severe storm prediction extremely limited, weather forecasting can be much improved. Meteorology and oceanography have been important driving forces in fluid mechanics research and in showing the need for ever more, bigger, and faster supercomputers. In the process of studying weather, subjects such as chaos theory and the butterfly effect have been born. Other new areas of technological interest, from ink jet printers to the cleaning of silicon wafers for the manufacture of semiconductors to flows in plants and animals, will continue to arise and attract our attention. Turbulence theory, long the bane of researchers, has made substantial progress in the twentieth century but still has more unknowns than knowns.

In the proverbs of Solomon, the bible states that "There be three things which are too wonderful for me, yea four of which I know not." It goes on to say that two of these are "The way of an eagle in the air," a clear reference to aerodynamics, and "The way of a ship in the midst of the sea," a reference to naval architecture. While our understanding of these two topics has increased over the years, every year new applications arise that challenge our understanding and tell us we need to know more. Turbulence is a prime example of this. In life, Saint Peter, the key keeper at the gates of heaven, was a commercial fisherman and hence is considered the patron saint of fishermen. As a simple fisherman, Peter was familiar with the concept of turbulence and no doubt by now has reduced it to a simple theory. When they meet St. Peter, many fluid mechanicians will agree with their colleague who said that the first thing he will ask Saint Peter is, "Explain turbulence to me!"

References on the History of Fluid Mechanics

Birkhoff, G., *Hydrodynamics, A Study in Logic, Fact, and Similitude*, Princeton University Press, Princeton, NJ, 1950.

Daintith, J., S. Mitchell, and E. Tootill, editors, *Biographical Encyclopedia of Scientists*, Facts on File, New York, 1981.

Garbrecht, G., *Hydraulics and Hydraulic Research: An Historical Review*, Gower Pub., Aldershot, United Kingdom, 1987.

Gillispie, C. C., editor in chief, *Dictionary of Scientific Biography*, Scribner, New York, 1980.

Hart, I. B., *The Mechanical inventions of Leonardo da Vinci*, Chapman & Hall, London, 1925.

MacKendrick, P., *Roman France*, St. Martin's, New York, 1972.

Morton, H. V., *The Fountains of Rome*, Macmillan, New York, 1966.

Rouse, H., and S. Ince, *History of Hydraulics*, Iowa State University, Ames, 1957; Dover, New York, 1963.

Rouse, H., *Hydraulics in the United States 1776–1976*, Iowa Institute of Hydraulic Research, Iowa City, 1976.

Van Dyke, M., J. V. Wehausen, and J. L. Lumley, *Annual Reviews of Fluid Mechanics*, vols. 1 and following, Annual Reviews Inc., Palo Alto, CA, 1969—. The first article in each volume usually is of historical interest.

von Kármán, T., and L. Edson, *The Wind and Beyond: Theodore von Kármán Pioneer in Aviation and Pathfinder in Space*, Little, Brown, Boston, 1967.

Design of a Pump System

In the main body of this text we have concentrated on the general theory of fluid mechanics and its application to the analysis of given situations. The design process utilizes these analytic methods, but also is concerned with broader issues. In this appendix we look at a class of piping systems to see the design process and how broad design issues are met.

The basic general requirements of fluid and ventilation systems are set in terms of delivery requirements, which usually are the flow rates and pressures at various points in the system. To make these requirements more specific, the engineer first develops a set of system diagrams. These diagrams give a schematic representation of the system and establish all system characteristics. These characteristics include applicable operating specifications and data necessary for the ordering of all system components. They are developed very early in the preliminary design stage to allow early ordering of components requiring long delivery times, and to define what should be included in the general arrangement drawings, which will be developed at a later time. These diagrams reflect the applicable specifications of the system owner, insurance underwriters, engineering societies, and applicable local, state, and federal codes.

1. System Diagrams

The following describes typical contents of system diagrams. The sequence indicates the approximate order in which design decisions and selections are made.

1. A schematic arrangement drawing is made that indicates the operational features of the system and shows what components are included. It should show methods for isolation of the various portions of the system to allow for maintenance or failure of a component, crossovers, and system redundancies that allow flexibility in the use of the system.
2. Flow rates and design pressures should be shown at each junction point in the system. If temperature is important in the system or is necessary for operation specifications, it should also be indicated at the required points.

3. Engineering specifications should be given for all pumps. These should include the following: head, flow rate, physical size, and speed. Motor specifications should include power, speed, and frame size.

4. Sizing of all piping should be included. Pipe diameters are set by the flow requirements and the maximum velocities that are acceptable to avoid erosion. Table F.1 shows some recommended and limiting velocities for various situations. These values can be used as a starting point for the design, but the diameters of the pipe may have to be increased if pressure drop or velocity in a given pipe is excessive. If the flow involves slurries, or if there are heavy solid particles in the flow, it is necessary to avoid low-velocity regions, since heavy particles can settle out and eventually cause blockage of the pipe. It is also necessary to determine the wall thickness, termed the *pipe schedule*. This is set by the pressure requirements in the line to ensure that there is no danger of the pipe bursting. Stress calculations, including effects of thermal stresses, should be made to determine needed wall thickness.

5. The type, location, and orientation of all valves is specified. Valve type is determined by the considerations given in Table F.2.

6. The type and the location of all special components is next indicated on the system diagrams. This list includes strainers and filters. In some cases strainers are placed before filters to prevent large particles from clogging the filter and shortening its life. Filter cartridges should be specified to indicate the largest particle size that is allowed to pass through the filter. Cartridge load buildup with age, which decreases the size of particles that can pass through the filter, must also be allowed for.

7. The type and the location of all instrumentation and control is shown. Instrumentation can be broadly considered as being of either the monitoring or control type. It may be appropriate to install some instrumentation fittings during construction so that portable instrumentation may be attached to monitor conditions during periodic testing and tuning of the system. Some of the types of measurements that may be made for monitoring and control follow.

a. *Pressure*. Made by gages for visual monitoring, or by transducers for remote control or monitoring, to record pressure change across components such as pumps and filters. The gages could indicate positive pressure, differential pressure, or vacuum pressure.

b. *Fluid levels*. Typically manometers are used for visual monitoring. Float switches, electrical resistance, and capacitance gages are used for remote monitoring and control.

c. *Flow rates*. The choice of a flow meter depends on whether an electrical output signal is desired, or whether a visual method can be used. The back pressure or fouling potential of the meter is also a consideration.

d. *Temperature*. Moderate temperatures can be measured with a thermometer if visual monitoring is sufficient. Capillary tube instruments may also be used for low-temperature measurements and control. Thermocouples or resistive devices are used for high-temperature measurements, or when remote reading is indicated.

e. *Salinity cells*. For installations using saltwater for cooling it may be necessary to monitor any leakage of saltwater into other parts of the system. This can be done

Table F.1. Fluid design velocities

Service	Fluid velocity (ft/s)	
	Nominal[a]	Maximum
Condensate pump suction	$\sqrt{d}$	3
Condensate pump discharge	$3\sqrt{d}$	8
Condensate drains	$0.3\sqrt{d}$	1
Hot-water suction	$\sqrt{d}$	3
Hot-water discharge	$3\sqrt{d}$	8
Feedwater suction	$1.3\sqrt{d}$	4
Feedwater discharge	$4\sqrt{d}$	10
Cold freshwater suction	$3\sqrt{d}$	15
Cold freshwater discharge	$3\sqrt{d}$	20
Lube-oil service pump suction	$5\sqrt{d}$	4
Lube-oil discharge	$\sqrt{d}$	6
Fuel-oil service suction	$2\sqrt{d}$	4
Fuel-oil service discharge	$1.5\sqrt{d}$	6
Fuel-oil transfer suction	$\sqrt{d}$	6
Fuel-oil transfer discharge	$2\sqrt{d}$	15
Diesel-oil suction	$2\sqrt{d}$	7
Diesel-oil discharge	$5\sqrt{d}$	12
Hydraulic-oil suction	$1.5\sqrt{d}$	8
Hydraulic-oil discharge	$8\sqrt{d}$	20
Seawater suction	$3\sqrt{d}$	15[b]
Seawater discharge	$5\sqrt{d}$	15[b]
Steam, high pressure	$50\sqrt{d}$	200
Steam exhaust, 215 psig	$75\sqrt{d}$	250
Steam exhaust, high vacuum	$75\sqrt{d}$	330

[a]d is the pipe internal diameter in inches.
[b]ft/s for galvanized steel pipe.

Table F.2. Valve-type selection

Function	Type of valve
Isolation capability for maintenance or failure of a component or section of the system	Usually a butterfly, gate, or ball valve
Coarse flow control, with coarse adjustment capability	Globe valve
Fine flow control, with fine adjustment capability	Needle valve
Reverse flow protection	Check valve
Pressure relief protection	Relief valve
Vacuum relief protection	Vacuum breakers or check valves
Pressure reduction	Reducing valves or orifices
Flow reduction (fixed amount)	Orifices
Vents and drains	Usually fixed after general arrangements are completed

with a conductivity meter. Other chemical factors such as pH may also require monitoring.

8. Sizing and materials should be specified for all components. This includes the piping, fittings, tanks, pits, heaters, cooling towers, etc. The fabrication specifications for welding, screwed fittings, flanged fittings, bend radii, etc., should also be included. Special instructions such as slope, orientation of components, etc., should also be noted on the drawings.

9. A testing procedure should be developed along with the development of the system diagram. These tests should include hydrostatic tests, acceptance tests, operations tests, emergency, casualty, operational, and safety tests and procedures.

2. Arrangement Drawings

Arrangement drawings are made after the physical plant characteristics have been decided upon and selection of the site for the piping system has been made. These drawings are usually made long after the long lead-time components have been ordered. The actual physical arrangement and position of each part of the system should be shown on these drawings.

These drawings are used for the actual fabrication and installation of the system. Some of the items that should be considered during this development period follow:

1. Minimum life cycle costs. This is not necessarily the same as minimum initial cost.
2. Personnel access to valves, instrumentation, sensing points, and equipment during system operation.
3. Personnel access to equipment for maintenance. This should allow space for equipment needed for the removal of partial or entire sections of the system,

including the removal of tube bundles in heat exchangers, repair of valve seats, pump replacement, etc.

4. Pipe material and path. The path of the pipe circuit should be laid out in such a manner that the following requirements are met: flow into system components is smooth to avoid erosion; straight sections of sufficient length are allowed upstream of instrumentation; cavitation is avoided by ensuring that pressures are elevated above vapor pressure.

5. All instrumentation should be clearly visible to operators and maintenance personnel.

6. Adequate drainage of the system should be provided for.

7. Consideration should be made for access for welding and installation of components during erection of the system, and also for inspecting all or part of the system during the acceptance phase.

8. Adequate anchors, hangers, and mountings for all piping and components should be provided for.

9. Sufficient flexibility should be built into the system to allow for thermal expansion.

For complicated systems it may be appropriate to construct a model of the system. This model is built from the system diagrams and the arrangement drawings, and could be either a physical model or a computer model. The fabrication and detail drawings would then be made by scaling the model. This method eliminates possible interference problems in the field and allows the construction personnel to see what the final installation should look like before construction is started on any portion of any part of the system.

3.   System Pressure Loss Calculations

One of the most important parts of sizing a piping system is the calculation of the pressure drop in the system due to viscous friction. Fluid properties such as viscosity and density should be determined for the expected temperature range.

An outline of the methodology used in sizing a piping system and selecting the pump follows.

1. For a required flow select a pipe diameter that will give velocities within the acceptable ranges indicated in Table F.1.

2. Determine the pressure losses that occur in the various pipes and fittings by the methods of Chapter 7. Adjust pipe diameters in any part of the system where pressure losses are excessive.

3. Check the pressure differential across components to determine if they are acceptable. Certain components are adversely effected by incorrect pressure differentials. As an example, a filter or strainer must have a minimum pressure differential in order to have flow through the cartridge or strainer. It also may have a maximum allowable pressure differential to avoid blowing out of the cartridge. Relief and

check valves may also require a minimum pressure differential to prevent chatter or to ensure sealing.

4. Develop a head-discharge curve for the system.

5. Select a pump by matching the head-discharge characteristics of the system with that of the pump. Check that the system and pump Reynolds numbers are not too disparate if you are using fluids other than water. If they are, consult with the pump manufacturer to find their recommendation. For fluids other than water, consult with the manufacturer to ensure that the pump case, impeller material, bearings, and all seals and gaskets are chemically compatible with your liquid. This consultation is a must!

6. Check the effects of possible off-design operating conditions on the system. Include the effects of pipe aging and the buildup of deposits on pipe walls.

Some Suggestions for Design Problems

Students are often asked to work on a design problem in the beginning fluids course, or perhaps in a later design course. Your instructor may have some explicit suggestions based on the instructor's own interests, or on available facilities. In the event that you select your own problem, the following are some suggested areas for you to consider. Many of them have attracted some of the best engineers and scientists in the world. Some progress has been made in some of the areas, but much remains to be done. Fresh approaches from fresh minds are needed. It is unlikely that you will be able to give a complete solution to any of them in the time available to you, so you should isolate a portion of the problem that interests you. It's your turn now!

In the production-line filling of bottles with liquids (soft drinks, beer, shampoo, liquid soap, milk, etc.), a problem that often arises is the foaming of the contents as they enter the container. Such foaming causes spilling of the liquid, resulting in cleanup and material costs. One solution of course is to insert just a little liquid, then let it rest until the foam has receded, then repeat the process. This is a costly solution. The most desirable solution would be to avoid the foaming in the first place, but yet to fill at a reasonably fast pace. Suggest solutions, and verify with lab-scale tests.

Ink jet printers for computers are of two kinds, drop-on-demand and continuous jet. In both cases the ink is directed to the paper by first giving the ink an electrical charge and then passing it between parallel plates. The electric field between the plates is adjusted to steer the drop to the right place, much as electrons are steered in the picture-tube of a television set or oscilloscope. In the drop-on-demand printers, a reservoir containing an orifice is pulsed to eject a drop when needed. In the continuous jet technology, a thin jet exits the orifice, then breaks up into droplets due to jet instability (see the book by H. Lamb referenced in Chapter 12). The droplets are then directed either to the paper or to a reservoir, as needed. The reservoir ink is filtered and then recycled. Both of these technologies have their usage. The continuous jet can be economically used when the printing rate needs are high, such as a central bank that processes many checks hourly for return to the client banks. The drop-on-demand system is more economical for the home computer. The ink is a high-cost item, since its viscosity and surface tension must be closely controlled. This usually requires that its temperature be controlled. Also, it must be kept free from contaminants, which usually means filtration. The orifice must be very small and absolutely free of imperfections at its sharp edge to ensure that the drop separates cleanly from the nozzle. All this contributes to the cost, complexity, and maintenance of the system. (The orifice and ink system can be the smallest—and most expensive—component in the printer.) Design a water jet system that models such ink jet technology, and test the system reusing the water over a period of time.

Jets of liquid at very high velocities have been found useful for cutting many materials. For instance the technique has been tried in Russian coal mines, where it proved much superior to mechanical cutting, in that no coal dust was introduced into the air, and it was not as wasteful of materials as other cutting methods. The procedure generally used is to pressurize the water to very high pressures, so that as it exits a nozzle the water leaves at velocities about 25–50% that of the sonic speed of water (1,000 m/s). Suggest ways as to how to achieve this high pressure to attain these velocities. (**Warning:** Should you decide to perform experiments with such jets, be *extremely* careful in dealing with them, since they will also cut soft tissue very efficiently. Tales abound about people who try to find the location of leaks in the exhaust manifolds of diesel engines by running hands along the manifold. This leads to fingers and hands of reduced size!)

The branch of fluid mechanics called *fluidics* (see reference in Chapter 12) uses simple fluid elements to construct computers using fluids rather than electronics. Simple amplifiers and logical blocks can be constructed and used as building blocks to control the direction of fluid jets (used in the steering of rocket engines), count items on a production line, or perform many other similar duties. Air and water are the fluids most commonly used. They have the advantage of providing an output power that can perform physical tasks—an electronic computer would require an interface to amplify the power. Show how you can use these units to build oscillators, digital displays, units for counting the output of a production line, or similar tasks.

Oil spills are a form of environmental pollution that has received a great deal of media attention, yet there are no completely satisfactory way of dealing with them. Absorbing the oil requires later disposal of the absorbent, often done by burning. Barriers work only if the oil slick is relatively thin and the current is slow. (Some work by the author suggests that barriers fail catastrophically as the Froude number based on barrier depth and current speed approaches 1.) In high currents the barriers turn a slick into highly dispersed droplets, making recovery hopeless. Burning of a slick is usually impossible and undesirable, considering the resulting environmental pollution, since the volatile portion of the oil rapidly evaporates and the residue is almost impossible to ignite. (In the breakup of the tanker Torrey Castle off the southwest coast of England, the Royal Air Force bombed the slick in a futile attempt to ignite it.) Suggest ways of dealing with oil slicks, both in channels containing runoff water from parking lots, and on streams and lakes. The same technique will likely not work in all situations. Perform small benchtop experiments.

A problem associated with the preceding one is the separation of immiscible fluids such as oil and water. Filters are a possibility, but they are usually limited to very low flow rates. Settling tanks are another possibility, but they are usually of very large physical size. In many applications (e.g., the bilge water on ocean-going ships) neither of these is practical. Suggest and test alternative methods.

Chimneys have been used for hundreds of years to exhaust heated—and frequently contaminated—gases into the atmosphere. The traditional argument was that "dilution is a solution to pollution." While that phrase has a glib sound to it, the usual result was putting the pollutant in someone else's backyard. Complaints were often met by making the chimney higher, resulting in that "someone" being someone farther from the source of the pollution. Several technical design problems arise from this problem. Are there alternatives to chimneys? How does one monitor the pollutants as they exit the chimney? How does one monitor the pollutant plume downstream from the chimney? Can such a plume be modeled in a wind tunnel? Can such a plume be monitored on a computer?

An aftermath of the Cold War is the disposal of chemical and nuclear weapons. The consequences of exploding these in the open air are too horrendous to be contemplated. Burning at extremely high temperatures is being tried, but the resulting exhaust gases are of questionable safety. Suggest alternative solutions.

Plating plants and painting shops have one thing in common—both exhaust undesirable contaminants. The acid-laden air from a plating process can have dire effects on the neighborhood (and the neighbors), and the volatiles from traditional hydrocarbon-base paints can violate EPA rules. Automobile manufacturers have shifted to water-base paints to solve this problem. Can you suggest alternative solutions for dealing with particle-laden air?

In many regions of the world water is in short supply, especially for agriculture in arid regions. One technology that has been successfully utilized of irrigation is a drip system, whereby the water is delivered directly to the desired plant by small-diameter flexible tubing. These systems regulate the pressure with orifices, and control flow rates by the use of small diameters and small exit orifices. Drip systems avoid evaporation losses such as those associated with sprays. A difficulty with a drip system, however, is the build-up of mineral deposits at the system exits. This is particularly a problem where water is "hard," and when the system is used only for several hours a week. Develop ways for making such systems self-cleaning.

Electricity-generating plants require vast amounts of water for condenser-cooling purposes. Often rivers or lakes are used as a water source and heat sink. A problem associated with this is that the water intake naturally has a current that results in fish and debris being sucked into the intake pipe and also the pumps. This is damaging to the pumps, not to say what it does to the fish! Much effort has been devoted to screens, fences, and even workers with nets and rakes, but the problem remains largely unsolved. It is no small problem and expense for both the utility or the consumer. Suggest better ways to deal with the problem.

Nuclear power plants must protect against failure of the system, for if the rods are stuck in the wrong position the normal cooling system cannot handle the thermal transfer. The standard solution is to have a large holding tank that can be flooded rapidly. These tanks usually have high-speed pumps in the intake lines, the pumps being raised above the reservoir floor. This requires that the pumps have the capacity to raise the water to the pump level. While this is not always a problem, it has been noticed that if a small vortex forms (much like that which forms at the center of a washbasin when you drain it) at the lip of the intake pipe, it can be sucked into the pump. This can cause the pump to stop pumping, defeating its reason for being. Perform simple laboratory experiments with transparent tubing and a small pump to see how such vortex formation can be suppressed.

It is frequently necessary to mix dry chemicals with liquids. While this could be done in tanks with stirring mechanisms, it is less expensive to mix it "on-the-fly" by injecting the powder into a flowing stream of water. This is an example of an inductor, and is used in many applications: by dentists to remove saliva from a patient's mouth, to spray weed killers onto a lawn, to pass interior air past sensors in automobile thermostatically controlled air conditioning systems, and others. Construct such a system, and examine both analytically and experimentally the capabilities of your system.

Wildfires are unfortunately of frequent occurrence in the West and elsewhere during dry summers, resulting in the loss of homes. This can be exacerbated by roofing materials such as cedar shingles that dry out rapidly and become highly flammable. Preinstallation treatment can avoid this to some extent, but experience has shown that in a few years such treatment can become valueless. Some inventors have found water-absorbent chemicals that if sprayed on roofs can retain large amounts of water. To be convenient, and thus widely used, they must be capable of being sprayed onto the roof. Two problems exist for such systems. (1) If water from the taps is to be used to spray the chemical, there must be a means of getting the dry chemical into the water stream in a metered amount. (See previous suggestion.) (2) If tap water cannot be used for some reason, a separate pump must be used. This pump must be inexpensive enough to keep the system economical, yet provide sufficient head to get a very viscous—and likely non-Newtonian—liquid at least 20 ft into the air. How would you design such a pump?

Home plumbing systems are often an undesirable source of noise. There is noise associated with water running in pipes in the wall, and flow through faucets and flush valves. Risers in the plumbing system guard against the most severe water hammer, but they can become water-filled over time and therefore useless. Determine the source of noise in valves and piping systems, and test alternate designs that might alleviate this.

Mixing of hot and cold water in the shower can be a difficult procedure, since slight changes in the system pressure can have pronounced effects on flow rates in the cold line. Design a valve system whereby temperatures can be set accurately and water temperature controlled within comfortable limits. The mixing valve should provide a temperature change linearly proportional to how much the valve was turned.

The use of water fountains has been a popular way of enhancing outdoor surroundings since Roman times. Typical fountains shoot continuous jets of water. Design a pump that will shoot "packets" of water several inches long and still follow the trajectory of a continuous jet.

Design a support for traffic signs that will minimize the motion of a sign in a strong wind. (On a windy day, stop sign supports can be observed to twist and sway due to vortex shedding past the sign. The amplitude of this motion is enhanced if the shedding frequency is near a natural frequency of the support.)

Repeat the previous suggestion for suspended traffic lights.

Design a wind-powered land vehicle. (*Hint*: Sails on sailboats are nothing more than flexible airfoils. The vehicle could use a fixed airfoil for propulsion, and a small engine for low-wind situations.)

Design a wind anemometer whose output is linearly proportional to wind speeds. Most propellers will turn at rates that vary nonlinearly with wind speeds. A proper propeller shape, perhaps a flat blade like a popsicle stick, may provide a more linear response.

Design a means for feeding granular material (gravel, stones, earth, dust, flour, etc.) through a funnel without clogging or otherwise blocking the flow. (The flow of solids takes on many of the characteristics of fluid flow, but has its own unique challenges.)

Glossary

Acceleration, advective acceleration, convective acceleration, local acceleration, temporal acceleration   Acceleration in fluid mechanics normally uses an Eularian description. The temporal, or local, acceleration is the time-derivative term, and the advective, or convective, acceleration is the derivative following the fluid particle.

Accumulator   A device, typically a gas-filled bag or air column, used to absorb pressure pulses.

Admissible roughness   The roughness height equal to the laminar sublayer thickness.

Air lift pump   Pumping action is induced by the rise of gas bubbles introduced at the column bottom.

Autocorrelation   A correlation between a function and itself.

Bernoulli equation, Bernoulli terms   The mechanical terms (pressure, elevation, and kinetic energy heads) in a work-energy formulation of Newton's law.

Blade element analysis   Analysis of flow through the rotor of a centrifugal machine.

Boundary layer   The layer of fluid near a solid boundary (sometimes also in the interior of a flow) in which the flow properties such as velocity change rapidly across the layer.

Body forces   Forces that act throughout a volume, such as gravity.

British gravitational system   The system of units commonly used in the U.S. Britain has largely converted to SI units.

Bucket (pitcher) pump   A pump with a handle that when pressed down lowers the pressure below a plunger, pulling water up through a one-way valve.

Buffer zone   A region in the turbulent boundary layer between the laminar sublayer and the logarithmic zones.

Bulbous bow   A rounded nose put on the bow of a ship below the waterline to produce a canceling effect on the ship's bow wave.

Bulk modulus   A measure of the compressibility of a substance, defined as the amount of incremental pressure needed to produce an incremental change on the density.

Buoyancy force   A body force equal to the weight of fluid displaced by the body.

641

Capillary tubes   Very-small-diameter tubes in which the flow of a fluid normally would be laminar.

Cavitation   Local reduction of the pressure in a liquid to vapor pressure. At that point the fluid vaporizes.

Center of pressure   The location of the resultant force due to a pressure distribution.

Coanda effect   The naturally occurring deflection of a jet toward an adjacent wall at moderate Reynolds numbers.

Constitutive equation   The relationship between the stress and the geometric quantities such as rate of deformation and strain. Examples are Newton's law of viscosity and Hooke's law.

Continuum   The modeling of a substance as a continuous material, with no molecules, atoms, and the like taken into consideration.

Control volume, control surface   A user-selected volume enclosed by a surface used in the analysis of flowing substances such as fluids.

Correlation   A number that shows the degree of relationship between quantities.

Couette flow   Flow between parallel plates or concentric cylinders caused by movement of the boundaries.

Critical, subcritical, and supercritical flow   Classifications for open channel flow based on the Froude number being equal to, less than, or greater than one.

Critical point   The maximum combination of pressure, density, and temperature at which the liquid and vapor states can coexist.

Deadhead conditions   A pump operating at zero discharge.

Density, mass density   The mass per unit volume of a substance. This is the inverse of specific volume.

Dimensionless parameter   A combination of quantities that has no physical dimensions.

Discharge   The amount of flow through an area per unit time.

Dispersive waves   Waves whose wave speed depends on the wavelength. Such waves change shape as they propagate.

Displacement thickness   A measurement of the thickness of the boundary layer based on continuity considerations.

Downstream   In the direction of flow of the current or stream.

Draft tube   The pipe at the exit of a turbine.

Drag coefficient   A dimensionless ratio of drag forces to momentum forces.

Drag crisis A narrow region of Reynolds numbers for which the drag coefficient suddenly decreases before increasing again. Occurs because of the shift of the boundary layer from laminar flow to turbulent flow.

Eddy viscosity An approximation to the turbulent Reynolds stresses, modeling them in the form where they are proportional to the viscous stresses.

Energy grade line A plot of pressure plus elevation plus kinetic energy heads along a pipeline.

Entropy A measure of the irreversibility of a process or flow.

Euler equations A differential equation formulation of the momentum equations that includes pressure and gravity forces but not viscous forces.

Euler number One-half the pressure coefficient.

Eulerian description A description of the kinematics of a flow where one follows a fluid particle for an infinitesimal amount of time. This is the most commonly used description in fluid mechanics.

Fick's law A constitutive relationship between velocity and pressure gradient for flow in a porous media.

Field quantities Quantities such as velocity, pressure, density, temperature that can vary from point to point in a flow field.

Fluidics Mechanical devices that operate with only the fluid moving. Control of the motion is accomplished with no other moving mechanical parts.

Flume An open channel used for studying flows or carrying water to hydraulic machinery.

Form (shape) drag The pressure drag due to body shape, often strongly affected by the flow in the wake.

Francis turbine A reaction turbine, where the moment is due to the momentum caused by the fluid striking the vanes in the rotor.

Free surface Usually the surface between a liquid and a gas (e.g., water and air), but also could refer to an interface between two liquids.

Froude number A dimensionless ratio of the momentum forces to the gravity force.

Gear (lobe) pump A positive displacement pump where the fluid is forced between two gears, a rotating lobe and a wall, etc., to produce a pumping action.

Headrace The channel leading from a reservoir to a penstock tube for delivery to a turbine.

Hydraulic diameter An approximation to pipe diameter for noncircular cross sections.

Hydraulic grade line A plot of pressure and elevation heads along a pipeline.

Hydraulic ram   An impulse pump that relies on elevation difference for pumping power.

Hydrostatic   Relating to forces due solely to the weight of the fluid.

Ideal flow, inviscid flow, frictionless flow   Flows in which viscosity plays a negligible role.

Incompressible flow   Flows in which the density of a fluid particle does not change appreciably as the particle moves through the flow.

Internal energy   Energy due to the molecular motion of the fluid. Internal energy changes in a liquid are usually negligible.

Irrotational flow, potential flow   Flow in regions where the vorticity is zero.

Isobars   Lines of constant pressure.

Jet pump   A pump that is basically a converging nozzle, as in a venturi meter.

Kaplan turbine   An impulse turbine that operates effectively at a lower head than does a Francis turbine.

Kinematic viscosity   The absolute viscosity of a fluid divided by its mass density.

Kurtosis   A mathematical quantity that measures the peakedness of a set of data.

Lagrangian description   A description of the motion that follows individual particles in the flow.

Laminar flow   Flow in a well-defined orderly manner.

Lapse rate   The rate of change of temperature with elevation in the earth's atmosphere.

Lift coefficient   The dimensionless ratio of lift forces to momentum forces.

Mach number   The dimensionless ratio of momentum forces to compressibility forces.

Maneuvering basin   A basin of water used for testing the maneuverability of ship models. The models are frequently radio-controlled.

Manometer   A transparent tube used for measuring pressure differences.

Mass discharge   The mass per unit time passing through a surface.

Material derivative, substantial derivative   The time derivative used for obtaining acceleration from velocity when using an Eulerian description of the flow.

Metacenter, metacentric height   The point determined by a line drawn vertically from the center of buoyancy of a body to an originally vertical line drawn through the center of mass of the body.

Minor losses   Losses in a piping system due to fittings.

Mixing length   A model for turbulence originally based on an analogy to molecular physics for the interaction of molecules.

Moment coefficient   A dimensionless ratio of moments to moment of momentum.

Momentum flow rate   The mass rate of flow times the velocity.

Momentum thickness   A measure of boundary layer thickness based on momentum considerations.

Moody diagram   A plot of pipe friction factor versus Reynolds number for various pipe roughnesses.

Navier–Stokes equations   A differential equation formulation of the momentum equations that includes pressure, gravity, and viscous forces.

Newtonian fluids   Fluids that obey Newton's law of viscosity, where stress is proportional to rate of deformation of the fluid.

Non-Newtonian fluids   Fluids that do not obey Newton's law of viscosity.

Nonuniform flow   Flow in which the velocity varies over an area.

No-slip boundary condition   The adherence of a fluid to a solid boundary, so that the fluid velocity equals the boundary velocity.

NPSH (net positive suction pressure)   The pressure needed at the inlet of a centrifugal pump to ensure that the flow does not cavitate inside the pump.

Path line   The path followed by an individual fluid particle.

Pelton turbine   A simple impulse turbine using the momentum of a jet to produce a moment.

Penstock tubes   The pipe connecting a turbine to the headrace.

Poiseuille flow   Flow in a tube or pipe caused by a pressure gradient or gravity.

Pressure, absolute pressure, gage pressure   The force per unit area normal to a surface. Gage pressure is a pressure difference.

Pressure coefficient   The dimensionless ratio of pressure forces to momentum forces.

Pressure prism   A graphical concept of the pressure distribution acting on a plane surface.

Rates of deformation   The normal rates of deformation are the rates of change per unit length of distances parallel to the various axes. The shear rates of deformation are the angular rates of change in planes containing the various axes.

Repeating variables   The variables that appear in more than one dimensionless parameter when performing a dimensional analysis.

Reynolds number   The dimensionless ratio of momentum forces to viscous forces.

Reynolds stresses   Terms that arise from the convective acceleration when separating velocities into mean and disturbance quantities.

Rigid-body motion   Motion where there is no relative motion between adjacent fluid particles, so that the fluid acts as if it were a solid body.

Rotational flow   Flow with vorticity.

Scalars   Quantities that at a point do not depend on direction. Also called zero-order tensors.

Screw pump   A positive displacement pump originally designed by Archimedes for lifting water.

Separation, separation point   A point in a boundary layer where reverse flow occurs, causing sudden thickening of the boundary layer.

Shaft work   The work done by or on a fluid system, usually by mechanical devices such as pumps and turbines.

SI (Le Système International d'Unités)   The system of units used throughout most of the world.

Similitude   Matching of geometry (geometrical similitude) and forces (dynamic similitude) in model studies.

Skewness   A mathematical quantity that measures the departure from symmetry of a set of data.

Skin friction   The force due to viscous stresses acting on a body.

Slider bearing   A bearing where one face moves with respect to the other. Frequently the surfaces have a slight incline with respect to each other.

Slope adverse, critical, horizontal, mild, steep   Comparison criteria for open channel flow depending on the steepness and sign of the slope of the channel bottom.

Sonic speed, speed of sound   The speed at which a sound wave (pressure wave) travels in a quiescent fluid.

Specific energy   The energy per unit mass.

Specific gravity   The specific weight of a fluid with respect to water (for liquids) or air (for gases).

Specific weight   The weight of a fluid per unit volume.

Stagnation point   A point where the velocity is zero.

Stall   A point of sudden drop in the lift force as the angle of attack is increased. Caused by separation of the boundary layer.

Statically equivalent   The forces and moments of the given system are replaced by a single concentrated force such that the sum of forces and moments are the same for both systems.

Steady flow   Flow that does not vary with time.

Streak line   A line caused by introduction of dye at a given point in the flow.

Streamline   A line to which the velocity vector is tangent at every point on the line.

Stream surface, stream tube   A surface made up of a bundle of streamlines.

Stress   Force per unit area.

Strouhal number   The dimensionless ratio of unsteady forces to momentum forces.

Surface forces   Forces acting on the surface of a region.

Surface tension   The force acting at the interface of two or more fluids due to the attraction between molecules in the two fluids.

Surfactants   Chemicals that change the surface tension of a liquid.

System, surroundings   A system is a selected region containing the flow of interest. The surroundings are the entire exterior surrounding the system.

Tailrace   The pipe or channel leading from the draft tube at the exit of a turbine to the river or channel bed.

Tensors   Quantities whose components transform with a change of coordinates according to a specific law. Scalars (e.g., temperature, density, pressure) are zero-order tensors, vectors (e.g., force, velocity) are first-order tensors. Examples of second-order tensors are stress, strain, rate of deformation, and moments of inertia.

Towing basin or tank   A testing facility where a model is towed either on the surface of water or submerged to determine drag and other forces.

Tsunamis   Waves on the ocean surface caused usually by underwater earthquakes or landslides.

Turbulent flow, turbulent bursts   Nonorderly, chaotic flow. Turbulent bursts are intermittent creations of turbulence in the boundary layer.

Turbulent fluctuations   The difference between the actual velocity and its mean value at a point in the flow.

Turbulent intensity   The autocorrelation of the turbulent velocity components.

Uniform flow   Flow that does not vary with position over a given area.

Unsteady flow   Flow that varies with time.

Upstream   In the direction of the source of the flow.

Vapor pressure   The pressure at a given temperature at which a liquid becomes vapor.

Vectors   A first-order tensor. Components transform according to the parallelogram rule. Examples are force and velocity.

Viscosity, absolute viscosity, kinetic viscosity   The proportionality constant between stress and rate of deformation in the constitutive equation of Newton.

Viscous sublayer   A very thin layer in the vicinity of the wall in turbulent flow.

Vorticity   A measure of the rate at which a point in the flow would rotate if it suddenly could be changed to the solid state.

Wake   The flow behind a body when either the body or the fluid is in motion.

Water power   The power required by a fluid system for given operating conditions.

Water tunnel   A facility for moving water past a body. The water equivalent of a wind tunnel.

Weber number   A dimensionless ratio of momentum forces to surface tension forces.

Wicket gates   Adjustable vanes at the entrance to a turbine.

Wind tunnel   An experimental facility for testing models in air flow. The tunnel may be closed or open, and the flow subsonic or supersonic.

Answers to Even-numbered Problems

Chapter 1

1.2 (a).
1.4 (a) 13.368 lb/ft^3, (b) 1.936 slugs/ft^3, (c) 2.05 lb-s/ft^2, (d) 1.059 ft^2/s, (e) 0.005 lb/ft, (f) 32×10^4 psi.
1.6 (e).
1.8 (a) 2,399 kPa, (b) 21.68 kPa, (c) 85.30 psig, (d) −4.90 psig.
1.10 (a) 0.25 psi, (b) 0.11 psi additional.
1.12 101.82 psi.
1.14 6,287.6 psi.
1.16 376.2 lb.
1.18 13.09 kg/m^3.
1.20 $\rho = 0.00222$ slugs/ft^3, SG = 0.91.
1.22 373.6 lb.
1.24 5,355 N.
1.26 0.288 lb.
1.28 −0.2098 mm kerosene, −0.4464 mm water.
1.30 77,834 kPa.
1.32 786 knots.
1.34 373.5 m/s.
1.36 94.33°C.
1.38 2.34 kPa.
1.40 0.0215 N/m.
1.42 213.3 Pa.
1.44 106.7 Pa.
1.46 0.009 lb/ft.
1.48 (a) 2.09 mm, (b) 5.8×10^{-4} N, (c) 11.6×10^{-4} N.
1.50 (b).
1.52 6.99×10^{-5} N-m.
1.54 0.17 ft/s.
1.56 (a) 3 lb, (b) 2.016×10^{-3} lb, (c) 0.114 lb.
1.58 0.509 N-m.

1.60   0.0075 ft^3/s/ft.

Chapter 2

2.2 $p_D = 174.7$ lb/ft^2, $p_C = 124.8$ lb/ft^2, $p_B = 125.1$ lb/ft^2, $p_A = -62.1$ lb/ft^2.
2.4   374.4 lb/ft^2.
2.6   5,241.6 lb/ft^2.
2.8   −139.8 lb/ft^2.
2.10   3.53 psi.
2.12   3.54 mm.
2.14   13.48 mm.
2.16   120 mm.
2.20   2.16 mm.
2.22   3,271.5 lb/ft, −3,536.7 lb/ft, 6.558 ft.
2.24   11.2 kPa, 15.1 kPa, 4,600 N.
2.26   21,840 lb, 13,650 lb.
2.28   12,542 lb.
2.30   18,432 lb, 73,728 lb-ft.
2.34   −9,511 lb/ft, 10,982.4 lb/ft, 6,532.8 lb/ft.
2.36   3,650.4 lb/ft, −873.6 lb/ft, 24,429.6 lb-ft/ft.
2.38   3,529.9 lb/ft, 2.640 ft.
2.40   2.067 ft.
2.42   655.2 lb/ft, −873.6 lb/ft, 1.571 ft.
2.44   3,120 lb/ft, −3,458.6 lb/ft, 22,380.8 lb-ft/ft.
2.46   2,470.9 lb/ft, 2.559 ft.
2.48   3.754 ft.
2.50   2,341.1 lb.
2.52   6,240 lb.
2.54   156 lb/ft, −1,225.2 lb/ft, 4.96 ft below center, 0.632 ft right of A.
2.56   1,123.2 lb-ft.
2.58   4,704.8 lb, 1,045.5 lb.
2.60   3,057.6 lb, −2,379.8 lb, 3.435 ft.
2.62   1,747.2 lb/ft, −2,032.1 lb/ft.
2.64   4,875 lb, 19,684.1 lb.
2.66   97.85 ft^3.
2.68   0°, 21.45°.
2.70   50.75°C.
2.72   0.002 m^3, 18,000 kg/m^3.
2.74   1.348 N.
2.76   0.160 m.
2.78   0.076 m above free surface.
2.80 $2g/3$.
2.82 $p_A = -93.6$ lb/ft^2, $p_B = -124.8$ lb/ft^2, $p_C = 249.6$ lb/ft^2, $p_D = 280.8$ lb/ft^2.
2.84   @ (0, 0) $p = 124.8$ lb/ft^2, @ (0, 4) $p = 0$.

2.86 2.5 g.

2.88 7.885 ft.

2.90 $p_A = 156$ lb/ft^2, $p_B = 31.2$ lb/ft^2.

2.92 45°, 3.43 Pa, 50 mm, 350 mm.

2.94 $g/3$.

2.96 124.8 lb, 187.2 lb.

2.98 5.95 rad/s, 187.2 lb/ft^2.

2.100 0.146 lb/ft^2.

2.102 $z = -0.0417 + 0.426\ r^2$.

2.104 1.18 ft.

2.106 138.1 rad/s, 260 lb/ft^2.

Chapter 3

3.2 $x = \text{constant} \times (y - 1.25)^{-1/8}$.

3.4 4.516×10^{-3} m^3/s, 3.974 kg/s, 38.97 N/s.

3.6 2.326 slugs/ft^3.

3.8 −15.67 lb, 15.67 lb.

3.10 1.3 ft^3/s.

3.12 59.9 lb.

3.14 0.0572 slugs/ft^3.

3.16 −16.63 lb, 19.20 lb.

3.18 36.67 m/s, 15.02 kN, 11.51 kN, −440.8 kN-m/s.

3.20 −222.7 lb, 78.8 lb, −2,168.6 lb-ft/s.

3.22 0.86 lb, 43 lb-ft/s.

3.24 12.68 ft/s.

3.26 357 mm.

3.28 2.01 ft/s, 245.1 lb/ft^2.

3.30 0.056 ft^3/s, 0.109 slugs/s.

3.32 10.87 m/s, 2.67×10^{-3} m^3/s.

3.34 −418.8 lb, −1,053.2 lb.

3.36 −722 lb, 193.5 lb.

3.38 −88.5 lb, −123.9 lb.

3.40 0.363 m^3/s.

3.42 $p_A = -374.4$ lb/ft^2, $p_B = 624$ lb/ft^2, $p_A = -729.4$ lb/ft^2.

3.44 112.8 ft^3/s.

3.46 36.69 ft, 40.26 ft.

3.48 4.08 ft/s, −10.95 lb/ft^2, 9.17 ft/s, −76.12 lb/ft^2, 2.28 lb, −4.63 lb.

3.50 4.77 ft/s, 19.1 ft/s, −42.94 lb/ft^2, 588.1 lb, 1,043 lb.

3.52 0.908 ft.

3.54 −16.41 lb/ft^2, 1,256.3 lb, 1,521.3 lb.

3.56 2.55 m/s, 7.07 m/s, 1.83×10^5 Pa, 15,738 N, −42,028 N.

3.58 −4,080 lb, 1,320 lb, −5,400 lb.

3.60 −12,040 N, −2,900 N, $-34.75\mathbf{i} + 24.75\mathbf{j}$ m/s, $7.50\mathbf{i} + 30.31\mathbf{j}$ m/s.

3.62 −2,437 lb, 653 lb, 194.9 ft/s @ 157.4°, −1,625 lb, 435 lb.
3.64 −1,652 lb, 684 lb, −5,351 lb, 2,217 lb, 45.8 lb, 27.4 lb.
3.66 −1,045 lb, 603 lb, 127.6°.
3.68 −129.6 lb, 483.8 lb, −46.7 lb, 174.2 lb, 18.1°.
3.70 −233.1 lb, 7.3 lb, −50**j** ft/s, 2.
3.72 52.57 ft^2.
3.74 90.9 slugs/s, 189.7 ft/s^2.
3.76 114.3 rad/s.
3.78 $(10 − 3s + 2s^2) \times (−3 + 4s)$, $−1.94 \times (10 − 3s + 2s^2) \times (−3 + 4s)$.
3.80 22.25 ft/s, 624 lb/ft^2.
3.82 60 ft/s, 844.5 lb/ft^2, 848.9 lb/ft^2, −667.4 lb, 163.8 lb.

Chapter 4

4.2 (a) $\psi = −0.5(x^2 + y^2)\Omega + $ constant,
(b) $\psi = −\Omega \ln (x^2 + y^2)^{0.5} + $ constant,
(c) $\psi = A \tan^{-1} (y/x) + $ constant,
(d) $\psi = −yA/(x^2 + y^2) + $ constant.
4.4 (a) $\psi = yx^2 − x \sin y^2 + $ constant,
(b) $Q = 18 − 3 \sin 4$.
4.6 $v_y = ky/(x^2 + y^2)$,
$\psi = k \tan^{-1} y/x + $ constant.
4.8 $U + m/w, w/(1 + m/wU)$.
4.10 $\psi = 3y + (3/\pi) \tan^{-1} y/x$, $r = (1 − \theta/\pi)/\sin \theta$, 8.73 lb/ft^2, $\theta = 1.5866$ radians.
4.12 $p/\rho = a \cos \theta \, dU/dt + 0.125U^2 [1 + 3 \cos^2\theta] + f(t)$, $2\rho a^3\pi/3$.
4.14 (a) $\phi = x^2/2 + (y + z)x − y^2/2 + yz + 3z + $ constant, (b) rotational, (c) rotational.
4.18 (a) no, (b) yes, (c) no, (d) no.
4.20 $dU/dt = (\rho_{sphere} − \rho_{fluid})g/(\rho_{sphere} + 0.5\rho_{fluid})$
4.22 $\Omega rz/d, 0.5\rho(\Omega rz/d)^2$.
4.26 $(iKe^{i\Omega t}/\rho\Omega)\{1 + [\sin ip(y/d − 1) − \sin (−ipy/d)]/\sin ip\}$
4.28 $f'' + (ra/\nu − 1/r)f ' = 0$, $\exp (−r^2a/\nu)$.
4.30 $v_r = 3U(r/h)[(z/h) − (z/h)^2]$, $v_z = U[−3(z/h)^2 + 2(z/h)^3]$.

Chapter 5

5.2 $\omega D/V, \mu/\rho VD, c/V$.
5.4 F_{drag}/pl^2.
5.6 $V\mu/D^2\gamma, \Delta pD/\mu V, \tau\rho D^2/\mu^2$.
5.8 $\mu/\rho VD, gD/V^2, F/\rho V^2D^2$.
5.10 $\nu/Vl, gl/V^2, p/\rho V^2, F_D/\rho V^2l^2$.
5.12 $\mu b/\rho Q, pb^4/\rho Q^2$.
5.14 $\nu D/Q, \gamma D/\Delta p, Q_{model}/Q_{prototype} = 2, \Delta p_{model}/\Delta p_{prototype} = 0.13$.
5.16 $C_D = $ drag force/inertia force, $C_p = $ pressure force/inertia force

5.18 (a) Froude, (b) Reynolds, (c) Froude,
(d) Reynolds, (e) Reynolds.

5.20 $p = 0.5\rho a^2 \Omega^2 C_p$(Froude).

5.22 60 ft/s, force ratio = 1.

5.24 1.96×10^{-4} lb, 0.0144 ft.

5.26 7.529.

5.28 3.33×10^{-4} N, 3.19 mm.

5.30 1,224.9 miles/hour.

5.32 Drag coefficient, Reynolds number, $\omega D/V$.

5.34 2.5 ft, 20 ft/s, 332.6 lb.

5.36 14.3 to 42.9 ft/s, power ratio = 1.86.

5.38 2,500 rpm, torque ratio = 0.02.

5.40 4.8 rpm, 0.896 lb-ft.

5.42 2 miles/hour, 2.15×10^{-8} lb-s/ft^2, 2,174,000 lb.

5.44 pressure ratio = 0.04.

5.46 8.75×10^{-6} ft^2/s, discharge ratio = 1/32.

5.48 $V_{model}/V_{prototype} = 0.338$, $\rho_{package\ model} = 475 \times (\rho_{package\ prototype} + 0.901)$.

5.50 0.625 ft^3/s, 0.391 psi, 400 hp.

5.52 2.25 ft/s, 300 lb-ft.

5.54 $\Omega = \sqrt{(g/h)}\, f(h/w)$.

Chapter 6

6.2 $u = \gamma(2yb - y^2)/2\mu$, $\gamma b^3/3\mu$.

6.4 $4u_{max}wb/3$, $16\rho u_{max}^2 bw/15$, $12\rho u_{max}^3 bw/35$.

6.6 $U_{top} = (\partial p/\partial x)b^2/6\mu$

6.8 zero pressure gradient, 4.41 lb/ft^2, 4.41 lb/ft^2, 631 ft^3/s/ft.

6.10 3.65×10^{-4} ft, 0.0114 lb/ft^2.

6.12 $Q/w = -\gamma b^3/3\mu + Vb$, $\gamma b^2/3\mu$, $\sqrt{(\mu V/\gamma)}$.

6.14 216.4 lb/ft^2/ft, 3.15 lb.

6.18 $(\Omega_0 r_0^2 - \Omega_i r_i^2)/(r_0^2 - r_i^2)$, $(\Omega_0 - \Omega_i)r_0^2 r_i^2/(r_0^2 - r_i^2)$.

6.20 $\pi\mu\Omega a^3[2h/(b - a) + a/2d]$

6.22 70.2 lb, 3.75×10^{-5} m.

6.24 2.21×10^{-2} lb/ft^2/ft.

6.26 0.178 in.

6.28 1.487×10^{-3} ft^3/s.

6.30 leg reading p_{max} depressed 0.178 ft. compared to the other leg.

6.32 0.019 m/s, 1.54 kPa, 9.17×10^{-3} N.

6.34 0.05 m, 2.999 Pa.

6.38 1.837×10^{-10} ft^3/s.

6.40 0.1 m.

6.42 $\delta^2 = (30\nu L/7U_0)[1 - (1 + x/L)^{-14}]$.

6.44 $\delta^2 = (5\nu L/U_0)[(1 + x/L)^{14} - (1 + x/L)^2]$.

6.46 $x/L = 0.123$.

6.48 $x/L = 0.1583$.
6.50 $x/L = 0.6398$.

Chapter 7

7.2 0.021.
7.4 40.29 ft, 2.27 ft additional.
7.6 10,484 ft, -3.153 lb/ft^2/ft, -3.756 lb/ft^2/ft.
7.8 2.93 ft.
7.10 11.6 hp.
7.12 8.51×10^4, 0.024, 39.2 hp.
7.14 3.25 ft.
7.16 14.24 hp.
7.18 2.867 lb/ft^2, 0.0459 ft.
7.20 6,233 lb/ft^2 above the entering pressure.
7.22 0.15 ft^3/s, 106 ft.
7.24 16.7 ft/s.
7.26 2.55 ft^3/s, 0, -908 lb/ft^2.
7.28 0.084 ft^3/s.
7.30 3.85×10^{-4} m^3/s.
7.32 0.254 m^3/s.
7.34 0.040 ft^3/s.
7.36 0.55 ft^3/s, 0.40 ft^3/s.
7.38 7.44×10^{-4} m^3/s.
7.40 18.93 ft/s.
7.42 0.297 m.
7.44 0.282 m.
7.46 0.728 m.
7.48 0.0101 ft, 2.35×10^{-4} ft, 3.73×10^{-6} ft.
7.50 10.77 lb.
7.52 53.9, 1.5.
7.54 4.77 kg/m^3.
7.56 2.041×10^{-19} coulomb.
7.58 0.948 m, 7.41 N, 9.08 N.
7.60 86.1 ft/s, 9.58 hp, 42.9 hp.

Chapter 8

8.2 20 ft/s, 29,578 lb-ft/s, 122,155 lb.
8.4 61,804 lb. 39,924 lb-ft/s.
8.6 264.5 ft^3/s, 20.59 ft.
8.8 10,252 lb, 11, 833 lb-ft/s.
8.10 5,518,000 lb.
8.12 3,091 lb-ft/s per unit width, 2,810 lb.

8.14 9,245 lb, no.
8.16 0.0427 m, 1.078 m/s, or 0.0881 m, 0.522 m/s.
8.18 0.267 ft, 6.077 ft/s, or 0.773 ft, 2.097 ft/s.
8.20 0.197 m.
8.22 0.28 ft.
8.24 8.858 m/s, 1.
8.26 0.0356 ft, 2.25 ft/s, and 0.1055 ft, 0.758 ft/s.
8.28 0.486 m, 1.54 m/s, 1.623 kW.
8.30 74.7 lb/ft width.
8.32 14.63 ft, 20 ft/s, 5.818 ft, 2 ft/s.
8.34 3,744 lb, 7,487 lb.
8.36 9.82 ft/s, 9.82 ft/s.
8.38 4 ft, 6.253 ft/s.
8.40 $r = h_2/h_1$, $c^2 = (2gh_1/3)(r^2 + r + 1)/r^2(r + 1)$.
8.42 6.316 ft/s.
8.44 0.896.
8.46 (a) subcritical, (b) and (c) supercritical.
8.48 20% reduction in normal depth, no change in critical depth.
8.50 0.789 m.
8.52 6.19 m.
8.54 $0.5\gamma R_h \sin \theta$, 1.51 lb/ft^2.
8.56 2.82 m.

Chapter 9

9.2 46.3 milliseconds, 1.049×10^{-4}.
9.6 30 km, 1.838.
9.8 704.8 ft/s.
9.10 267.5 K.
9.12 12.78 psia, 277.8° R, 1, 1.185 in^2.
9.14 1.912 inches.
9.16 95.1 psia, 551.6° R, 373 ft/s, 0.27, 1.
9.18 33.8 kPa.
9.20 1.893 atmospheres, 0.0182 slugs/s.
9.22 1.147 slugs/s, 184.91 psia.
9.24 285.9 K, 0.809, 212.9 m/s.
9.26 0.145, 501.8° R.
9.28 $480.4 \leq p_{exit} \leq 500$ kPa and $p_{exit} = 31.8$ kPa, 0.101 kg/s.
9.30 329.2 K, 150 kPa, 0.103 kg/s.
9.32 0.9375, 0.0935.
9.34 1, 416.4° R, 1,000.1 ft/s, 0.43, 481.8° R, 176.1 psia, 1.86, 295.3° R, 31.7 psia.
9.36 0.0168 slugs/ft^3, 1, 1.77, 18.22 psia, 63.6 psia.
9.38 0.800.

9.40 1 cm^2, 2.1, 0.236 kg/s.

9.42 16.4 K, 3,032.8 K, 1.196 kPa, 89.10 kPa.

9.44 $\delta = 22.97°$, $\beta = 64.7°$.

9.46 4.906 ft.

9.48 1.724 ft.

9.50 647.7 (m/s)^2.

Chapter 10

10.2 0.511 to 1.115.

10.4 ν, 1.725×10^{-3} ft^2/s.

10.6 $\mu = 32dT/\pi\Omega D^4$.

10.8 $\mu = gD^2(\rho_{sphere} - \rho_{fluid})/18V$.

10.10 1.68, 10.59.

10.12 62.8 ft^3.

10.14 0.44.

10.16 $Q = w(H - h)\sqrt{(2gh)}$.

Chapter 11

11.2 25.45 in, 168.7 ft^3/s.

11.4 No.

11.6 $\Omega_{cutoff} = 3,315$ rpm, Francis turbine.

11.8 93.8%

11.10 90 rpm, 680 ft^3/s, 7,600 hp.

11.12 $H = 40 - Q/3 - Q^2 - 2Q^3/3$.

11.14 0, 4.31, 6.81, 4.08 hp.

11.16 122.5, 116.4, 91.9, 36.8 ft, 0, 1.75, 3.5, 5.25 thousands of gpm.

11.18 2.387 ft^3/s.

11.20 11.20 hp.

11.22 54.6%, 13.4%.

11.24 33.2 ft.

11.26 22.26°, 78.21°, 64.2 lb-ft.

Index